Grassland Management: Progress and Trends

Grassland Management: Progress and Trends

Editor: Elias Bennett

R CALLISTO
REFERENCE
www.callistoreference.com

Callisto Reference,
118-35 Queens Blvd., Suite 400,
Forest Hills, NY 11375, USA

Visit us on the World Wide Web at:
www.callistoreference.com

ISBN: 978-1-63239-817-8 (Hardback)

Cataloging-in-publication Data

Grassland management : progress and trends / edited by Elias Bennett.
    p. cm.
Includes bibliographical references and index.
ISBN 978-1-63239-817-8
1. Grassland--Management. 2. Grassland ecology. 3. Grassland conservation. I. Bennett, Elias.
QH541.5.P7 G73 2017
577.4--dc23

Table of Contents

Preface...VII

Chapter 1 **A Multi-Scale Distribution Model for Non Equilibrium Populations Suggests Resource Limitation in an Endangered Rodent** 1
William T. Bean, Robert Stafford, H. Scott Butterfield, Justin S. Brashares

Chapter 2 **Community Structure of Skipper Butterflies (Lepidoptera, Hesperiidae) along Elevational Gradients in Brazilian Atlantic Forest Reflects Vegetation Type Rather than Altitude** 10
Eduardo Carneiro, Olaf Hermann Hendrik Mielke, Mirna Martins Casagrande, Konrad Fiedler

Chapter 3 **Effects of Grazing Regimes on Plant Traits and Soil Nutrients in an Alpine Steppe, Northern Tibetan Plateau** 21
Jian Sun, Xiaodan Wang, Genwei Cheng, Jianbo Wu, Jiangtao Hong, Shuli Niu

Chapter 4 **Bird Communities and Biomass Yields in Potential Bioenergy Grasslands** 30
Peter J. Blank, David W. Sample, Carol L. Williams, Monica G. Turner

Chapter 5 **Defining Landscape Resistance Values in Least-Cost Connectivity Models for the Invasive Grey Squirrel: A Comparison of Approaches using Expert-Opinion and Habitat Suitability Modelling** 40
Claire D. Stevenson-Holt, Kevin Watts, Chloe C. Bellamy, Owen T. Nevin, Andrew D. Ramsey

Chapter 6 **Historical, Observed, and Modeled Wildfire Severity in Montane Forests of the Colorado Front Range** 51
Rosemary L. Sherriff, Rutherford V. Platt, Thomas T. Veblen, Tania L. Schoennagel, Meredith H. Gartner

Chapter 7 **Vertical Profiles of Soil Water Content as Influenced by Environmental Factors in a Small Catchment on the Hilly-Gully Loess Plateau** 68
Bing Wang, Fenxiang Wen, Jiangtao Wu, Xiaojun Wang, Yani Hu

Chapter 8 **Global Validation of a Process-Based Model on Vegetation Gross Primary Production using Eddy Covariance Observations** 80
Dan Liu, Wenwen Cai, Jiangzhou Xia, Wenjie Dong, Guangsheng Zhou, Yang Chen, Haicheng Zhang, Wenping Yuan

Chapter 9 **Spatial and Temporal Variations of Ecosystem Service Values in Relation to Land use Pattern in the Loess Plateau of China at Town Scale** 92
Xuan Fang, Guoan Tang, Bicheng Li, Ruiming Han

Chapter 10   **Effects of Warming and Clipping on Ecosystem Carbon Fluxes across Two Hydrologically Contrasting Years in an Alpine Meadow of the Qinghai-Tibet Plateau** 105
Fei Peng, Quangang You, Manhou Xu, Jian Guo, Tao Wang, Xian Xue

Chapter 11   **Effect of Abandonment on Diversity and Abundance of Free-Living Nitrogen-Fixing Bacteria and Total Bacteria in the Cropland Soils of Hulun Buir, Inner Mongoli** 119
Huhe, Shinchilelt Borjigin, Yunxiang Cheng, Nobukiko Nomura, Toshiaki Nakajima, Toru Nakamura, Hiroo Uchiyama

Chapter 12   **Effects of Traditional Flood Irrigation on Invertebrates in Lowland Meadows** 129
Jens Schirmel, Martin Alt, Isabell Rudolph, Martin H. Entling

Chapter 13   **Comparison of Phenology models for Predicting the Onset of Growing Season over the Northern Hemisphere** 138
Yang Fu, Haicheng Zhang, Wenjie Dong, Wenping Yuan

Chapter 14   **How Ecosystem Services Knowledge and Values Influence Farmers' Decision-Making** 150
Pénélope Lamarque, Patrick Meyfroidt, Baptiste Nettier, Sandra Lavorel

Chapter 15   **The Soil Biota Composition along a Progressive Succession of Secondary Vegetation in a Karst Area** 166
Jie Zhao, Shengping Li, Xunyang He, Lu Liu, Kelin Wang

Chapter 16   **Impacts of Land Cover Data Selection and Trait Parameterisation on Dynamic Modelling of Species' Range Expansion** 175
Risto K. Heikkinen, Greta Bocedi, Mikko Kuussaari, Janne Heliölä, Niko Leikola, Juha Pöyry, Justin M. J. Travis

Chapter 17   **The Effects of Timing of Grazing on Plant and Arthropod Communities in High-Elevation Grasslands** 189
Stacy C. Davis, Laura A. Burkle, Wyatt F. Cross, Kyle A. Cutting

Chapter 18   **Impacts of Diffuse Radiation on Light use Efficiency across Terrestrial Ecosystems Based on Eddy Covariance Observation in China** 203
Kun Huang, Shaoqiang Wang, Lei Zhou, Huimin Wang, Junhui Zhang, Junhua Yan, Liang Zhao, Yanfen Wang, Peili Shi

Chapter 19   **Modeling Pollinator Community Response to Contrasting Bioenergy Scenarios** 212
Ashley B. Bennett, Timothy D. Meehan, Claudio Gratton, Rufus Isaacs

Chapter 20   **Evidence for Frozen-Niche Variation in a Cosmopolitan Parthenogenetic Soil Mite Species (Acari, Oribatida)** 222
Helge von Saltzwedel, Mark Maraun, Stefan Scheu, Ina Schaefer

Permissions

List of Contributors

Index

Preface

Grassland management can be defined as human intervention in regions dominated by grass vegetation to achieve predetermined goals. This book on grassland management includes the study and classification of flora and fauna found in grasslands, their assessment and management. This extensive book on grasslands sheds light on the constitution of ecosystems and conservation techniques for sustainable development. Some of the diverse topics covered in this book address the varied branches that fall under this field. It brings forth some of the most innovative concepts and elucidates the unexplored aspects of grassland management. Conservationists, microbiologists and forest preservationists will find the book helpful. The readers would gain knowledge that would broaden their perspective about grassland management.

The main aim of this book is to educate learners and enhance their research focus by presenting diverse topics covering this vast field. This is an advanced book which compiles significant studies by distinguished experts. This book addresses successive solutions to the challenges arising in the area of application, along with it; the book provides scope for future developments.

It was a great honour to edit this book, though there were challenges, as it involved a lot of communication and networking between me and the editorial team. However, the end result was this all-inclusive book covering diverse themes in the field.

Finally, it is important to acknowledge the efforts of the contributors for their excellent chapters, through which a wide variety of issues have been addressed. I would also like to thank my colleagues for their valuable feedback during the making of this book.

Editor

A Multi-Scale Distribution Model for Non-Equilibrium Populations Suggests Resource Limitation in an Endangered Rodent

William T. Bean[1]*, Robert Stafford[2], H. Scott Butterfield[3], Justin S. Brashares[4]

1 Humboldt State University, Arcata, California, United States of America, **2** California Department of Fish and Game, Los Osos, California, United States of America, **3** The Nature Conservancy, San Francisco, California, United States of America, **4** Department of Environmental Science, Policy and Management, University of California, Berkeley, California, United States of America

Abstract

Species distributions are known to be limited by biotic and abiotic factors at multiple temporal and spatial scales. Species distribution models, however, frequently assume a population at equilibrium in both time and space. Studies of habitat selection have repeatedly shown the difficulty of estimating resource selection if the scale or extent of analysis is incorrect. Here, we present a multi-step approach to estimate the realized and potential distribution of the endangered giant kangaroo rat. First, we estimate the potential distribution by modeling suitability at a range-wide scale using static bioclimatic variables. We then examine annual changes in extent at a population-level. We define "available" habitat based on the total suitable potential distribution at the range-wide scale. Then, within the available habitat, model changes in population extent driven by multiple measures of resource availability. By modeling distributions for a population with robust estimates of population extent through time, and ecologically relevant predictor variables, we improved the predictive ability of SDMs, as well as revealed an unanticipated relationship between population extent and precipitation at multiple scales. At a range-wide scale, the best model indicated the giant kangaroo rat was limited to areas that received little to no precipitation in the summer months. In contrast, the best model for shorter time scales showed a positive relation with resource abundance, driven by precipitation, in the current and previous year. These results suggest that the distribution of the giant kangaroo rat was limited to the wettest parts of the drier areas within the study region. This multi-step approach reinforces the differing relationship species may have with environmental variables at different scales, provides a novel method for defining "available" habitat in habitat selection studies, and suggests a way to create distribution models at spatial and temporal scales relevant to theoretical and applied ecologists.

Editor: Stephanie S. Romanach, U.S. Geological Survey, United States of America

Funding: Funding was provided by The Nature Conservancy, U.S. Department of Agriculture, Bureau of Land Management, California Department of Fish and Game, and the A. Starker Leopold Wildlife Graduate Student Fund. The funders had no role in study design, data collection and analysis, decision to publish, or preparation of the manuscript.

Competing Interests: The authors have declared that no competing interests exist.

* Email: bean@humboldt.edu

Introduction

Species distribution models (SDMs) have become a cornerstone of theoretical (e.g., [1]) and applied (e.g., [2]) ecological research [3,4]. In these models, species occurrence data and environmental correlates are used to define the limits of a species distribution ([4]). However, understanding and predicting the relationship between environmental resources and species distributions is complicated by the temporal and spatial scale of analysis, with most SDMs aimed at mapping range-wide associations using abiotic climatic factors. Perhaps because of the broad temporal and spatial scale at which these analyses are conducted, most recommendations suggest that SDMs operate best for populations at equilibrium (e.g.,[5]). By contrast, most species, especially those of conservation concern, are rarely, if ever, at equilibrium [6].

Guisan and Thuiller [7] provide a framework for modeling species distributions at disparate scales. At broad (*e.g.*, biogeographic) spatial and temporal scales, species' distributions tend to be limited primarily by abiotic factors [8]. At finer spatial and temporal scales, species are limited by local community interactions such as resource factors, dispersal, predation, and competition. Guisan and Thuiller's work suggests a multi-step approach to modeling. That is, they encourage practitioners to first define a species' range-wide distribution, and then model limiting factors within that area to better understand relationships with environmental factors at finer spatial or temporal scales. Echoing Hutchinson [9], Guisan and Thuiller [7] refer to the broad-scale, bioclimatic range as the "potential distribution", and they define the "realized distribution" as the bioclimatic range filtered through dispersal, disturbance, and biotic interactions.

Guisan and Thuiller's [7] research closely parallels work in the field of habitat selection. Johnson [10] defined habitat selection as a strictly hierarchical process, with first-order selection occurring at the level of the physical or geographical range, second-order selection determining the home range, and so on. While Johnson [10] described habitat selection at each scale as a decision-based process by the individual animal, and Guisan and Thuiller [7] formulate it as an environmental filtering process, both clearly suggest that local occurrences are separately constrained within a higher hierarchical biogeographic distribution. As Wiens et al. [11] demonstrated in their landmark study of shrubsteppe birds, not only are these hierarchical levels of habitat selection distinct, animals may select habitat in contrasting directions at different spatial or temporal scales. Environmental factors that predict habitat selection at macro scales (*e.g.* vegetation, cover, temperature, rainfall) may have little predictive value at finer scales, or may even be correlated with selection in opposite directions.

Habitat selection studies have long recognized this problem of temporal or spatial scale incompatibility [12], and resource selection studies frequently examine habitat selection at multiple scales [13]. Despite the long history of research on habitat selection, the problem of defining "available" habitat has been a common and recurring one [14]. Typically, researchers use some measure of a home range, a buffer around used points, or some meaningful political or biological boundary [12]. We suggest that a more appropriate definition of available habitat would follow the well-understood construction of hierarchical habitat selection. That is, a study of habitat selection at multiple scales should follow the theory of Guisan and Thuiller [7] and Johnson [10] by explicitly modeling habitat selection at each hierarchical stage.

Guisan and Thuiller's [7] multi-step approach has been used to model distribution limited by dispersal [15], and habitat type [16], but to our knowledge has not been used to examine the role of resource availability. Resource availability has long been hypothesized as a key factor limiting species' distributions [17], and recent work has supported this (e.g., [18], [19], [20]). In particular, the temporal dynamics of resource availability can be critical to fine-scale distribution modeling in either space or time. While species at broad spatial and temporal scales may be considered at equilibrium, managers are frequently tasked with understanding shifts in distribution at much finer time intervals, such as between years or even seasons [21,22]. At such temporal scales, variability of resources can greatly impact species distributions, particularly where the presence of a species is positively or negatively related with resource availability [23].

Recent advances in remote sensing techniques have allowed for estimates of resource abundance at fine temporal scales [24]. In particular, the Normalized Difference Vegetation Index (hereafter "NDVI") has been used as a reliable estimate of biomass in grassland systems [25], and population dynamics in herbivores have been shown to be correlated with NDVI (e.g., [26], [27], [28]). Recent work has shown that NDVI can be a useful predictor of distribution in large herbivores [29].

In this study, we created a multi-step species distribution model for the giant kangaroo rat (*Dipodomys ingens*, hereafter "GKR"). The GKR is an endangered rodent endemic to southern-central California [30]. GKRs are believed to be limited to areas with loamy soils, flat or gently rolling hills, and to areas with mean annual precipitation no greater than approximately 30 cm [31,32]. First, we estimated the potential distribution (or first-order habitat selection) of the GKR using population-wide occurrence data and static environmental predictor variables (slope, soil particle size, and six climatic variables relating to temperature and precipitation) using the machine-learning method Maxent [33]. Maxent represents an ideal method for modeling the "potential distribution" because it assumes the most uniform distribution of a species' occurrence across the study area, minimally constrained by the provided environmental correlates. Maxent is a presence-background model, and in fact its authors suggest the results may represent the species' potential distribution [33]. Over broader spatial scales and longer time scales, we

predicted a negative relationship between GKR presence and precipitation.

We used this model of potential distribution to define available habitat in order to understand finer scale temporal dynamics in GKR population extent (i.e., annual changes in the "realized" distribution). These temporal models incorporated a suite of primary productivity estimates based on the NDVI. In particular, we predicted, based on previous research [30,32,34], that GKR presence would show a positive correlation with resource abundance within their potential distribution, possibly with a time lag reflecting a delayed demographic response of GKR to resource availability. Due to the GKR's strong association between population demographics and precipitation, other factors that may also limit population extent (e.g., predation and competition) were not considered in these models.

Methods

Study site and focal species

The GKR is a state and federally endangered, burrowing, granivorous rodent endemic to deserts grasslands of California, USA [32]. Once widespread in the western San Joaquin Valley, habitat loss from agriculture and other development have severely restricted its range to a half-dozen populations in and around the California Coast Range [30]. The GKR is considered both a keystone species and an ecosystem engineer [35,36]. As grasses begin to senesce in April, GKRs remove all herbaceous vegetation from the top of their burrows [31,37]. This behavior results in clear circles of bare soil, 2–7 m in diameter, where GKRs are present. Aerial surveys have therefore been a useful tool in mapping GKR population extent in years of high primary productivity [38].

This study is primarily focused on the Carrizo Plain National Monument (hereafter "Carrizo"), an area that contains the largest remaining population of GKRs. Carrizo represents the largest representative landscape of San Joaquin Valley annual grassland [39]. Carrizo experiences variable precipitation (mean = 20 cm, sd = 10 cm) that contributes directly to variability in primary productivity, which in turn may drive dramatic annual changes in GKR distribution [32]. Based on aerial surveys, GKR population extent in Carrizo was estimated to expand more than 50% between 2001 and 2006 [38]. Understanding the role of primary productivity in driving these changes is crucial to biodiversity management for this endangered ecosystem. Both the size of the GKR population and its management and monitoring history make Carrizo an ideal study site for examining the role of resource availability on species distributions.

Other factors that often limit a species' realized distribution – predation, parasitism, competition and dispersal – were not believed to be limiting factors for GKR in Carrizo. Within the study area, the open and flat topography, coupled with GKR reproductive habits allow for rapid dispersal. Within Carrizo, GKR appear to be competitively dominant [31,36]. Because of these features of their ecology, GKR distribution was less likely to be affected by dispersal or competition and, thus, the GKR was a good species for testing models of realized distribution based solely on resource abundance.

This study was carried out in strict accordance with the recommendations in the Guidelines of the American Society of Mammalogists for the Use of Wild Mammals in Research. The protocol was approved by the Animal Care and Use Committee at the University of Californa, Berkeley (R304).

GKR Distribution

We obtained estimates of GKR distribution from three sources: (1) historical occurrence records from the Global Biodiversity Information Facility (GBIF); (2) contemporary trapping sites throughout GKR range; and (3) aerial surveys of GKR population extent within Carrizo.

Occurrence records were downloaded from the GBIF using the *dismo* package [40] in R [41], and limited to points collected since 1950 (N = 38). We obtained an additional 185 records of GKR presence or absence from trapping conducted in 2010 and 2011. 157 points were selected randomly throughout GKR range, and trapped for three nights with five traps [42]. Eight additional sites were stratified across a range of habitat suitability values from a preliminary distribution model constructed in 2008, and a final 20 presence points were obtained from ongoing trapping in the center of Carrizo [36]. Of the 185 sites trapped, 120 were occupied in either 2010 or 2011, and thus included in the range-wide potential distribution model. Additional details on trapping methodology are provided in Bean et al. [42].

In 2001, 2006, 2010 and 2011, we conducted Carrizo-wide aerial flight surveys in late summer to estimate GKR extent. Using 800 m wide transects with two observers (i.e., monitoring 400 m on each side) and a global positioning system (GPS), we mapped the total extent of active burrows. GPS points were recorded whenever the observers entered or left areas of observable GKR activity. These points were then connected as lines and buffered 400 m on each side to create an estimate of total extent. These surveys were shown to be a reliable estimate of GKR population extent in a given year [38].

Potential Distribution Modeling

We created a multi-step model to estimate GKR distribution. We first used Maxent to estimate the potential GKR distribution with range-wide occurrence data (museum records and our trapping data] and static environmental variables. Second, we used logistic regression to estimate limits to the potential distribution based on local resource abundance (Fig. 1).

We used the software package Maxent to estimate GKR potential distribution [33]. Maxent uses a maximum entropy approach to estimate the most uniform distribution of a species' occurrence across the study area, minimally constrained by the provided environmental correlates. Maxent is a presence-background model, and therefore may better model the species' potential distribution [33].

To estimate the potential distribution for GKR, we selected a suite of environmental variables believed to limit GKR distribution range-wide. We obtained 19 climate layers [43] frequently used in distribution modeling as independent variables [44]. Bioclim layers are estimated as mean conditions from 1950 to 2000. We limited the variables to six we believed sufficient in describing GKR distribution, and that had limited correlation with each other. These included annual mean temperature (BIO1); annual precipitation (BIO12); minimum temperature of the coldest month (BIO6); precipitation of the driest month (BIO14); and precipitation of the driest quarter (BIO17). In addition, we used slope [45], and soil particle size derived from the SSURGO database [46]. Soil particle size was converted to raster format using ArcGIS 9.2, and all inputs were analyzed at 30 s resolution (the coarsest resolution of all predictor variables). Soil particle size was classified as categorical, with the rest classified continuous.

The output of this initial Maxent distribution model was a map of habitat suitability (Fig. 1), with each 30 s cell representing an index of suitability. To convert the map from a continuous suitability distribution to a binary map of potential distribution, we selected a threshold, above which cells were classified as potential GKR distribution and below which cells were classified as outside potential GKR distribution. A number of methods have been proposed for selecting thresholds [47–49]. However, the optimal thresholds recommended in previous work focused on best predicting overall presence or absence for a species. In this case, we were interested in defining the maximum potential distribution for the species. Therefore, in order to err on the side of inclusiveness, we selected a threshold (0.059) that included 99% of presence points from the modeled potential distribution.

Realized Distribution Modeling

Having produced an estimate of the potential distribution for GKR, we then examined the effects of resource availability on GKR realized distribution for four study years. Because the GKR relies on grass seeds as a food resource, we expected a positive correlation between primary productivity and GKR presence. GKRs dry and cache most of the seeds they collect in underground chambers [37], so GKR presence in a given area may lag primary productivity for a year or more.

To create a spatially explicit measure of primary productivity in Carrizo we acquired 16-day composites (250 m × 250 m) of NDVI measured by the Moderate Resolution Imaging Spectroradiometer ("MODIS") platform [50]. The NDVI is calculated as

$$(NIR - R) / (NIR + R) \qquad \text{(Equation 1)}$$

where NIR represents spectral reflectance within the near infrared band (841–876 nm), and R represents the visible red band (620–670 nm). Values approaching -1.0 tend to represent areas with water, while areas greater than 0 and approaching 1.0 tend to represent areas of photosynthetic activity [51]. Pre-processed 16-day composites of NDVI measured from MODIS have been shown to better measure primary productivity than single measures. These composites correlate well with biomass in grassland systems [25].

We created a suite of generalized linear models (GLM) to predict GKR presence using the NDVI for each year (2001, 2006, 2010 and 2011) [13]. We examined two drivers of GKR presence: first, and of primary interest, we tested the effect of primary productivity (i.e., resource abundance) on GKR presence. Second, we tested if GKR presence in the previous year would also be a significant predictor of GKR presence in the current year. The independent variables included in the model to evaluate these predictions represented resource abundance in the current or previous year, or were proxies of GKR presence in the previous year (Table 1). These hypotheses were first tested independently before being included in the suite of models (Fig. 2).

First, to estimate resource abundance, we used the highest recorded NDVI value for a given growing season (November through May, the typical rainy season in the Mediterranean climate of coastal California) as an estimate of primary productivity for that 250 m × 250 m cell. In estimating distribution for GKR in 2006, for example, we estimated primary productivity in the previous year as the peak NDVI from November to May, 2004–2005; and primary productivity in the current year as peak NDVI from November to May, 2005–2006. NDVI can be inflated by soil moisture if the soil is visible [51] and NDVI appeared to peak approximately 1–2 weeks before the typical peak growth in Carrizo, suggesting that soil moisture was influencing NDVI measurement. However, precipitation and aboveground biomass are correlated and, despite the lag in measurements, peak NDVI has repeatedly been shown to correlate strongly with peak aboveground biomass in grasslands [24,52].

1. Model GKR range-wide potential distribution / "available" habitat

Soil   Bio1

Bio 12   Bio 14   Bio 17

Bio 6   Slope

Trapping &
Museum Records

0   500   1,000 Km

2. Define "available" habitat as areas with suitable long-term climate, topography, and soil

0   150   300 Km

Select threshold   Clip to surveyed area

3. Model annual changes in GKR realized distribution as response to NDVI

a. Generate random points from used versus available/potential areas each year
b. Estimate GKR response to NDVI

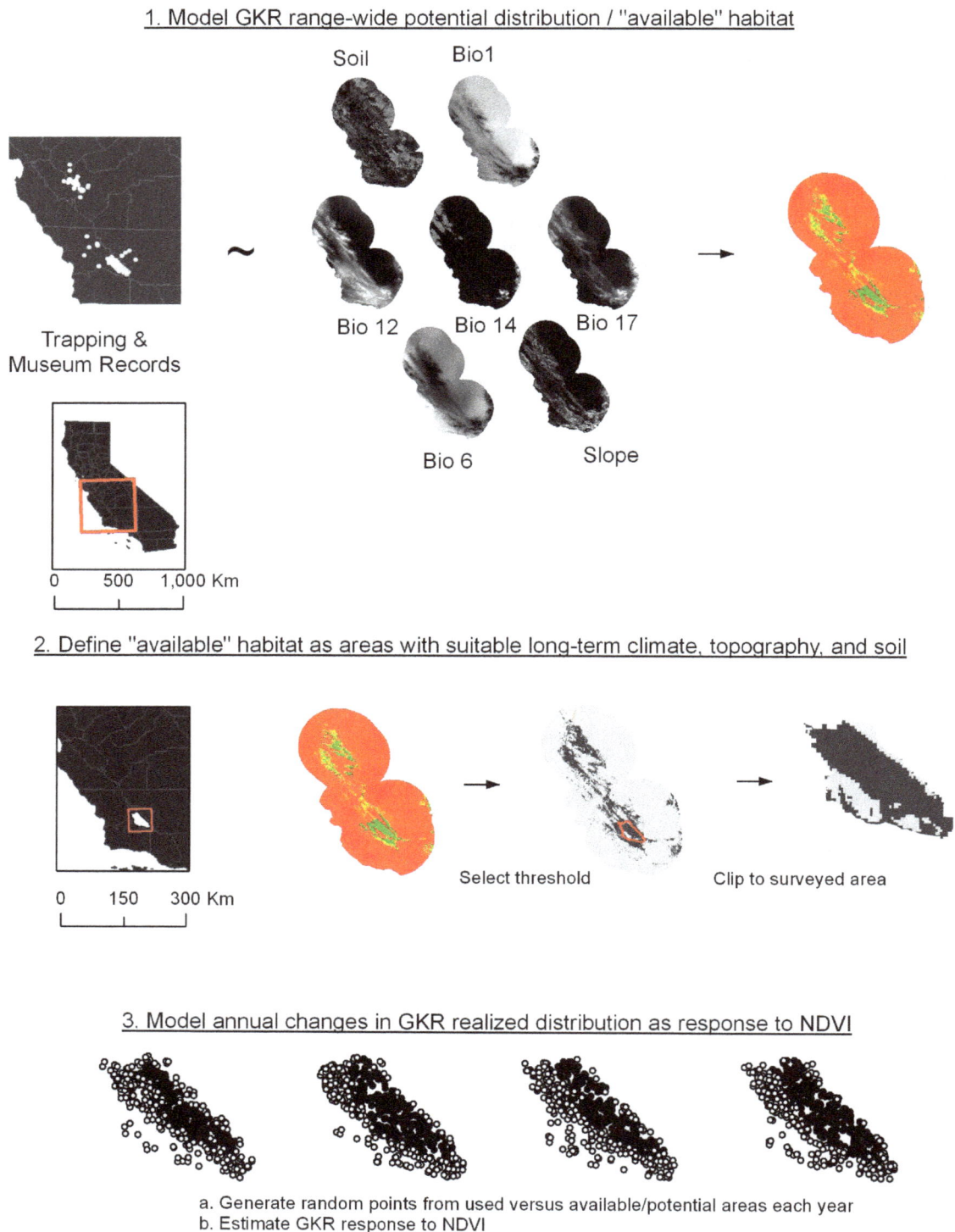

Figure 1. Flow chart of multi-step modeling approach. Here we present a multi-method, multi-scale approach to estimating species distributions. In the first step, Maxent is used to relate contemporary trapping and historical museum records with static environmental variables. The result is a model of potential distribution at a range-wide scale. Predictor variables included soil particle size ("Soil"), annual mean temperature ("Bio1"), minimum temperature of coldest month ("Bio 6"), annual precipitation ("Bio 12"), precipitation of driest month ("Bio 14"), precipitation of driest quarter ("Bio 17"), and slope. We then selected a threshold to define all available habitat for GKR, with the 99% Maxent value for training data used as the threshold. Finally, within the potential habitat in the Carrizo Plain National Monument, we examined annual changes in population extent based on aerial surveys and driven by measures of resource availability (NDVI).

Table 1. Hypothesized relationships between estimates of primary productivity (NDVI) and the local presence of the giant kangaroo rat.

Candidate Predictors	Hypothesized Mechanism
Maximum NDVI$_{T1,T0}$	Estimate of primary productivity, a bottom-up limitation on GKR presence (with potential one year lag)
Minimum NDVI$_{T0}$	Proxy for GKR presence in previous year
NDVI slope during plant senescence$_{T0}$	Proxy for GKR presence in previous year (GKR remove vegetation more quickly than it senesces)

T1 represents the Normalized Difference Vegetation Index data from the same year as the rat distribution was estimated; T0 is data from the previous year.

For three of the four years of surveys, no estimate was available for GKR presence in the previous year. Instead of a direct estimate from aerial surveys, it was therefore necessary to create proxies of GKR presence in the previous year. Because GKR clear their burrow mounds of vegetation, we assumed that GKR would have a direct effect on the NDVI after peak green up. First, we assumed that later in the summer, the areas with GKR would have lower plant biomass than areas without GKR. We therefore included the

lowest measured NDVI value from later in the year (April to December) as a proxy for GKR presence, assuming a negative correlation between the two (i.e., areas with GKR would have lower minimum NDVI). Second, we assumed that GKR removed vegetation from around their burrows faster than vegetation naturally senesced. To estimate vegetation removal by GKR, we measured the slope of NDVI decline from its peak. We subtracted the NDVI value from one time step (i.e. 16 days) after peak from

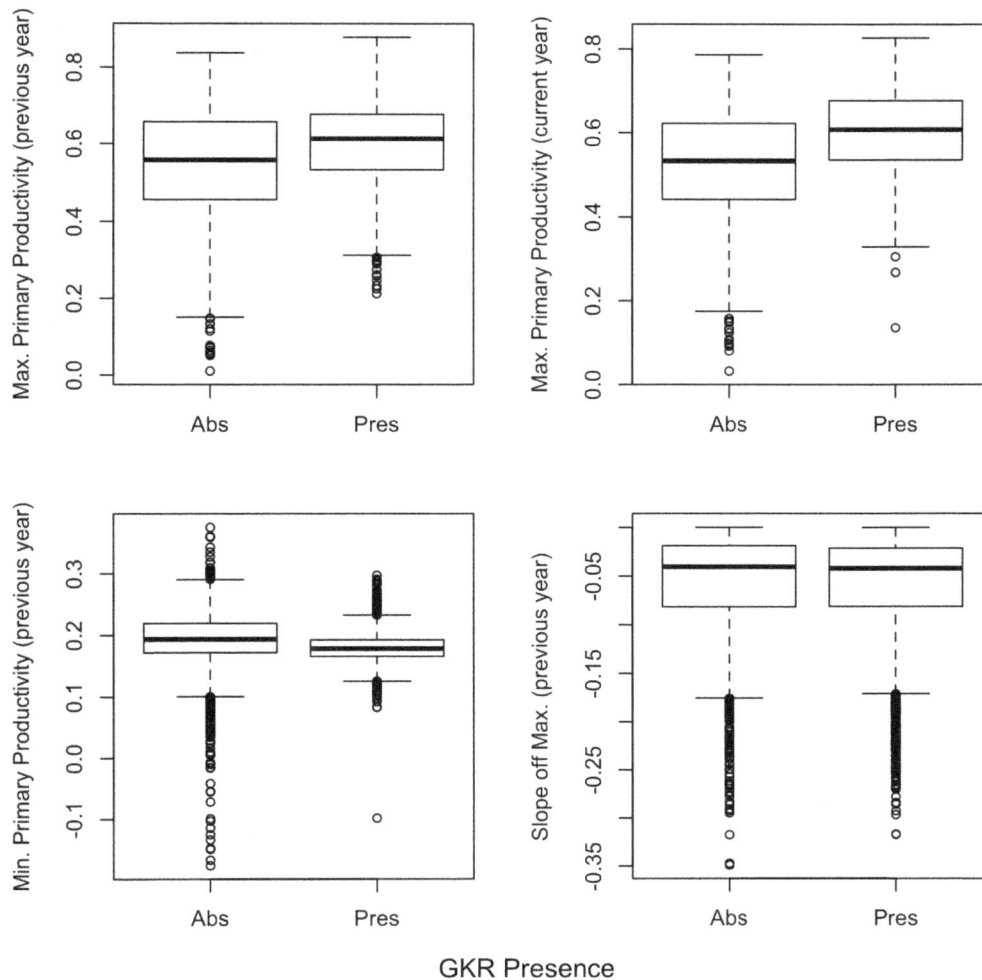

GKR Presence

Figure 2. Relationships between primary productivity (measured as NDVI) and GKR presence. GKR were expected to have a positive relationship with maximum primary productivity in the previous and current year; a negative relationship with the minimum primary productivity measured in the previous year; and a negative relationship with the rate of decrease of primary productivity in the previous year. Relationships are shown from 500 random points estimated from aerial surveys in 2011. All differences were significant (t-test, p<0.05).

the peak NDVI value, and again hypothesized that a larger difference would suggest GKR activity. These two measurements (minimum NDVI and NDVI slope) were used as proxies for GKR presence in an area, and in effect represent a null model of GKR distribution: if current GKR distribution could be predicted solely from the prior year's presence, plant biomass would not be considered a factor limiting the realized GKR distribution.

Although individual GKR burrows ($\sim$27–36 m^2) represent a small fraction of a single MODIS pixel (250 m$\times$250 m), the heterogeneity of the landscape supports analyses at this scale. The density of the GKR burrows, and the strong difference in signal between the perturbed bare soil on burrow and dried grass off burrow, suggest that a mixed pixel with GKR activity has a significantly different signal than one without GKR activity.

GKR distribution models were ranked using Akaike Information Criteria (AIC) [53]. Models were created for all of the presence points, with year included as a fixed effect. For each year of the model, we used 500 random points from the potential distribution, 250 within GKR realized distribution and 250 outside active areas.

The accuracy of the best model (as identified using AIC) was assessed with the PresenceAbsence package in R [54]. For each model, we calculated a threshold to test predicted presence and absence points for each model. Each threshold was set to the observed prevalence [48]. We then calculated the percent correctly classified (PCC), Cohen's kappa, sensitivity, specificity, and the true skill statistic (TSS, sensitivity + specificity -1), a prevalence-independent measure of accuracy [55]. Testing data was obtained in two ways: for all four years, we randomly selected 500 new points from the aerial surveys in each year, in the same manner as the training data. We also used the set of 105 GKR trapping points in Carrizo collected in 2010 and 2011 to test the models in those years.

Results

Potential Distribution Modeling

As expected, GKR potential distribution is limited to a narrow band of habitat on the western San Joaquin Valley and nearby Coast Ranges (Fig. 1). The most important variables in predicting GKR distribution included precipitation of the driest quarter, precipitation of the driest month, and minimum temperature of the coldest month (Table 2). Surprisingly, annual precipitation was not an important predictor of GKR distribution. Instead, precipitation in the driest month and driest quarter were more important predictors. Probability of GKR presence was highest in areas where the driest month received a mean of 0 mm precipitation. Similarly, probability of GKR presence was highest in areas where the driest quarter received a mean of 4 mm precipitation. GKRs inhabited areas that have a narrow band of mean annual temperatures between 14° and 16°C.

The area classified in the Maxent model as the potential distribution of GKR closely resembled the combined distribution from 2001, 2006, 2010 and 2011 (Fig. 3). However, there were portions of Carrizo in the northwest and southeast classified as suitable that were not part of the realized distribution in any of the years monitored. AUC for the potential distribution model was 0.98.

Realized Distribution Modeling

In the best model of GKR realized distribution, population extent was positively related to primary productivity in both the previous and current year, suggesting a strong influence of bottom-up regulation on GKR distribution (Table 3). GKR presence in the previous year also was an important predictor of GKR presence in the current year. Both proxies of prior GKR presence performed as expected: GKR distribution was negatively correlated to both minimum NDVI and the slope from the peak NDVI from the previous year. Realized distribution model accuracy from the testing data was "useful" (AUC = 0.74). The threshold was set at 0.50 (as expected, due to the prevalence of the model data [48]). Using the aerial surveys as testing data, model sensitivity was 0.70 and specificity 0.71. The model Kappa score and the true skill statistic (TSS) were 0.40. The best model correctly classified 70.1% of all test points as inside or outside the GKR's estimated realized distribution. Using the trapping data (obtained independently of the training data), sensitivity = 0.65; specificity = 0.66; Kappa = 0.29 and TSS = 0.29, while 65.8% of all test points were correctly classified.

Discussion

This study joins a growing body of literature that attempts to use ecological theory on limits to population extent and species ranges to inform, interpret and advance species distribution models (e.g. [1,16,56,57]). Specifically, we presented a technique of multi-step modeling to define a species' potential and realized distribution, and in doing so explored the relationship between primary productivity and animal distribution.

Consistent with theory on potential and realized distributions [4,7], our results showed that the potential distribution of GKR was larger than any of the distributions observed in the four years of aerial surveys. In other words, there were areas within Carrizo that should have been suitable for GKR, but monitoring documented them as uninhabited. This result supports conclusions of Guisan and Thuiller [7], and Grinnell [17] and Hutchinson [9] before them, who suggested that species' distributions are limited by more than fixed environmental conditions, a fundamentally important concept for distribution modelers and ecologists.

The fact that distribution models built only on static bioclimatic factors may poorly estimate realized distributions has several important implications for how these models are applied to questions in biodiversity conservation. For example, distribution models are often relied upon to project the impact of climate change on species' distributions. Without incorporating mechanisms that limit the focal species' realized distribution, these models are likely over-estimating the range of conditions within which the species will survive and reproduce under different climate change scenarios [58,59]. At the same time, we may be ignoring important local ecological processes by examining patterns of distribution at range-wide spatial scales [60]. In this study, GKR potential distribution was limited to areas with little to no rain in the driest months of the year; therefore, future increases in precipitation might be expected to reduce GKR distribution. However, we found that within the area of potential distribution, GKR were positively correlated with primary productivity. It is possible, then, that an increase in mean annual precipitation would decrease the potential distribution range-wide (where agriculture has already rendered suitable habitat uninhabitable), but increase the realized distribution within a core conservation area.

Within the potential distribution, we found that the best model of factors limiting the realized distribution of GKR showed a clear, positive correlation between primary productivity (measured as peak NDVI) and the presence of GKR in the previous year. This result conforms to recent findings from studies of other species that show rapid changes in distribution in response to temporal and

Table 2. Variable importance for range-wide model of giant kangaroo rat distribution reported by Maxent.

Variable	Percent Contribution	Permutation Importance
Precipitation of Driest Quarter	33.5	29.8
Annual Mean Temperature	21.1	0.4
Precipitation of Driest Month	16.2	33.2
Minimum Temperature of Coldest Month	15.8	22.9
Slope	5.8	5.8
Annual Precipitation	3.9	6.8
Soil Particle Size	3.7	1

spatial variability of NDVI (*e.g.*, Mongolian gazelles (*Procapra gutturosa*; [19,20]); and African buffalo (*Syncerus caffer*; [61,62]).

A key step in our approach required a selection of threshold to convert a continuous model of habitat suitability at the range-wide scale to a binary presence-absence map of potential distribution (or "available habitat" *sensu* Johnson [1980]). An alternative approach that might prove useful to explore would utilize the continuous distribution of habitat suitability as an informative prior in modeling habitat selection at finer scales. However, it is unclear whether animals select habitat in this manner. We suggest that, for GKRs, climate, soil, and topography serve as a simple filter to defining the potential distribution. That is, for example, either a GKR can construct a burrow in a particular soil type or it can't – we do not expect GKRs to have a continuous response to resource abundance in relation to soil particle size. Nevertheless, additional research on the relationship across habitat selection at multiple scales is warranted.

This study of GKR distribution in Carrizo, while conducted at a relatively small spatial scale, focused on the temporal dynamics of species' distributions. Niche and distribution theory tend to assume a species is at equilibrium, but this study and others (e.g., [21,63]) show that for many applications, considering the temporal dynamics of a species' distribution is essential. Although the importance of non-static suitability models in grassland systems has been recognized [64], the difficulties in addressing such variability have thus far limited research in this area [65].

This study focused specifically on resource abundance as a limiting factor for GKR.

While the approach presented here combining distribution models at different scales allows new insights, it is not without its shortcomings. One particular problem is our inability to identify the "true" potential distribution. By its very nature, it may be impossible to know a species' potential distribution; in fact the potential distribution may only be a theoretical construct. We can only measure the realized distribution and estimate the potential distribution from those measurements. This issue is highlighted regularly in the invasive species modeling literature. Species that appear to have a limited distribution in their native range often show a spectacular ability to live in "unsuitable" conditions when introduced to new areas (e.g., [66,67]). In these cases, the species' realized distribution in its native range is so limited by competition, dispersal, and other ecological factors that any estimate of its potential distribution will be woefully inadequate for predicting the spread of a species. Oftentimes, ecological limits to the realized distribution may be correlated with environmental conditions, thereby preventing true knowledge of the species' limits of its potential distribution. In this case, additional steps (*e.g.* physiological tests) may be required to estimate its potential distribution.

As for GKR's competitive dominance, the relationship between precipitation limitation and competition may be impossible to untangle. The *Heteromyidae* in general appear to have evolved to claim a desert-grassland niche unfilled by other small mammals. The observed relationship between dry summer months and GKR presence may be as much related to the *lower* limit for larger rodents (*e.g.*, the California ground squirrel, *Otospermophilus beecheyi*) than an upper limit for GKR. Again, this illustrates the conceptual difficulty surrounding niche theory, but the temporal mechanisms outlined in this study ought to remain relevant. GKR display differing responses to precipitation at range-wide and local

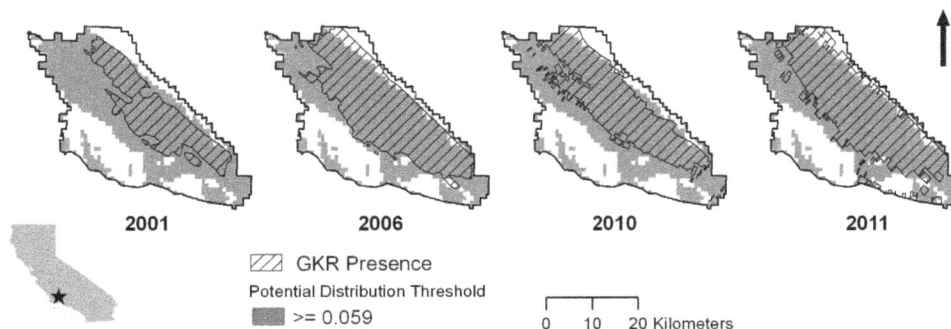

Figure 3. Results of GKR distribution mapping and potential distribution modeling. Hatched polygons show areas of GKR activity in 2001, 2006, 2010 and 2011. Dark grey areas indicate the thresholded potential distribution for GKR from a range-wide Maxent model using presence points from confirmed GKR trapping locations and museum records. A fixed model is unsuitable for predicting annual changes in population extent.

Table 3. Logistic regression of GKR presence in relation to NDVI.

Model	AIC	ΔAIC	w_i
$-1.26+6.11*MaxNDVI_{T1}+4.15*MaxNDVI_{T0} - 23.79*MinNDVI_{T0} - 3.19*SlopeNDVI_{T0} -0.25*year2006 - 0.55*year2010+0.44*year2011$	2,425.8	0	1
$MaxNDVI_{T1}+MaxNDVI_{T0}+MinNDVI_{T0}+year$	2,439.3	13.50	0
$MaxNDVI_{T1}+MinNDVI_{T0}+SlopeNDVI_{T0}+year$	2,452.7	26.91	0
$MaxNDVI_{T1}+MinNDVI_{T0}+year$	2,461.5	35.72	0
$MaxNDVI_{T0}+MinNDVI_{T0}+SlopeNDVI_{T0}+year$	2,502.9	77.05	0
$MaxNDVI_{T0}+MinNDVI_{T0}+year$	2,514.7	88.94	0
$MaxNDVI_{T1}+year$	2,623.9	198.15	0
$MaxNDVI_{T1}+SlopeNDVI_{T0}+year$	2,624.8	199.03	0
$MaxNDVI_{T1}+MaxNDVI_{T0}+year$	2,625.2	199.35	0
$MaxNDVI_{T1}+MaxNDVI_{T0}+SlopeNDVI_{T0}+year$	2,626.1	200.34	0
$MaxNDVI_{T0}+year$	2,702.4	276.56	0
$MaxNDVI_{T0}+SlopeNDVI_{T0}+year$	2,703.6	277.82	0

The full model performed the strongest. In this model, GKR presence is positively correlated with peak primary productivity in the current year and previous year, and negatively correlated with minimum NDVI and NDVI slope in the previous year. This suggests that the best predictor of GKR presence in a given year is a positive correlation with resource abundance over two years, and presence in the area the previous year.

scales. This fact is a crucial finding for those interested in modeling ecologically relevant species' distributions.

Incorporating detailed mechanisms into species distribution models, at ecologically relevant scales and informed by ecological theory is an important next step in the field of spatial ecology. We have presented an approach to estimating a species' potential distribution and address questions about the ecological limits to its realized distribution. We presented further evidence that non-equilibrium populations are often limited not just by fixed, environmental conditions, but also other ecological conditions that vary spatially and temporally. Such research will be important as distribution modeling moves from the "how" to the "why."

References

1. Anderson BJ, Arroyo BE, Collingham YC (2009) Using distribution models to test alternative hypotheses about a species' environmental limits and recovery prospects. Biological Conservation 142: 488–499. doi:10.1016/j.biocon.2008.10.036.
2. Kremen C, Cameron A, Moilanen A, Phillips SJ (2008) Aligning conservation priorities across taxa in Madagascar with high-resolution planning tools. Science. doi:10.1126/science.1155193.
3. Franklin J (2013) Species distribution models in conservation biogeography: developments and challenges. Diversity and Distributions 19: 1217–1223. doi:10.1111/ddi.12125.
4. Guisan A, Zimmermann NE (2000) Predictive habitat distribution models in ecology. Ecological Modelling 135: 147–186.
5. Peterson AT, Soberon J, Pearson RG, Anderson RP, Martinez-Meyer E, et al. (2011) Ecological Niches and Geographic Distributions (MPB-49). Princeton University Press. 1 pp. doi:10.2307/j.ctt7stnh.
6. Hanski I (1999) Metapopulation ecology. Oxford; New York: Oxford University Press.
7. Guisan A, Thuiller W (2005) Predicting species distribution: offering more than simple habitat models. Ecology Letters 8: 993–1009. doi:10.1111/j.1461-0248.2005.00792.x.
8. Soberon J, Peterson T (2005) Interpretation of Models of Fundamental Ecological Niches and Species' Distributional Areas. Biodiversity Informatics. Available: https://journals.ku.edu/index.php/jbi/article/view/4/2.
9. Hutchinson GE (1957) Concluding remarks.: Cold Spring Harbor Symposia on Quantitative Biology. 13 pp.
10. Johnson DH (1980) The comparison of usage and availability measurements for evaluating resource preference. Ecology 61: 65–71.
11. Wiens JA, Rotenberry JT, Van Horne B (1987) Habitat occupancy patterns of North American shrubsteppe birds: the effects of spatial scale. Oikos 48: 132–147.
12. Thomas DL, Taylor EJ (2006) Study designs and tests for comparing resource use and availability II. Journal of Wildlife Management. doi:10.2193/0022-541X(2006)70[324:SDATFC]2.0.CO;2.
13. Boyce M, McDonald L (1999) Relating populations to habitats using resource selection functions. Trends in Ecology & Evolution 14: 268–272.
14. Garshelis DL (2000) Delusions in habitat evaluation: measuring use, selection, and importance. In: Fuller TK, Boitani L, editors. Research Techniques in Animal Ecology. New York, NY. pp. 111–157.
15. Pulliam HR (2000) On the relationship between niche and distribution. Ecology Letters 3: 349–361. doi:10.1046/j.1461-0248.2000.00143.x.
16. Pearson RG, Dawson TP, Liu C (2004) Modelling species distributions in Britain: a hierarchical integration of climate and land-cover data. Ecography 27: 285–298. doi:10.1111/j.0906-7590.2004.03740.x.
17. Grinnell J (1917) Field Tests of Theories Concerning Distributional Control. The American Naturalist 51: 115–128. doi:10.2307/2456106.
18. Pettorelli N, Vik JO, Mysterud A, Gaillard J-M, Tucker CJ, et al. (2005) Using the satellite-derived NDVI to assess ecological responses to environmental change. Trends in Ecology & Evolution 20: 503–510. doi:10.1016/j.tree.2005.05.011.
19. Ito TY, Miura N, Lhagvasuren B, Enkhbileg D, Takatsuki S, et al. (2006) Satellite tracking of Mongolian gazelles (Procapra gutturosa) and habitat shifts in their seasonal ranges. Journal of Zoology 269: 291–298. doi:10.1111/j.1469-7998.2006.00077.x.
20. Mueller T, Olson KA, Fuller TK (2008) In search of forage: predicting dynamic habitats of Mongolian gazelles using satellite-based estimates of vegetation productivity. Journal of Applied Ecology 45: 649–658.
21. Bissonette JA, Storch I (2007) Temporal dimensions of landscape ecology: wildlife responses to variable resources. New York, NY: Springer.
22. Basille M, Fortin D, Dussault C, Ouellet JP (2013) Ecologically based definition of seasons clarifies predator–prey interactions. Ecography 36: 220–229.

Acknowledgments

We thank J. Hurl, K. Sharum and L. Saslaw for logistical field support; S. Beissinger, C. Burton, K. Fiorella, C. Golden, C. Gurney, M. Kelly, W. Lidicker, L. Prugh, E. Rubidge, and S. Sawyer for comments on multiple drafts. We are indebted to the thoughtful comments of M. Basille and three anonymous reviewers on earlier versions of this manuscript.

Author Contributions

Conceived and designed the experiments: WB JB RS HSB. Performed the experiments: WB RS. Analyzed the data: WB. Wrote the paper: WB JB RS HSB.

23. Falcucci A, Ciucci P, Maiorano L (2009) Assessing habitat quality for conservation using an integrated occurrence-mortality model. Journal of Applied Ecology 46: 600–609.

24. Pettorelli N (2013) The Normalized Difference Vegetation Index. Oxford University Press. 1 pp.

25. Kawamura K, Akiyama T, Yokota H, Tsutsumi M, Yasuda T, et al. (2005) Monitoring of forage conditions with MODIS imagery in the Xilingol steppe, Inner Mongolia. International Journal of Remote Sensing 26: 1423–1436. doi:10.1080/01431160512331326783.

26. Andreo V, Lima MP, Provensal C (2009) ... Polop, J.(2009) Population dynamics of two rodent species in agroecosystems of central Argentina: intra-specific competition, land-use, and climate effects. 10 pp.

27. Hamel S, Garel M, Bianchet MF (2009) Spring Normalized Difference Vegetation Index (NDVI) predicts annual variation in timing of peak faecal crude protein in mountain ungulates. Journal of Applied Ecology 46: 582–589.

28. Cao L, Cova TJ, Dennison PE (2011) Using MODIS satellite imagery to predict hantavirus risk. Global Ecology and Biogeography 20: 620–629.

29. Pettorelli N, Ryan S, Mueller T, Bunnefeld N, Jędrzejewska B, et al. (2011) The Normalized Difference Vegetation Index (NDVI): unforeseen successes in animal ecology. Climate Research 46: 15–27. doi:10.3354/cr00936.

30. Williams DF (1992) Geographic distribution and population status of the giant kangaroo rat, Dipodomys ingens (Rodentia, Heteromyidae). In: Williams DF, Byrne S, Rado TA, editors. Endangered and sensitive species of the San Joaquin Valley, California their biology, management, and conservation. Sacramento: Endangered and sensitive species of the San Joaquin Valley. pp. 301–327. doi:10.1111/j.1365-2656.2011.01930.x/full.

31. Grinnell J (1932) Habitat Relations of the Giant Kangaroo Rat. Journal of Mammalogy 13: 305–320. doi:10.2307/1374134.

32. Williams DF, Kilburn KS (1991) Dipodomys ingens. Mammalian Species: 1–7.

33. Phillips SJ, Anderson RP, Schapire RE (2006) Maximum entropy modeling of species geographic distributions. Ecological Modelling 190: 231–259.

34. Hawbecker AC (1944) The giant kangaroo rat and sheep forage. The Journal of Wildlife Management 8: 161–165.

35. Goldingay RL, Kelly PA, Williams DF (1997) The Kangaroo Rats of California: Endemism and Conservation of Keystone Species. 3: 47.

36. Prugh LR, Brashares JS (2012) Partitioning the effects of an ecosystem engineer: kangaroo rats control community structure via multiple pathways. Journal of Animal Ecology 81: 667–678. doi:10.2307/41496035.

37. Shaw WT (1934) The ability of the giant kangaroo rat as a harvester and storer of seeds. Journal of Mammalogy 15: 275–286. doi:10.2307/1374514.

38. Bean WT, Stafford R, Prugh LR, Scott Butterfield H, Brashares JS (2012) An evaluation of monitoring methods for the endangered giant kangaroo rat. Wildlife Society Bulletin 36: 587–593. doi:10.1002/wsb.171.

39. Germano DJ, Rathbun GB (2012) Effects of grazing and invasive grasses on desert vertebrates in California. The Journal of Wildlife Management 76: 670–682. doi:10.2307/41519406.

40. Hijmans RJ, Phillips SJ, Leathwick JR, Elith J (n.d.) Package "dismo." Available: http://cran.r-project.org/web/packages/dismo/index.html.

41. R Core Team (2012) R A language and environment for statistical computing. Available: http://www.R-project.org/.

42. Bean WT, Prugh LR, Stafford R, Butterfield HS, Westphal M, et al. (2014) Species distribution models of an endangered rodent offer conflicting measures of habitat quality at multiple scales. Journal of Applied Ecology 51: 1116–1125. doi:10.1111/1365-2664.12281.

43. Hijmans RJ, Cameron SE, Parra JL, Jones PG, Jarvis A (2005) Very high resolution interpolated climate surfaces for global land areas. International Journal of Climatology 25: 1965–1978.

44. Hijmans RJ, Graham CH (2006) The ability of climate envelope models to predict the effect of climate change on species distributions. Global Change Biology 12: 2272–2281.

45. United States Geological Survey (n.d.) Shuttle Radar Topography Mission.

46. Soil Survey Staff, Natural Resources Conservation Service, United States Department of Agriculture (2012) Soil Survey Geographic (SSURGO) Database. Available: http://sdmdataaccess.nrcs.usda.gov/.

47. Bean WT, Stafford R, Brashares JS (2012) The effects of small sample size and sample bias on threshold selection and accuracy assessment of species distribution models. Ecography. doi:10.1111/j.1600-0587.2011.06545.x.

48. Liu C, Berry PM, Dawson TP, Pearson RG (2005) Selecting thresholds of occurrence in the prediction of species distributions. Ecography 28: 385–393.

49. Liu C, White M, Newell G (2013) Selecting thresholds for the prediction of species occurrence with presence-only data. Journal of Biogeography 40: 778–789.

50. Carroll ML, M DC, Sohlberg RA, Townshend JR (n.d.) 250m MODIS Normalized Difference Vegetation Index.

51. Huete A, Didan K, Miura T, Rodriguez EP, Gao X (2002) Overview of the radiometric and biophysical performance of the MODIS vegetation indices. Remote Sensing of Environment 83: 195–213.

52. Butterfield HS, Malmström CM (2009) The effects of phenology on indirect measures of aboveground biomass in annual grasses. International Journal of Remote Sensing 30: 3133–3146.

53. Burnham KP, Anderson DR (2002) Model selection and multimodel inference: a practical information-theoretic approach. New York: Springer-Verlag.

54. Freeman EA, Moisen G (2008) PresenceAbsence: An R package for presence absence analysis. Journal of Statistical Software 23.

55. Allouche O, Tsoar A, Kadmon R (2006) Assessing the accuracy of species distribution models: prevalence, kappa and the true skill statistic (TSS) - ALLOUCHE - 2006 - Journal of Applied Ecology - Wiley Online Library. Journal of Applied Ecology 43: 1223–1232.

56. Rushton SP, Ormerod SJ, Kerby G (2004) New paradigms for modelling species distributions? - Rushton - 2004 - Journal of Applied Ecology - Wiley Online Library. Journal of Applied Ecology 41: 193–200.

57. Austin M (2007) Species distribution models and ecological theory: a critical assessment and some possible new approaches. Ecological Modelling 200: 1–19.

58. Pearson RG, Dawson TP (2003) Predicting the impacts of climate change on the distribution of species: are bioclimate envelope models useful? Global Ecology and Biogeography 12: 361–371.

59. Franklin J (2010) Moving beyond static species distribution models in support of conservation biogeography. Diversity and Distributions 16: 321–330. doi:10.2307/40604232.

60. Morin X, Thuiller W (2009) Comparing niche- and process-based models to reduce prediction uncertainty in species range shifts under climate change. Ecology 90: 1301–1313.

61. Ryan SJ, Knechtel CU, Getz WM (2006) Range and habitat selection of African buffalo in South Africa. Journal of Wildlife Management 70: 764–776. doi:10.2193/0022-541X(2006)70%5B764:RAHSOA%5D2.0.CO;2.

62. Viña A, Bearer S, Zhang H, Ouyang Z, Liu J (2008) Evaluating MODIS data for mapping wildlife habitat distribution. Remote Sensing of Environment 112: 2160–2169.

63. Suárez-Seoane S, García de la Morena EL, Morales Prieto MB, Osborne PE, de Juana E (2008) Maximum entropy niche-based modelling of seasonal changes in little bustard (Tetrax tetrax) distribution. Ecological Modelling 219: 17–29. doi:10.1016/j.ecolmodel.2008.07.035.

64. Fryxell JM, Wilmshurst JF, Sinclair ARE (2004) Predictive Models of Movement by Serengeti Grazers. Ecology 85: 2429–2435. doi:10.1890/04-0147.

65. Fernandez Gimenez ME, Allen Diaz B (1999) Testing a non-equilibrium model of rangeland vegetation dynamics in Mongolia. Journal of Applied Ecology 36: 871–885. doi:10.1046/j.1365-2664.1999.00447.x.

66. Beaumont LJ, Gallagher RV, Thuiller W, Downey PO, Leishman MR, et al. (2009) Different Climatic Envelopes among Invasive Populations May Lead to Underestimations of Current and Future Biological Invasions. Diversity and Distributions 15: 409–420. doi:10.1111/j.1472-4642.2008.00547.x.

67. Steiner FM, Schlick-Steiner BC, VanDerWal J, Reuther KD, Christian E, et al. (2008) Combined modelling of distribution and niche in invasion biology: a case study of two invasive Tetramorium ant species. Diversity and Distributions 14: 538–545. doi:10.1111/j.1472-4642.2008.00472.x.

Community Structure of Skipper Butterflies (Lepidoptera, Hesperiidae) along Elevational Gradients in Brazilian Atlantic Forest Reflects Vegetation Type Rather than Altitude

Eduardo Carneiro[1]*, **Olaf Hermann Hendrik Mielke**[1], **Mirna Martins Casagrande**[1], **Konrad Fiedler**[2]

1 Laboratório de Estudos de Lepidoptera Neotropical, Zoology Department, UFPR. Curitiba, Paraná, Brasil, **2** Division of Tropical Ecology & Animal Biodiversity, University of Vienna, Vienna, Austria

Abstract

Species turnover across elevational gradients has matured into an important paradigm of community ecology. Here, we tested whether ecological and phylogenetic structure of skipper butterfly assemblages is more strongly structured according to altitude or vegetation type along three elevation gradients of moderate extent in Serra do Mar, Southern Brazil. Skippers were surveyed along three different mountain transects, and data on altitude and vegetation type of every collection site were recorded. NMDS ordination plots were used to assess community turnover and the influence of phylogenetic distance between species on apparent community patterns. Ordinations based on ecological similarity (Bray-Curtis index) were compared to those based on phylogenetic distance measures (MPD and MNTD) derived from a supertree. In the absence of a well-resolved phylogeny, various branch length transformation methods were applied together with four different null models, aiming to assess if results were confounded by low-resolution trees. Species composition as well as phylogenetic community structure of skipper butterflies were more prominently related to vegetation type instead of altitude per se. Phylogenetic distances reflected spatial community patterns less clearly than species composition, but revealed a more distinct fauna of monocot feeders associated with grassland habitats, implying that historical factors have played a fundamental role in shaping species composition across elevation gradients. Phylogenetic structure of community turned out to be a relevant additional tool which was even superior to identify faunal contrasts between forest and grassland habitats related to deep evolutionary splits. Since endemic skippers tend to occur in grassland habitats in the Serra do Mar, inclusion of phylogenetic diversity may also be important for conservation decisions.

Editor: M. Alex Smith, University of Guelph, Canada

Funding: Funding was provided by CNPq and CAPES to EC as a doctoral scholarship. OHHM and MMC are granted with "Produtividade de Pesquisa"; see (http://www.cnpq.br/web/guest/apresentacao13). The funders had no role in study design, data collection and analysis, decision to publish, or preparation of the manuscript.

Competing Interests: The authors have declared that no competing interests exist.

* Email: carneiroeduardo@hotmail.com

Introduction

Since the early observations by Forster [1] and von Humboldt [2], species turnover across altitudinal gradients has matured into an important paradigm of community ecology. High mountain chains are usually selected for analyses, mainly because their wide elevational ranges increase the power of detecting patterns in community structure (see [3,4]). Geologically older mountain chains whose peaks are considerably lower as a consequence of long exposition to erosion forces have received less attention, since effects of elevation on species diversity or species composition might not be as evident. In such places, historical events may play an additional fundamental role in assemblage turnover in addition to altitude. More recently, the integration of phylogenetic structure into community ecology has yielded important insights as to how historical factors influence community structure [5].

The Serra do Mar is located near the south-eastern coast of Brazil and represents a good example of an old mountain chain in the southern hemisphere, where animal and plant communities might have been structured through both ecological and historical processes. The start of its rising processes has been dated at about 90 mya, soon after the splitting of Gondwana into its two biggest daughter continents, viz. South America and Africa [6]. Despite its moderate altitudinal amplitude today (peaks generally between 800–1500 m), distinct climatic and pedological shifts occur along these rather short elevational gradients, together with a high endemism rate in certain plant taxa such as Bromeliaceae, Orchidaceae and Gesneriaceae [6,7].

Environmental dimensions change in a predictable manner with increasing altitude in the Serra do Mar. Mean annual temperature decreases by an average of 0.5–0.6 K per 100 m altitude, while soil depth decreases and wind intensity increases [8]. Correlated to those abiotic factors and modulated through superimposed effects of anthropogenic land-use, four different vegetation types can be observed on these mountains: Atlantic rain forest, early succession flora, cloud forest, and grasslands [7]. While Atlantic rain forest prevails in the lowlands (up to 1100 or 1400 m, depending on the mountain), grasslands are generally restricted to altitudes above

1300 m. These grasslands are relict vegetation ecosystems which had been far more widespread in south and southeast Brazil during dry cold periods of the late Quaternary [9]. Since then, this vegetation has become increasingly restricted to mountain tops, due to forest expansion, with many animal species related to this kind of ecosystem being included in red lists [10,11]. Early successional flora is not only found after human interference, but also due to natural disturbance events such as forest fires or landslides [7,8].

Communities of host-specific herbivores, whose life histories are intrinsically related to the vegetation of their habitats, may either respond directly to abiotic environmental gradients, or alternatively their species turnover may rather track biotic changes in vegetation types [12,13]. The challenge remains to disentangle whether abiotic (climate, area size) or biotic conditions (vegetation) are more important as drivers of their community composition and species turnover along elevational gradients [14,15], and how phylogenetic history contributes to understanding contrasting patterns between taxa [5,16].

Various groups of nocturnal Lepidoptera have frequently served as targets to study elevational diversity patterns in the tropics [17–19], yet surprisingly few quantitative butterfly studies do exist from elevational gradients on tropical mountains [20,21]. Especially, the few studies addressing phylogenetic structure in butterfly communities along elevational gradients are extra-tropical [22,23], where faunas are distinctly less diverse (but see [24] for a recent example on a rich Neotropical moth assemblage). Skipper butterflies (Lepidoptera, Hesperioidea) might also have the potential to address elevational patterns and underlying processes, but thus far no specific case study on this somehow elusive family exists from mountain ranges anywhere in the tropics.

Recent studies revealed contrasting patterns of phylogenetic community structure along environmental gradients, i.e. different assemblages respond to environmental or biotic factors depending on their specific requirements, evolutionary history and biogeography [25–28]. Skippers might have particular potential to reveal how vegetation types and altitude influence phylogenetic composition along altitudinal gradients, since major skipper lineages are conservative and contrasting in relation to their larval food plant affiliations [29]. While Hesperiinae larvae feed exclusively on monocotyledonous plants, Pyrginae larvae are bound to various families of dicotyledonous plants, mainly Fabaceae, Malvaceae and Rutaceae [29–31]. In addition, skippers are the only group of butterflies commonly observed throughout all sections of altitudinal gradients in south Brazil (pers. observ.), and they comprise far

Table 1. Two different sets of sample unit delimitations used to analyze skipper assemblages in the Serra do Mar (Brazil)[1].

Samples 1 (m)	Mountain	Veg. Type	Samples 2 (m)	Mountain	Veg. Type	Altitude Class
1000–1100	Anhangava	FOR	998–1060	Anhangava	FOR	low
1100–1200	Anhangava	FOR+ESV	1061–1122	Anhangava	FOR	low
1200–1300	Anhangava	ESV+GRA	1123–1206	Anhangava	ESV	medium
1300–1400	Anhangava	GRA	1207–1289	Anhangava	ESV	medium
1400–1500	Anhangava	GRA	1290–1364	Anhangava	GRA	medium*
900–1000	Araçatuba	FOR+ESV	1365–1440	Anhangava	GRA	high
1000–1100	Araçatuba	ESV	912–938	Araçatuba	FOR	low
1100–1200	Araçatuba	GRA	939–1019	Araçatuba	ESV	low
1200–1300	Araçatuba	GRA	1020–1099	Araçatuba	ESV	low
1300–1400	Araçatuba	GRA	1100–1175	Araçatuba	GRA	low
1400–1500	Araçatuba	GRA	1176–1250	Araçatuba	GRA	medium
1500–1600	Araçatuba	GRA	1251–1325	Araçatuba	GRA	medium
1600–1700	Araçatuba	GRA	1326–1400	Araçatuba	GRA	medium
900–1000	Caratuva	FOR	1401–1475	Araçatuba	GRA	high
1000–1100	Caratuva	FOR+ESV	1476–1550	Araçatuba	GRA	high*
1100–1200	Caratuva	ESV	1551–1625	Araçatuba	GRA	high
1200–1300	Caratuva	ESV	1625–1682	Araçatuba	GRA	high
1300–1400	Caratuva	FOR	980–1031	Caratuva	FOR	low
1400–1500	Caratuva	FOR+GRA	1032–1083	Caratuva	FOR	low
1800–1900	Caratuva	GRA	1084–1158	Caratuva	ESV	low
			1159–1233	Caratuva	ESV	medium
			1234–1306	Caratuva	ESV	medium
			1307–1362	Caratuva	FOR	medium
			1363–1418	Caratuva	FOR	medium
			1419–1488	Caratuva	GRA	high
			1800–1860	Caratuva	GRA	high*

[1]Samples 1: delimited only by altitude; Samples 2: delimited by vegetation type and altitude. Each location is assigned to mountains, elevational belts and vegetation types. Note that the delimitation by altitude plus vegetation increases the number of sample units. Abbreviations: FOR: forest; ESV: early successional vegetation; GRA: grassland.

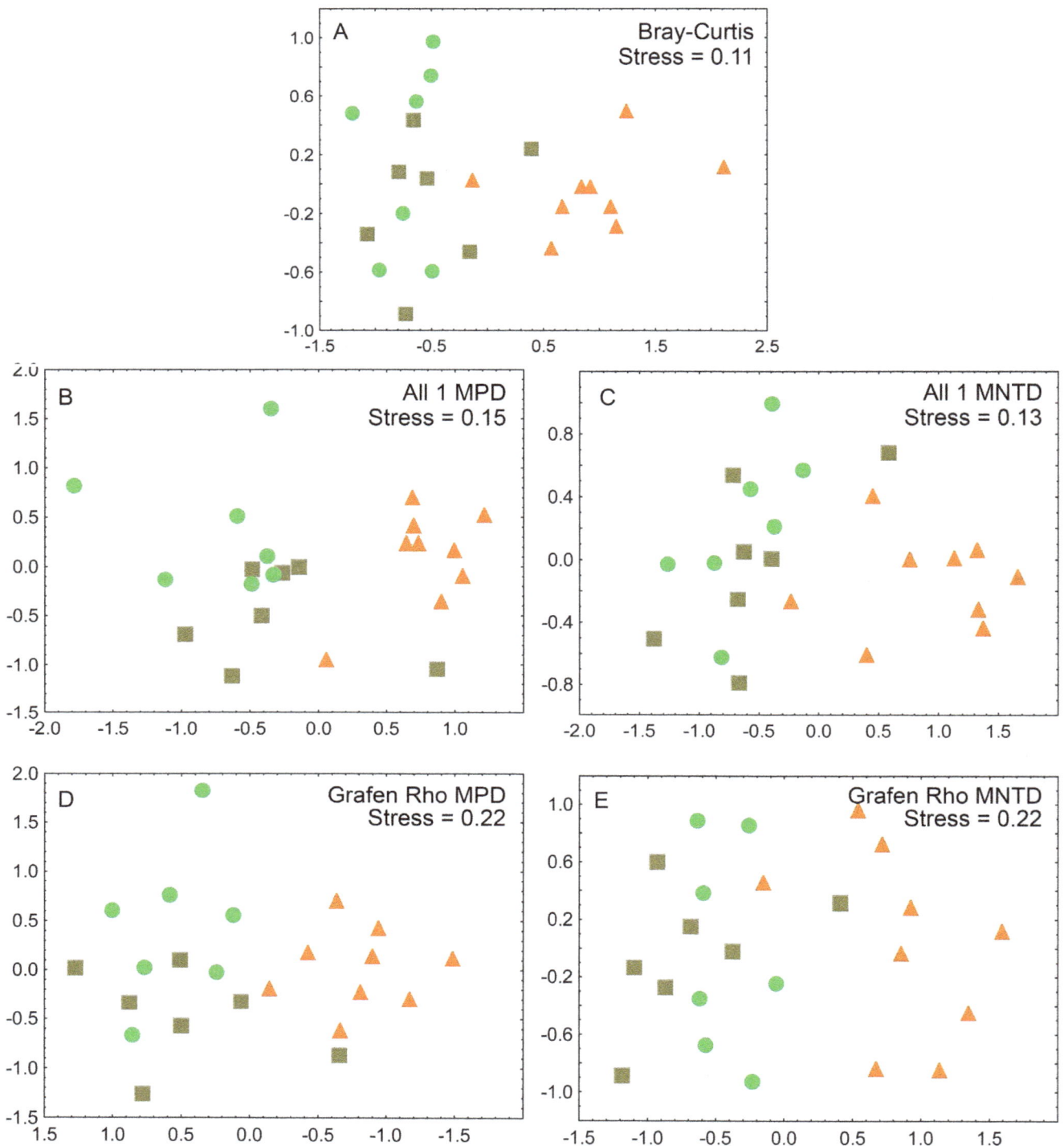

Figure 1. NMDS ordination plots of Hesperiidae assemblages along elevational gradients in Serra do Mar, Brazil. Ordination patterns were assessed based on Bray-Curtis similarities of species lists (A), and compared to two phylogenetic distance indexes (MPD and MNTD) using equal branch lengths (All 1) and Grafen's Rho transformation method (charts B–E). Samples are scored according to altitude and vegetation types. Assemblages are basically ordered along the first axis from low (left) to high elevations (right). Symbols: green circles (forest), brown squares (early successional vegetation), orange triangles (grassland). Stress values indicate goodness of fit of two-dimensional representations to the underlying distance matrices.

larger numbers of grassland endemics when compared to Papilionoidea families in the region [32].

The present study aimed to elucidate the structure of Hesperiidae assemblages along altitudinal gradients of moderate extension, using ecological and phylogenetic measures of community similarity, on three different mountains, in relation to different vegetation types. Specifically, the following hypotheses were tested: 1. Assemblages of skippers are structured according to vegetation type as well as altitude; 2. Phylogenetic similarity is more informative for community responses to altitude and vegetation than ecological similarity (e.g. species composition); 3. Vegetation has a stronger influence on skipper assemblages than altitude; 4. Subfamilies of skippers show discordant community patterns because they differ in life-history traits governing their habitat preferences.

Methods

Study location and sampling methods

Three mountains located in the Serra do Mar of Paraná state, Brazil were sampled in this study: Anhangava (25°23′30″S; 49°00′15″W), Araçatuba (25°54′07″S; 48°59′37″W) and Caratuva mountain (25°13′30″S; 48°51′40″W). The three locations sampled are embedded in protected areas regulated by the state environmental agency (IAP/PR). For location details see [33]. Capture of specimens and their transport to the laboratory for subsequent identification were permitted by licences n° 59.08 (IAP/PR) and 14.595-1 (IBAMA/Sisbio). Hesperiidae specimens were captured with insect nets during up- and down-walking of transects. Each transect was walked 11 times from 2009 to 2011, between 9:00 and 16:00 h. All recorded specimens were immediately labeled according to the elevation (measured to the nearest 10 m using a Garmin 60Cx GPS device) and vegetation type of their sampling locality. In all, 1578 records of 155 species make up the data on which all analyses are based. No endangered species were recorded in this study.

On the mountains in Serra do Mar up to four different vegetation types are present above 900 m: montane forest, cloud forest, early successional vegetation, and grassland. Montane forest refers to a well-developed, tall grown (up to 10 m), vertically stratified Atlantic Rainforest located throughout the slopes of Serra do Mar. Cloud forest stands are more dense, lower in growth (tree height from 3–7 m) with less well defined strata. Successional vegetation and grasslands both lack a canopy stratum. Succession vegetation is dominated by bracken fern *Pteridium aquilinum* (Dennstaedtiaceae), and it is often located where fire or human impact have occurred recently [8]. Grassland sites present a varied floral composition, dominated by several species of Poaceae and Cyperaceae, but occasionally with scattered low trees and shrubs [7].

Although these vegetation types are in part related to altitude (e.g. grasslands are located on mountain tops), their ranges also vary according to a mountain's relief, soil depth, or biogeographical history [7]. This peculiarity allowed us to examine whether vegetation or altitude *per se* plays the more important role on community differentiation. Therefore, skipper samples were delimited and analyzed in two different ways: first, the whole altitudinal range on each mountain was divided into belts of 100 m elevational extent, without considering changes of vegetation types within; second, samples were again delimited by altitude, but in addition the prevalent vegetation type was superimposed to delimit sample sites (Table 1). As a consequence, these latter operational units were not equally sized according to elevational bands, but varied from 50 to 75 m extension each. Because only two specimens of skippers were collected in cloud forest, this vegetation type was excluded from all analyses.

Ecological and phylogenetic community structure analyses

NMDS ordinations were used to search for both ecological and phylogenetic structure in skipper assemblages along elevational gradients. This methodology enables the recognition of spatial gradients across communities through comparisons of pairwise similarities, or distances, between all samples [34]. Brehm & Fiedler [35] evaluated different ordination methods for identifying elevational gradients with incompletely sampled communities and concluded that different techniques performed quite similarly. Furthermore, NMDS has the advantage of fixing *a priori* the number of dimensions to be considered for analysis, and to be grounded on rank statistics which renders this ordination method very robust [36,37].

In a first series of NMDS explorations, based on Bray-Curtis matrix similarities, it was assessed whether inclusion, or exclusion, of hilltopping species (i.e. where adult butterflies aggregate at mountain tops for mate location [38]), or the segregation of samples only by altitude (100 m belts), or by altitude plus vegetation type (50–75 m belts), would affect ordination patterns, as already shown for species richness patterns [33]. Since both these factors indeed influenced the ordination of assemblages (Fig. S1), in the subsequent main series of analyses all hilltopping species were omitted and sample sites were classified according to elevation plus vegetation type. Species that exhibit hilltopping behavior are listed in Carneiro et al. [33]. Because hilltopping species were quite numerous, two samples, one from the 1800–1900 m band on Caratuva and the 1476–1550 m band on Araçatuba, became too small and therefore had to be excluded

Table 2. Spearman rank correlation coefficients *r* (plus associated *p*-values) between altitude of sample sites and the site scores along the two ordination axes extracted from NMDS ordinations[1].

	NMDS Axis 1		NMDS Axis 2	
	r	*p*	*r*	*p*
Bray-Curtis	**0.67**	**0.001**	0.31	0.148
MPD All1	**0.54**	**0.008**	0.07	0.758
MPD Grafen's Rho	**0.53**	**0.009**	0.06	0.792
MNTD All1	**0.54**	**0.008**	**0.60**	**0.002**
MNTD Grafen's Rho	**0.49**	**0.019**	**0.49**	**0.017**

[1]Different sets of Hesperiidae assemblages were considered with different measures of species or phylogenetic composition, sampled along altitudinal gradients in Serra do Mar, Paraná, Brazil. 'All 1' refers to equal branch lengths assigned to the tree topology while 'Grafen's Rho' refers to Grafen's branch length transformation method [45]. Correlations that remain significant after applying a table-wide false discovery rate approach are printed in bold face.

Table 3. Pairwise Spearman rank correlation coefficients r (plus associated p-values) for the similarity matrices extracted by ecological and phylogenetic metrics[1].

	Bray-Curtis	All 1 MPD	All 1 MNTD	Grafen Rho MPD
Bray-Curtis	-			
All 1 MPD	0.509 / **<0.001**	-		
All 1 MNTD	0.849 / **<0.001**	0.397 / **0.001**	-	
Grafen Rho MPD	0.608 / **<0.001**	0.787 / **0.001**	0.498 / **<0.001**	-
Grafen Rho MNTD	0.877 / **<0.001**	0.357 / **0.001**	0.957 / **0.001**	0.522 / **<0.001**

[1]Ecological ordination was measured with Bray-Curtis similarity matrix, while four phylogeny-based distance matrices (MPD and MNTD) were calculated using two branch-length options each. 'All 1' refers to equal branch lengths assigned to the tree topology while 'Grafen's Rho' refers to Grafen's Rho branch length transformation method [45]. Correlations that remain significant after applying a table-wide false discovery rate approach are printed in bold face.

altogether. As shown before [33], neither vegetation type nor altitude influenced the efficiency of sampling skippers. Therefore, differences in sample coverage are not expected to affect our results.

The Bray-Curtis similarity index was used to measure ecological similarities of assemblages. Complementarily, the incidence-based Chao-Soerensen index [39] was also calculated, but this did not yield any deviant patters (data not shown). Additionally, two indexes were calculated to assess phylogenetic distances between assemblage samples: the Mean Pairwise Distance (MPD) and the Mean Nearest Neighbor Distance (MNTD) [5]. Both indices were compared because they provide different perspectives of phylogenetic similarities, as an overall pattern of relatedness (MPD) or as how closely related species can be (MNTD) [40]. Additionally, MNTD is more affected by changes at the terminal branches of a phylogeny, whereas MPD is more sensitive to changes at the basis of a phylogeny [40].

To calculate measurements of phylogenetic structure, a phylogenetic hypothesis of the sampled taxa is obviously required. Skipper phylogeny is still very imperfectly resolved [29,41]. Based on the latest phylogenetic approach [29], a tentative community supertree was constructed. Topology of high rank taxa was recovered after [29]. Groups (G) and subgroups (SG) stated by Evans [42] were maintained only when they did not conflict with the current tree topology [29]. Species were clumped according to their respective genera (Fig. S2). When genera or species sampled in our study had not been included in the phylogenetic analyses, they were replaced by their closest tribal affiliates, using expert taxonomical arrangement as best possible surrogate of phylogenetic support [40,43]. Because branch length estimates were not available, and even node ages and fossil calibrations are still largely questioned in butterfly systematics [44], equal branch lengths were arbitrarily assigned to the tree topology (which is the same approach as the nodal distance method; [40]). We then compared these results with branch lengths using Grafen's Rho transformation (r = 0.5) [45] (Fig. S2). Additional branch length transformation methods were also calculated (e.g. Pagel's and Nee's methods: [46,47]), but results did not systematically differ from those presented below (data not shown).

To ascertain that phylogenetic structure of communities obtained from the supertree and branch transformation techniques were different from what would be expected at random, we correlated the empirical phylogenetic distance matrices with matrices produced by four distinct null models (means from 1000 randomizations) through Spearman matrix rank correlation tests (procedure RELATE in PRIMER: [48]). These null models included the shuffling of species labels across the phylogeny (null 0), randomization of species from the sample pool (null 1), randomization of species from the phylogeny pool (null 2), or by swapping versions of sample/species matrix (see [49,50] for details). The choice of an adequate null model is relevant, since their different assumptions might lead to contrasting results [51]. Because our phylogenetic trees do not contain branch length information based on genetic differences, we opted for challenging the results obtained through all these null models and then interpret results.

To test whether taxonomic scale may be informative to community similarities between different altitude and vegetation types, the analyses was first performed for the whole family Hesperiidae, and then separately for communities composed only of Pyrginae+Pyrrhopyginae species (i.e. dicot feeders, hereafter called Pyrginae); or only of Hesperiinae (i.e. monocot feeders). *Urbanus teleus* (Hübner, 1821) was included in the Hesperiinae analyses, since this species has confirmed host-plant records only in

Table 4. Spearman matrix rank correlation coefficients r (999 permutations) between phylogenetic distance matrices MPD (mean pairwise distance) and MNTD (mean nearest neighbor distance and four types of null models: 0–3 [49,50][1].

	Model 0		Model 1		Model 2		Model 3	
	r	p	r	p	r	p	r	p
Hesperiidae								
MPD								
All 1	**0.41**	**0.001**	0.05	0.725	0.02	0.580	0.22	0.030
Grafen Rho	0.06	0.239	0.05	0.303	0.05	0.300	0.04	0.648
MNTD								
All 1	**0.83**	**0.001**	**0.48**	**0.001**	**0.48**	**0.001**	**0.50**	**0.001**
Grafen Rho	**0.17**	**0.013**	**0.27**	**0.003**	**0.27**	**0.001**	**0.37**	**0.001**
Hesperiinae								
MPD								
All 1	**0.60**	**0.001**	0.00	0.524	0.06	0.782	−0.23	0.996
Grafen Rho	0.02	0.376	0.05	0.248	−0.02	0.671	−0.09	0.935
MNTD								
All 1	**0.69**	**0.001**	**0.33**	**0.001**	**0.33**	**0.001**	**0.33**	**0.001**
Grafen Rho	**0.28**	**0.001**	**0.18**	**0.003**	**0.28**	**0.001**	**0.20**	**0.002**
Pyrginae								
MPD								
All 1	**0.28**	**0.013**	0.11	0.209	−0.08	0.753	**0.36**	**0.023**
Grafen Rho	0.18	0.081	−0.32	0.973	−0.05	0.658	−0.07	0.665
MNTD								
All 1	**0.44**	**0.001**	**0.24**	**0.018**	0.18	0.097	**0.31**	**0.001**
Grafen Rho	0.19	0.091	0.18	0.092	0.14	0.153	0.17	0.097

[1]Two different branch length transformation methods were applied with the mean of 999 randomly generated matrices, according to four null models assumptions (see Methods section). 'All 1': equal branch lengths assigned to the tree topology; 'Grafen's Rho': Grafen's branch length transformation method [45]. Significant correlations here indicate that observed sample matrices are not substantially different from random expectations. Null model tests were applied to the three taxa Hesperiidae, Hesperiinae and Pyrginae. Correlations that remain significant after applying a table-wide false discovery rate approach are printed in bold face.

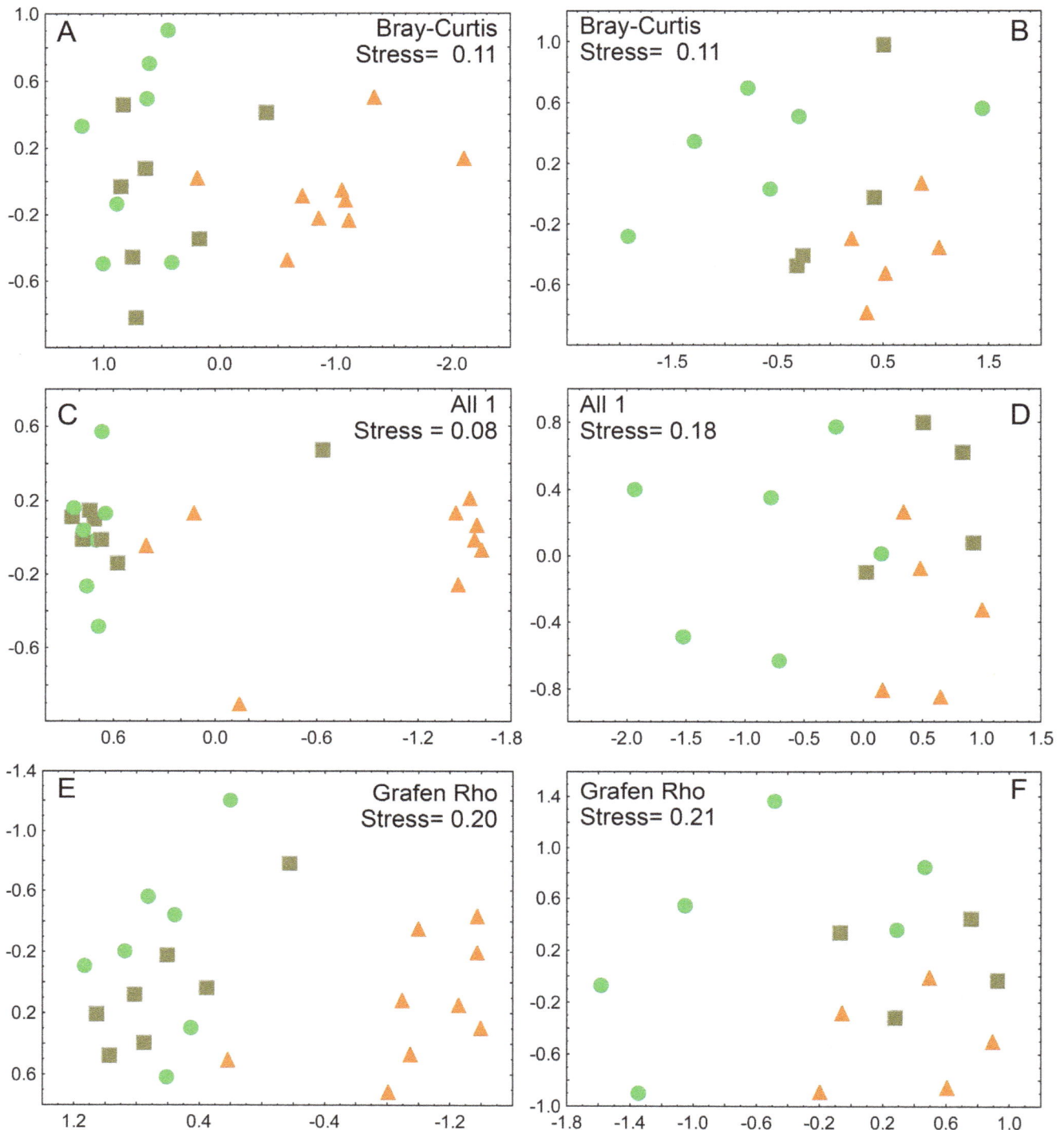

Figure 2. NMDS ordination plots of Hesperiinae (left panels, with monocot-feeding larvae) and Pyrginae (right panels, with larvae feeding on various dicot families) assemblages along elevational gradients in Serra do Mar, Paraná, Brazil. Ordinations are based on ecological (Bray-Curtis: A, B) and phylogenetic (MPD) similarity indexes. Two branch length transformations were applied to obtain MPD: 'All 1' (all branches set to unity: C, D) and Grafen's Rho model (E, F). Samples are partitioned according to altitude and vegetation types. Symbols: green circles (forest), brown squares (early successional vegetation), orange triangles (grassland). Stress values indicate goodness of fit of two-dimensional representations to the underlying distance matrices.

Poaceae [29], disregarding a probably misleading record from Fabaceae [31]. Pyrginae and Hesperiinae similarity matrices were thereafter compared across the same common sampling sites by

Spearman matrix rank correlation, aiming to test whether taxa with such contrasting hostplant associations differ in their ecological and phylogenetic similarity patterns in response to

Table 5. Two-way ANOSIM results (R statistics and associated p-values), evaluating the effects of vegetation type and altitude on ecological similarities (Bray-Curtis) and phylogenetic distances (MPD) of skipper assemblages[1].

	Vegetation		Altitude	
	R^2	p	R^2	p
Hesperiidae				
Bray-Curtis	**0.47**	**0.004**	**0.10**	**0.001**
All1	0.19	0.13	0.13	0.81
Grafen Rho	**0.35**	**0.006**	0.01	0.5
Hesperiinae				
Bray-Curtis	**0.46**	**0.005**	0.05	0.328
All1	**0.49**	**0.004**	0.13	0.18
Grafen Rho	**0.39**	**0.013**	0.08	0.279
Pyrginae				
Bray-Curtis	**0.64**	**0.003**	0.76	0.03
All1	0.08	0.298	0.39	0.852
Grafen Rho	0.06	0.507	0.40	0.856

[1]Analyses were performed at three different taxonomic levels (entire family Hesperiidae, and two major subfamilies), and for two different branch length transformation methods: 'All 1': equal branch lengths assigned to the tree topology; 'Grafen's Rho': Grafen's branch length transformation method [45]. Results printed in bold remained significant after applying a false discovery rate approach.

elevation and vegetation change. Vegetation and altitude were classified as factors and tested in two-way ANOSIM analyses, where R statistics values served as a measure of effect size. Vegetation was divided into forest, early successional vegetation, and grassland, while the elevational categories included in this analyses were 'low' (900 m–1150 m), 'medium' (1150 m–1400 m), and 'high' (1400 m–1650 m; see Table 1).

Bray-Curtis similarity indexes, NMDS ordinations, RELATE tests and two-way ANOSIM were calculated with PRIMER 6.1.13 [48]. Phylogenetic trees were drawn and branch lengths put in and transformed using Mesquite 2.72 [52], including the PDAP Package [53]. Phylogenetic community distances were calculated using COMDIST and COMDISTNT functions available in PHYLOCOM 4.2 [54]. To avoid spurious significance resulting from multiple tests, a "False Discovery Rate" approach [55] was taken [56]. All test results passing adjusted criteria were assigned with an asterisk (*).

Results

Ecological and phylogenetic community structure at family level

The total skipper fauna collected along all transects comprised of 1578 specimens representing 155 species. Spatial resolution revealed by NMDS ordinations uncovered a nearly even gradient-like pattern mainly represented by the first ordination axis, along which communities were ordered from high to low altitudes (Fig. 1a, 'Bray-Curtis' in Table 2). However, assemblages from grassland sites (from elevations between 1100–1650 m elevation) became segregated from all others, whereas skipper assemblages from forest and early successional vegetation tended to be spatially unordered along both axes. Therefore, an elevational gradient was not evident when looking exclusively to points within each vegetation type, indicating that elevation is secondary to vegetation type in shaping species composition.

When skipper assemblages were ordinated by their phylogenetic distances in the MPD model, spatial patterns were remarkably similar, and NMDS representations achieved almost equal goodness-of-fit (Figs. 1b–e). Application of various branch length transformations did not alter the general pattern observed in NMDS ordinations, but two-dimensional representations had poorer fit (i.e. higher stress values and lower correlations of ordination axis scores with altitude) than if assuming equal branch lengths set to unity. Ordinations based on MPD and MNTD measures, respectively, showed similar patterns with regard to group clustering. Axis 2 significantly correlated with altitude in ordinations based on MNTD, but not for those using MPD (Table 2). Spearman rank correlations between the ecological Bray-Curtis similarity matrix and the four phylogeny-based distance matrices (MPD and MNTD with two branch-length options each) were always highly significant ($p \leq 0.001$), but MNTD index showed higher coefficient values with Bray-Curtis ecological index (0.84 to 0.87), compared to MPD (0.50 to 0.60) (Table 3). Strikingly, phylogenetic indexes showed the lowest coefficient values of correlation between each other (0.35 to 0.52).

Differences between MPD and MNTD distance measures were particularly obvious with regard to null-model tests (Table 4). The MNTD matrix correlated strongly with random matrices created under any of the four null models compared. In contrast, correlations between the MPD matrix and null models were only significant (and much weaker so) when all branch lengths were assigned to one, i.e. in null models 0 and 3 (shuffling of phylogenetic terminal labels and swapping species between samples, respectively). Therefore, MNTD results cannot be reliably interpreted as relating to ecological conditions and we restricted further considerations to MPD measures.

Effects of vegetation, altitude and taxon scale: Pyrginae vs. Hesperiinae

When analyzed separately, the two subfamilies Hesperiinae and Pyrginae revealed important differences in their patterns of assemblage similarities (Fig. 2). The NMDS ordination of Hesperiinae rendered a similar configuration as the entire family (Fig. 2a). Grassland skipper assemblages were distinctly set apart, with no clear differentiation between communities associated with forest and early successional vegetation. Pyrginae assemblages, in contrast, revealed a more distinct grouping of forest sites, as opposed to non-forested grassland and early successional sites (Fig. 2b). Ordinations based on phylogenetic distances were qualitatively similar, but tended to show less segregation between vegetation types, especially for the Pyrginae (Figs. 2c–f). These observations are supported by ANOSIM tests. Vegetation type was the main factor responsible for governing Hesperiidae and Hesperiinae assemblages, rather than altitude *per se* (Table 5). *R* values for comparisons based on phylogenetic distance measures were generally lower than those for comparisons of species composition. This was particularly pronounced in the subfamily Pyrginae.

Discussion

Ecological and phylogenetic community structure

Hesperiidae assemblages differed greatly in their species composition across elevational transects in Serra do Mar, even though sampling occurred over a moderate altitudinal range of 500–700 m extension. Although some herbivorous insect taxa are especially sensitive to abiotic gradients related to altitude [57,58], vegetation may exert more direct and distinct effects on species turnover. This was clearly the case with skipper assemblages in our study. Species turnover along elevational gradients has regularly been demonstrated to occur in many groups of plants and animals (for tropical butterflies and moths, e.g. [18,21]). Yet those studies often required far more extensive altitudinal ranges to uncover species turnover patterns.

In unconstrained ordinations, ecological species composition patterns were strikingly similar to those inferred from phylogenetic distances such as MPD and MNTD. In contrast to the number of studies investigating ecological species turnover along elevational gradients, fewer studies thus far attempted to address phylogenetic turnover in community compositions. This is due to the lack of robust phylogenetic hypotheses for most invertebrate groups, especially in tropical biota. In our case, the finding that MNTD distance patterns were not different from random expectations, opposite to MPD, might reflect the inaccuracy of our phylogenetic tree, in which terminal branches are less well established (presence of many polytomies) than is the resolution among higher level groups [29,41]. Skipper phylogeny has been addressed only recently, and molecular information from many genera, especially of Neotropical origin, is still completely lacking. Nevertheless, MPD data allowed us to reveal almost the same spatial pattern as abundance-weighted species turnover. This corroborates that phylogenetic distance information can be relevant also in groups with substantial uncertainty about their phylogenetic relationships [40,59].

Even though different null models, as expected, showed divergent results when applied to skippers [49,51], a non-random general pattern with regard to phylogenetic community distances was clearly observed. In general, various branch length transformation methods did not massively affect the outcome of ordination analyses, but the assumption of equal branch lengths reduced the information content to levels of random relationships between samples. Therefore, trees without genetically founded branch length estimates can still furnish consistent results [40], once different indexes are used and compared to a variety of null models to verify *a priori* whether phylogenetic inaccuracy may have obscured ecologically relevant aspects of phylogenetic community structure.

Effects of vegetation, altitude and taxon scale: Pyrginae vs. Hesperiinae

While numerous studies have elucidated how elevational gradients influence community similarities, only a few specifically addressed the relevance of vegetation types in relation to merely abiotic gradients on animal assemblages [12,60–62]. This integration, however, is essential, since high altitude habitats often represent different vegetation types which not necessarily are concordant to climatic change with elevation alone [14]. In cases where vegetational and abiotic dimensions were related to butterfly or moth assemblages, altitude (and climate) usually emerged as the single best predictor of faunal composition or richness instead of habitat type [12,60,62]. Accordingly, vegetation type at most emerged as modulating patterns of elevational faunal change [61,63,64].

In contrast, vegetation type played the major role with regard to community differentiation in skipper butterflies of Serra do Mar, instead of altitude *per se*. Along elevational transects of moderate extent, factors beyond mere laps rate may gain higher relative importance, as temperature differences are less prominent than on high mountain ranges like the Andes or Himalaya [4]. Accordingly, the presence of distinct vegetation types should attain higher weight in elevation gradients studies, addressing not only a novel ecological constraint (e.g. humidity and solar incidence are clearly contrasting between grasslands and forests), but also evolutionary dimensions, such as insect-plant interactions.

Life history traits are particularly related to the diversification of Hesperiidae. The two major subfamilies are rather conservative in larval food plant affiliations with either monocot or dicot plant families [29]. Hesperiinae (exclusive monocot feeders) are a more recent lineage than all other dicot feeders, with almost twice the global species richness of Pyrginae [29]. Hence diversification rate in this group must be distinctly higher, in contrast to the relatively low diversification rate estimated for the entire family [65]. Another unrelated group of monocot feeding butterflies (viz. Satyrinae) displays a similar pattern of speciation, and the expansion of grassland habitats around the world has been linked to this "explosive diversification" [66]. The high number of Hesperiinae species endemic to grasslands ecosystems in south Brazil reinforces this statement.

Hesperiinae assemblages of grassland sites were more distinctly clustered when analyses were based on phylogenetic distances rather than abundance-weighted species lists. This may hint towards a historical relationship with grassland ecosystems through evolutionary time. In other words, skipper species in this highly distinct vegetation type represent lineages with unique evolutionary history [16]. Consequently, environmental filtering caused by altitudinal climate shifts seems to play only a secondary role in structuring those assemblages.

Implications for Conservation

Conservation of natural grasslands ecosystems is of transcontinental concern [67]. Atlantic forests are biodiversity hot-spots not only because of high species richness of organisms, but also because of high numbers of endemics [68]. Because mountain grassland patches in the Serra do Mar are inserted amongst this forest landscape, endemic organisms of grasslands are routinely counted to this ecosystem, especially because fine scale distribution of grassland patches is hardly represented in distribution maps. Grassland ecosystems embedded in Atlantic Forest are phylogenetically linked to unique skipper assemblages, in analogy to high-altitude grassland butterflies in the European Alps [23]. The reduction of grassland ecosystems in Atlantic Forest is historically related to climate fluctuations, but nowadays these ecosystems are also threatened by occasional anthropogenic fires, invasive exotic grasses and extensive tourism [69]. The present findings highlight the need of conserving these high altitude habitats also from the perspective of a unique skipper butterfly fauna. Although difficult to identify, Neotropical skippers comprise rich assemblages (compared for example to frugivorous butterflies), and are suitable sensitive biological indicators in Atlantic Forest [68]. Hence, further taxonomic and ecological studies into this family are desired, since for most skipper species we still lack information on hostplant associations, phylogenetic relationships, and geographical distribution records. Even, quite a number of unknown species remain to be described.

Supporting Information

Figure S1 Explorative NMDS ordination plots of Hesperiidae assemblages along elevational gradients in Serra do Mar, Brazil. Ordination patterns were first assessed based on Bray-Curtis similarities, with samples partitioned into 100 m altitudinal bands (a), into 100 m altitudinal bands, but excluding hilltopping species (b), and partitioned according to altitude and vegetation types (excluding hilltopping species) c). Arrows indicate mountain summit samples.

Figure S2 Phylogenetic relationships of Hesperiidae (Insecta, Lepidoptera) species recorded in Serra do Mar, Paraná, Brazil. Topology of high rank taxa was recovered after Warren et al. [29]. Groups (G) and subgroups (SG) stated by Evans [42] were maintained only when not conflicting with the topology published in Warren et al. [29]. Species were clumped according to its respectively genera. As no branch lengths are still available for skipper phylogeny, equal branch lengths (above) and Grafen's Rho transformation (below) were arbitrarily assigned to quantify phylogenetic differences between species.

Acknowledgments

The authors are thankful for the critical review of earlier manuscript versions by André V. L. Freitas (UNICAMP), Emygdio L. A. M. Filho (UFPR), Gilson R. P. Moreira (UFRGS), Maurício O. Moura (UFPR) and two anonymous reviewers.

Author Contributions

Conceived and designed the experiments: EC OHHM MMC KF. Performed the experiments: EC. Analyzed the data: EC KF. Contributed reagents/materials/analysis tools: EC OHHM MMC KF. Wrote the paper: EC OHHM MMC KF.

References

1. Forster JR (1778) Observations made during a voyage round the world on physical geography, natural history, and ethic philosophy. London: G. Robinson. 676 p. Available: http://www.biodiversitylibrary.org/item/106982.
2. von Humboldt A (1849) Aspects of nature, in different lands and different climates; with scientific elucidations. Philadelphia: Lea and Blanchard. 480 p. Available: http://www.biodiversitylibrary.org/item/62179.
3. Rahbek C (2005) The role of spatial scale and the perception of large-scale species-richness patterns. Ecol Lett 8: 224–239. doi:10.1111/j.1461-0248.2004.00701.x.
4. Nogués-Bravo D, Araújo MB, Romdal T, Rahbek C (2008) Scale effects and human impact on the elevational species richness gradients. Nature 453: 216–219. doi:10.1038/nature06812.
5. Webb CO, Ackerly DD, McPeek MA, Donoghue MJ (2002) Phylogenies and community ecology. Annu Rev Ecol Syst 33: 475–505. doi:10.1146/annurev.ecolsys.33.010802.150448.
6. Almeida FFM de, Brito Neves BB de, Dal Ré Carneiro C (2000) The origin and evolution of the South American Platform. Earth-Sci Rev 50: 77–111. doi:10.1016/S0012-8252(99)00072-0.
7. Roderjan CV, Galvão F, Kuniyoshi YS (2002) As unidades fitogeográficas do estado do Paraná. Ciênc E Ambiente 26: 693–712.
8. Roderjan CV, Struminsky E (1992) Caracterização e proposta de manejo da Serra da Baitaca - Quatro Barras. Curitiba: FUPEF/FBPN. 121 p.
9. Behling H (2002) South and southeast Brazilian grasslands during Late Quaternary times: a synthesis. Palaeogeogr Palaeoclimatol Palaeoecol 177: 19–27. doi:10.1016/S0031-0182(01)00349-2.
10. Mikich SB, Bernils RS (2004) Livro vermelho da fauna ameaçada no estado do Paraná. Curitiba: Instituto Ambiental do Paraná. 763 p.
11. Machado AMB, Drummond GM, Paglia AP (2008) Livro vermelho da fauna brasileira ameaçada de extinção. Brasília: MMA/Fundação Biodiversitas. 160 p.
12. Gutiérrez D (1997) Importance of historical factors on species richness and composition of butterfly assemblages (Lepidoptera: Rhopalocera) in a northern Iberian mountain range. J Biogeogr 24: 77–88. doi:10.1111/j.1365-2699.1997.tb00052.x.
13. Axmacher JC, Brehm G, Hemp A, Tünte H, Lyaruu HVM, et al. (2009) Determinants of diversity in afrotropical herbivore insects (Lepidoptera: Geometridae): plant diversity, vegetation structure or abiotic factors? J Biogeogr 36: 337–349. doi:10.1111/j.1365-2699.2008.01997.x.
14. Lomolino MV (2001) Elevation gradients of species-density: historical and prospective views. Glob Ecol Biogeogr 10: 3–13. doi:10.1046/j.1466-822x.2001.00229.x.
15. Machac A, Janda M, Dunn RR, Sanders NJ (2011) Elevational gradients in phylogenetic structure of ant communities reveal the interplay of biotic and abiotic constraints on diversity. Ecography 34: 364–371. doi:10.1111/j.1600-0587.2010.06629.x.
16. Graham CH, Fine PVA (2008) Phylogenetic beta diversity: linking ecological and evolutionary processes across space in time. Ecol Lett 11: 1265–1277. doi:10.1111/j.1461-0248.2008.01256.x.
17. Novotny V, Weiblen GD (2005) From communities to continents: beta diversity of herbivorous insects. Ann Zool Fenn 42: 463–475.

18. Beck J, Chey VK (2007) Beta-diversity of geometrid moths from northern Borneo: effects of habitat, time and space. J Anim Ecol 76: 230–237. doi:10.1111/j.1365-2656.2006.01189.x.
19. Muñoz A, Amarillo-Suárez Á (2010) Altitudinal variation in diversity of Arctiidae and Saturniidae (Lepidoptera) in a Colombian cloud forest. Rev Colomb Entomol 36: 292–299.
20. Pyrcz TW, Wojtusiak J (2002) The vertical distribution of pronophiline butterflies (Nymphalidae, Satyrinae) along an elevational transect in Monte Zerpa (Cordillera de Mérida, Venezuela) with remarks on their diversity and parapatric distribution. Glob Ecol Biogeogr 11: 211–221. doi:10.1046/j.1466-822X.2002.00285.x.
21. Pyrcz TW, Wojtusiak J, Garlacz R (2009) Diversity and distribution patterns of Pronophilina butterflies (Lepidoptera: Nymphalidae: Satyrinae) along an altitudinal transect in north-western Ecuador. Neotrop Entomol 38: 716–726. doi:10.1590/S1519-566X2009000600003.
22. Hawkins BA (2010) Multiregional comparison of the ecological and phylogenetic structure of butterfly species richness gradients. J Biogeogr 37: 647–656. doi:10.1111/j.1365-2699.2009.02250.x.
23. Pellissier L, Ndiribe C, Dubuis A, Pradervand J-N, Salamin N, et al. (2013) Turnover of plant lineages shapes herbivore phylogenetic beta diversity along ecological gradients. Ecol Lett 16: 600–608. doi:10.1111/ele.12083.
24. Brehm G, Strutzenberger P, Fiedler K (2013) Phylogenetic diversity of geometrid moths decreases with elevation in the tropical Andes. Ecography 36: 1247–1253. doi:10.1111/j.1600-0587.2013.00030.x.
25. Cavender-Bares J, Keen A, Miles B (2006) Phylogenetic structure of Floridian plant communities depends on taxonomic and spatial scale. Ecology 87: 109–122.
26. Swenson NG, Enquist BJ, Pither J, Thompson J, Zimmerman JK (2006) The problem and promise of scale dependency in community phylogenetics. Ecology 87: 2418–2424.
27. Cadotte MW, Borer ET, Seabloom EW, Cavender-Bares J, Harpole WS, et al. (2010) Phylogenetic patterns differ for native and exotic plant communities across a richness gradient in Northern California. Divers Distrib 16: 892–901. doi:10.1111/j.1472-4642.2010.00700.x.
28. Hoiss B, Krauss J, Potts SG, Roberts S, Steffan-Dewenter I (2012) Altitude acts as an environmental filter on phylogenetic composition, traits and diversity in bee communities. Proc R Soc B Biol Sci 279: 4447–4456. doi:10.1098/rspb.2012.1581.
29. Warren AD, Ogawa JR, Brower AVZ (2009) Revised classification of the family Hesperiidae (Lepidoptera: Hesperioidea) based on combined molecular and morphological data. Syst Entomol 34: 467–523. doi:10.1111/j.1365-3113.2008.00463.x.
30. Beccaloni G, Hall SK, Viloria AL, Robinson GS (2008) Catalogue of the hostplants of the neotropical butterflies = Catálogo de las plantas huésped de las mariposas neotropicales. Zaragoza: Sociedad Entomológica Aragonesa.
31. Robinson GS, Ackerly PR, Kitching IJ, Beccaloni GW, Hernández LM (2010) A database of the World's Lepidopteran hostplants. Available: http://www.nhm.ac.uk/hosts. Accessed 18 August 2010.

32. Dolibaina DR, Mielke OHH, Casagrande MM (2011) Butterflies (Papilionoidea and Hesperioidea) from Guarapuava and vicinity, Paraná, Brazil: an inventory based on records of 63 years. Biota Neotropica 11: 341–354. doi:10.1590/S1676-06032011000100031.

33. Carneiro E, Mielke OHH, Casagrande MM, Fiedler K (2014) Skipper Richness (Hesperiidae) along elevational gradients in Brazilian Atlantic Forest. Neotrop Entomol 43: 27–38. doi:10.1007/s13744-013-0175-8.

34. Legendre P, Legendre L (2013) Numerical ecology. Amsterdam; Boston: Elsevier.

35. Brehm G, Fiedler K (2004) Ordinating tropical moth ensembles from an elevational gradient: a comparison of common methods. J Trop Ecol 20: 165–172. doi:10.1017/S0266467403001184.

36. Minchin PR (1987) An evaluation of the relative robustness of techniques for ecological ordination. Vegetatio 69: 89–107. doi:10.1007/BF00038690.

37. Clarke KR (1993) Non-parametric multivariate analyses of changes in community structure. Aust J Ecol 18: 117–143. doi:10.1111/j.1442-9993.1993.tb00438.x.

38. Pe'er G, Saltz D, Münkemüller T, Matsinos YG, Thulke H-H (2013) Simple rules for complex landscapes: the case of hilltopping movements and topography. Oikos 122: 1483–1495. doi:10.1111/j.1600-0706.2013.00198.x.

39. Chao A, Chazdon RL, Colwell RK, Shen T-J (2005) A new statistical approach for assessing similarity of species composition with incidence and abundance data. Ecol Lett 8: 148–159. doi:10.1111/j.1461-0248.2004.00707.x.

40. Webb (2000) Exploring the phylogenetic structure of ecological communities: an example for rain forest trees. Am Nat 156: 145–155. doi:10.1086/303378.

41. Warren AD, Ogawa JR, Brower AVZ (2008) Phylogenetic relationships of subfamilies and circumscription of tribes in the family Hesperiidae (Lepidoptera: Hesperioidea). Cladistics 24: 642–676. doi:10.1111/j.1096-0031.2008.00218.x.

42. Evans WH (1955) A catalogue of the American Hesperiidae: indicating the classification and nomenclature adopted in the British Museum (Natural History). London: British Museum (Natural History). 552 p.

43. Sanderson MJ, Purvis A, Henze C (1998) Phylogenetic supertrees: Assembling the trees of life. Trends Ecol Evol 13: 105–109. doi:10.1016/S0169-5347(97)01242-1.

44. Wahlberg N, Wheat CW, Peña C (2013) Timing and patterns in the taxonomic diversification of Lepidoptera (butterflies and moths). PLoS ONE 8: e80875. doi:10.1371/journal.pone.0080875.

45. Grafen A (1989) The phylogenetic regression. Philos Trans R Soc Lond B Biol Sci 326: 119–157. doi:10.1098/rstb.1989.0106.

46. Pagel MD (1992) A method for the analysis of comparative data. J Theor Biol 156: 431–442. doi:10.1016/S0022-5193(05)80637-X.

47. Purvis A (1995) A composite estimate of primate phylogeny. Philos Trans R Soc Lond B Biol Sci 348: 405–421. doi:10.1098/rstb.1995.0078.

48. Clarke KR, Warwick RM (2009) PRIMER for Windows. Primer-E Ltd.

49. Gotelli N (2000) Null model analysis of species co-occurrence patterns. Ecology 81: 2606–2621. doi:10.2307/177478.

50. Webb CO, Ackerly DD, Kembel SW (2008) Phylocom: software for the analysis of phylogenetic community structure and trait evolution. Bioinformatics 24: 2098–2100. doi:10.1093/bioinformatics/btn358.

51. Kembel SW, Hubbell SP (2006) The phylogenetic structure of a Neotropical forest tree community. Ecology 87: 86–99. doi:10.1890/0012-9658(2006)87[86:TPSOAN]2.0.CO;2.

52. Maddison WP, Maddison DR (2001) Mesquite, a modular system for evolutionary analysis, Available: http://mesquiteproject.org.

53. Midford PE, Garland Jr P, Maddison WP (2009) PDAP:PDTREE package for Mesquite. Available: http://mesquiteproject.org/pdap_mesquite/.

54. Webb CO, Ackerly DD, Kembel SW (2011) Phylocom: software for the analysis of phylogenetic community structure and trait evolution. Available: http://www.phylodiversity.net/phylocom.

55. Benjamini Y, Hochberg Y (1995) Controlling the false discovery rate: A practical and powerful approach to multiple testing. J R Stat Soc Ser B Methodol 57: 289–300. doi:10.2307/2346101.

56. Pike N (2011) Using false discovery rates for multiple comparisons in ecology and evolution. Methods Ecol Evol 2: 278–282. doi:10.1111/j.2041-210X.2010.00061.x.

57. Whittaker RH (1952) A study of summer foliage insect communities in the Great Smoky Mountains. Ecol Monogr 22: 1–44. doi:10.2307/1948527.

58. Hodkinson ID (2005) Terrestrial insects along elevation gradients: species and community responses to altitude. Biol Rev 80: 489–513. doi:10.1017/S1464793105006767.

59. Swenson NG (2009) Phylogenetic resolution and quantifying the phylogenetic diversity and dispersion of communities. PLoS ONE 4: e4390. doi:10.1371/journal.pone.0004390.

60. Storch D, Konvicka M, Benes J, Martinková J, Gaston KJ (2003) Distribution patterns in butterflies and birds of the Czech Republic: separating effects of habitat and geographical position. J Biogeogr 30: 1195–1205. doi:10.1046/j.1365-2699.2003.00917.x.

61. Axmacher JC, Fiedler K (2008) Habitat type modifies geometry of elevational diversity gradients in geometrid moths (Lepidoptera Geometridae) on Mt Kilimanjaro, Tanzania. Trop Zool 21: 243–251.

62. Illán JG, Gutiérrez D, Wilson RJ (2010) Fine-scale determinants of butterfly species richness and composition in a mountain region. J Biogeogr 37: 1706–1720. doi:10.1111/j.1365-2699.2010.02314.x.

63. Lien VV, Yuan D (2003) The differences of butterfly (Lepidoptera, Papilionoidea) communities in habitats with various degrees of disturbance and altitudes in tropical forests of Vietnam. Biodivers Conserv 12: 1099–1111. doi:10.1023/A:1023038923000.

64. Pellissier L, Pradervand J-N, Pottier J, Dubuis A, Maiorano L, et al. (2012) Climate-based empirical models show biased predictions of butterfly communities along environmental gradients. Ecography 35: 684–692. doi:10.1111/j.1600-0587.2011.07047.x.

65. Heikkilä M, Kaila L, Mutanen M, Peña C, Wahlberg N (2011) Cretaceous origin and repeated tertiary diversification of the redefined butterflies. Proc R Soc B Biol Sci 279: 1093–1099. doi:10.1098/rspb.2011.1430.

66. Peña C, Wahlberg N, Weingartner E, Kodandaramaiah U, Nylin S, et al. (2006) Higher level phylogeny of Satyrinae butterflies (Lepidoptera: Nymphalidae) based on DNA sequence data. Mol Phylogenet Evol 40: 29–49. doi:10.1016/j.ympev.2006.02.007.

67. Suttie JM, Reynolds SG, Batello C (2005) Grasslands of the World. Food & Agriculture Org. 548 p.

68. Brown KS, Freitas AVL (2000) Atlantic forest butterflies: indicators for landscape conservation. Biotropica 32: 934–956. doi:10.1111/j.1744-7429.2000.tb00631.x.

69. Martinelli G (2007) Mountain biodiversity in Brazil. Braz J Bot 30: 587–597. doi:10.1590/S0100-84042007000400005.

Effects of Grazing Regimes on Plant Traits and Soil Nutrients in an Alpine Steppe, Northern Tibetan Plateau

Jian Sun[1], Xiaodan Wang[2]*, Genwei Cheng[2], Jianbo Wu[2], Jiangtao Hong[2], Shuli Niu[1]

1 Synthesis Research Centre of Chinese Ecosystem Research Network, Key Laboratory of Ecosystem Network Observation and Modelling, Institute of Geographic Sciences and Natural Resources Research, Chinese Academy of Sciences, Beijing, China, 2 The key laboratory of mountain surface processes and eco-regulation, Institute of Mountain Hazard and Environment, Chinese Academy of Sciences, Chengdu, China

Abstract

Understanding the impact of grazing intensity on grassland production and soil fertility is of fundamental importance for grassland conservation and management. We thus compared three types of alpine steppe management by studying vegetation traits and soil properties in response to three levels of grazing pressure: permanent grazing (M1), seasonal grazing (M2), and grazing exclusion (M3) in the alpine steppe in Xainza County, Tibetan Plateau. The results showed that community biomass allocation did not support the isometric hypothesis under different grassland management types. Plants in M1 had less aboveground biomass but more belowground biomass in the top soil layer than those in M2 and M3, which was largely due to that root/shoot ratios of dominant plants in M1 were far greater than those in M2 and M3. The interramet distance and the tiller size of the dominant clonal plants were greater in M3 than in M1 and M2, while the resprouting from rhizome buds did not differ significantly among the three greezing regimes. Both soil bulk density and soil available nitrogen in M3 were greater than in M1 at the 15–30 cm soil depth ($P=0.05$). Soil organic carbon and soil total nitrogen were greater in M3 than in M1 and M2 ($P=0.05$). We conclude that the isometric hypothesis is not supported in this study and fencing is a helpful grassland management in terms of plant growth and soil nutrient retention in alpine steppe. The extreme cold, scarce precipitation and short growing period may be the causation of the unique plant and soil responses to different management regimes.

Editor: Cheng-Sen Li, Institute of Botany, China

Funding: This research was jointly funded by the Knowledge Innovation Program of the Chinese Academy of Sciences (KZCX2-XB3-08), CAS Strategic Priority Research Program (XDA05050602), and the Open Fund of Key Laboratory of Ecosystem Network Observation and Modelling (110301A1PA). The funders had no role in study design, data collection and analysis, decision to publish, or preparation of the manuscript.

Competing Interests: The authors have declared that no competing interests exist.

* Email: wxd@imde.ac.cn

Introduction

Alpine grasslands make up the dominant ecosystem occupying approximately 94% of Northern Tibet [1]. The natural environment of the region is extremely harsh, and the alpine steppe, a fragile ecosystem, is extremely susceptible to the impacts of human activities [2]. It suffers from overgrazing, deforestation, and the harvesting of numerous herbs commonly used in traditional medicines [3–5]. Studies examining the response of above- and belowground biomass, the root/shoot ratio, the morphological characteristics of dominant plants, and the soil properties to human disturbance offer important insights that can contribute to adopting the most effective approach to grassland management in an alpine steppe, in which it is particularly difficult to recover from ecosystem degradation due to the region's long period of frost and relatively short growing season [3,6].

Although grazing and fencing, both of which have a substantial affect on vegetation traits and soil properties [7–9], are the most prevalent management regimes for grasslands worldwide, and although the effects of herbivores on soil properties in (sub)alpine ecosystems have recently been reported [10,11], knowledge about plants and soils in response to grassland management regimes (i.e., fencing and grazing) in the Tibetan Plateau remains limited due to

an extremely difficult geographic situation [9,12]. With regard to soil properties, it has been documented that grazing depresses soil carbon storage by changing the plant biomass and composition of a Tibetan alpine meadow [13]. In contrast, Shi et al. [14] found grazing exclusion to decrease soil organic carbon storage in an alpine grassland of the Tibetan Plateau, while another report suggested that seasonal grazing might enrich soil nutrients [15]. Such conflicting results indicate that different grazing intensities may have varying impacts on soil properties.

Understanding the influence of different management types on grassland production is essential for improving grassland conservation and management [16]. Previous studies have yielded varying results for aboveground biomass changes. Grazing thus increased [17] or decreased [18] aboveground production in different cases. The fencing optimization hypothesis posits that fencing significantly enhances above- and belowground biomass by means of carbon reallocation to plant growth and promotes increases in soil nutrient concentrations [19,20]. Belowground biomass has also been shown to be affected by grazing, with some studies indicating that plants reduce the proportion of aboveground parts and allocate more biomass to belowground parts, so as to geminate and resist environmental stress (i.e., grazing pressure) [21]. Conversely, more belowground biomass was

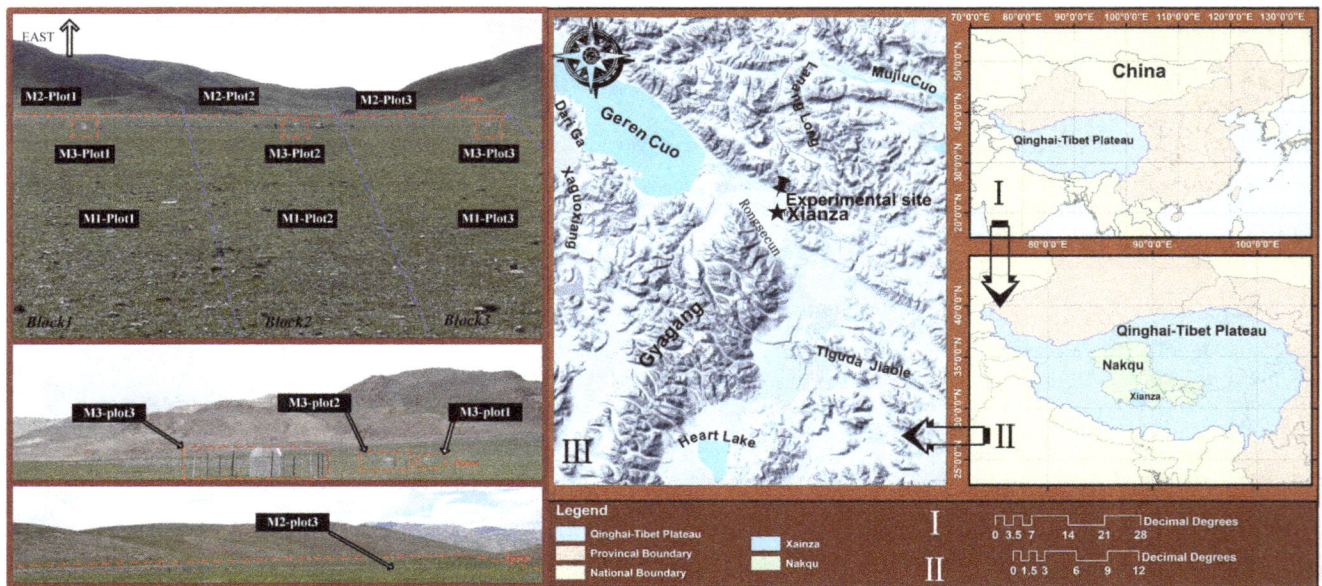

Figure 1. The experiment sites in alpine steppe, northern Tibetan Plateau.

allocated to the surface layer than the subsurface of the soil profile with increasing grazing pressure [22]. For instance, the root to shoot ratio in the grazing pattern was significantly higher than for the mowing and fencing patterns in temperate grassland [23].

The root:shoot ratio provides the basis for understanding the response or adaptive strategies of plants to environmental stress [24,25], and reflects how plants respond to different selection pressures [26,27]. Theory predicts the ratio of roots to shoots to respond isometrically across different individuals and community types under varying environmental conditions [28–30]. While biomass allocation has been widely examined from individual to community to ecosystem levels [31–35], few studies compare the biomass partitioning of dominant alpine species under different grassland management types.

Aside from biomass, the morphological characteristics of clonal plants are also regarded as a sensitive index for assessing the possible effect of grazing pressure [36,37]. The traditional theory is that a clone with a guerrilla-like foraging strategy should be able to locate and exploit favorable patches in which competitors are absent. In a habitat in which potential competitors are more homogeneously distributed, a phalanx clone exhibits a "consolidation" strategy and may be able to persist and monopolize locally available resources [38]. However, we know little about the responses of the morphological characteristics of clonal plants to grazing pressure in alpine steppes.

In this study, we examined plant traits and soil properties in response to different grassland management strategies in the Northern Tibetan Plateau. The specific aims are: (1) to reveal the impact of different grassland managements on plant traits and soil properties and to test the grazing optimization hypothesis; (2) to test the isometric partitioning theory on different alpine species under varying grassland management patterns; and (3) to explore the effects of grazing stress on clonal growth.

Materials and Methods

Study area

Our study was conducted in the Alpine Steppe Experiment Station (N 30°57′, E 88°42′, 4675 m a.s.l.) in Xainza County, Northern Tibet (Fig. 1). The annual mean air temperature of this region is 0°C, while the mean air temperature ranges from − 10.1°C in January and 9.6°C in July. The average annual precipitation is 300 mm. There is no absolute frost-free season, and the frosty period can last up to 279.1 days [39]. The region's soil is classified as steppe soil with sandy loam [40]. Vegetation coverage of the alpine steppe is approximately 20% and is dominated by *Stipa purpurea* (Bunch grass) and *Carex moorcrofii* (sedge family, creeping rhizomes) (Fig. 2A). Other common species include *Stellera chamaejasme Linn.* (Bunch weeds), *Oxytropis glacialis* (Bunch weeds), and *Leontopodium alpinum* (Fig. 2B). All of these plants are perennial herbs.

Grazing treatments and experimental plots

There were three grazing treatments in this study. i.e, M1 (grazing): The pasture is always used for grazing. M2 (seasonal grazing): The pasture has been used for grazing in the non-growing seasons but not in the growing season (from May to September) since 2007. M3 (grazing exclusion): Herbivores have been excluded from the pasture since 2010.

The study area was divided into three blocks. Each block contained three plots, the eastern plot having a grazing intensity of M2, the middle plot a grazing intensity of M3, and the western plot a grazing intensity of M1 (Fig. 1). The dimensions of each M3 fenced plot were 6×6 m; the dimensions for each M2 plot were 200×500 m. Three M1 plots were located outside of the fenced area in each block, with each plot having dimensions of 200×500 m. In each plot, a quadrat (30 cm×30 cm) was used to measure biomass and take samples. The estimated average livestock density in both M1 and M2 was approximately 120 sheep units per km². (The "sheep unit" is the most frequently used unit of measurement for evaluating the carrying capacity of pasture areas in China and includes figures for cattle, which are converted

Figure 2. The vegetation composition diagram is in the study area. The Vegetation coverage of the alpine steppe was about 20%, which are dominated by *Stipa purpurea* and *Carex moorcrofii* (panel A) and the accompanying species are *Stellera chamaejasme Linn.*, *Leontopodium alpinum*, *Oxytropis microphylla* (panel B) and so on.

to sheep; thus, e.g., one cow is equivalent to five sheep based on the average weight of a cow) [41].

Ethics statement

No specific permits were required for the samples collected from any of the sites, and the field studies did not involve endangered or protected species.

Biomass measure and sample taking

We established a 30 cm×30 cm quadrat in each of the sampling plots to harvest aboveground biomass (AGB) in late August of 2012. After harvesting the aboveground biomass, we took soil samples by auger with a 5-cm core diameter at two depths (0–15 cm and 15–30 cm) to estimate belowground biomass (BGB). According to previous research, most belowground biomass is located in the top 30 cm layer [42]. Soil samples were also taken at depths of 0–15 cm and 15–30 cm in order to measure soil

elements. After being air-dried and sieved (2 mm mesh), the soil samples were carefully handpicked to extract the surface organic materials and fine roots for an analysis of soil chemical properties. Soil nutrients (including soil organic carbon, soil total nitrogen content, available nitrogen, total phosphorus and available phosphorus) were determined using standard protocol [43].

In order to estimate the root:shoot ratio of the dominant species (*S. purpurea*, *C. moorcrofii*, *O. glacialis*, and *L. alpinum*), we took three soil blocks in each sampling plot of different management regimes. The patch was excavated with a spade with a diameter of 20 cm to a depth of 30 cm, which was selected based on the root morphological traits of the species [42]. The root samples obtained from the sites were immediately placed in a cloth bag and then soaked in water to remove the residual soil by means of a 0.5 mm sieve. The roots of each sample were carefully separated from soil and other belowground materials first. Then individuals of the all species were carefully separated from other plant roots. Finally, we

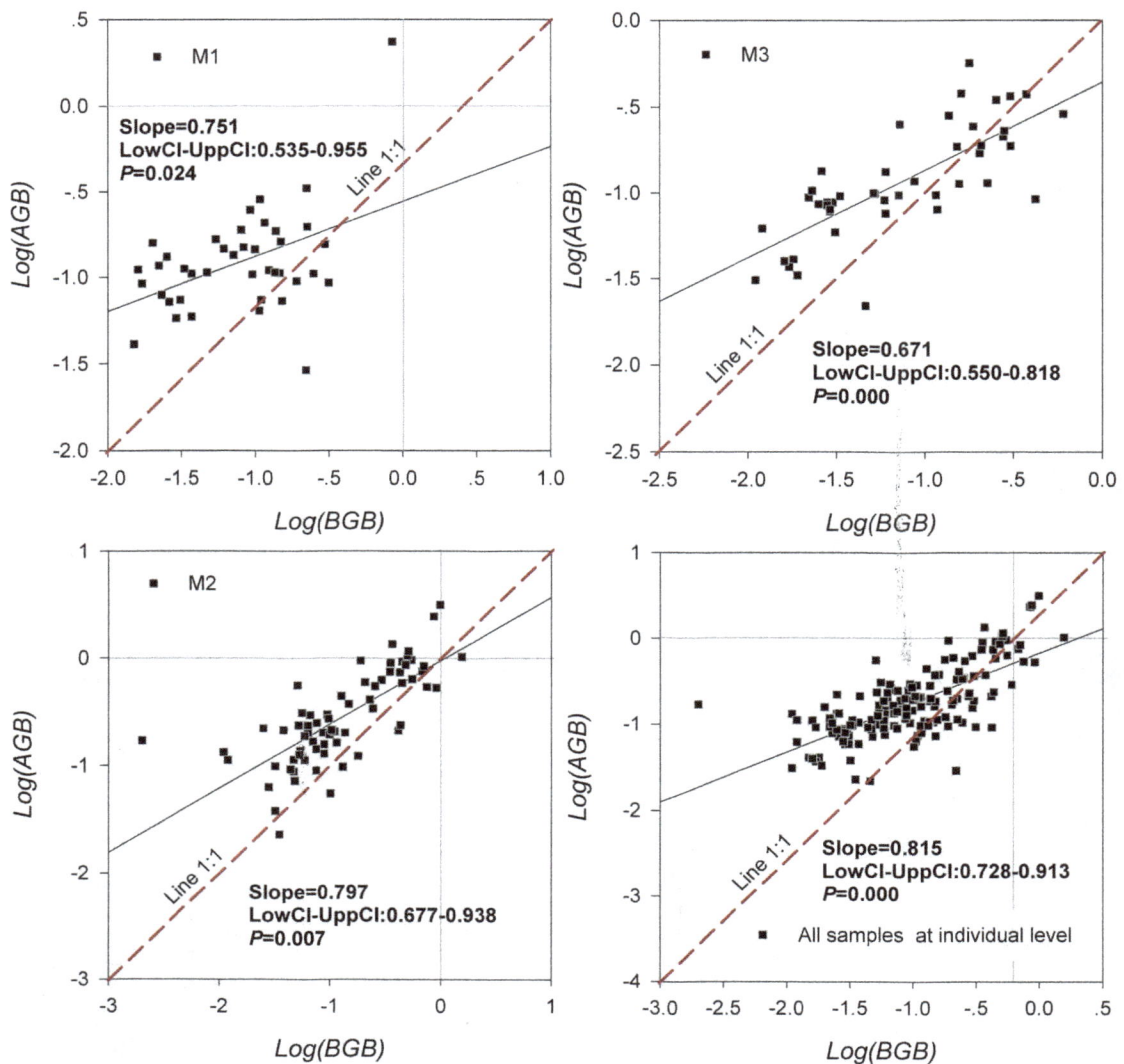

Figure 3. Relationships between AGB and BGB under grazing (M1), seasonal grazing (M2), and grazing exclusion (M3) regimes.

separated the roots and shoots of each individual. Individual biomass (shoots and roots) and community biomass (AGB and BGB) was oven-dried at 65°C until a constant weight was achieved. Individual biomass data were used to analyze allometric functions under different management regimes (with a total of 40, 45, and 68 individuals in M1, M2, and M3, respectively).

The plant morphology of *C. moorcrofii* is regarded as a fairly good indicator that reflects the different grassland management regimes [44]. We thus sampled *C. moorcrofii.* from other 30 cm×30 cm quadrats at each of the 9 plots. The samples were then taken back to the laboratory, where the morphological characteristics (plant height, interramet distance, number of rhizome buds, and tillering number) were immediately measured. Special attention was paid to avoid destroying the structural integrity of the plant throughout this process.

Statistical analysis

We performed a standardized major axis (SRMA) regression to examine whether AGB and BGB scale isometrically at an individual level across all samples, and then explored their

relationships under different management types. The regression relationship of the form $LogAGB = \alpha + \beta LogBGB$ was used to describe the allometric relationship between AGB and BGB, where x is BGB, y is AGB, α is the intercept, and β is the scaling slope [25,28,45,46]. The scaling slope and y-intercept of the allometric function were determined using the software package titled "Standardized Major Axis Tests & Routines Version 2.0" [47]. If a 95% confidence interval of the scaling slope covered 1.0, the relationship between AGB and BGB was considered to be isometric.

One-way ANOVA was used to test differences in soil properties and biomes (AGB, BGB, and R/S) among the different treatments, and the Tukey test was used to distinguish differences at a $P = 0.05$ level.

Results

Biomass partitioning

The relationship between BGB and AGB under different grazing pressures was characterized by the linear function $LogAGB = \alpha + \beta LogBGB$ (Fig. 3). The slopes of the allometric

Table 1. Aboveground biomass (AGB), belowground biomass (BGB), and their ratios (R/S) under different grassland managements.

Management Types	Category	M1		M2		M3	
		Mean	STDEV	Mean	STDEV	Mean	STDEV
AGB (g·cm^{-2})		48.09	12.39	51.09	35.48	62.25	3.21
BGB (g·cm^{-2})	0–15 cm	312.92	65.6	238.07	53.95	247.16	38.96
	15–30 cm	11.95	7.64	6.77	5.64	14.72	10.27
	Ratio(0–15 cm/15–30 cm)	26.19	8.59	35.17	9.57	16.79	3.79
	0–30 cm	324.87	58.18	244.84	59.59	261.88	48.67
R/S		7.37	3.58	5.79	2.23	4.2	0.67

Note: The differences of biomes under the different grassland management type were insignificant at 0.05 level.

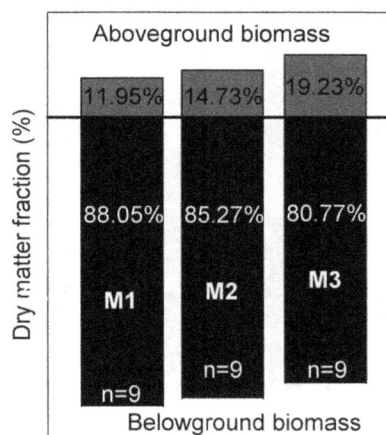

Figure 4. Biomass fractions (% of total biomass) in herbaceous plants in M1, M2 and M3 of Tibetan Plateau.

relationship were 0.751 (for M1), 0.797 (for M2), 0.671 (for M3), and 0.815 (for all samples at an individual level), respectively. All of the slopes differed significantly from 1 at $P = 0.05$.

Responses of AGB, BGB, and R/S to different management types

AGB in M1 was slightly but not significantly lower than in M2 and M3 (Table 1). In contrast, BGB of M1, M2, and M3 amounted to 324.87, 244.84, and 261.88 g·m^{-2}, respectively. The ratio of root to shoot in M1 was also higher than in M2 and M3. The root biomass was mainly distributed in a soil depth of 0–15 cm, while the root biomass in a soil depth of 15–30 cm only accounted for 2.77–5.62% of total BGB. The ratios of BGB in the 0–15 cm-layer to BGB in the 15–30 cm-layer of M1, M2, and M3 were 26.19, 35.17, and 16.79, respectively.

To investigate the characteristics of biomass allocation further, we analyzed the dry matter fraction of M1, M2, and M3 in the Tibetan Plateau (Fig. 4). The results indicated that the roots of total plants in the Tibetan Plateau amounted to 88.05% (in M1), 85.27% (in M2), and 80.77% (in M3).

Traits of clonal plants under different management types

C. moocroftii was tallest in M3 and shortest in M1 ($P<0.05$, Fig. 5A). Both interramet distance (Fig. 5C) and the tiller number (Fig. 5D) in M3 were significantly larger than in M1 and M2. No significant differences were found in the number of rhizome buds among M1, M2, and M3 treatments ($P<0.05$, Fig. 5B).

Soil properties in response to management types

At a soil depth of 0–15 cm, soil bulk density, soil organic carbon, total nitrogen, and available nitrogen were non-significant among M1, M2, and M3 (Table 2). At a soil depth of 15–30 cm, both soil bulk densiy and soil available nitrogen in M3 were greater than in M1 at a $P = 0.05$ level. Soil organic carbon and soil total nitrogen in M3 were greater than in M1, and M2 ($P = 0.05$).

Discussion

Biomass partitioning

Based on the results of our SMA analysis, biomass allocation at the community level did not agree with the isometric hypothesis under different grassland management types. Plants consistently sense changes in their environment and often allocate a greater

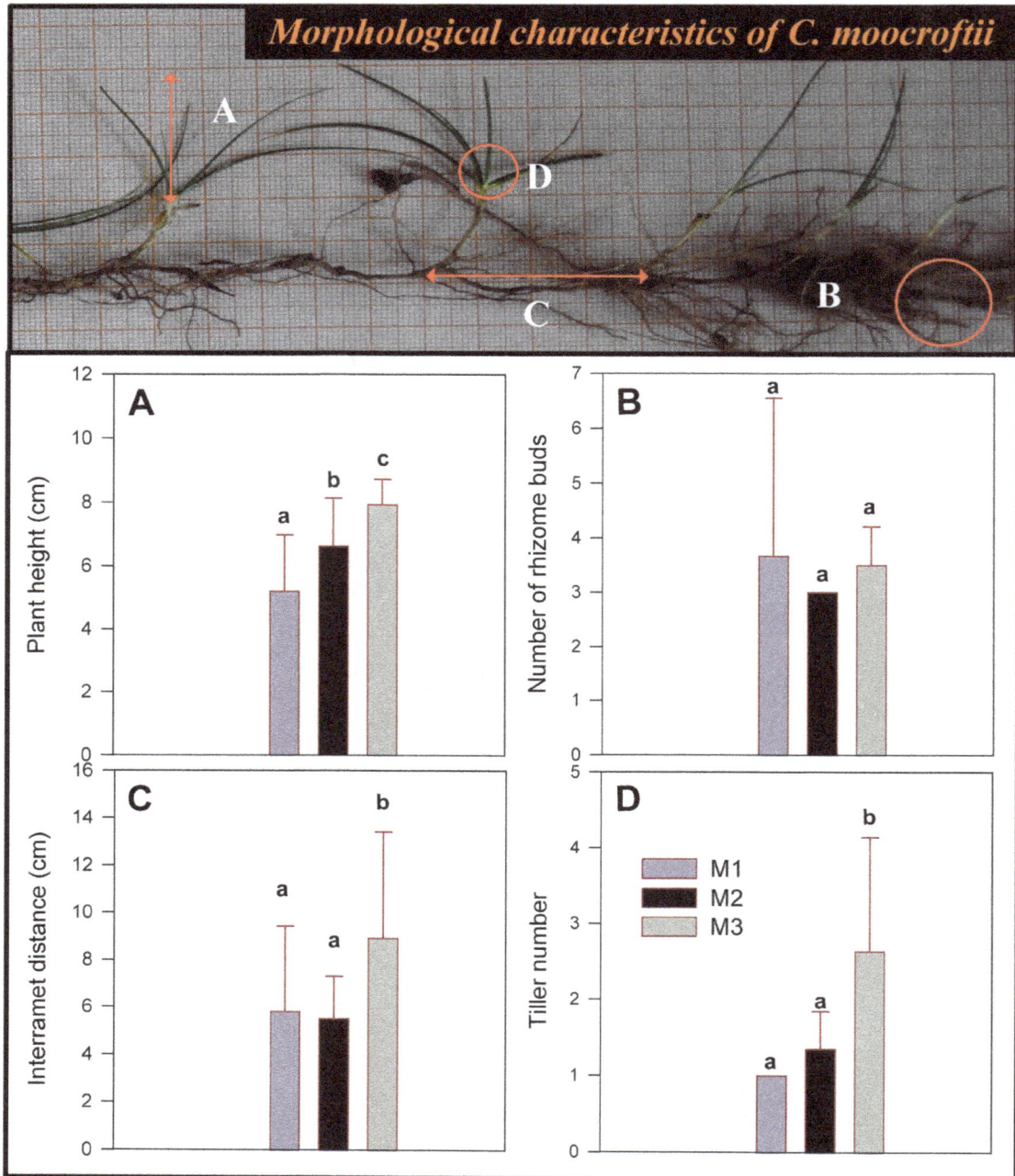

Figure 5. The morphological characteristics of *C. moocroftii* **in the different grassland management types.** A, represent plant height, B, number of rhizome buds, C, interramet distance, and D tiller number.

proportion of their biomass to the root system when water or mineral elements are scarce [48]. Moreover, roots have also been found to store carbohydrates in alpine grasslands [49]. In the alpine steppe, greater plant biomass allocation to belowground reflects the plants' response to the harsh alpine environment (limited precipitation and low temperatures) for survival. Another reason may be the relatively narrow variation in plant size in that region. Moreover, many species in the Tibetan alpine grasslands do not have typical hierarchical branching structures [42], which did not meet the assumption that stem length scales isometrically with respect to root length. This characteristic is clearly reflected in

the root morphology structure shown in Fig. 2B and Fig. 5. Biomass allocation in these areas may thus support the allometric biomass partitioning hypothesis, rather than the isometric allocation hypothesis.

Responses of AGB, BGB, and R/S to different management types

AGB in the grazing plots was found to be lower than in the non-grazing plots, which is in agreement with the findings of Medina-Roldan et al. [50]. We suggested that alpine grassland production might be decreased to some degree by grazing as a result of the

Table 2. Soil properties under the different grassland managements.

Management Types	Depth	M1 Mean	M1 STDEV	M2 Mean	M2 STDEV	M3 Mean	M3 STDEV
Soil bulk density (g·cm⁻³)	**0-15 cm**	1.71	0.09	1.48	0.08	1.60	0.11
Soil organic carbon (g·kg⁻¹)		10.09	3.38	12.72	2.45	11.02	0.85
Soil total nitrogen(g·kg⁻¹)		1.12	0.26	1.54	0.40	1.19	0.11
Soil available Nitrogen(mg·kg⁻¹)		62.96	24.14	84.65	17.24	81.61	7.14
Soil total phosphorous(g·kg⁻¹)		0.34	0.03	0.33	0.04	0.28	0.02
Soil available phosphorous(mg·kg⁻¹)		3.91	0.72	3.83	0.16	3.39	0.29
Soil bulk density(g·cm⁻³)	**15-30 cm**	1.76 **a**	0.04	1.58 **ab**	0.09	1.48 **b**	0.14
Soil organic carbon (g·kg⁻¹)		5.21 **a**	0.87	5.81 **a**	0.25	10.16 **b**	2.58
Soil total nitrogen(g·kg⁻¹)		0.64 **a**	0.13	0.57 **a**	0.04	1.07 **b**	0.27
Soil available Nitrogen(mg·kg⁻¹)		32.36 **a**	4.87	40.57 **ab**	8.51	68.01 **b**	17.37
Soil total phosphorous(g·kg⁻¹)		0.27	0.01	0.24	0.04	0.37	0.09
Soil available phosphorous(mg·kg⁻¹)		2.95	0.54	3.25	0.61	3.33	0.11

Note: The bold lower case represents the result of variance analysis at 0.05 level.

consumption effect of the livestock. The higher BGB in M1 than M2 and M3 in our alpine steppe conflict with the results of Cheng et al. [51], who found grazing to significantly decrease belowground biomass in the Loess Plateau. Gao et al. [52] also reported significantly less belowground biomass under conditions of heavy grazing in comparison with a non-grazing site in Inner Mongolia. The conflicting results between our study and previous studies in other areas probably stem from differences in environmental conditions. In the extreme cold region of alpine grassland, plants reduce the proportion of aboveground parts and allocate larger amounts of biomass to belowground parts in order to germinate and resist grazing pressures [21]. More BGB in the top soil layer of M1 than M2 and M3 suggests that grazing results in more belowground biomass being allocated to the topsoil [53], possibly due to the alpine environment inhibiting root growth in ungrazed plots. We did not find an ameliorating effect of reduced grazing (M2) or grazer exclusion (M3), suggesting that the effects of grazing cessation may only be detectable in the long term, as proposed by Steffens et al. [54].

The root:shoot ratio (R/S) in M1 was much larger than in M2 and M3, suggesting that plants allocate more to BGB than AGB in order to maximize resources for optimal growth in the grazed environment [21]. We compared the dry matter fraction of M1, M2, and M3 in the Tibetan Plateau (Fig. 4) with dry matter fractions in Argentina, Bolivia, Ecuador, the Arctic, and the Alps [49], and we found that the proportion of BGB was much higher in our study sites than for plants in a semi-arid grassland ecosystem (Bolivia), mountain grassland ecosystem (Argentina), and a humid mountain grassland ecosystem (Ecuadorian Andes). The main reason may be associated with the comparatively slow depletion of carbohydrates in roots, resulting from low respiration rates in the extremely cold winter, and the slower root turnover in colder environments [55]. Meanwhile, long-term grazing results in a higher BGB fraction in the Tibetan Plateau than for plants in other cold alpine environments and the Arctic.

The clonal plant in response to different management types

Clonal plants defend against large herbivores by regulating their morphological characteristics [56]. Clonal plant *C. moocroftii* in M3 was notably taller than in M1 and M2 because it was not bitten by livestock, and they can plastically respond to environmental heterogeneity by placing ramets and changing tiller size in favorable sites [57,58]. We found the interramet distance and tiller number in M3 to be greater than for plants in M1 and M2. Tillers may produce more daughter tillers as a result of increased nutrient availability in grazing exclusion. *C. moocroftii* thus responds to grazing by decreasing the tiller number [37]. The results also indicate that *C. moocroftii* may have both phalanx and guerrilla strategies. The phalanx strategy involves the production of a compact structure of closely spaced ramets (i.e., M1), and the guerrilla strategy involves the production of loosely arranged and more widely spaced ramets in order to seek out more soil nutrients in a relatively suitable habitat (i.e., M3).

Clonal growth by means of lateral roots is regarded as a typical trait of opportunistic species growing in disturbed habitats, as such growth potentially produces a large number of buds on lateral roots [59]. However, our results implied that different grassland management systems did not affect resprouting from rhizome buds, which suggests that reprouting from rhizome buds is not a major adjustment stragegy in response to grazing intensity. This finding is probably due to the short time period since the implementation of fencing and seasonal grazing.

Soil properties in response to different management types

Grazing intensity is one of the most important factors influencing soil properties [50,60]. It has been documented that the trampling action of grazing animals impacts the soil by increasing bulk density [61] and mechanical resistance, and reducing porosity, water infiltration, and aggregate stability [62], and native perennial cover and litter cover, with the consequence of changing the soil nutrient concentrations [63]. In our study, bulk density was found to increase with increasing grazing intensity, and all soil organic carbon, total nitrogen, and available nitrogen contents in M3 were slightly higher than in M1 and M2. We thus report that grazing exclusion may improve soil N availability. Arevalo et al. [64] found an increase in soil phosphorus content in response to increased goat grazing pressure in pastures. In contrast, our results indicated that soil total phosphorous and available phosphorous contents in M1 did not differ significantly from those of M2 and M3. The extreme cold, limited precipitation and short growing period may lead to no obvious change in phosphorus content.

Conclusions

By studying plant traits and soil nutrients in response to grazing intensity in an alpine steppe, we found that the isometric hypothesis is not supported in this particular region. The extreme cold, limited amount of precipitation and short growing period may lead to the plants' and soils' unique to different management types. Overall, the study suggests that the implementation of grazing exclusion played a positive role in the sustainable development of the alpine steppe region.

Acknowledgments

The authors are grateful to two anonymous reviewers all gave very helpful editorial comments.

Author Contributions

Conceived and designed the experiments: XDW GWC. Performed the experiments: JS JBW JTH. Analyzed the data: JS. Contributed reagents/materials/analysis tools: JS GWC JTH. Wrote the paper: JS SLN.

References

1. Lu XY, Fan JH, Yan Y, Wang XD (2013) Responses of Soil CO_2 Fluxes to Short-Term Experimental Warming in Alpine Steppe Ecosystem, Northern Tibet. Plos One 8.

2. Gao QZ, Li Y, Wan YF, Qin XB, Jiangcun WZ, et al. (2009) Dynamics of alpine grassland NPP and its response to climate change in Northern Tibet. Climatic Change 97: 515–528.

3. Sun J, Cheng GW, Li WP, Sha YK, Yang YC (2013) On the Variation of NDVI with the Principal Climatic Elements in the Tibetan Plateau. Remote Sensing 5: 1894–1911.

4. Wang GX, Cheng GD, Shen YP, Qian J (2003) Influence of land cover changes on the physical and chemical properties of alpine meadow soil. Chinese Science Bulletin 48: 118–124.

5. Sun J, Cheng GW, Li WP (2013) Meta-analysis of relationships between environmental factors and aboveground biomass in the alpine grassland on the Tibetan Plateau. Biogeosciences 10: 1707–1715.

6. Yu HY, Luedeling E, Xu JC (2010) Winter and spring warming result in delayed spring phenology on the Tibetan Plateau. Proceedings of the National Academy of Sciences of the United States of America 107: 22151–22156.

7. Zheng SX, Ren HY, Li WH, Lan ZC (2012) Scale-Dependent Effects of Grazing on Plant C: N: P Stoichiometry and Linkages to Ecosystem Functioning in the Inner Mongolia Grassland. Plos One 7.

8. Wu GL, Liu ZH, Zhang L, Chen JM, Hu TM (2010) Long-term fencing improved soil properties and soil organic carbon storage in an alpine swamp meadow of western China. Plant and Soil 332: 331–337.

9. Wang XD, Yan Y, Cao YZ (2012) Impact of historic grazing on steppe soils on the northern Tibetan Plateau. Plant and Soil 354: 173–183.

10. Haynes AG, Schutz M, Buchmann N, Page-Dumroese DS, Busse MD, et al. (2014) Linkages between grazing history and herbivore exclusion on decomposition rates in mineral soils of subalpine grasslands. Plant and Soil 374: 579–591.

11. Risch AC, Haynes AG, Busse MD, Filli F, Schutz M (2013) The Response of Soil CO2 Fluxes to Progressively Excluding Vertebrate and Invertebrate Herbivores Depends on Ecosystem Type. Ecosystems 16: 1192–1202.

12. Fan YJ, Hou XY, Shi HX, Shi SL (2013) Effects of grazing and fencing on carbon and nitrogen reserves in plants and soils of alpine meadow in the three headwater resource regions. Russian Journal of Ecology 44: 80–88.

13. Sun DS, Wesche K, Chen DD, Zhang SH, Wu GL, et al. (2011) Grazing depresses soil carbon storage through changing plant biomass and composition in a Tibetan alpine meadow. Plant Soil and Environment 57: 271–278.

14. Shi XM, Li XG, Li CT, Zhao Y, Shang ZH, et al. (2013) Grazing exclusion decreases soil organic C storage at an alpine grassland of the Qinghai-Tibetan Plateau. Ecological Engineering 57: 183–187.

15. Whitford WG, Steinberger Y (2012) Effects of seasonal grazing, drought, fire, and carbon enrichment on soil microarthropods in a desert grassland. Journal of Arid Environments 83: 10–14.

16. Yang XH, Guo XL, Fitzsimmons M (2012) Assessing light to moderate grazing effects on grassland production using satellite imagery. International Journal of Remote Sensing 33: 5087–5104.

17. Hilbert DW, Swift DM, Detling JK, Dyer MI (1981) Relative Growth-Rates and the Grazing Optimization Hypothesis. Oecologia 51: 14–18.

18. Lkhagva A, Boldgiv B, Goulden CE, Yadamsuren O, Lauenroth WK (2013) Effects of grazing on plant community structure and aboveground net primary production of semiarid boreal steppe of northern Mongolia. Grassland Science 59: 135–145.

19. Wu GL, Du GZ, Liu ZH, Thirgood S (2009) Effect of fencing and grazing on a Kobresia-dominated meadow in the Qinghai-Tibetan Plateau. Plant and Soil 319: 115–126.

20. Pandey CB, Singh JS (1992) Influence of Rainfall and Grazing on Belowground Biomass Dynamics in a Dry Tropical Savanna. Canadian Journal of Botany-Revue Canadienne De Botanique 70: 1885–1890.

21. Ma WL, Shi PL, Li WH, He YT, Zhang XZ, et al. (2010) Changes in individual plant traits and biomass allocation in alpine meadow with elevation variation on the Qinghai-Tibetan Plateau. Science China-Life Sciences 53: 1142–1151.

22. Burke IC, Lauenroth WK, Riggle R, Brannen P, Madigan B, et al. (1999) Spatial variability of soil properties in the shortgrass steppe: The relative importance of topography, grazing, microsite, and plant species in controlling spatial patterns. Ecosystems 2: 422–438.

23. Liu M, Liu GH, Wu X, Wang H, Chen L (2014). Vegetation traits and soil properties in respones to utilization patterns of grassland in Hulun Buir City, Inner Mongolia, China. Chinese Geographical Science, 24(4):471–478.

24. Xie JB, Tang LS, Wang ZY, Xu GQ, Li Y (2012) Distinguishing the Biomass Allocation Variance Resulting from Ontogenetic Drift or Acclimation to Soil Texture. Plos One 7.

25. Yang YH, Luo YQ (2011) Isometric biomass partitioning pattern in forest ecosystems: evidence from temporal observations during stand development. Journal of Ecology 99: 431–437.

26. Baruch Z (1994) Responses to Drought and Flooding in Tropical Forage Grasses.1. Biomass Allocation, Leaf Growth and Mineral Nutrients. Plant and Soil 164: 87–96.

27. Sanchez-Vilas J, Bermudez R, Retuerto R (2012) Soil water content and patterns of allocation to below- and above-ground biomass in the sexes of the subdioecious plant Honckenya peploides. Annals of Botany 110: 839–848.

28. Enquist BJ, Niklas KJ (2002) Global allocation rules for patterns of biomass partitioning in seed plants. Science 295: 1517–1520.

29. Niklas KJ (2005) Morphogenesis and biomechanics: The roles of mechanical perturbation and other environmental stresses in plant development and evolution. Evolving Form and Function: Fossils and Development: 25–41.

30. Cheng DL, Niklas KJ (2007) Above- and below-ground biomass relationships across 1534 forested communities. Annals of Botany 99: 95–102.

31. McConnaughay KDM, Coleman JS (1999) Biomass allocation in plants: Ontogeny or optimality? A test along three resource gradients. Ecology 80: 2581–2593.

32. Shipley B, Meziane D (2002) The balanced-growth hypothesis and the allometry of leaf and root biomass allocation. Functional Ecology 16: 326–331.

33. Reich PB, Oleksyn J, Wright IJ, Niklas KJ, Hedin L, et al. (2010) Evidence of a general 2/3-power law of scaling leaf nitrogen to phosphorus among major plant groups and biomes. Proceedings of the Royal Society B-Biological Sciences 277: 877–883.

34. McCarthy MC, Enquist BJ (2007) Consistency between an allometric approach and optimal partitioning theory in global patterns of plant biomass allocation. Functional Ecology 21: 713–720.

35. Enquist BJ (2002) Universal scaling in tree and vascular plant allometry: Toward a general quantitative theory linking plant form and function from cells to ecosystems. Tree Physiology 22: 1045–1064.

36. Tolvanen A, Schroderus J, Henry GHR (2001) Age- and stage-based bud demography of Salix arctica under contrasting muskox grazing pressure in the High Arctic. Evolutionary Ecology 15: 443–462.

37. Jonsdottir IS (1991) Effects of Grazing on Tiller Size and Population-Dynamics in a Clonal Sedge (Carex-Bigelowii). Oikos 62: 177–188.

38. Cheplick GP (1997) Responses to severe competitive stress in a clonal plant: Differences between genotypes. Oikos 79: 581–591.

39. Lu XY, Fan JH, Yan Y, Wang XD (2011) Soil water soluble organic carbon under three alpine grassland types in Northern Tibet, China. African Journal of Agricultural Research 6: 2066–2071.

40. Wang XD, Li MH, Liu SZ, Liu GC (2006) Fractal characteristics of soils under different land-use patterns in the and and semiarid regions of the Tibetan Plateau, China. Geoderma 134: 56–61.

41. Xie YZ, Wittig R (2004) The impact of grazing intensity on soil characteristics of Stipa grandis and Stipa bungeana steppe in northern China (autonomous region of Ningxia). Acta Oecologica-International Journal of Ecology 25: 197–204.

42. Wang LA, Niu KC, Yang YH, Zhou P (2010) Patterns of above- and belowground biomass allocation in China's grasslands: Evidence from individual-level observations. Science China-Life Sciences 53: 851–857.

43. Bao SD (2000) Soil chemical anlysis of agriculture. Beijing: Chinese Agriculture Press.

44. Hu JY (2007) Morphological and structural traits and their ecological adaptation of two dominant species in alpine grassland in Tibetan Plateau. Master's degree thesis. Institute of botany, Chinese academy of sciences.

45. Yang YH, Fang JY, Ji CJ, Han WX (2009) Above- and belowground biomass allocation in Tibetan grasslands. Journal of Vegetation Science 20: 177–184.

46. Li HT, Han XG, Wu JG (2006) Variant scaling relationship for mass-density across tree-dominated communities. Journal of Integrative Plant Biology 48: 268–277.

47. Warton DI, Wright IJ, Falster DS, Westoby M (2006) Bivariate line-fitting methods for allometry. Biological Reviews 81: 259–291.

48. Hermans C, Hammond JP, White PJ, Verbruggen N (2006) How do plants respond to nutrient shortage by biomass allocation? Trends in Plant Science 11: 610–617.

49. Patty L, Halloy SRP, Hiltbrunner E, Korner C (2010) Biomass allocation in herbaceous plants under grazing impact in the high semi-arid Andes. Flora 205: 695–703.

50. Medina-Roldan E, Paz-Ferreiro J, Bardgett RD (2012) Grazing exclusion affects soil and plant communities, but has no impact on soil carbon storage in an upland grassland. Agriculture Ecosystems & Environment 149: 118–123.

51. Cheng J, Wu GL, Zhao LP, Li Y, Li W, et al. (2011) Cumulative effects of 20-year exclusion of livestock grazing on above- and belowground biomass of typical steppe communities in arid areas of the Loess Plateau, China. Plant Soil and Environment 57: 40–44.

52. Gao YZ, Giese M, Lin S, Sattelmacher B, Zhao Y, et al. (2008) Belowground net primary productivity and biomass allocation of a grassland in Inner Mongolia is affected by grazing intensity. Plant and Soil 307: 41–50.

53. Rodriguez MA, Brown VK, Gomezsal A (1995) The Vertical-Distribution of Belowground Biomass in Grassland Communities in Relation to Grazing Regime and Habitat Characteristics. Journal of Vegetation Science 6: 63–72.

54. Steffens M, Kolbl A, Totsche KU, Kogel-Knabner I (2008) Grazing effects on soil chemical and physical properties in a semiarid steppe of Inner Mongolia (PR China). Geoderma 143: 63–72.

55. Gill RA, Jackson RB (2000) Global patterns of root turnover for terrestrial ecosystems. New Phytologist 147: 13–31.

56. Smit C, Bakker ES, Apol MEF, Olff H (2010) Effects of cattle and rabbit grazing on clonal expansion of spiny shrubs in wood-pastures. Basic and Applied Ecology 11: 685–692.

57. Yu FH, Dong M, Krusi B (2004) Clonal integration helps Psammochloa villosa survive sand burial in an inland dune. New Phytologist 162: 697–704.

58. Gao Y, Xing F, Jin YJ, Nie DD, Wang Y (2012) Foraging responses of clonal plants to multi-patch environmental heterogeneity: spatial preference and temporal reversibility. Plant and Soil 359: 137–147.

59. Sosnova M, van Diggelen R, Klimesova J (2010) Distribution of clonal growth forms in wetlands. Aquatic Botany 92: 33–39.

60. Gao YH, Schumann M, Chen H, Wu N, Luo P (2009) Impacts of grazing intensity on soil carbon and nitrogen in an alpine meadow on the eastern Tibetan Plateau. Journal of Food Agriculture & Environment 7: 749–754.

61. Evans CRW, Krzic M, Broersma K, Thompson DJ (2012) Long-term grazing effects on grassland soil properties in southern British Columbia. Canadian Journal of Soil Science 92: 685–693.

62. Zhou ZC, Gan ZT, Shangguan ZP, Dong ZB (2010) Effects of grazing on soil physical properties and soil erodibility in semiarid grassland of the Northern Loess Plateau (China). Catena 82: 87–91.

63. Yates CJ, Norton DA, Hobbs RJ (2000) Grazing effects on plant cover, soil and microclimate in fragmented woodlands in south-western Australia: implications for restoration. Austral Ecology 25: 36–47.

64. Arevalo JR, Chinea E, Barquin E (2007) Pasture management under goat grazing on Canary Islands. Agriculture Ecosystems & Environment 118: 291–296.

4

Bird Communities and Biomass Yields in Potential Bioenergy Grasslands

Peter J. Blank[1]*, **David W. Sample**[2], **Carol L. Williams**[3], **Monica G. Turner**[1]

1 Department of Zoology, University of Wisconsin, Madison, Wisconsin, United States of America, **2** Wisconsin Department of Natural Resources, Madison, Wisconsin, United States of America, **3** Wisconsin Energy Institute, University of Wisconsin, Madison, Wisconsin, United States of America

Abstract

Demand for bioenergy is increasing, but the ecological consequences of bioenergy crop production on working lands remain unresolved. Corn is currently a dominant bioenergy crop, but perennial grasslands could produce renewable bioenergy resources and enhance biodiversity. Grassland bird populations have declined in recent decades and may particularly benefit from perennial grasslands grown for bioenergy. We asked how breeding bird community assemblages, vegetation characteristics, and biomass yields varied among three types of potential bioenergy grassland fields (grass monocultures, grass-dominated fields, and forb-dominated fields), and assessed tradeoffs between grassland biomass production and bird habitat. We also compared the bird communities in grassland fields to nearby cornfields. Cornfields had few birds compared to perennial grassland fields. Ten bird Species of Greatest Conservation Need (SGCN) were observed in perennial grassland fields. Bird species richness and total bird density increased with forb cover and were greater in forb-dominated fields than grass monocultures. SGCN density declined with increasing vertical vegetation density, indicating that tall, dense grassland fields managed for maximum biomass yield would be of lesser value to imperiled grassland bird species. The proportion of grassland habitat within 1 km of study sites was positively associated with bird species richness and the density of total birds and SGCNs, suggesting that grassland bioenergy fields may be more beneficial for grassland birds if they are established near other grassland parcels. Predicted total bird density peaked below maximum biomass yields and predicted SGCN density was negatively related to biomass yields. Our results indicate that perennial grassland fields could produce bioenergy feedstocks while providing bird habitat. Bioenergy grasslands promote agricultural multifunctionality and conservation of biodiversity in working landscapes.

Editor: Sarah C. Davis, Ohio University, United States of America

Funding: Funding for this project came from the Wisconsin Department of Natural Resources Federal Wildlife Restoration Grant W-160-P. The funders had no role in study design, data collection and analysis, decision to publish, or preparation of the manuscript.

Competing Interests: The authors have declared that no competing interests exist.

* Email: blankpj@gmail.com

Introduction

Interest in bioenergy is increasing in the U.S. due to concerns about climate change, energy independence, air and water quality, and other issues [1,2]. Most biofuels in the U.S. are currently made from corn grain [2], but biofuels derived from cellulose (e.g., cellulosic ethanol) are an important part of the national renewable energy policy [3]. Using perennial grasslands to produce bioenergy feedstocks could help meet national cellulosic bioenergy goals and could promote multifunctionality of working lands by producing agricultural commodities and ecological benefits [4,5,6,7,8]. However, grassland production for bioenergy is still in its infancy [9], and the degree to which grassland production and conservation goals can be aligned is uncertain. Notably, the influence of potential bioenergy grassland crop types on biodiversity is not well understood.

Among potential bioenergy crops, low-input high-diversity grassland fields, such as native prairies, could provide greater biodiversity value and ecosystem services compared to high-input low-diversity fields, such as intensively managed annual row crops [10,11,12]. Grasslands planted with native warm-season (C_4)

grasses, such as switchgrass (*Panicum virgatum*) and big bluestem (*Andropogon gerardii*), have high bioenergy potential [13,14] and could provide valuable wildlife habitat [7,15]. Grassland birds, which have experienced substantial population declines in recent decades [16], may particularly benefit if diverse grasslands are used for bioenergy production [10,17]. Bird use of perennial bioenergy grasslands could be influenced by vegetation characteristics of the fields [18,19,20,21,22] and by landscape context around the fields [23,24,25,26,27]. To date, few studies have assessed the value of potential bioenergy grasslands for breeding birds, and among the studies on the potential effects of biomass crops on biodiversity, many lack appropriate study designs and spatial replication, making inferences from these studies limited [28].

In addition to its effects on birds, plant diversity in perennial grasslands may also affect biomass yields and therefore bioenergy potential [7]. Tilman et al. [12] found that greater plant diversity was associated with higher energy yields in experimental plots in Minnesota, but this research has been contradicted by several studies [e.g., 7,29]. For example, greater plant diversity was associated with lower biomass yields in Conservation Reserve

Program (CRP) grasslands in the northeast U.S. [30]. Compared to monocultures, diverse plant mixtures have greater variation in chemical content and physical properties that lower bioenergy conversion efficiency and add cost to conversion processes [30,31]. Because it is unclear if industrial-scale bioenergy production will call for monotypic or mixed-species (i.e., polycultures) production systems [29,32], more information is needed about biomass productivity in potential grassland bioenergy crops [2]. An improved understanding of the biomass potential of grassland crops will also allow tradeoffs between biomass yields and biodiversity to be evaluated [7,33].

We studied bird community assemblages in potential bioenergy crops along a gradient of plant diversity ranging from intensively managed cornfields and grass monocultures, to more diverse grass-dominated fields (grasslands with >50% live vegetation in grass), to diverse forb-dominated fields (grasslands with <50% live vegetation in grass). In each grassland field we also measured vegetation characteristics and biomass yields. We asked: (1) How do bird communities differ among cornfields and potential bioenergy grasslands? (2) How do vegetation and landscape context influence bird communities in potential bioenergy grasslands? (3) How do biomass yields differ among potential bioenergy grassland types? (4) What are the tradeoffs between producing biomass and providing bird habitat in perennial grassland fields? A main objective of the study was to develop recommendations for improving the habitat quality of future grassland biomass production fields for grassland birds. This study is one of the few to assess how bird communities are influenced by vegetation characteristics and landscape context of potential grassland bioenergy crops, and the first to quantitatively evaluate the tradeoffs between biomass production and bird habitat in perennial grassland fields.

Methods

Study Area

Our study was conducted in southern Wisconsin, U.S. (Fig. 1), a predominantly agricultural landscape representative of much of the Midwest. Climate of the region is continental (warm humid summers and cold winters), with annual precipitation averaging 86 cm and occurring largely from May to September [34]. However, the region experienced a severe drought in the summer and fall of 2012 [35]. Topography is generally flat to rolling, with gentle hills and some shallow depressions containing wetlands. Soils are primarily composed of prairie-derived Mollisols and forest-derived Alfisols [36]. Agricultural lands are dominated by corn, soybean, alfalfa, small grains, and livestock pastures, but the region also includes forests, restored grasslands, lakes, wetlands, and a densely populated urban area (Madison, Wisconsin). Southern Wisconsin is well suited for this study because the central U.S. is predicted to have high local biomass production of switchgrass under future climate scenarios [37], and several grassland bird species nesting in the study area are listed as Species of Greatest Conservation Need (SGCN) in Wisconsin [38].

Study sites

We established study sites on 30 grassland fields and 11 cornfields (Fig. 1; Tables S1 and S2). At present, there are few grassland fields grown for biomass feedstocks in Wisconsin [9,39], and most are small agronomic research plots unsuitable for field-scale research on bird communities. Therefore, we used existing perennial grassland fields to represent grasslands that could be harvested for biomass [as in 15,40]. Among the grassland fields, five were grass monocultures managed for seed production and 25

were planted for conservation purposes and were not commercially harvested. Average field sizes were 22 ha for cornfields (SE = 7 ha) and 19 ha for grassland fields (SE = 3 ha). All of the grassland fields we studied were planted with warm-season grasses because these grasses are considered to have high bioenergy potential [13,14,30]. For example, switchgrass is considered a valuable bioenergy feedstock source because of its perennial growth, relatively easy establishment, ability to grow on marginal agricultural lands, low nutrient requirements, high yield potential, and drought resistance [13,41]. We only included grassland fields that had no significant woody cover because we assumed that bioenergy production fields harvested annually would have little to no woody cover.

The five grass monocultures were identified after a thorough search for commercial-scale grass monocultures in Wisconsin. The grass monoculture fields were intensively-managed, seed production fields in which seeds are harvested in autumn and later sold for native grass plantings. They included one switchgrass field, two big-bluestem fields, one indiangrass (*Sorghastrum nutans*) field, and one field with switchgrass on one side of the field and indiangrass on the other (the two grasses on this field were not grown as a mixture, therefore we considered it a monoculture). Three of these grass monocultures, owned by Agrecol Corporation in Evansville WI, are currently being used to produce bioenergy feedstocks: the seed waste (hulls and straw) from their seed cleaning operation is used to produce grass biomass pellets that are burned to heat their production facilities and have been sold for residential and commercial heating [42]. To our knowledge, our study is the first analysis of the potential effects of grassland bioenergy production on bird communities to include study fields that are actually used to produce biomass feedstocks for bioenergy.

We classified the 25 conservation grassland fields as either grass-dominated (>50% of live vegetation cover in grass; $n = 14$) or forb-dominated (<50% of live vegetation cover in grass; $n = 11$), in part based on Garlock et al. [31] who classified cellulosic ethanol feedstocks into grass- and forb-dominated samples. Switchgrass, indiangrass, and big bluestem were the most common grasses among grassland sites (Table S1). Commonly encountered forbs among grass- and forb-dominated fields included wild bergamot (*Monarda fistulosa*), pinnate prairie coneflower (*Ratibida pinnata*), blackeyed susan (*Rudbeckia hirta*), Canada goldenrod (*Solidago canadensis*), wholeleaf rosinweed (*Silphium integrifolium*), and compassplant (*Silphium laciniatum*; Table S1). Thus, the grass- and forb-dominated grassland fields represent a range of grass-to-forb cover ratios that may be harvested for bioenergy in the future if cellulosic feedstock sources other than grass monocultures are desired.

Study sites were located on private farms, nonprofit conservation areas, U.S. Fish and Wildlife Service Waterfowl Production Areas (WPAs), and Wisconsin Department of Natural Resources (DNR) wildlife areas. Permission for the research was granted from each of the private farm owners and from the Wisconsin DNR for work on the wildlife areas. The nonprofit conservation areas and the WPAs were open to the public and no permits or approvals were necessary to work on those properties. Three of the grassland fields on private lands were enrolled in the CRP.

Bird surveys

We surveyed one 100-m radius bird point count circle in each field. The point was randomly located within the field and at least 100 m from the field edge. All points were ≥350 m apart. We surveyed 17 fields in 2011 and 31 fields in 2012. Seven fields were surveyed in both years (one cornfield and six grassland fields). Within each year, fields were surveyed three times between 30

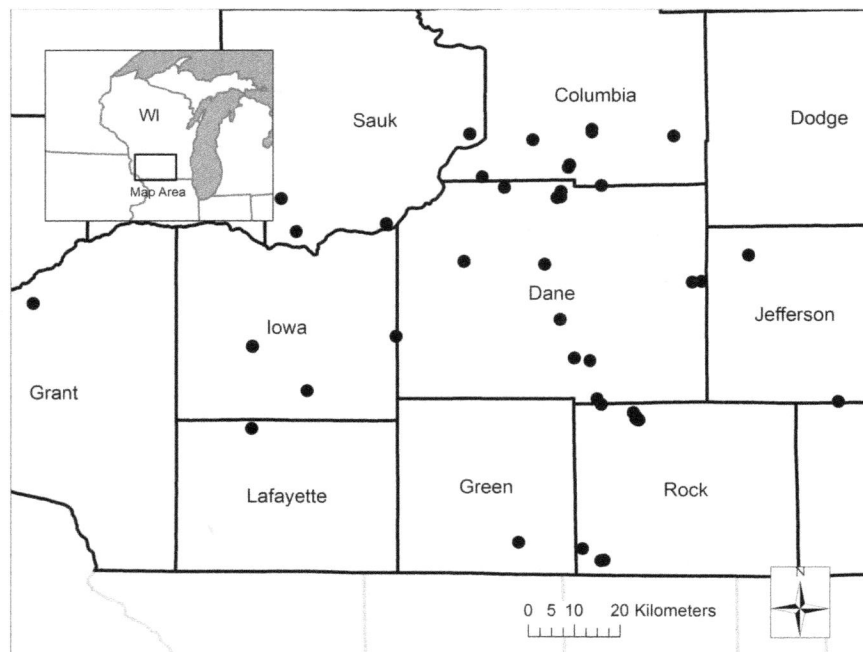

Figure 1. Study sites (solid circles) and counties in southern Wisconsin, USA.

May and 13 July. Each point count was 10 min in duration and all birds seen or heard were recorded (Dataset S1). All surveys were between sunrise and four hours after sunrise and were not conducted during rain, heavy fog, or sustained winds >16 km/hr. One observer conducted each survey. All surveys were conducted by PJB or E.R. Keyel. Because the bird surveys were strictly observational and did not involve bird handling, no animal research or ethical approvals were required to conduct this research.

Vegetation Surveys

We surveyed vegetation composition and structure in all grassland fields in the same year in which they were surveyed for birds (Dataset S2). In 2011 vegetation surveys were conducted between early and mid-August, and in 2012 surveys were conducted between late-June and mid-July. We established three vegetation transects evenly spaced within each 100-m radius bird point count circle. The middle transect went down the center of the circle and the outer transects averaged 60 m from the middle transect. We surveyed vegetation in four 0.5-m^2 Daubenmire plots [43] on each transect, totaling 12 plots per circle. Plots were evenly spaced and approximately 44 m apart along each transect. Within each plot we visually estimated the percent cover of total canopy (live + residual vegetation), live grasses (warm- and cool-season), live forbs, and each individual live plant species. We estimated plant species richness by counting the total number of live plant species per plot. Litter depth was measured at three evenly spaced positions in each plot. We measured vertical vegetation density from visual obstruction measurements, taken in the four cardinal directions, of a modified Robel pole when the pole was viewed from a distance of four m and a height of 1.5 m [44]. Average litter depth and vertical density per plot were calculated from the multiple measurements in each plot. Percent of live vegetation in grass was calculated by dividing the percent cover of live grass by the sum of the percent cover of live grass and forbs.

Landscape analysis

We superimposed the point-count locations onto the 2011 cropland data layer [45] in ArcMap 10 [46]. Land-cover classes were reclassified as cropland, grassland/pasture, forest, wetland, low development, high development, barren land, and shrubland. Grassland and pasture were combined into one category called grassland [as in 47]. We evaluated R^2 values from univariate models testing the influence of the proportion of agriculture and grassland in 1-, 2-, 3-, 4-, and 5-km radius landscapes around each point on several bird response variables (e.g., total bird density, SGCN density). These results indicated that the 1-km radius landscape scale was equal to or better than other landscape scales at explaining variation in bird responses to landscape factors. Aerial photographs of the study sites also indicated that 1-km radius landscapes around the points encompassed the dominant land-cover types near the study fields. Therefore, we used a 1-km radius landscape scale for further statistical analyses of landscape metrics (Dataset S2), which is a scale that has been related to grassland bird community structure in other studies (e.g., [24]).

Biomass and Gross Bioenergy Yields

We measured fall biomass yield in October 2012 by recording leaf area index (LAI) and predicting biomass yield with an allometric equation we developed for our study area. We chose October because it is when farmers will harvest biomass according to best management practices that conserve nutrients [33]. In 20 grassland fields, we measured LAI in 12 plots laid out in the same arrangement as the summer vegetation survey plots (see above) with an AccuPAR LP-80 Ceptometer (Decagon Devices, Inc., Pullman, WA). From nine of these fields, aboveground biomass was harvested at 10–15 cm stubble height [13,30] in four 0.5-m^2 plots ($n = 36$ harvested plots) chosen to span a range of biomass yields. Biomass was dried at 60°C for at least five days and then weighed. We related harvested aboveground biomass to LAI from 35 of the 36 plots (one plot was omitted due to measurement error):

$$Biomass\ yield\ (kg/ha) = 156.5 + 1603.9\ (LAI) \qquad (1)$$

Based on this equation ($R^2 = 0.72$), average fall biomass yields were predicted for all 20 grassland fields on which LAI was measured (Dataset S2).

To predict fall 2012 biomass yield in five grassland fields that were mowed before we could sample LAI and in the 11 grassland fields surveyed in 2011, we developed an equation to relate October 2012 biomass yield to seven vegetation variables measured in summer 2012 [percent cover of canopy (Canopy), total grasses (Grass), warm-season grasses (WSG), cool-season grasses (CSG), and forbs (Forbs); vertical density (Robel; min. height of Robel pole visibility, dm), and litter depth (LD, dm)]:

$$
\begin{aligned}
Biomass\ yield\ (Kg/ha) = \\
-678.9 - 3.6\ (Canopy) + 71.2\ (Grass) \\
-21.6\ (WSG) + 9.3\ (CSG) \\
-6.4\ (Forbs) + 669.3\ (Robel) + 62.4\ (LD)
\end{aligned} \qquad (2)
$$

This equation ($R^2 = 0.72$) was used to predict fall biomass yields in the remaining 16 fields (Dataset S2).

Statistical Analyses

We developed models for three community-level bird metrics (bird species richness, total bird density, and density of SGCNs) and the density of the four most commonly detected bird species. Recently, methods have been developed that incorporate detection probabilities into estimates of species richness and densities of animal populations [e.g., 48,49,50]. However, after evaluating many different model structures that incorporate detection probabilities, we were unable to find models that adequately fit the data. Therefore, we used unadjusted counts as the response variables in our models. We believe that detection rates were relatively high and were consistent among fields because all of our counts were conducted under favorable weather conditions with low wind and no precipitation, surveys were conducted by two skilled observers (PJB and E. R. Keyel), it was very rare to observe birds before or after the counts that were not detected during the counts, and the vegetation was short enough so that most birds detected were observed by sight and sound. However, we recognize that detection may have been <100% [51] and that our counts may be biased low, and as such our counts are indices of abundance. Thus, in this paper we estimate relative and not absolute effects of potential bioenergy crop fields and landscape attributes on bird metrics.

All analyses were performed in R 3.0.1 [52]. For all bird models, we fit generalized linear mixed models with the lmer function in the lme4 package [53], specified a Poisson distribution, and included site and year as random effects. Because there were very few birds in cornfields, we were unable to fit models that included cornfields as a factor level. Differences in bird metrics among grassland field types were tested by specifying field type as a fixed factor and including the percent cover of agriculture and grassland within 1 km as covariates to account for landscape-level effects.

We used an information theoretic approach [54] to evaluate the influence of vegetation characteristics and landscape context on each of the bird metrics. This approach allows for inference from multiple models of the same response variable, produces a weight of evidence for each candidate model, and incorporates model selection uncertainty into parameter estimates. The full (global) model of each bird metric included three vegetation explanatory variables (Forbs, WSG, and Robel) and two landscape explanatory variables at the 1-km radius scale [percent cover of agriculture (Ag) and (Grassland)]. $Robel^2$ was also included to test for non-linear relationships with vertical density, because previous research suggests that bioenergy grasslands may be suitable for the widest array of grassland birds at intermediate levels of vegetation density [40]. Among the vegetation variables measured, we did not include plant species richness, grass cover, and total canopy cover because they were highly correlated with other vegetation variables ($r \geq 0.7$), and did not include litter depth and cool-season grass cover because they had highly skewed distributions. The proportion of forest in the landscape was omitted because it was highly correlated with proportion of agriculture ($r = -0.7$), and we did not include other land-cover classes because they had highly skewed distributions or comprised a small proportion of the study landscapes. We tested all possible combinations of the six retained explanatory variables for each of the bird response variables, including a null model with no covariates. The predictor variables were centered and standardized before entering them into the models to improve the interpretability of the regression coefficients (Schielzeth, 2010). We tested the assumptions of the global models by plotting the residuals against the fitted values and each covariate in the full model to ensure that the residuals did not spread as the fitted and covariate values increased [55]. All of the global models met the assumptions. We also produced model-averaged estimates, and 95% unconditional confidence intervals, of continuous explanatory variables with the modavg function in the AICcmodavg package [56]. If the unconditional confidence intervals of the model-averaged parameter estimates did not overlap zero, we interpreted this as strong support that the predictor variable was related to the response.

Differences in fall biomass yields among grassland field types were also evaluated by using a mixed model with the lmer function, with field type as a fixed factor and site and year as random factors. We evaluated relationships between fall biomass yields and summer vegetation variables with the lme function in the nlme package [57] by using biomass yield as a response variable and each vegetation variable as an explanatory variable in separate models. We predicted bird metrics as a function of Robel, and included $Robel^2$ in the prediction models because model-averaged parameter estimates indicated that $Robel^2$ was strongly related to several bird metrics. Because Robel was strongly correlated with biomass yield (Biomass), we also predicted bird metrics as a function of Biomass and $Biomass^2$. We did not include quadratic terms in the prediction models of SGCN density because there was little evidence that those terms improved the models. Forbs and Grassland were included as covariates and were held constant at their means in the bird prediction models because these variables were found to consistently influence bird response. One grass monoculture field with a high estimated biomass yield in 2011 relative to other fields (12.02 Mg/ha) was omitted from the biomass prediction models because it was a statistical outlier that had a high influence on the results.

Results

Grasses represented 98%, 66%, and 32% of the live vegetation cover in grass monoculture, grass-dominated, and forb-dominated fields, respectively (Table 1). Litter depth was close to zero in grass monocultures, and vertical density was similar between grass monocultures and forb-dominated fields. Across all grassland fields, forb cover was positively correlated with plant species richness ($r = 0.8$) and negatively correlated with live vegetation in grass ($r = -0.8$). The 1-km radius landscapes around the grassland sites were primarily composed of agriculture (mean = 36%, range: 3 to 64%), grasslands (mean = 30%, range: 7 to 75%), forests (mean = 17%, range: 1 to 58%) and wetlands (mean = 10%, range: 0 to 32%).

Twenty-nine bird species were detected among all study sites (Table S3), including six species in cornfields, eight in grass monocultures, 19 in grass-dominated fields, and 20 in forb-dominated fields. Thirteen species were habitat generalists that commonly occur in grasslands, 11 species were grassland obligates, and 10 species are considered SGCNs in Wisconsin. Of the SGCNs, we observed none in cornfields, three in grass monocultures, eight in grass-dominated fields, and seven in forb-dominated fields. Three habitat generalists [Red-winged Blackbirds (*Agelaius phoeniceus*), Song Sparrows (*Melospiza melodia*), and Common Yellowthroats (*Geothlypis trichas*)] and one grassland obligate species [Dickcissel (*Spiza americana*)] had the greatest overall densities and were detected at the most sites. Cornfields and grass monocultures had the lowest bird densities for most species compared to grass- and forb-dominated fields.

Among grassland fields, total bird density was greatest in forb-dominated fields, lower in grass-dominated fields, and lowest in grass monocultures (Fig. 2). Species richness was greater in forb-dominated fields than grass monocultures, and the densities of Red-winged Blackbirds, Song Sparrows, and Common Yellowthroats were greater in forb-dominated fields than grass monocultures and grass-dominated fields. However, SGCN density was lowest in forb-dominated fields and Dickcissel (also a SGCN) density was lower in forb-dominated fields than in grass-dominated fields.

Both local vegetation and landscape variables were related to bird metrics, as indicated by the most supported models (Table S4) and model-averaged parameter estimates (Table S5). Forb cover was consistently positively related to bird species richness and the density of total birds and Red-winged Blackbirds. Warm-season grass cover was negatively related to Red-winged Blackbird and Common Yellowthroat densities. SGCN density was negatively associated vertical vegetation density. The square of vertical vegetation density was negatively associated with the density of total birds, Red-winged Blackbirds, Song Sparrows, and Common Yellowthroats, indicating unimodal relationships with these bird metrics and vegetation density. The amount of grassland habitat within 1 km was positively related to bird species richness and the density of total birds, SGCNs, Red-winged Blackbirds, Common Yellowthroats, and Dickcissels.

Fall biomass yields on grassland fields ranged from 1.24 to 12.02 Mg/ha. Biomass yield (mean ± SE) in grass monocultures (6.4 ± 1.03 Mg/ha) was greater than grass-dominated (3.58 ± 0.85 Mg/ha, $t = 3.28$, $P = 0.003$) and forb-dominated (4.67 ± 0.86 Mg/ha, $t = 2.02$, $P = 0.05$) fields. Yields were similar between grass-dominated and forb-dominated fields ($t = 1.7$, $P = 0.10$). Fall biomass yield was positively related to summer canopy cover ($t = 7.3$, $P < 0.001$) and vertical density ($t = 14.12$, $P < 0.001$; Fig. 3), but was not related to plant species richness ($t = 0.83$, $P = 0.44$) or other summer vegetation characteristics. Among the six grassland fields for which we estimated biomass yields in both 2011 and 2012, yields were greater in 2011 (7.22 ± 1.20 Mg/ha) than in 2012 (4.14 ± 0.95 Mg/ha), suggesting a large yearly change in biomass yield, most likely due to the severe drought experienced in Wisconsin in the summer of 2012 [35].

Predicted total bird density had unimodal relationships with vertical vegetation density and biomass yield and declined above approximately 4 Mg biomass/ha (Fig. 4). We found similar unimodal trends for Red-winged Blackbird and Common Yellowthroat densities. Predicted SGCN density was negatively related to both vertical vegetation density and biomass yield.

Discussion

Perennial, warm-season grasslands representing potential bioenergy crops provided habitat for a variety of grassland birds, including 10 SGCNs in Wisconsin. Bird densities for most species were greater in perennial grassland field types than cornfields. Recent increases in commodity prices have created incentives for farm owners to convert grasslands into corn [5,47]. Our results concur with previous studies that suggest that an increase in corn production to meet bioenergy demand would be detrimental to grassland bird populations [4,10,15,17]. Bird species of conservation concern may experience greater negative impacts of increased corn production than more common species [4]. In general, further conversion of grasslands to annual row crops would decrease wildlife habitat, increase carbon emissions, and harm water quality [10].

Table 1. Descriptive statistics of vegetation variables in the three grassland field types[a].

Variable	Grass monoculture		Grass-dominated		Forb-dominated	
	Mean	SE	Mean	SE	Mean	SE
Plant species richness (# species/0.5 m²)	1.5	0.2	5.2	0.5	6.2	0.6
Total canopy cover (%) (live + residual)	54.2	7.8	38.6	3.4	47.9	4.2
Total grass cover (%) (live)	53.5	8.1	21.6	2.1	12.3	1.6
Warm-season grass cover (%) (live)	39.7	8.3	16.2	2.7	16.6	1.8
Cool-season grass cover (%) (live)	1.7	1.7	15.5	2.9	13.5	2.6
Live vegetation in grass (%)	98.4	0.6	66.0	3.5	31.6	3.2
Forbs (%) (live)	0.9	0.3	11.0	1.6	29.1	4.0
Vertical density (min. height of Robel pole visibility, dm)	6.1	1.3	3.6	0.4	6.6	0.5
Litter depth (cm)	0.1	0.1	6.5	1.5	6.7	1.6

[a]Data are from 11 sites in August 2011 and 25 sites in July 2012. Sample sizes: grass monoculture ($n = 5$), grass-dominated ($n = 14$), forb-dominated ($n = 11$). For six sites measured in both years, the mean was calculated across years.

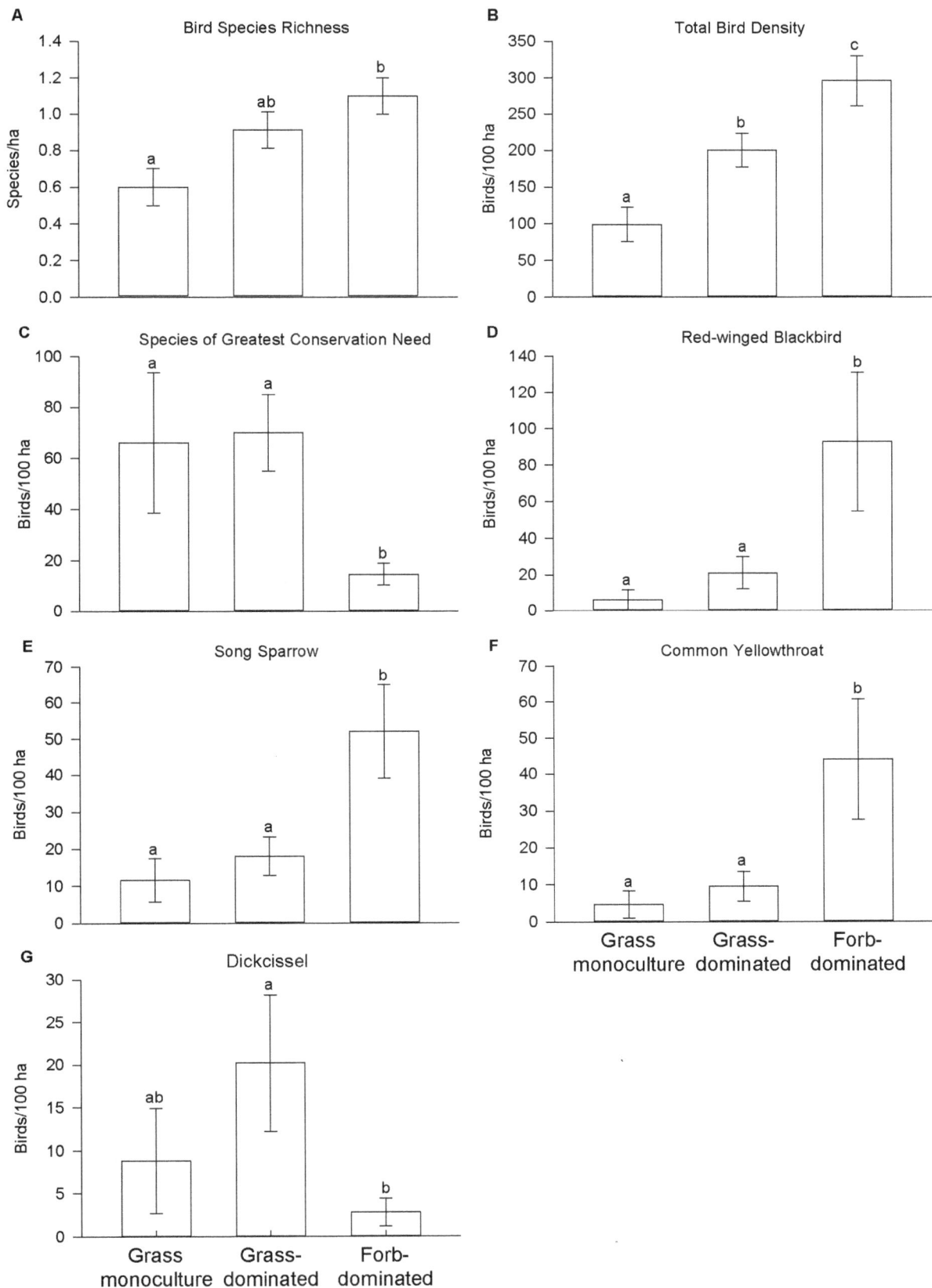

Figure 2. Predicted means of bird metrics in the three grassland field types. Error bars are ±1 SE. Bars with the same letter above them are not significantly different.

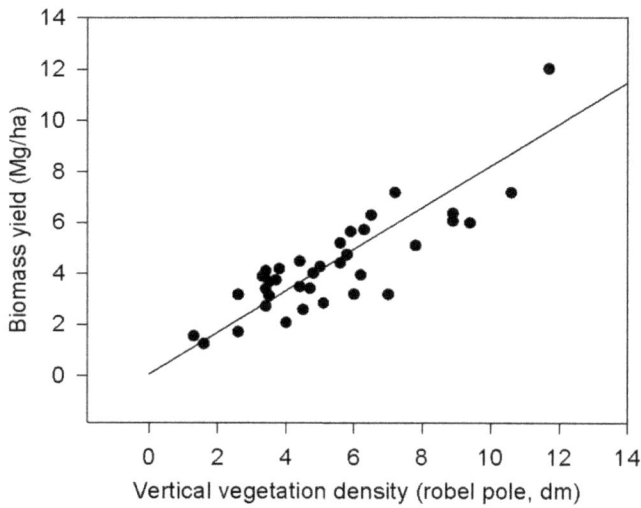

Figure 3. Relationship between biomass yield and vertical vegetation density in grassland fields. Predicted (solid line) and observed values (closed circles) are shown.

Among the field types we studied, most bird metrics were greater in forb-dominated fields but were similar between grass monocultures and grass-dominated fields. Forb cover was also positively associated with bird species richness and total bird

density. Because forb cover was positively correlated with plant species richness, we infer that most birds were attracted to sites with greater plant species richness and a higher ratio of forbs to grasses. These results support research suggesting perennial bioenergy grasslands with greater plant diversity and forb cover would provide better bird habitat than less diverse crops [10,17,40]. Diverse grasslands provide food for herbivores and nectivores, which may increase insect diversity and in turn provide additional food resources for birds [10,40,58]. Diverse grasslands could also increase populations of insect crop-pollinators and pest natural enemies [7,10,59]. However, feedstocks with lower grass cover and greater plant diversity may have lower bioenergy conversion efficiencies and therefore may not be as profitable as less diverse feedstocks [29,30,31]. Additionally, grassland fields that start out with high plant diversity may decrease in diversity over time due to repeated annual harvests or dominance by some grass species [29].

In our study, predicted SGCN and Dickcissel densities were lowest in forb-dominated fields compared to less diverse grasslands. These trends in SGCN and Dickcissel densities were likely driven by the habitat structure of the fields. Most grassland SGCNs in Wisconsin prefer grasslands with low to moderate vertical density and are more responsive to habitat structure than to plant species composition [22,60]. Average vertical density was greater in forb-dominated fields than in grass monocultures or grass-dominated fields. Thus, the forb-dominated fields were likely too tall and dense for most SGCNs. SGCN density in grass monocultures may have been inflated because one grass mono-

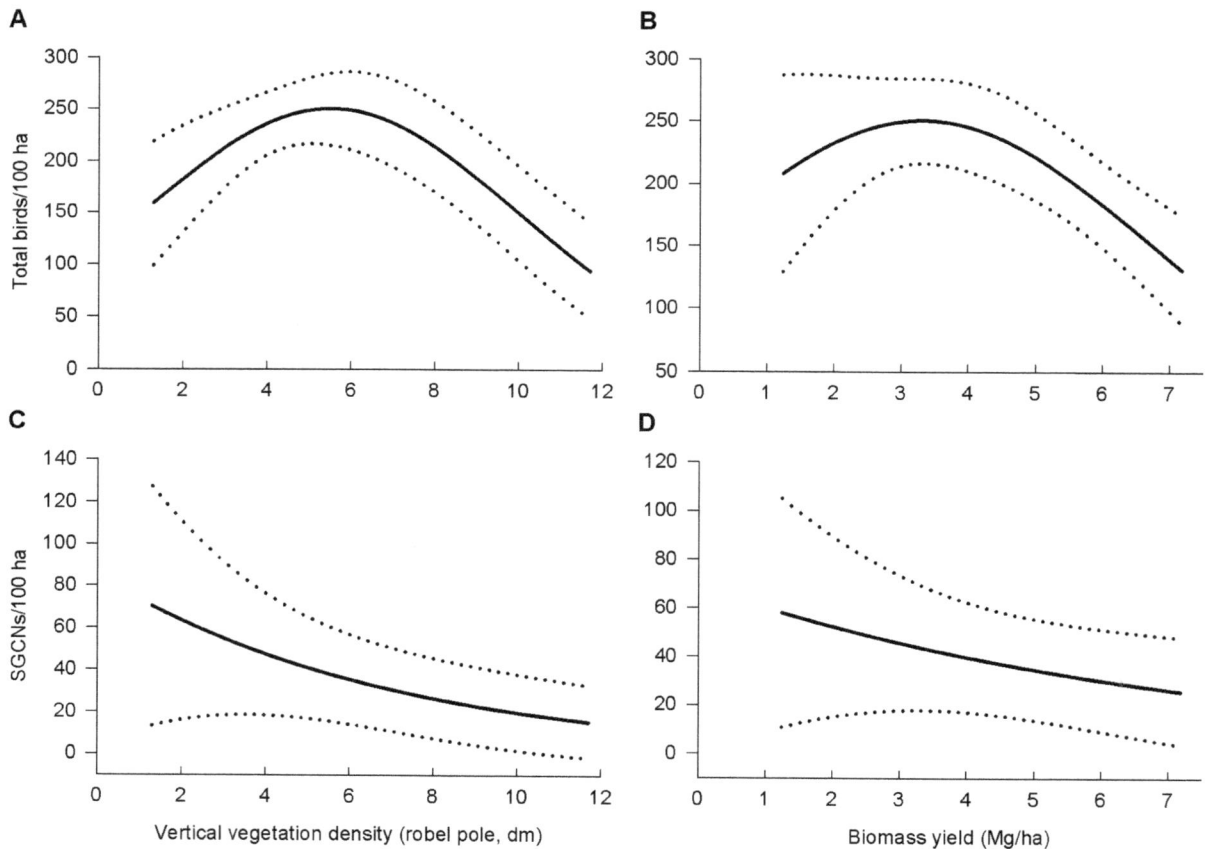

Figure 4. Predicted relationships of bird densities with vertical vegetation density and biomass yields. Total bird density (a–b) and SGCN density (c–d) are shown. Dotted lines are 95% confidence intervals.

culture field had been colonized by a few common milkweed (*Asclepias syriaca*) plants that were popular perches for Dickcissels, and another grass monoculture field was adjacent to a florally diverse field that appeared to attract SGCNs to the area. Further, grass-dominated fields contained an average of 11% forbs, which can be enough to accommodate the needs of many SGCNs [60].

Habitat amount generally has stronger effects on biodiversity than habitat configuration [61]. The amount of grassland habitat in the landscapes surrounding our study sites was positively associated with most bird metrics, suggesting grassland bioenergy fields would be more beneficial for grassland birds if they are near other grassland parcels [10,17,25,62]. We did not find strong evidence of a relationship between most bird metrics and the amount of agriculture in the landscape. Agriculture and forest cover were negatively correlated, suggesting that the bird communities in grasslands were not strongly influenced by forest cover in the landscape. This contrasts with other studies that have found that grassland birds are negatively influenced by the proportion of forest in the landscape [15,20,40,63,64], and may be explained by the fairly low average amount of forest cover (17%) in the agriculture-dominated landscape of our study area.

Based on estimates of potential grassland biomass yields from research in the upper Midwest, the biomass yields from the fields in our study represent plausible, future, commercial-scale, grassland biomass yields [13,42]. Our results agree with others who suggest that grass monocultures would produce greater biomass yields than grass-forb mixtures [29,32,41,65]. Notably, biomass yield across all of our grassland sites was not related to plant species richness, contrasting with other studies that have suggested such a relationship [e.g.,12,30]. Thus, we found that although grass monocultures had the greatest biomass yields, increasing plant diversity in general in perennial grassland fields may have little effect on biomass yields. The frequency, timing, and amount of grassland biomass harvests will also impact biomass yields and bird habitat quality and should be considered when developing harvesting strategies [2,10,21,29,66].

Our analyses relating breeding bird metrics to biomass yields focuses attention on the possible tradeoffs between maintaining bird habitat and producing grassland biomass. Our results support previous suggestions that bioenergy production fields managed for the greatest biomass yield would be of lesser value to bird communities than lower-yielding fields [5,10,40]. Our analyses relating bird metrics to biomass yields may have been influenced by several factors. For example, our biomass yield estimates were obtained following a record-setting drought in 2012, so yields in more average growing seasons could be substantially greater and may influence bird response. Additionally, total bird density was dominated by the most common species (e.g., Red-winged Blackbirds and Common Yellowthroats), and SGCN density was dominated by Dickcissels, therefore trends for those metrics may be driven by a few species. We note that our analyses relate breeding bird densities measured in summer to biomass yields measured in October; this temporal distinction is important because grassland biomass will be harvested well after the breeding season.

Of particular importance for conservation is the impact that grassland biomass fields could have on SGCN populations. We found that SGCN density was negatively associated with vertical vegetation density and biomass yields, suggesting that tall and dense grassland bioenergy fields managed for the greatest biomass yields would be of lesser value to imperiled grassland bird species. If future perennial grass production is designed to maximize biomass yields by creating monocultures of tall and dense grasses, although some habitat generalists (e.g., Song Sparrow) or species

that tolerate dense grasses (e.g., Sedge Wren [*Cistothorus platensis*]) may adapt [17,67], biodiversity will generally be reduced [7]. We agree with Webster et al. [7:457] who suggest that extracting "optimal rather than maximal quantities of biomass" may be necessary to maintain or enhance biodiversity.

In this study we used species richness and bird densities to test for habitat preferences among birds that utilize grassland habitats. We acknowledge that greater richness and densities of birds does not necessarily indicate higher quality habitat for birds and that reproductive success may differ among grassland biomass crops [68,69]. Reproductive success in grassland biomass fields may vary with differences in predation, brood parasitism, food availability, and other factors. These factors will depend on both the local habitat characteristics and the landscape context around the crops. Management intensity and the timing of biomass harvests could also affect reproductive success, and nest losses could be minimized if sustainable harvesting practices were followed [33].

There is a growing understanding that working agricultural lands will need to produce commodities as well as provide environmental benefits [8,70,71]. Although recent profitability analyses indicate that cellulosic bioenergy crops such as switchgrass and mixed grasses would be less profitable than corn as a biofuel feedstock [72], these studies do not include the costs of biodiversity loss, soil loss, and other societal and ecological impacts associated with conventional, annual row crop agriculture. With appropriate financial incentives, farm owners could be motivated to use diverse grassland bioenergy plantings, and sacrifice some biomass yield, in order to support greater agricultural multifunctionality and meet conservation goals [8,10,17].

Supporting Information

Table S1 Field types, common plant species, and estimated biomass yields of grassland study sites in southern Wisconsin.

Table S2 Soil characteristics, determined from the SSURGO Database [73], of grassland study sites in southern Wisconsin.

Table S3 Mean bird densities (birds/100 ha) in cornfields and the three grassland field types, and the total number of sites where each species was detected, in 2011–2012.

Table S4 Top models with $\Delta \, AIC_c \leq 2.0$ for each of the seven bird metrics. Predictor variables in the full (global) models of each response variable included % cover of Forbs (Forbs) and warm-season grasses (WSG), vertical vegetation density (Robel), Robel², and % agriculture (Ag) and grassland (Grassland) within 1 km.

Table S5 Model-averaged parameter estimates and 95% unconditional confidence limits of explanatory variables included in models of the seven bird metrics. Predictor variables in the full (global) models of each response variable included % cover of Forbs (Forbs) and warm-season grasses (WSG), vertical vegetation density (Robel), Robel², and % agriculture (Ag) and grassland (Grassland) within 1 km. Confidence intervals that do not overlap zero are bolded.

Dataset S1 Abundance of individual bird species during point counts at each study site. See Metadata worksheet for definition of variables.

Dataset S2 Vegetation characteristics, biomass yields, and landscape attributes of grassland study sites. See Metadata worksheet for definition of variables.

Acknowledgments

We thank E. R. Keyel, B. Blum, and T. G. Whitby for their assistance in the field, and G. Oates and G. Sanford for their assistance with the biomass yield measurements. We appreciate helpful comments on this manuscript from T. Meehan and R. Graves. We are also grateful to the land owners and managers who allowed us to work on their properties.

Author Contributions

Conceived and designed the experiments: PJB DWS CLW MGT. Performed the experiments: PJB. Analyzed the data: PJB. Contributed reagents/materials/analysis tools: DWS. Contributed to the writing of the manuscript: PJB DWS CLW MGT.

References

1. Sagar AD, Kartha S (2007) Bioenergy and sustainable development? Annual Review of Environment and Resources 2007: 131–167.
2. U.S EPA (2011) Biofuels and the environment: the first triennial report to congress. U.S. Environmental Protection Agency, Office of Research and Development, National Center for Environmental Assessment, Washington, DC. EPA/600/R-10/183F.
3. U.S Congress (2007) H.R. 6–110th Congress: Energy Independence and Security Act. http://www.govtrack.us/congress/bills/110/hr6. Accessed 19 July 2013.
4. Fletcher Jr RJ, Robertson BA, Evans J, Doran PJ, Alavalapati JR, et al. (2010) Biodiversity conservation in the era of biofuels: risks and opportunities. Frontiers in Ecology and the Environment 9: 161–168.
5. Groom MJ, Gray EM, Townsend PA (2008) Biofuels and biodiversity: principles for creating better policies for biofuel production. Conservation Biology 22: 602–609.
6. Paine LK, Peterson TL, Undersander D, Rineer KC, Bartelt GA, et al. (1996) Some ecological and socio-economic considerations for biomass energy crop production. Biomass and Bioenergy 10: 231–242.
7. Webster CR, Flaspohler DJ, Jackson RD, Meehan TD, Gratton C (2010) Diversity, productivity and landscape-level effects in North American grasslands managed for biomass production. Biofuels 1: 451–461.
8. Jordan N, Warner KD (2010) Enhancing the multifunctionality of US agriculture. BioScience 60: 60–66.
9. Williams CL, Charland P, Radloff G, Sample D, Jackson RD (2013) Grass-shed: Place and process for catalyzing perennial grass bioeconomies and their potential multiple benefits. Journal of Soil and Water Conservation 68: 141A–146A.
10. Fargione JE, Cooper TR, Flaspohler DJ, Hill J, Lehman C, et al. (2009) Bioenergy and wildlife: threats and opportunities for grassland conservation. BioScience 59: 767–777.
11. Meehan TD, Hurlbert AH, Gratton C (2010) Bird communities in future bioenergy landscapes of the Upper Midwest. Proceedings of the National Academy of Sciences 107: 18533–18538.
12. Tilman D, Hill J, Lehman C (2006) Carbon-negative biofuels from low-input high-diversity grassland biomass. Science 314: 1598–1600.
13. Miesel JR, Renz MJ, Doll JE, Jackson RD (2012) Effectiveness of weed management methods in establishment of switchgrass and a native species mixture for biofuels in Wisconsin. Biomass and Bioenergy 36: 121–131.
14. Mulkey V, Owens V, Lee D (2008) Management of warm-season grass mixtures for biomass production in South Dakota USA. Bioresource technology 99: 609–617.
15. Werling BP, Dickson TL, Isaacs R, Gaines H, Gratton C, et al. (2014) Perennial grasslands enhance biodiversity and multiple ecosystem services in bioenergy landscapes. Proceedings of the National Academy of Sciences 111: 1652–1657.
16. Askins RA, Chávez-Ramírez F, Dale BC, Haas CA, Herkert JR, et al. (2007) Conservation of grassland birds in North America: understanding ecological processes in different regions. Ornithological Monographs 64: 1–46.
17. Robertson BA, Rice RA, Sillett TS, Ribic CA, Babcock BA, et al. (2012) Are agrofuels a conservation threat or opportunity for grassland birds in the United States? The Condor 114: 679–688.
18. Gill DE, Blank P, Parks J, Guerard JB, Lohr B, et al. (2006) Plants and breeding bird response on a managed Conservation Reserve Program grassland in Maryland. Wildlife Society Bulletin 34: 944–956.
19. McCoy TD, Kurzejeski EW, Burger LWJ, Ryan MR (2001) Effects of conservation practice, mowing, and temporal changes on vegetation structure on CRP fields in northern Missouri. Wildlife Society Bulletin 29: 979–987.
20. Robertson BA, Doran PJ, Loomis ER, Robertson JR, Schemske DW (2011) Avian use of perennial biomass feedstocks as post-breeding and migratory stopover habitat. PloS one 6: e16941.
21. Roth AM, Sample DW, Ribic CA, Paine L, Undersander DJ, et al. (2005) Grassland bird response to harvesting switchgrass as a biomass energy crop. Biomass and Bioenergy 28: 490–498.
22. Sample DW (1989) Grassland birds in southern Wisconsin: habitat preference, population trends, and response to land use changes. M.Sc. Thesis, University of Wisconsin, Madison.
23. Cunningham MA, Johnson DH (2006) Proximate and landscape factors influence grassland bird distributions. Ecological Applications 16: 1062–1075.
24. Fletcher RJ, Koford RR (2002) Habitat and landscape associations of breeding birds in native and restored grasslands. Journal of Wildlife Management 66: 1011–1022.
25. Ribic CA, Guzy MJ, Sample DW (2009) Grassland bird use of remnant prairie and Conservation Reserve Program fields in an agricultural landscape in Wisconsin. American Midland Naturalist 161: 110–122.
26. Ribic CA, Koford RR, Herkert JR, Johnson DH, Niemuth ND, et al. (2009) Area sensitivity in North American grassland birds: patterns and processes. The Auk 126: 233–244.
27. Ribic CA, Sample DW (2001) Associations of grassland birds with landscape factors in southern Wisconsin. American Midland Naturalist 146: 105–121.
28. Dauber J, Jones MB, Stout JC (2010) The impact of biomass crop cultivation on temperate biodiversity. Global Change Biology 2: 289–309.
29. Griffith AP, Epplin FM, Fuhlendorf SD, Gillen R (2011) A comparison of perennial polycultures and monocultures for producing biomass for biorefinery feedstock. Agronomy Journal 103: 617–627.
30. Adler PR, Sanderson MA, Weimer PJ, Vogel KP (2009) Plant species composition and biofuel yields of conservation grasslands. Ecological Applications 19: 2202–2209.
31. Garlock RJ, Bals B, Jasrotia P, Balan V, Dale BE (2012) Influence of variable species on the saccharification of AFEX pretreated biomass from unmanaged fields in comparison to corn stover. Biomass and Bioenergy 37: 49–59.
32. Anderson-Teixeira KJ, Duval BD, Long SP, DeLucia EH (2012) Biofuels on the landscape: Is "land sharing" preferable to "land sparing"? Ecological Applications 22: 2035–2048.
33. Hull S, Arntzen J, Bleser C, Crossley A, Jackson R, et al. (2011) Wisconsin sustainable planting and harvest guidelines for nonforest biomass. Joint collaboration of the Wisconsin Department of Natural Resources, the University of Wisconsin, and the Wisconsin Department of Agriculture, Trade, and Consumer Protection.
34. NOAA (2013) Milwaukee/Sullivan, unique local climate data. National Oceanic and Atmospheric Administration, National Weather Service.
35. NOAA (2013) 2012 Wisconsin yearly weather summary. National Oceanic and Atmospheric Administration, National Weather Service.
36. USDA (1978) Soil survey of Dane County, Wisconsin. U.S. Department of Agriculture, Soil Conservation Service, Wisconsin.
37. Behrman KD, Kiniry JR, Winchell M, Juenger TE, Keitt TH (2013) Spatial forecasting of switchgrass productivity under current and future climate change scenarios. Ecological Applications 23: 73–85.
38. Wisconsin Department of Natural Resources (2005) Wisconsin's strategy for wildlife species of greatest conservation need. Madison, WI, USA.
39. Runge TM, Porter PA (2013) Not quite converted: how fuel prices both started and stopped a biomass heating plant project. Biofuels 4: 493–500.
40. Robertson BA, Dora PJ, Loomis LR, Robertson J, Schemske DW (2011) Perennial biomass feedstocks enhance avian diversity. Global Change Biology Bioenergy 3: 235–246.
41. Schmer MR, Vogel KP, Mitchell RB, Perrin RK (2008) Net energy of cellulosic ethanol from switchgrass. Proceedings of the National Academy of Sciences 105: 464–469.
42. Porter PA, Barry J, Samson R, Doudlah M (2008) Growing Wisconsin energy: a native grass pellet bio-heat roadmap for Wisconsin. Agrecol Corporation, 2918 Agriculture Drive, Madison, WI 53718, USA.
43. Daubenmire R (1959) A canopy-coverage method of vegetational analysis. Northwest Science 33: 43–64.
44. Robel RJ, Briggs JN, Dayton AD, Hulbert LC (1970) Relationship between visual obstruction measurements and weight of grassland vegetation. Journal of Range Management 23: 295–297.
45. USDA (2011) National Agricultural Statistics Service Cropland Data Layer. Available at http://nassgeodata.gmu.edu/CropScape/. USDA-NASS, Washington, DC.
46. ESRI (2011) ArcGIS Desktop: Release 10. Redlands, CA.
47. Wright CK, Wimberly MC (2013) Recent land use change in the Western Corn Belt theatens grasslands and wetlands. Proceedings of the National Academy of Sciences 110: 4134–4139.
48. Karanth KK, Nichols JD, Sauer JR, Hines JE, Yackulic CB (2013) Latitudinal gradients in North American avian species richness, turnover rates and extinction probabilities. Ecography 37: 626–636.

49. Reese GC, Wilson KR, Flather CH (2013) Program SimAssem: software for simulating species assemblages and estimating species richness. Methods in Ecology and Evolution 4: 891–896.

50. Fiske I, Chandler RB (2011) UNMARKED: an R package for fitting hierarchical models of wildlife occurrence and abundance. Journal of Statistical Software 43: 1–23.

51. Diefenbach DR, Brauning DW, Mattice JA (2003) Variability in grassland bird counts related to observer differences and species detection rates. Auk 120: 1168–1179.

52. R Core Development Team (2013) R: A language and environment for statistical computing. R Foundation for statisitical computing, Viennam Austria. http://www.R-project.org/.

53. Bates D, Maechler M, Bolker B (2012) lme4: Linear mixed-effects models using S4 classes/R package version 0.999999-0.

54. Burnham KP, Anderson DR (2002) Model selection and multimodel inference: a practical information-theoretic approach. New York: Springer.

55. Zuur AF, Ieno EN, Walker NJ, Saveliev AA, Smith GM (2009) Mixed effects models and extensions in R. New York: Springer.

56. Mazerolle MJ (2013) AICcmodavg: Model selection and multimodel inference based on (Q)AIC(c). R package version 1.28.

57. Pinheiro J, Bates D, DebRoy S, Sarkar D, and the R Development Core Team (2013) nlme: linear and nonlinear mixed effects models. R package version 3.1-111.

58. McIntyre NE, Thompson TR (2003) A comparison of conservation reserve program habitat plantings with respect to arthropod prey for grassland birds. American Midland Naturalist 150: 291–301.

59. Werling BP, Meehan TD, Robertson BA, Gratton C, Landis DA (2011) Biocontrol potential varies with changes in biofuel–crop plant communities and landscape perenniality. GCB Bioenergy 3: 347–359.

60. Sample DW, Mossman MJ (1997) Managing habitat for grassland birds: a guide for Wisconsin. Wisconsin Department of Natural Resouces, Madison, WI. PUBL-SS-925-97.

61. Fahrig L (2003) Effects of habitat fragmentation on biodiversity. Annual Review of Ecology, Evolution, and Systematics: 487–515.

62. Davis SK, Fisher RJ, Skinner SL, Shaffer TL, Brigham RM (2013) Songbird abundance in native and planted grassland varies with type and amount of grassland in the surrounding landscape. Journal of Wildlife Management 77: 908–919.

63. Helzer CJ, Jelinski DE (1999) The relative importance of patch area and perimeter-area ratio to grassland breeding birds. Ecological Applications 9: 1448–1458.

64. Robertson BA, Landis BA, Sillett TS, Loomis ER, Rice RA (2013) Perennial agroenergy feedstocks as en route habitat for spring migratory birds. Bioenergy Research 6: 311–320.

65. Wang D, Lebauer DS, Dietze MC (2010) A quantitative review comparing the yield of switchgrass in monocultures and mixtures in relation to climate and management factors. GCB Bioenergy 2: 16–25.

66. Murray LD, Best LB (2003) Short-term bird response to harvesting switchgrass for biomass in Iowa. Journal of Wildlife Management 67: 611–621.

67. Murray LD, Best LB, Jacobsen TJ, Braster ML (2003) Potential effects on grassland birds of converting marginal cropland to switchgrass biomass production. Biomass and Bioenergy 25: 167–175.

68. Van Horne B (1983) Density as a misleading indicator of habitat quality. Journal of Wildlife Management 47: 893–901.

69. Winter M, Faaborg J (1999) Patterns of area sensitivity in grassland nesting birds. Conservation Biology 13: 1424–1436.

70. Matson PA, Vitousek PM (2006) Agricultural intensification: will land spared from farming be land spared for nature? Conservation Biology 20: 709–710.

71. Perfecto I, Vandermeer J (2010) The agroecological matrix as alternative to the land-sparing/agriculture intensification model. Proceedings of the National Academy of Sciences 107: 5786–5791.

72. James LK, Swinton SM, Thelen KD (2010) Profitability analysis of cellulosic energy crops compared with corn. Agronomy Journal 102: 675–687.

73. USDA (2013) Web soil survey. Natural Resources Conservation Service, Washington, DC.

5

Defining Landscape Resistance Values in Least-Cost Connectivity Models for the Invasive Grey Squirrel: A Comparison of Approaches Using Expert-Opinion and Habitat Suitability Modelling

Claire D. Stevenson-Holt[1]*, Kevin Watts[2], Chloe C. Bellamy[3], Owen T. Nevin[4], Andrew D. Ramsey[5]

1 Centre for Wildlife Conservation, University of Cumbria, Ambleside, Cumbria, United Kingdom, 2 Centre for Ecosystems, Society and Biosecurity, Forest Research, Farnham, Surrey, United Kingdom, 3 Centre for Ecosystems, Society and Biosecurity, Forest Research, Roslin, Midlothian, United Kingdom, 4 School of Medical and Applied Sciences, Central Queensland University, Gladstone, Queensland, Australia, 5 School of Biological and Forensic Sciences, University of Derby, Derby, Derbyshire, United Kingdom

Abstract

Least-cost models are widely used to study the functional connectivity of habitat within a varied landscape matrix. A critical step in the process is identifying resistance values for each land cover based upon the facilitating or impeding impact on species movement. Ideally resistance values would be parameterised with empirical data, but due to a shortage of such information, expert-opinion is often used. However, the use of expert-opinion is seen as subjective, human-centric and unreliable. This study derived resistance values from grey squirrel habitat suitability models (HSM) in order to compare the utility and validity of this approach with more traditional, expert-led methods. Models were built and tested with MaxEnt, using squirrel presence records and a categorical land cover map for Cumbria, UK. Predictions on the likelihood of squirrel occurrence within each land cover type were inverted, providing resistance values which were used to parameterise a least-cost model. The resulting habitat networks were measured and compared to those derived from a least-cost model built with previously collated information from experts. The expert-derived and HSM-inferred least-cost networks differ in precision. The HSM-informed networks were smaller and more fragmented because of the higher resistance values attributed to most habitats. These results are discussed in relation to the applicability of both approaches for conservation and management objectives, providing guidance to researchers and practitioners attempting to apply and interpret a least-cost approach to mapping ecological networks.

Editor: Benjamin Lee Allen, University of Queensland, Australia

Funding: This project was funded by the Forestry Commission GB and the National School of Forestry at the University of Cumbria. The funders had no role in study design, data collection and analysis, decision to publish, or preparation of the manuscript.

Competing Interests: The authors have the following competing interest: This work was funded by the Forestry Commission GB and National School of Forestry at the University of Cumbria.

* Email: claire.stevenson@cumbria.ac.uk

Introduction

Effective biodiversity conservation within fragmented landscapes often requires the modelling of connectivity to define the extent of the problem, target conservation activities and to evaluate the impacts of landscape change [1]. Connectivity is defined as the degree to which the landscape facilitates or impedes species movement among resource patches [2]. A landscape consists of a complex, often dynamic, heterogeneous mixture of habitats and land uses which may impact on important ecological processes, such as species movement, habitat selection and survival, and influence behavioural and physiological responses [2–5]. The study of the impacts of the matrix on species movement, known as functional connectivity [6], is now the subject of much research within modified and fragmented landscapes [7]. Assessing functional connectivity is commonly used to aid conservation strategies by identifying potential movement pathways across fragmented landscapes for species of conservation concern [8–10]. It has also been used to help predict the potential dispersal and movement of invasive species to aid species management by identifying areas to target resources [11,12].

Geographical Information System (GIS), raster-based least-cost analysis techniques are often used to assess functional connectivity by modelling the impact of permeability of the surrounding landscape matrix on species movement [10]. It has been used in conservation [8–10] and invasive species management contexts [11,12]. For example, the population expansion of the grey squirrel (*Sciurus carolinensis*) in Britain, following its first introduction in 1876 [13], has had negative effects upon the forestry industry and native biodiversity [14–16]. In particular, it has occurred simultaneously with the decline and replacement of native red squirrel (*Sciurus vulgaris*) populations through resource competition and disease [14–16]. Therefore, an understanding of how grey squirrels utilise and move through the landscape is essential for effective red squirrel conservation and grey squirrel management. By using least-cost modelling it is possible to identify the potential dispersal areas, in addition to the most probable dispersal corridors, to assess the extent of spread [11]. Developing these models involves defining a species' 'core' or 'source' habitat and assigning resistance values to the surrounding landscape features, based on the actual or perceived impact to species movement at a particular resolution [17]. A cell with a high resistance value is used to represent an area that an individual is unlikely to traverse under typical conditions because of high energy, mortality, or other ecological costs [18]. Using information on a species' maximum dispersal distance, the area around a core habitat patch that is accessible to a species can be mapped with a simple Euclidean buffer. The permeability buffer zone is then taken into account so that the buffer is compressed or stretched according to the cumulative resistance scores assigned to the underlying landscape features. Overlapping buffers therefore signify connections where the species is assumed to be able to move between core habitat patches, forming a functionally connected habitat network.

It is widely acknowledged [4,18,19] that a critical step in least-cost modelling is defining resistance values for each type of landscape feature. Beier et al. [19] highlighted three ranked choices for estimating landscape resistance values with the first being the most highly ranked option: 1. empirical animal movement data, genetic distance or rates of inter-patch movements; 2. animal occurrence, density or fitness; 3. literature review and expert opinion. Ideally, resistance values should be informed and parameterised with independent field data, such as extensive mark release recapture studies, actual movement data from radio-telemetry or Global Positioning System (GPS) studies [11,20], data from experimental studies to record movement through different land cover types [21], or inferred movement data from landscape genetics [9]. However, as these resistance values are species and landscape specific, there is an understandable shortage of such empirical data [22]. Zeller et al. [23] reviewed the different types of data used to parameterize least-cost models and concluded that expert-opinion and occurrence data are most often used. However, they also suggest that comparative studies on the data used to derive resistance values are needed.

Although the use of expert-opinion to parameterise least-cost models is seen as subjective and out performed by values informed by empirical data [24], many studies utilise this type of information to parameterise models [3,12,25]. The use of expert-opinion may be appropriate in some cases, such as where there is a particular shortage of empirical data, an urgency to act, or a focus on general principles, focal species or particular species traits. However, in an attempt to make the setting of landscape resistance values less biased and more data-driven, some researchers [26–31] are starting to utilise species distribution models, such as MaxEnt [32], to parameterise least-cost connectivity models (defined as option 2 by Beier [19]). This study uses MaxEnt, a species distribution model which utilises maximum entropy principles to predict a species' use of a landscape based upon occurrence data and a selected set of environmental predictors [32]. The habitat suitability indices provided by the models can then be used in calculations [26–31] to create least-cost connectivity models. Given that resistance values informed by empirical data are ranked higher [19] and seen to outperform expert-opinion values [24], it is hypothesised that the HSM-informed values will produce a more accurate least-cost network

than expert-opinion data. The aim of this study is to investigate how expert-derived resistance values compare against values informed by habitat suitability modelling (HSM). The results of this study provide guidance to researchers and practitioners on the suitability of these approaches for informing management and research objectives relating to both species of conservation concern and invasive species spread.

Materials and Methods

Ethical statement

Ethical clearance for this study was approved by the University of Cumbria Ethics Committee, ref 09/17. This was a desk based study with no field work required. Therefore, research permits and licences were not required.

Study site

To compare expert-derived resistance values against HSM-informed values, grey squirrel within the county of Cumbria UK (Figure 1), are used as the study species. Whilst six large woodlands in Cumbria are designated red squirrel refuge reserves (Figure 1), the grey squirrel remains throughout the county. A number of previous studies have used expert-derived least-cost models to define habitat connectivity for Britain's native red squirrels and invasive grey squirrels [33–36], providing expert-opinion on land cover resistance. In addition, Cumbria has an extensive collection of grey squirrel distribution records available with which to create HSM-informed data for comparison. Cumbria covers an area of 6,768 km² and has a sparse population of 490,000 people. The Lake District National Park is located in the centre of Cumbria and has legislation and planning restrictions to conserve the landscape. The National Park Authority are responsible for implementing legislation and planning decisions aimed at conserving the landscape and its species, which means that little has changed regarding land use during the time frame that the species presence data used within this study were recorded (2000–2009). The topography is varied with the Cumbrian Mountain range (≤978 m a.s.l.) that runs approximately west to east across the middle of the county. The majority of land at these higher elevations is used for grazing with little woodland habitat. However, at lower elevations there are numerous woodlands, and other semi-natural habitats, scattered within an agricultural matrix which may provide greater potential for squirrel movement.

Identifying least-cost networks

Land cover types from a highly accurate and up to date vector land cover map (Ordnance Survey Master Map) were reclassified into 21 broad land cover categories for Cumbria (Table 1). The map was rasterised at 10 m resolution to ensure accurate representation of narrow linear features, such as strips of woodland. All woodland patches were classed as core habitat as squirrels use these areas for nesting and breeding [37,38]. This map was then parameterised with five alternative expert-derived resistance sets from previous studies (Table 1). The resistance values given in the different studies varied substantially. An additional set of values was developed by the authors by refining Stevenson's [35] scores (referred to as new expert-derived), following a review of the literature and the ecological underpinning of the values that had been applied previously, as described below.

Coniferous, mixed and broadleaved woodland were all assigned the lowest resistance value of 1, as core habitat. Scrub, coppice, orchard, and garden were given relatively low resistance values because they often contain tree species and are commonly used by

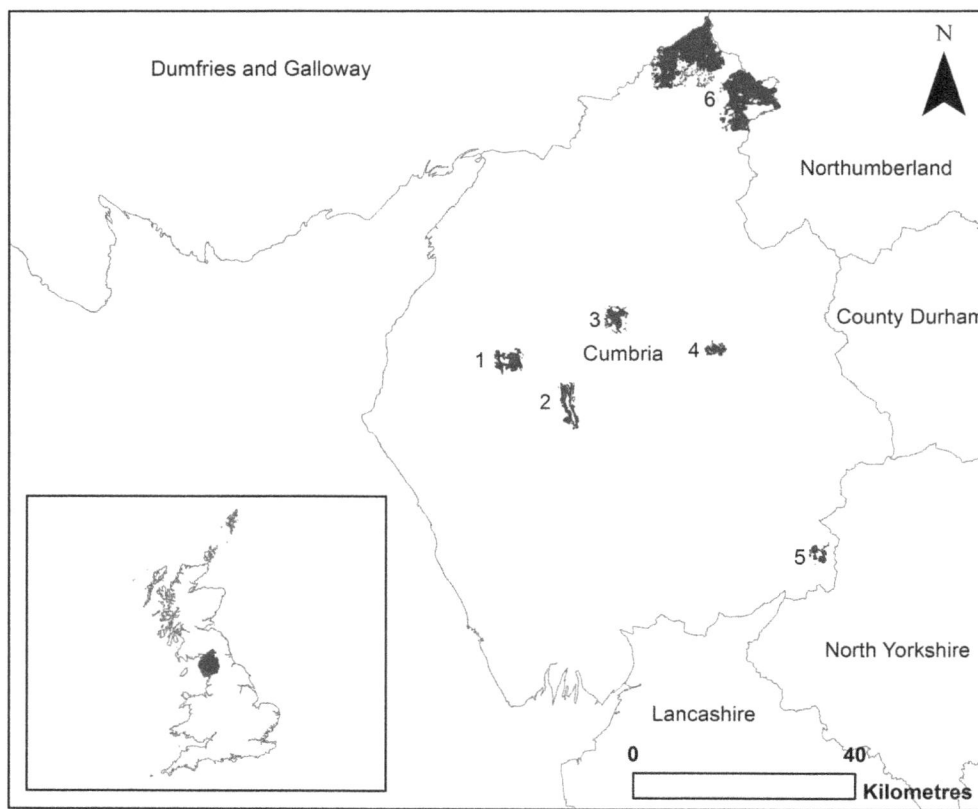

Figure 1. Map of red squirrel reserves in Cumbria and neighbouring counties with reference to its location in the UK. * 1. Whinlatter; 2. Thirlmere; 3. Greystoke; 4. Whinfell; 5. Garsdale/Mallerstang and 6. Kielder (Cumbria proportion of). Boundary lines were obtained through EDINA Digimap Ordnance Survey Service, http://digimap.edina.ac.uk/digimap/home.

grey squirrels for commuting [11,20]. Path, track, road and railway verge may also be used as commuting corridors, [13], but their use may confer higher mortality risks and therefore they were assigned a relatively high score. Improved/arable/amenity, rough grassland and heath were all attributed higher values still, as squirrel species tend to avoid open habitats [39]. Due to the threat of railways and the difficulty of moving over marsh, water, urban areas, buildings, and rocky areas like cliffs, the high scores assigned in previous studies were maintained.

Least-cost networks were created for each set of resistance values (Table 1) using the least-cost network process outlined in Watts et al. [10]. This network tool analysis utilises ArcView 9.1 and the Spatial Analyst extension (ESRI, Redlands, CA). The first step defined suitable patches of woodland habitat and generated a cost surface raster from the land cover map, by joining the resistance values (Table 1) to the 21 land cover classes. Secondly, the 'cost-distance' function in the Spatial Analyst toolbox was used to create a cost-distance surface between woodland patches. The resulting accumulated cost raster was then reclassified to a standardised maximum dispersal distance of 8 km to ensure comparability between the different resistance sets. The 'region group' function was used to define each discrete network, using an eight-cell rule so that touching cells, either adjacent and diagonally opposite, within the minimum distance of any given patch were considered part of the same network.

Deriving resistance scores from habitat suitability modelling

Records of grey squirrel presence were obtained from Save Our Squirrels (http://www.saveoursquirrels.org.uk/). These consisted of 2,281 verified sightings recorded year-round between 2000 and 2009 given by members of the public from both within woodland habitat (35%) and the wider landscape (65%). The grid references and type of habitat the sightings were recorded and verified by Save Our Squirrels. Sightings that were recorded outside of the grey squirrels known distribution range were also verified by contacting the recorder. The points outside of core woodland habitat are believed to relate to landscape use and movement, rather than indicating suitable foraging, breeding or nesting resources [37,38]. It is these non-woodland records that are used to infer the permeability of the landscape matrix using the habitat suitability modelling software, MaxEnt [40,32]. MaxEnt assigns each raster cell a Habitat Suitability Index (HSI) based on the environmental conditions at locations where a species has been recorded, using the maximum entropy method [41]. There are three output formats given by the MaxEnt programme: raw, cumulative and logistic; the most easily intuitive logistic HSI scores, which indicate the probability of occupancy ranging between 0–1 and assuming that this is 0.5 at an average site [40,32], were used in this study.

Both the species records and environmental data were prepared for modelling with MaxEnt. The squirrel data were filtered to remove locations recorded at a resolution of >100 m. Of the remaining 2,008 points, 842 squirrel presences recorded were

Table 1. Land cover resistance values based on previous least-cost modelling studies and resistance set based on expert opinion.

Land cover classification	Stevenson (2008)	Humphrey et al.(2007)	Williams (2008)	Verbeylen et al (2003) *	Gonzales (2000)	New expert-derived	Land cover ranking
Broadleaf	1	0	0	1	1	1	1
Mixed	1	0	0	1	1	1	2
Coniferous	1	0	0	1	1	1	3
Orchard	30	1	4	300	5	16	4
Scrub	1	2	10	10	5	16	18
Coppice	1	1	4	300	5	16	5
Garden	1	1	20	10	5	11	6
Improved/arable/amenity	30	10	20	800	8	40	20
Rough grassland	30	10	20	800	8	40	15
Heath	22	20	20	300	9	37	19
Path	16	50	20	800	1	27	9
Railway verge	12	10	20	800	4	27	11
Road verge	12	10	20	800	4	27	7
Marsh	76	50	50	800	9	91	14
Water	115	50	100	800	999	130	13
Urban	57	50	50	800	1	72	16
Railway	40	50	50	800	9	55	12
Road	12	50	30	800	9	27	8
Track	12	50	30	800	1	27	10
Building	1000	50	50	1000	999	1000	17
Rock	1000	50	50	1000	999	1000	21

*Note the resistance scores given in Verbeylen et al.'s (2003) study are for red squirrels in an urbanised matrix and are used for comparison.

within the matrix. A categorical land cover raster (gridded data map) was created using the same Ordnance Survey Master Map data and 21 broad land cover categories as previously described (Table 1). However, a coarser resolution of 100 m was used to match the spatial accuracy of the squirrel records. To ensure that linear habitat features were not under-represented, each land cover type was rasterised separately at a 10 m resolution and then aggregated to 100 m using the 'maximum' rule. These rasters were mosaicked using ranks that prioritised the classification of each 100 m square containing more than one land cover type (Table 1). All areas of woodland (core habitat), rocks and buildings (highly impermeable) were removed from the land cover map to prevent their incorporation in the model. In an effort to account for sampling bias towards accessible areas, a well-known and common characteristic of species data collected in an *ad-hoc* or non-systematic way [42], all areas over 500 m from a road, track or path were also removed from the map. This left a total of 665 squirrel records that fell within the remaining areas of the land cover map which were used to train and test the habitat suitability model. Each point (located in the south west of the 100 m grid square) was adjusted by 50 m east and 50 m north to locate each point in the centre of the grid square. This was to ensure that the points matched the 100 m raster landscape i.e. were within one cell, not potentially boarding four.

All models were run in MaxEnt Version 3.3.3k, using primarily default settings (regularisation multiplier = 1; duplicate occurrences removed; maximum number of background points = 10000, as used in Kramer-Schadt et al. [43]). Five-fold cross validation was used to calculate mean Area Under Roc Curve (AUC) and extrinsic omission rates (the average proportion of test points that fall outside the area predicted to be suitable), following use of the occupancy threshold rule that maximises the sum of test sensitivity and specificity (as recommended by Liu et al., [44]). Residual spatial autocorrelation (rSAC) can inflate measures of model performance [45–47] therefore Moran's correlograms were created (1 – predicted HSI for each species record; [48]) using the Spatial Analysis in Macroecology software program (SAM; [49]). Significance of Moran's I was calculated using a randomisation test with 9,999 Monte Carlo permutations, correcting for multiple testing.

The response curves, which showed the mean predicted probability of a species' presence (p; 0–1 scale) within each land cover type, were used to derive the resistance values for each land cover type. For both the new expert-derived and the HSM-informed values, woodland was given a value of 1, as permeable core habitat, and rock and building given values of 1000, as impermeable land cover types. The remaining land cover type values were inverted and standardised to the same scale as the new expert-derived values, (1–130; using 1-($p \times 130$)). These values were then used to identify least-cost networks using the same approach as applied to the new expert-derived resistance scores.

Comparing resistance scores and resulting habitat networks

An area-minimisation methodology was applied to select for the smallest network that captures the majority (≥90%) of the filtered distribution point data (n = 842). This methodology, derived during this study, was based on the principle that when managing invasive species, areas for control must be defined and defensible to provide successful management [50]. As the grey squirrel population continues to expand in the Cumbria study site, it is important that control efforts are targeted to provide effective management. By identifying habitat networks management can be targeted in these specific areas of the landscape. The larger the

habitat networks are the more widespread management would have to be. Therefore, the resistance set which produced networks that include a high proportion of distribution points but a small network area are regarded as the better networks as management can be targeted in these focused areas. In addition a chi square test was used to test whether a significant number of distribution points were within the networks when compared to random points.

The HSM-informed resistance scores and the resulting networks were compared to those created with the new expert-derived set selected by the area-minimisation criteria. A Wilcoxon signed ranks test was used to assess the relative difference between scores. The habitat networks produced were also measured and compared visually and using the distribution points. Distribution points that were within the new expert-derived networks but not within the HSM-informed networks were identified along with the land cover type they were in and vice versa.

Results

Habitat Suitability Model performance

The results from five-fold cross-validation test showed that the models performed well (Training sample size = 532; Test sample size = 133; Training AUC = 0.80±0.001; Test AUC = 0.78±0.04; Test gain = 0.70±0.19; Extrinsic omission rate = 0.23, $P<0.001$), indicating that land cover type provides useful information on the likelihood of grey squirrel presence. No significant residual spatial autocorrelation was found at any distance lag. Moran's I values were <0.05 and statistically insignificant at each distance lag, indicating that the residuals were not spatially autocorrelated.

Selecting an 'optimal model' from expert-derived resistance sets

There was considerable variation between the previous studies and new expert-derived resistance sets, with network area ranging from 78% to 15% of the total landscape area, containing between 99% and 32% of the squirrel point data (Figure 2). However, when the networks were tested against the occurrence data within the matrix all resulting networks contained significantly more distribution points than expected by chance (n = 842, Stevenson 2008, $\chi2 = 623$, df. = 1, p<0.001; Humphreys et al. 2007, $\chi2 = 238$, df. = 1, p<0.001; Williams 2008, $\chi2 = 357$, df. = 1, p< 0.001; Verbeylen et al. 2003, $\chi2 = 169$, df. = 1, p<0.001; Gonzales 2000, $\chi2 = 213$, df. = 1, p<0.001; new expert-derived, $\chi2 = 623$, df. = 1, p<0.001). Using the area-minimisation methodology, the new expert-derived resistance set was shown to have above 90% of sightings within the networks and the lowest networks area of 49% of the total landscape (Figure 2). This was therefore selected and used for further comparison with the HSM-informed networks.

Comparing expert-derived and HSM-informed networks

The HSM-informed network had significantly more grey squirrel distribution points within it than expected by chance (n = 842, $\chi2 = 836$, df. = 1, p<0.001). However, the new expert-derived network contained significantly more points than the HSM-informed network (n = 842, $\chi2 = 185$, df. = 1, p<0.001). The majority of land cover types were given higher resistance values using the HSM approach compared to those derived from the new expert-derived values, with relative differences ranging from 7–86% (Table 2). These differences were found to be statistically significant (n = 16, Wilcoxon signed ranks test, $p = 0.002$); water and coppice were the only habitats to be assigned an HSM-informed lower resistance value compared to those derived from the new expert-derived resistance set. The

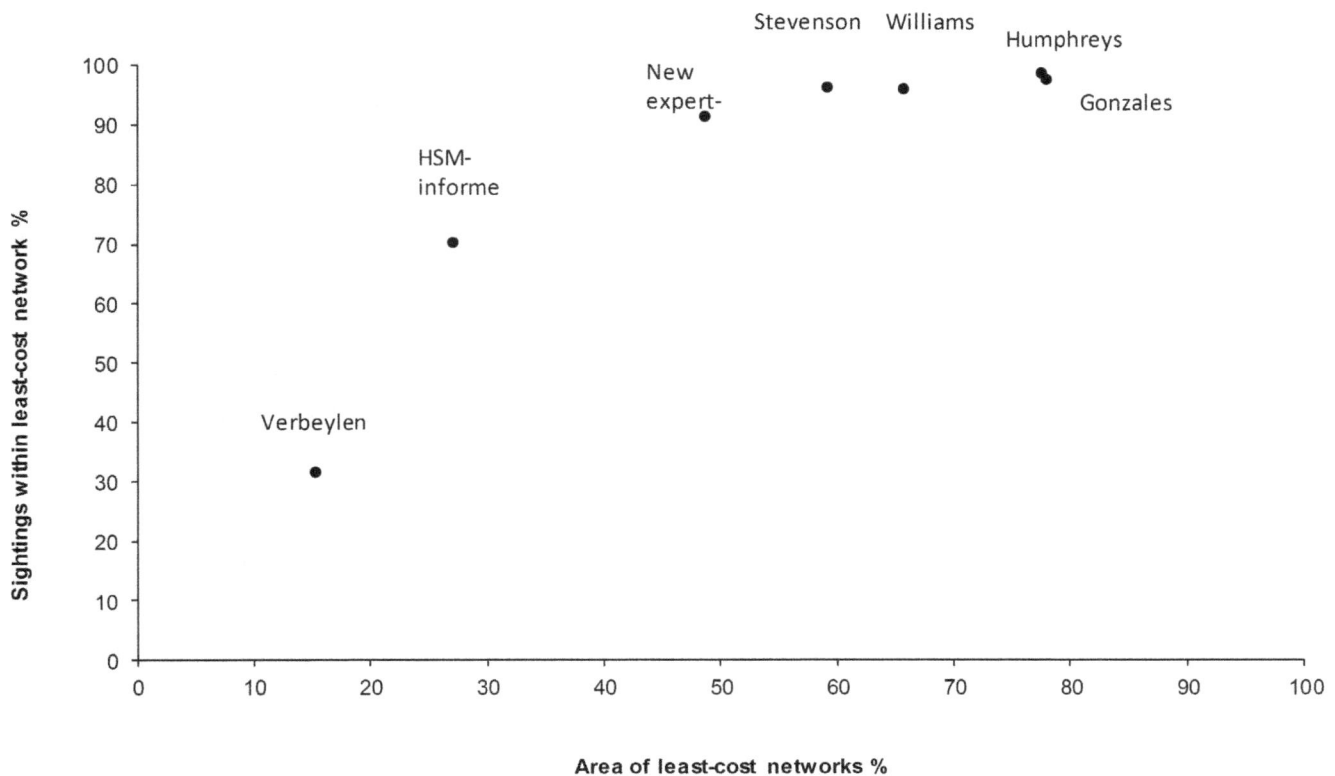

Figure 2. Comparison between expert-derived and Habitat Suitability Model-derived resistance values. Note: values that produce a network with >90% sightings points and the lowest network area is considered the best model for management.

largest differences in the resistance values assigned to habitats by the two approaches were between scrub, tracks, railways and railway verges (Table 2).

The least-cost model parameterised with the new expert-derived resistance values identified 738 discrete networks, although two of these cover substantial areas; habitat network 1 in the north and habitat network 2 in the south (Figure 3). The mean network size was 4.7 km^2 ($\pm$84.4). These networks accounted for 42% of the Cumbrian land cover (3,518 km^2) and appear to be separated by the land cover types within the Cumbrian Mountains. The HSM-informed resistance values generated comparatively smaller and more fragmented networks, owing to the higher resistance scores attributed to most habitat types. This network was 55% the size of the new expert-derived network (1,953 km^2; 34% of land cover) and sat almost entirely inside it, with only 0.2% extending beyond the expert-derived network, over areas of water. The mean network size was 0.3 km^2 ($\pm$5.0) and 5,840 separate networks were identified in Cumbria (Figure 3). Ten of these were relatively large (>20 km^2). The HSM-informed networks also indicated that networks in the north and south of the county were separated by the Cumbrian Mountains range. Both identified Grizdale Forest and surrounding woodlands as a large, well connected grey squirrel habitat network (Figure 3).

The smaller HSM-informed least-cost networks contained 592 (70%) of 842 species records within the habitat network (compared to 772 (92%) using new expert-derived) (Figure 4). As the HSM-informed scores were based upon the actual distribution data it was expected that the resulting networks would include a substantial amount of distribution points. The number of points outside of the HSM-informed least-cost networks was 250; of these

points missed by the HSM-informed network 180 were included within the new expert-derived networks. These 180 points were located in improved/arable/amenity land (77%), gardens (8%), rough grassland (6%), urban (3%), road (2%), road verge (1%), tracks (1%), marshland (1%), scrub (1%) or water (1%). The number of points outside of the new expert-derived networks was 70; of these points none were included within the HSM-informed least-cost networks.

Discussion

When estimating resistance values Beier [19] highlighted three ranked choices. Although using animal movement data, genetic distance or rates of inter-patch movements (option 1) is the preferable option to define resistance values, animal occurrence data (option 2) and/or literature review and expert opinion (option 3) may be the only information available to many researchers and conservationists trying to model functional connectivity in fragmented landscapes. In this study resistance values derived from expert-opinion have been compared to HSM-informed values. Both techniques identified least-cost networks that contained significantly more distribution points than would be expected by chance. However, differences occur between the degree of model assumptions and biases (based on the different types of data), resistance values for certain land cover types and the least-cost networks identified. This has implications for the reliability of using such data in meeting conservation and management objectives.

To derive a set of expert-opinion resistance values it is useful to compare previous resistance values from multiple sources, particularly if the studies have similar species and environmental

Table 2. Average probability of grey squirrel presence according to land cover type.

Habitat type	HSM p score	HSM-resistance score	New expert-derived resistance score	Difference between HSM and Expert-derived resistance scores
Scrub	0.10	117	16	0.86
Track	0.16	109	27	0.75
Railway	0.17	108	27	0.75
Railway verge	0.17	108	27	0.75
Path	0.34	86	27	0.69
Heath	0.17	108	37	0.66
Garden	0.73	32	11	0.66
Road	0.41	77	27	0.65
Improved/arable/ amenity	0.13	113	40	0.65
Rough grassland	0.13	113	40	0.65
Road verge	0.43	74	27	0.64
Orchard	0.77	30	16	0.47
Urban	0.29	92	72	0.22
Marsh	0.17	108	91	0.16
Broadleaf	N/A	1	1	0.00
Coniferous	N/A	1	1	0.00
Mixed	N/A	1	1	0.00
Building	N/A	1000	1000	0.00
Rock	N/A	1000	1000	0.00
Coppice	0.86	15	16	−0.07
Water	0.18	107	130	−0.21

p = mean predicted probability of presence according to habitat type.

conditions. The resistance values given in previous studies were highly variable, resulting in varied least-cost habitat network areas and number of distribution points within networks. Although the land cover resistance values given in these studies were for red or grey squirrels, the studies took place in different countries with different regional environmental conditions and large scale and inevitable differences in landscape composition and structure. This may account for the differences in values given and resulting networks. Verbeylen et al [3] in particular was focused on red squirrels and based in an urban area which is very different to the largely-rural and sparsely populated Cumbria. However by assessing the range of different resistance values given in these studies and additional literature on land cover use, the new expert-derived resistance set was created. The area-minimisation method suggests that these values appear to be the best set for management purposes in this area, capturing a high percentage of distribution points within the smallest network area.

The resistance values for the new expert-derived and HSM-informed least-cost models in this study were significantly different from one another. The HSM-informed model provided higher resistance values for most land cover types. The validity of HSM-informed least-cost models may be limited as the probability of occurrence in a particular land cover type does not always equate to the resistance of that land cover type during species movement [19]. In using distribution/occurrence data, certain land cover types may be undervalued when in reality they are used by the species. Conversely there will be land cover types that are overvalued. A key assumption of presence only modelling is that the data has come from random sampling or is representative of

the whole landscape [51]. It is questionable whether the degree of bias in presence data can be truly known [51]. Squirrels are well known to use scrub habitat and will use this and linear features to aid dispersal [13,52–54], yet scrub and railway verge (a linear feature) were given high HSM-informed resistance values due to a low number of distribution points. Of the distribution points missed by the HSM-informed networks but included within the new expert-derived networks, 77% were within improved/arable/ amenity land cover type. This suggests that the inverted HSM values for this land cover may be too high, and squirrels may be able to cross these hostile areas quickly and undetected. The dispersal distance used for both expert-derived model and the HSM-informed model were set at 8 km. Therefore, it is the higher resistance values given to certain land cover types using the inverted-HSM that led to the identification of smaller and more fragmented networks.

The HSM-informed networks were 45% smaller than the expert-derived networks and were spatially nested inside these networks. The smaller mean size of HSM-informed networks suggests that grey squirrel occurs in a highly fragmented and functionally unconnected landscape. Both models highlight the land cover types of the Cumbrian Mountains as a barrier to movement; the combination of relatively high elevation and intense grazing result in a lack of woodland in the area. Although, some individuals may attempt to cross the barrier, the lack of available habitat will impede dispersal subjecting individuals to high levels of predation and starvation. There are no recorded introductions of the grey squirrel into Cumbria [55,56] and therefore these animals have been able to spread to their present

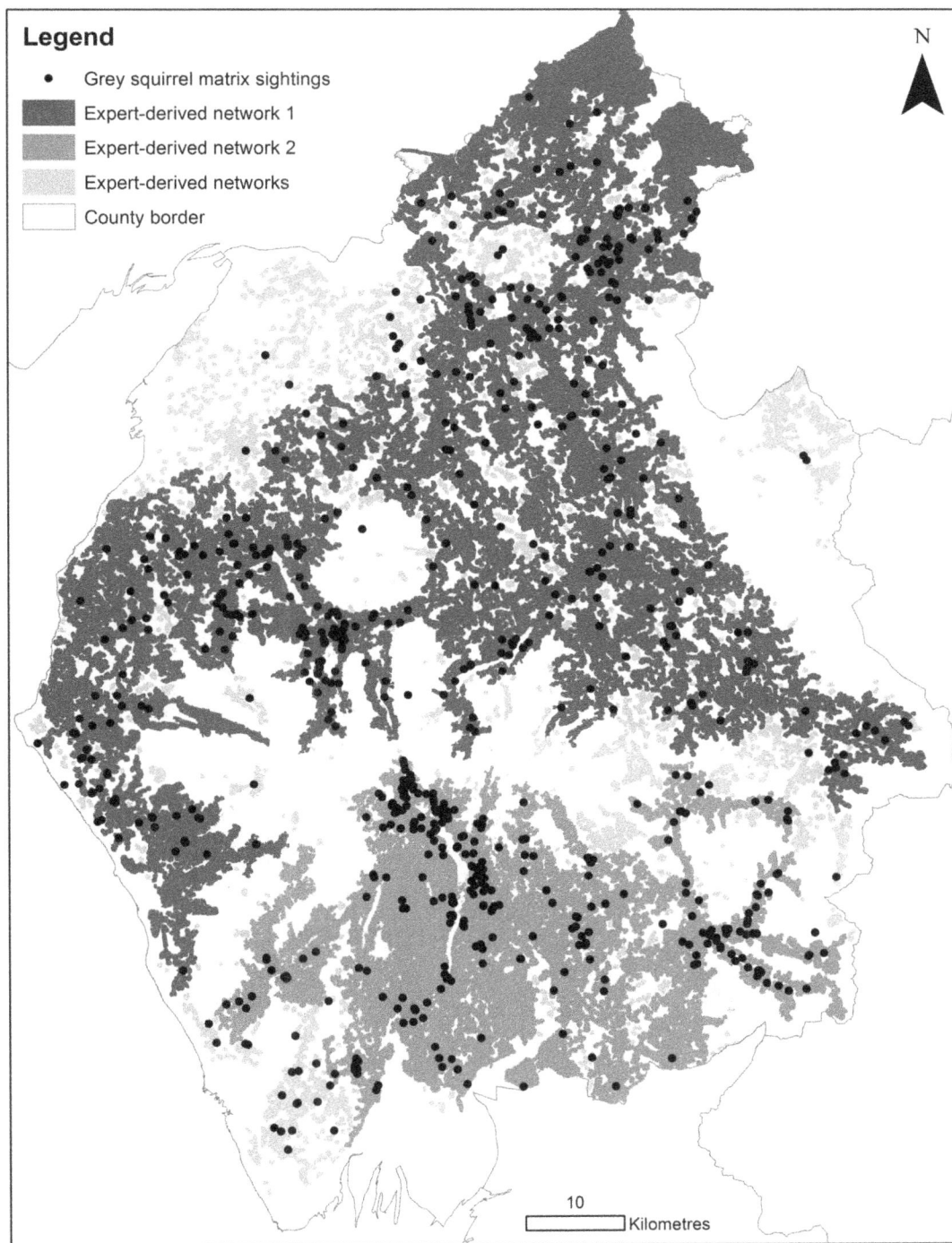

Figure 3. Grey squirrel least-cost habitat networks identified from expert-derived resistance values. Boundary lines were obtained through EDINA Digimap Ordnance Survey Service, http://digimap.edina.ac.uk/digimap/home.

distribution in the north and south of the county by natural means. The expert-derived model identified two large networks, one in the north and one in the south, suggesting a much more connected landscape.

Studies have suggested that expert-opinion based models perform less accurately than models informed by empirical data [24,57,58]. Given that HSM-informed networks are derived from known distribution data, these models could be interpreted as

identifying more precise areas in the landscape that are connected for a species. In comparison, the expert-derived networks include those areas where sighting have not been recorded but are judged by experts as permeable to the species during dispersal. Experts may overestimate the importance of certain land cover types erring on the side of caution and therefore rendering the model less accurate [24]. Where actions might require a more precise approach, such as identifying possible protected areas or sites for

Legend

- Grey squirrel matrix sightings
- HSM-derived network >20km2
- 5,830 habitat networks <20km2
- County border

N

10
Kilometres

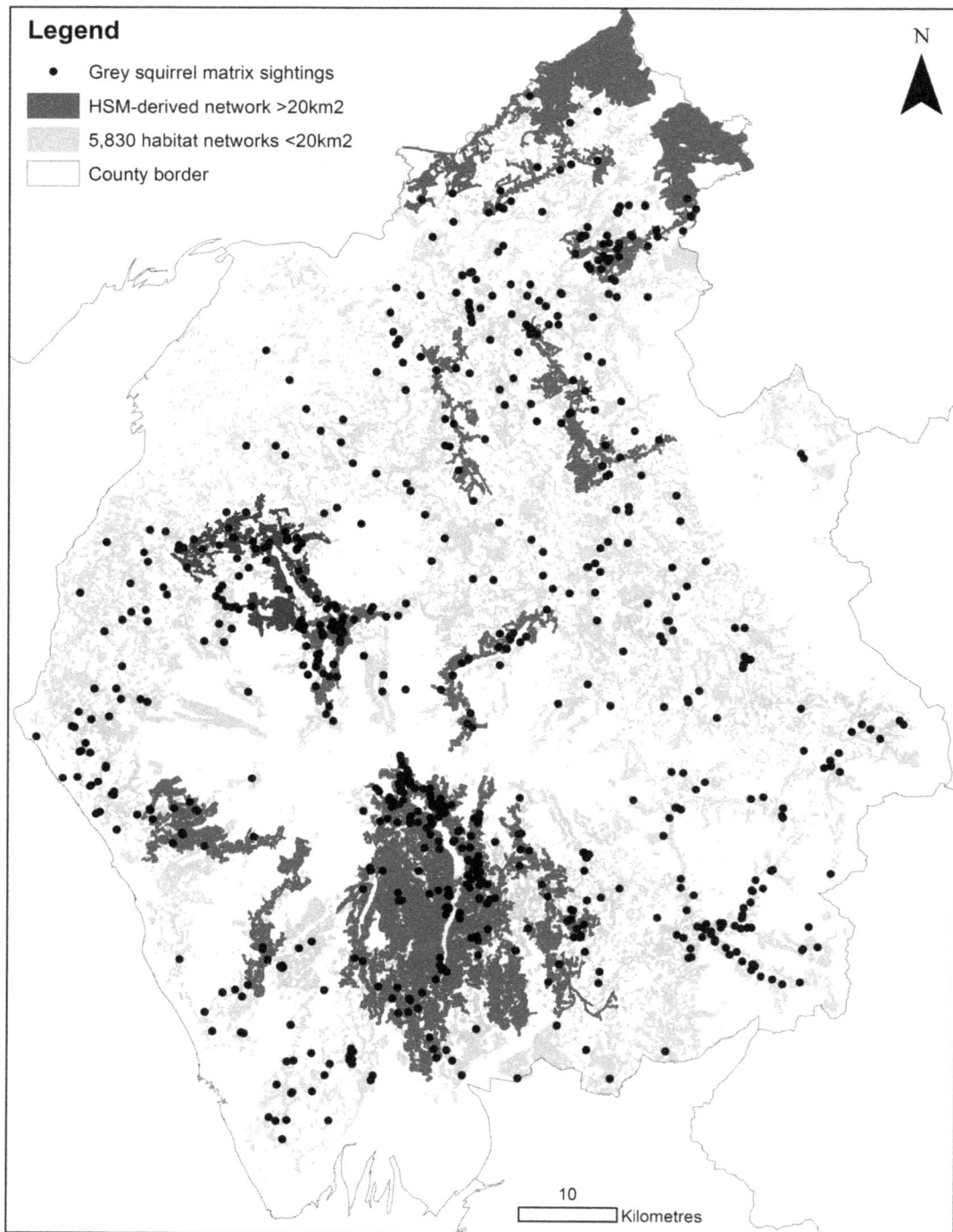

Figure 4. Highly fragmented grey squirrel least-cost habitat networks identified with Habitat Suitability Model-derived resistance values. Boundary lines were obtained through EDINA Digimap Ordnance Survey Service, http://digimap.edina.ac.uk/digimap/home.

an efficient and intensive control program, a HSM modelling approach would be appropriate. However, when assessing invasive species it is not just the most likely areas that a species will disperse to, but the entire possible range that needs identified. In an invasive species context, it may be more appropriate to apply a conservative less precise model, such as the expert-derived model, to enable all possible areas of dispersal to be included within the network.

In the case of invasive species the assessment of potential movement and impact is needed as soon as possible to aid management planning. This method is not dependent upon extensive species distribution data and can therefore be produced relatively quickly. Clevenger et al. [24] found that expert only derived resistance values had a weaker correlation with empirical-derived values than literature-derived values. Systematically collecting expert opinion, as promoted by Eycott et al [59], in combination with published data on land cover usage will enable resistance values to be assigned in the initial stages to give an indication of species movements whilst other empirical data is collected where possible. Adriensen et al [18] suggested that once a 'starter kit' of resistance values has been identified, sensitivity studies can be initiated and multiple alternative resistance sets can be tested [60]. Once species distribution data is collected, HSM-informed least-cost networks can be identified and used to aid the selection of most likely used sites to focus monitoring or eradication programs. It should not be assumed that using distribution data (option 2 in Beier et al. [19]) to identify resistance values is better or worse than using well developed expert-opinion (option 3 in Beier et al. [19]) as the choice of which method to use may depend upon the aims and objectives of the user and the appropriate precision of the approach.

This paper describes the first step towards developing least-cost habitat networks using *ad hoc* species records and a simple, land cover-based habitat suitability model. It is acknowledged, however, that species respond to their surrounding environment over a range of spatial scales and that both local and landscape features will affect both the suitability of the core habitat and the permeability of the surrounding matrix [5,61]. More complex models incorporating multiscale information on the terrain, built environment, and the composition, structure and arrangement of habitat patches are likely to provide more accurate and useful models [45], providing predictions at each location, rather than assuming consistent levels of permeability for a particular land cover type. This spatially explicit technique would enable landscape level decision making, improving our ability to identify important networks of habitat and enabling a targeted and informed approach to both conservation and infrastructural development.

Conclusion

Even though approaches to gather expert opinion are becoming more systematic and robust, it should not be seen as a blanket substitute for empirical data. Empirical data will continue to be important for studies on single species, where there is considerable uncertainty or where there is significant investment in time and money on conservation activities. Conservation planners must be aware of the subjectivity and pitfalls of the different types of data used in least-cost models, without any further validation or sensitivity testing of model values. If expert opinion is the only option available it should be used as a first step by systematically

combining multiple expert opinions and published data, but with the knowledge that further assessment of resistance values through sensitivity analysis and empirical data will be needed. Where distribution data is already available, the type of data collection and the subjective translation issues of over and under valuing land cover types must be assessed with expert knowledge or empirical data and explicitly stated in methodologies [51,62].

This study successfully compared expert-derived and HSM-informed resistance values used in least-cost modelling. Although the results of the models differed, both identified equally useful least-cost networks. For the grey squirrel in Cumbria, both expert-derived and HSM-informed networks have shown that there is a separation between north and south of Cumbria due to the land cover types and lack of habitat of the Cumbrian Mountain range. The expert-derived networks indicate a conservative less precise least-cost network that indicates the potential dispersal range of the grey squirrel and suggests that there may be multiple infiltration routes into the county from the north and south. This conservative expert-derived approach is useful when dealing with invasive or generalist species to identify the potential extend of spread. When assessing endangered or specialist species, or areas that are highly likely to contain target species, the HSM-informed network provides smaller precise networks. These precise networks should be used to inform targeted conservation to increase connectivity for species of conservation concern, or to inform targeted management to prevent the incursion of invasive species. The variable but acceptable precision of both expert-derived and HSM-informed least-cost networks highlights the need to consider data reliability and environmental context when deciding on the most appropriate management of invasive species.

Acknowledgments

The Authors would like to thank Dr Sallie Bailey for advice and comments, Phillip Handley for GIS advice and Simon O'Hare for sightings data. Country and county outlines in Figures 1, 3 and 4 and OSMM data were obtained through EDINA Digimap Ordnance Survey Service, http://digimap.edina.ac.uk/digimap/home.

Author Contributions

Conceived and designed the experiments: CDSH KW OTN ADR. Performed the experiments: CDSH KW CB. Analyzed the data: CDSH KW CB. Contributed reagents/materials/analysis tools: CDSH KW CB OTN ADR. Wrote the paper: CDSH KW CB OTN ADR.

References

1. Worboys G, Francis WL, Lockwood M (2009) Connectivity conservation management: A global guide (with particular reference to mountain connectivity conservation). London: Earthscan/James & James.
2. Taylor PD, Fahrig L, Henein K, Merriam G (1993) Connectivity is a vital element of landscape structure. Oikos 68: 571–573.
3. Verbeylen G, De Bruyn L, Adriaensen F, Matthysen E (2003) Does matrix resistance influence red squirrel (*Sciurus vulgaris* L. 1978) distribution in an urban landscape? Landscape Ecology 18: 791–805.
4. Spear SF, Balkenhol N, Fortin MJ, McRae BH, Scribner K (2010) Use of resistance surfaces for landscape genetic studies: Considerations for parameterization and analysis. Mol Ecol 19: 3576–3591.
5. Ricketts TH (2001) The matrix matters: Effective isolation in fragmented landscapes. The American Naturalist 158: 87–99.
6. Tischendorf L, Fahrig L (2000) On the usage and measurement of landscape connectivity. Oikos 90: 7–19.
7. Crooks KR (2006) Connectivity conservation. Cambridge: Cambridge Univ Pr.
8. Ferreras P (2001) Landscape structure and asymmetrical inter-patch connectivity in a metapopulation of the endangered Iberian lynx. Biol Conserv 100: 125–136.
9. Epps CW, Wehausen JD, Bleich VC, Torres SG, Brashares JS (2007) Optimizing dispersal and corridor models using landscape genetics. J Appl Ecol 44: 714–724.
10. Watts K, Eycott AE, Handley P, Ray D, Humphrey JW, et al. (2010) Targeting and evaluating biodiversity conservation action within fragmented landscapes: An approach based on generic focal species and least-cost networks. Landscape Ecology 25: 1305–1318.
11. Stevenson CD, Ferryman M, Nevin OT, Ramsey AD, Bailey S, et al. (2013) Using GPS telemetry to validate least-cost modeling of gray squirrel (*Sciurus carolinensis*) movement within a fragmented landscape. Ecology and Evolution 3: 2350–2361.
12. Gonzales EK, Gergel SE (2007) Testing assumptions of cost surface analysis- a tool for invasive species management. Landscape Ecology 22: 1155–1168.
13. Middleton AD (1930) The ecology of the American grey squirrel (*Sciurus carolinensis* gmelin) in the British isles. J Zool, Lond. 100: 809–843.
14. Gurnell J, Wauters LA, Lurz PWW, Tosi G (2004) Alien species and interspecific competition: Effects of introduced eastern grey squirrels on red squirrel population dynamics. J Anim Ecol 73: 26–35.
15. Gurnell J, Mayle B (2003) Ecological impacts of the alien grey squirrel (*Sciurus carolinensis*) in Britain. In Bowen CP, editor. MammAliens – A one day conference on the problems caused by non- native British mammals. London: Peoples Trust for Endangered Species/Mammals Trust UK. pp.40–45.
16. Kenward RE (1983) The causes of damage by red and grey squirrels. Mamm Rev 13: 159–166.
17. Sawyer SC, Epps CW, Brashares JS (2011) Placing linkages among fragmented habitats: Do least-cost models reflect how animals use landscapes? J Appl Ecol 48: 668–678.

18. Adriaensen F, Chardon JP, De Blust G, Swinnen E, Villalba S, et al. (2003) The application of 'least-cost' modelling as a functional landscape model. Landscape and Urban Planning 64: 233–247.

19. Beier P, Majka DR, Spencer WD (2008) Forks in the road: Choices in procedures for designing wildland linkages. Conserv Biol 22: 836–851.

20. Driezen K, Adriaensen F, Rondinini C, Doncaster CP, Matthysen E (2007) Evaluating least-cost model predictions with empirical dispersal data: A case-study using radio tracking data of hedgehogs (*Erinaceus europaeus*). Ecological Modelling 209: 314–322.

21. Stevens VM, Polus E, Wesselingh RA, Schtickzelle N, Baguette M (2005) Quantifying functional connectivity: Experimental evidence for patch-specific resistance in the natterjack toad (*Bufo calamita*). Landscape Ecol 19: 829–842.

22. Eycott AE, Stewart GB, Buyung-Ali LM, Bowler DE, Watts K, et al. (2012) A meta-analysis on the impact of different matrix structures on species movement rates. Landscape Ecol 27: 1263–1278.

23. Zeller KA, McGarigal K, Whiteley AR (2012) Estimating landscape resistance to movement: A review. Landscape Ecol 27: 777–797.

24. Clevenger AP, Wierzchowski J, Chruszcz B, Gunson K (2002) GIS-generated, expert-based models for identifying wildlife habitat linkages and planning mitigation passages. Conserv Biol 16: 503–514.

25. Chardon JP, Adriaensen F, Matthysen E (2003) Incorporating landscape elements into a connectivity measure: A case study for the speckled wood butterfly (*Pararge aegeris* L.). Landscape Ecology 18: 561–573.

26. Wang Y, Yang K, Bridgman CL, Lin L (2008) Habitat suitability modelling to correlate gene flow with landscape connectivity. Landscape Ecol 23: 989–1000.

27. Richards-Zawacki CL (2009) Effects of slope and riparian habitat connectivity on gene flow in an endangered Panamanian frog, *Atelopus varius*. Divers Distrib 15: 796–806.

28. Decout S, Manel S, Miaud C, Luque S (2010) Connectivity loss in human dominated landscape: Operational tools for the identification of suitable habitat patches and corridors on amphibian's population. Landscape International Conference IUFRO, Portugal.

29. Wang IJ, Summers K (2010) Genetic structure is correlated with phenotypic divergence rather than geographic isolation in the highly polymorphic strawberry poison-dart frog. Mol Ecol 19: 447–458.

30. Decout S, Manel S, Miaud C, Luque S (2012) Integrative approach for landscape-based graph connectivity analysis: A case study with the common frog (*Rana temporaria*) in human-dominated landscapes. Landscape Ecol 27: 267–279.

31. Howard A, Bernardes S (2012) A maximum entropy and least cost path model of bearded capuchin monkey movement in Northeastern brazil. Intergraph. Available: https://intergraphgovsolutions.com/assets/white-paper/A_Maximum_Entropy-2_Imagery.sflb.pdf Accessed 2013.

32. Phillips SJ, Anderson RP, Schapire RE (2006) Maximum entropy modeling of species geographic distributions. Ecol Model 190: 231–259.

33. Gonzales EK (2000) Distinguishing between modes of dispersal by introduced eastern grey squirrels (*Sciurus carolinensis*). MSc Thesis, University of Guelph.

34. Humphrey J, Smith M, Shepherd N, Handley P (2007) Developing lowland habitat networks in Scotland: Phase 2. Edinburgh: Forestry Commission.

35. Stevenson CD (2008) Modelling red squirrel population viability under a range of landscape scenarios in fragmented woodland ecosystems on the Solway plain, Cumbria. London: People's Trust for Endangered Species. Available http://insight.cumbria.ac.uk. Accessed 2013.

36. Williams S (2008) Red squirrel strongholds consultation. Edinburgh: Forestry Commission.

37. Lowe VPW (1993) The spread of the grey squirrel (*Sciurus carolinensis*) into Cumbria since 1960 and its present distribution. J Zool, Lond 231: 663–667.

38. Skelcher G (1997) The ecological replacement of red by grey squirrels. In: Gurnell J, Lurz P, editors. The conservation of red squirrels, *Sciurus vulgaris* L. London: People's Trust for Endangered Species. pp.67–78.

39. Nixon CM, McClain MW, Donohoe RW (1980) Effects of clear-cutting on grey squirrels. Journal of Wildlife Management 44: 403–412.

40. Phillips SJ, Dudik M (2008) Modeling of species distributions with maxent: New extensions and a comprehensive evaluation. Ecography 31: 161–175.

41. Phillips SJ, Dudík M, Elith J, Graham CH, Lehmann A, et al. (2009) Sample selection bias and presence-only distribution models: Implications for background and pseudo-absence data. Ecol Appl 19: 181–197.

42. Warton DI, Renner IW, Ramp D (2013) Model-based control of observer bias for the analysis of presence-only data in ecology. Plos One 8: e79168.

43. Schadt S, Knauer F, Kaczensky P, Revilla E, Wiegand T, et al. (2002) Rule-based assessment of suitable habitat and patch connectivity for the Eurasian lynx. Ecol Appl 12: 1469–1483.

44. Liu C, White M, Newell G (2013) Selecting thresholds for the prediction of species occurrence with presence-only data. J Biogeogr 40: 778–789.

45. Bellamy C, Scott C, Altringham J (2013) Multiscale, presence-only habitat suitability models: Fine-resolution maps for eight bat species. J Appl Ecol 50: 892–901.

46. Merckx B, Steyaert M, Vanreusel A, Vincx M, Vanaverbeke J (2011) Null models reveal preferential sampling, spatial autocorrelation and overfitting in habitat suitability modelling. Ecol Model 222: 588–597.

47. Veloz SD (2009) Spatially autocorrelated sampling falsely inflates measures of accuracy for presence-only niche models. J Biogeogr 36: 2290–2299.

48. De Marco P, Diniz-Filho JA, Bini LM (2008) Spatial analysis improves species distribution modelling during range expansion. Biol Lett 4: 577–580.

49. Rangel TF, Diniz-Filho JAF, Bini LM (2010) SAM: A comprehensive application for spatial analysis in macroecology. Ecography 33: 46–50.

50. Zalewski A, Piertney SB, Zalewska H, Lambin X (2009) Landscape barriers reduce gene flow in an invasive carnivore: Geographical and local genetic structure of American mink in Scotland. Mol Ecol 18: 1601–1615.

51. Yackulic CB, Chandler R, Zipkin EF, Royle JA, Nichols JD, et al. (2013) Presence-only modelling using MaxEnt: When can we trust the inferences? Methods in Ecology and Evolution 4: 236–243.

52. Taylor KD, Shorten M, Lloyd HG, Courtier FA (1971) Movements of the grey squirrel as revealed by trapping. J Appl Ecol 8: 123–146.

53. Fitzgibbon CD (1993) The distribution of grey squirrel dreys in farm woodland: The influence of wood area, isolation and management. J Appl Ecol 30: 736–742.

54. Wauters LA, Gurnell J, Currado I, Mazzoglio P (1997) Grey squirrel *Sciurus carolinensis* management in Italy - squirrel distribution in a highly fragmented landscape. Wildlife Biology 3: 117–123.

55. Shorten M (1954) Squirrels. London: Collins.

56. Shorten M (1957) Squirrels in England, Wales and Scotland, 1955. J Animal Ecol 26: 287–294.

57. Pearce J, Cherry K, Whish G (2001) Incorporating expert opinion and fine-scale vegetation mapping into statistical models of faunal distribution. J Appl Ecol 38: 412–424.

58. Seoane J, Bustamante J, Diaz-Delgado R (2005) Effect of expert opinion on the predictive ability of environmental models of bird distribution. Conserv Biol 19: 512–522.

59. Eycott AE, Marzano M, Watts K (2011) Filling evidence gaps with expert opinion: The use of Delphi analysis in least-cost modelling of functional connectivity. Landscape Urban Plann 103: 400–409.

60. Rayfield B, Fortin MJ, Fall A (2010) The sensitivity of least-cost habitat graphs to relative cost surface values. Landscape Ecol 25: 519–532.

61. Wiens JA (1989) Spatial scaling in ecology. Funct Ecol 3: 385–397.

62. Beier P, Majka DR, Newell SL (2009) Uncertainty analysis of least-cost modeling for designing wildlife linkages. Ecological Applications 19: 2067–2077.

Historical, Observed, and Modeled Wildfire Severity in Montane Forests of the Colorado Front Range

Rosemary L. Sherriff[1]*, **Rutherford V. Platt**[2], **Thomas T. Veblen**[3], **Tania L. Schoennagel**[3], **Meredith H. Gartner**[3]

1 Department of Geography, Humboldt State University, Arcata, California, United States of America, **2** Department of Environmental Studies, Gettysburg College, Gettysburg, Pennsylvania, United States of America, **3** Department of Geography, University of Colorado, Boulder, Colorado, United States of America

Abstract

Large recent fires in the western U.S. have contributed to a perception that fire exclusion has caused an unprecedented occurrence of uncharacteristically severe fires, particularly in lower elevation dry pine forests. In the absence of long-term fire severity records, it is unknown how short-term trends compare to fire severity prior to 20th century fire exclusion. This study compares historical (i.e. pre-1920) fire severity with observed modern fire severity and modeled potential fire behavior across 564,413 ha of montane forests of the Colorado Front Range. We used forest structure and tree-ring fire history to characterize fire severity at 232 sites and then modeled historical fire-severity across the entire study area using biophysical variables. Eighteen (7.8%) sites were characterized by low-severity fires and 214 (92.2%) by mixed-severity fires (i.e. including moderate- or high-severity fires). Difference in area of historical versus observed low-severity fire within nine recent (post-1999) large fire perimeters was greatest in lower montane forests. Only 16% of the study area recorded a shift from historical low severity to a higher potential for crown fire today. An historical fire regime of more frequent and low-severity fires at low elevations (<2260 m) supports a convergence of management goals of ecological restoration and fire hazard mitigation in those habitats. In contrast, at higher elevations mixed-severity fires were predominant historically and continue to be so today. Thinning treatments at higher elevations of the montane zone will not return the fire regime to an historic low-severity regime, and are of questionable effectiveness in preventing severe wildfires. Based on present-day fuels, predicted fire behavior under extreme fire weather continues to indicate a mixed-severity fire regime throughout most of the montane forest zone. Recent large wildfires in the Front Range are not fundamentally different from similar events that occurred historically under extreme weather conditions.

Editor: Ben Bond-Lamberty, DOE Pacific Northwest National Laboratory, United States of America

Funding: Data collection and analysis were funded by the National Science Foundation (awards BCS-0540928, 0541480, 0802667, 0541594 to TTV, RLS and RVP) and Humboldt State University Sponsored Programs Foundation and College of Arts, Humanities, and Social Sciences (awards to RLS). The funders had no role in the study design, decision to publish, or preparation of the manuscript.

Competing Interests: The authors have declared that no competing interests exist.

* Email: sherriff@humboldt.edu

Introduction

The social, environmental and fiscal costs of wildfire have escalated dramatically over the last few decades [1–2]. The costs associated with recent wildfires are particularly high in the arid mountain West, where residential structures abut or intermingle with wildland vegetation (Wildland-Urban Interface - WUI) and the exurban population has grown rapidly in recent decades [3]. Large fire events in the 1990s and early 2000s in the western U.S., particularly in lower elevation, relatively dry-pine forests, have contributed to widespread concern that fire exclusion has caused an unprecedented threat of uncharacteristically severe fires in these ecosystems [4–5]. Broad-scale monitoring of fire severity from satellite imagery since ca. 1984 shows a significant trend towards increased severity only in parts of the Southwest [6], yet findings are varied in other studies depending on the spatial scale, selected data types, and the location of study (e.g. Pacific West) [7–12]. In the absence of longer broad-scale records of fire severity, it is unknown how such short-term trends compare to fire severity prior to fire exclusion. In the context of debate about the potential effects of fire exclusion on modern fire regimes and their departures from historical fire regimes (i.e. prior to fire exclusion) [4,13–15], there is a critical need for research on whether the severity and other characteristics of modern fires depart from historical fire regimes for specific ecosystem types at broad landscape scales. The current study compares reconstructed historical fire severity, observed fire severity in recent fires and modeled fire severity from current fuel structures, and discusses the consequences of fire regime changes for management options under expected future climate in the Colorado Front Range.

Development of management goals and adaptation options in fire-prone ecosystems interfacing with the WUI is inextricably related to quantifying the range of variation of a set of ecological patterns and processes exhibited naturally or under human influences during a specified historical period [16–17]. This widely used approach, called the historical range of variability (HRV) framework uses historical ecological data to test hypotheses about the drivers and mechanisms of contemporary and future ecological change [18]. Management using the HRV framework assumes that ecosystem resilience (the ability to recover quickly) is reflected in observed ranges of past vegetation and fire dynamics. Retrospective ecological studies are considered essential for understanding likely consequences of climate change on future fire and landscape dynamics [18–19]. While there has been a shift away from using reference conditions as default for fire and forest management goals in relation to future climate change [19–20], HRV still provides the most viable framework for understanding the sensitivity of ecosystem resilience and ecological integrity to changes in fire regimes [21]. Ecological integrity has been defined as a measure of the composition, structure, and function of an ecosystem in relation to the system's natural or historical range of variation, as well as changes caused by humans [22–23]. A key research objective in the context of fire-prone landscapes in the U.S. West is the assessment of the historical role of fires of varying severities on the resilience and integrity of current ecosystems and their potential consequences for ecosystem services valued by humans [21].

Fuel reduction is currently the dominant management tool for reducing the likelihood of high-severity fire in the contexts of ecological restoration and mitigation of climate change impacts [24–26]. This dual approach (forest restoration and fire mitigation) assumes that the probability of severe fire occurrence has increased to uncharacteristic levels during decades of fire suppression in western forests [4,27–34]. Guided by this assumption, over 190 million acres of public land have been identified as "unnaturally dense" with an increased likelihood of catastrophic wildfires [24], [26]. In response to such concerns, both forest managers and communities have begun to develop strategies to alleviate the potential impacts of wildfire [35], and millions of hectares of forest lands have been treated in recent decades [36]. The goals of fuels reduction to decrease the likelihood of severe wildfires and restore historical forest structure and species composition are complementary in ecosystems where fuels and fire severity have increased, yet are incompatible elsewhere and threaten ecosystem integrity and ecosystem services [37–43].

In the Colorado Front Range, tree-ring evidence, historical landscape photographs, and General Land Office surveys demonstrate that the historical (i.e. pre-1920) fire regime of ponderosa pine and mixed-conifer forests included low-severity fires (i.e. non-lethal to large fire-resistant trees) as well as high-severity fires (i.e. killing >70% of canopy trees) [13,44–54]. There is a broad consensus that most of the montane zone of Colorado was characterized by fire regimes of mixed severity, including some component of high-severity fires [55]. However, a better

understanding of which habitats across this heterogeneous landscape were explicitly affected predominantly by low-severity or higher-severity fires, or a combination of both, is necessary for determining where woody fuels have become uncharacteristically abundant as a consequence of fire exclusion [37,56].

Thus, in this research we examine changes in fire regimes across 564,413 ha of the montane forest zone of the Colorado Front Range (Figure 1). We compile and analyze new and existing datasets to refine the spatial resolution and expand the geographic scope of retrodicted (reconstructed) historical fire regimes in comparison with present wildfire potential. We address two main questions about the montane forests of the Colorado Front Range: 1) What areas and landscape characteristics were characterized by an historical fire regime of predominantly low-severity or mixed-severity (which includes moderate- or high-severity fires)? 2) How does historical fire severity compare with observed severity of large wildfires (since 2000) and modeled potential wildfire behavior across the landscape? Finally, we discuss how these comparisons can inform fire mitigation and ecological restoration under expected climate change.

Methods

Study area

The study area is located in the north-central Front Range of Colorado, bounded by six counties and the extent of the montane zone between 1800–3000 m to the east and west (Figure 1). Within the lower to upper montane zones, the mean annual precipitation ranges are ~35.6 to 51 cm and the mean annual temperature ranges are 11°C to 2.4°C, respectively [57]. The lower montane zone (~1800 to 2200 m) comprises primarily pure ponderosa pine (*Pinus ponderosa*) on south-facing slopes and a mixture of ponderosa pine and Douglas-fir (*Pseudotsuga menziesii*) on north-facing slopes. The upper montane zone (~2200 to 3000 m) is composed of ponderosa pine stands on south-facing slopes and more dense stands of ponderosa pine and Douglas-fir on north-facing slopes along with lodgepole pine (*Pinus contorta*), aspen (*Populus tremuloides*) and dispersed limber pine (*Pinus flexilis*) trees at higher elevations. The LANDFIRE existing vegetation type (EVT) layer was used to delineate the montane study area (564,413 ha) within 1800–3000 m because it represents the most up-to-date and detailed cover type classification available across land ownerships in the region, which is proportionally represented by the following EVT types: ponderosa pine (24.2%); lodgepole pine (23.1%); mixed montane conifer forest and woodland types (49.2%); and intermixed with pixels classified as lower montane-foothill grassland (<0.1%) or shrubland (3.5%) (Figure 1; EVT types included in the study are listed in Table S1). However, the accuracy of the LANDFIRE EVT classification at field sites was relatively low (see RESULTS).

During the period from 1980 to 2011, 48 fires burned over 100,000 ha along the Front Range causing severe damage to property and infrastructure in the WUI [2,58–59]. These large fires can occur any time of year and typically burn when wind speeds are high and weather conditions are dry, which is common along the Front Range. During these types of conditions, fire suppression is typically ineffective and fires often escape initial suppression efforts [2].

Field sampling

We used a combination of existing datasets (141 sites) [52–53,60–62] and newly sampled datasets (91 sites) for a total of 232 sites with information on forest structure (tree age structure) and fire history (tree-ring fire-scar records) in montane forest types.

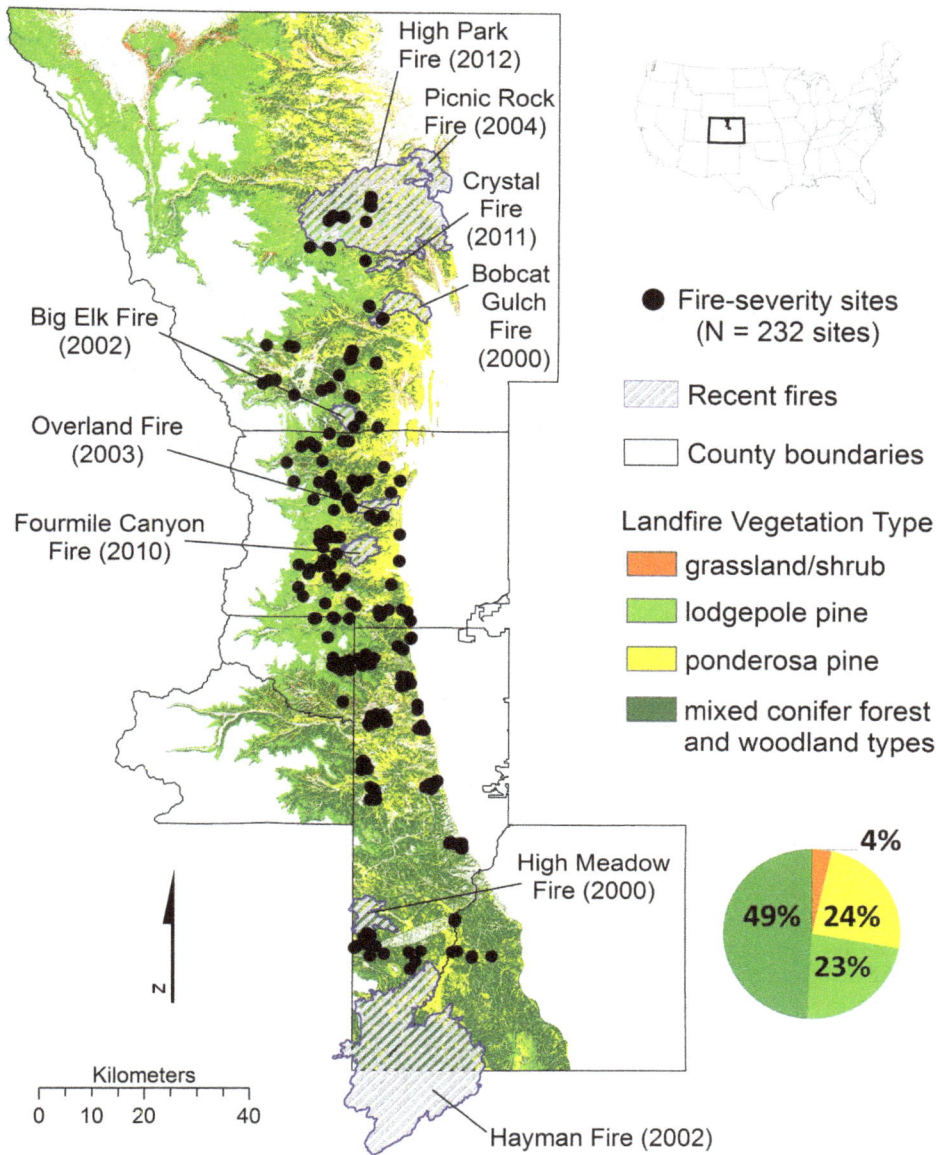

Figure 1. Study area in north-central Colorado, USA. The map includes the montane zone by cover type (LANDFIRE Existing Vegetation Type), 232 sites sampled for historical fire severity, and recent wildfires used for comparison and verification of fire behavior modeling. County boundaries (solid black lines) from north to south are of Gilpin and Clear Creek counties and the mountainous western regions of Larimer, Boulder, and Douglas counties, and Jefferson County.

The majority of sites were sampled in a stratified-random design and all sites were selected in relative proportion to their cover type within the study area, allowing inferences from our site-level datasets to reflect trends across the 564,413 ha montane study area. Because sites were dispersed over the broad regional area of montane forests we do not extrapolate the spatial spread of fire severity between sites, or determine if fires in the same year at multiple sites were from a single ignition or multiple ignition sources. We rejected sites with evidence of past logging or signs of other major anthropogenic disturbance (i.e. mining). Our sampling goal, similar to previous site-level studies of historic fire regimes in the region [51,53], was to identify the predominant structural influence of historical fires at each site (fire-severity regime as explained below). Understanding the historical fire regime and the effects on forest structure without major Euro-

American influences provides the natural range of conditions. The interactive effects of logging and/or other anthropogenic disturbances are important, but beyond the scope of our study. Laboratory (tree-ring) analysis of new sites followed the same procedures used in prior studies for analyzing fire-scar samples, stand age and size structure [51–53,60–62]. The criteria used to characterize the historical fire severity at each site were derived from the results of these prior studies and are described below (*Site-level classification of historical fire severity*).

Stand-level sites. At 120 of the 232 sites [51–53,60–62] (and unpublished new data), we sampled in areas of relatively consistent (unvarying) forest structure and physical environment with the extent of sampling area varying in size from 10 to 232 ha according to tree density and the extent of the homogeneous stand structure. At 44 of the stand-level sites, fire-scarred trees were systematically sampled [60,61], fire-scarred trees and fire scars on each fire-scarred tree were tallied, and forest structure was evaluated based on the same protocol used by Sherriff and Veblen [52] (see Tables 1–2). At the remaining 76 stand-level sites, the sampling goal was more intensive. In addition to systematically sampling fire-scarred trees throughout the site, at least 50 of the closest trees ($\geq$4 cm dbh) at a constant distance along randomly located transects were cored, and at least five of the largest and oldest characteristic live or dead trees were selectively cored to ensure large (and the oldest characteristic) trees were represented in our sample. In addition, forest structure (size, decay class, standing/down, distance from transect point), seedling (<30 cm height) and sapling (>30 cm height and <4 cm diameter) data were tallied by species along belt transects. To estimate the ages of trees too small to core (<4 cm in diameter) and to estimate the number of rings missed due to coring height (approximately 20 cm above the root-shoot boundary), 341 juvenile trees, reflecting the range of tree species within the sampling sites (in order of abundance - ponderosa pine, Douglas-fir, lodgepole pine, aspen and limber pine), were cut at sites of different elevation and aspect. Relatively open sites were selected to mimic post-fire growth conditions. Permits for sampling were issued by Rocky Mountain National Park, USDA Forest Service, and county and city open space land management agencies. Sampling did not involve protected or endangered species. For sites where seedlings were not collected, the median age-to-coring height was used from a site of similar elevation and aspect (across all sampled sites: median of 4 years – ponderosa pine; 9 years – Douglas-fir; 8 years – lodgepole pine; 1 year – aspen; 9 years – limber pine). To correct for years missed due to missing the pith, we followed Duncan's [63] procedure. Cores with more than 20 years estimated to pith were excluded.

Plot-level sites. The remaining 112 sites with information on forest structure and fire were smaller in size (3-ha plots) and randomly located from a 1-km grid throughout the CFR study area [52] (and new unpublished data). Sites with abundant logging or other human disturbances were rejected and an adjacent location was sampled. At each random plot-level site the number of fire-scarred trees and fire scars on each fire-scarred tree were tallied, fire-scarred trees were cored to estimate fire dates [64], forest structure was documented using an existing protocol [52] (see Tables 1–2), and approximately 10 cores were taken for reconstructing tree ages.

Analytical procedures

Site-level classification of historical fire severity. For this analysis, we use common and broad definitions of fire-severity regimes for montane forests of the Colorado Front Range that are relevant to forest structure and management (Table 1). Low-severity fire regimes are dominated by frequent (Mean Fire Interval – MFI <30 years), non-stand replacing fires within a stand ($\sim$100 ha) that leave multiple fire scars on individual trees throughout the stand and kill young seedlings and subcanopy trees while maintaining open, low-density stands of fire-resistant canopy trees. Mixed-severity regimes have varied fire effects that include low-severity, non-stand replacing fire to high-severity, stand- (or canopy) replacing fire both within stands and across landscapes, often in relation to topography [40,47–48,51–53,65,66]. High-severity fires occur less frequently (MFI >35–100+ years) than non-stand replacing fires, leave only small or no patches of pre-fire remnant trees, result in few to no fire scars on individual trees throughout the stand, and often initiate recruitment of a new cohort of canopy trees [51–53,55].

First, we classified the severity of fires at each (stand- and plot-level) site based on their influence on forest structure. We focused on fire dates recorded by fire scars on at least two trees per site (spreading fires) in the same year prior to 1915 (ca. effective fire exclusion). Next we used two primary metrics, along with the dates and frequency of spreading fires, for characterizing the severity of fires at each site – the percentage of remnant trees (proportion of the trees older than a spreading fire date) and the percentage of tree establishment (proportion of the trees that established within 40 years after a spreading fire). These criteria follow the approach described in prior studies in the region [51,53], correspond to classification of fire severity in other studies [65–66], and are summarized in Table 1. Specifically, individual fires were classified as low severity when $\geq$80% of the trees that were sampled had survived the fire (based on remnant tree ages) and $\leq$20% of the trees established following the fire. Fires were classified as moderate severity when 21–79% of the trees survived/established following the fire. Fires were classified as high severity when < 20% of the trees that were sampled had survived the fire and > 80% of the trees established following the fire. We excluded fires where no tree establishment followed within 40 years, assuming there was no structural impact or burning within the site [53]. Individual fires that occurred prior to subsequent mixed-severity fire, and were not associated with a pulse of establishment (>20% trees/site), were not included in the analysis of fire severity. We recognize that the estimate of fire effects for fires that occurred prior to a mixed-severity fire is less reliable because of subsequent

Table 1. Definitions of historical fire severity terms.

Term	Definition of fire effects
Historical high-severity fire	A fire that had high mortality of live, standing vegetation (<20% of the sampled trees survived the fire) and high tree establishment (>80% of the sampled trees) following the fire.
Historical low-severity fire	A fire that had low to no mortality of live, standing vegetation (>80% of the sampled trees survived) and low to no establishment (<20% of the sampled trees established following the fire).
Historical low-severity fire regime	Dominated by frequent (Mean Fire Interval <30 years), non-stand replacing fires within a stand ($\sim$100 ha) that leave multiple fire scars on individual trees throughout the stand and kill young seedlings and subcanopy trees while maintaining open, low-density stands of fire-resistant canopy trees.
Historical moderate-severity fire	A fire that had effects that were intermediate between low and high severity.
Historical mixed-severity fire regime	Varied fire effects that included low-severity, non-stand replacing fire to high-severity, stand- (or canopy) replacing fire both within stands and across landscapes, often in relation to topography.

These terms may have different meanings in the literature depending on the context in which they are used.

Table 2. Large fires in the montane study area from 2000–2012.

Fire	Year	Size (ha)*	Fire Severity (%)* Unburned/Low	Low	Moderate	High
Bobcat Fire	2000	3688	30.5	16.8	22.6	30.1
High Meadows Fire	2000	3884	21.7	37.8	32.0	8.5
Big Elk Fire	2002	1741	42.1	20.0	17.1	20.8
Hayman Fire	2002	52,016	21.1	12.7	22.5	43.7
Overland Fire	2003	1244	27.9	12.9	22.6	36.6
Picnic Rock Fire	2004	3626	27.6	37.7	22.6	12.1
Four Mile Fire	2010	2285	11.0	25.6	34.3	29.1
Crystal Fire	2011	914	22.9	38.5	26.9	11.7
High Park Fire	2012	34,905	20.8	29.7	26.9	22.5

* Total area burned and fire severity percentages within each fire perimeter (MTBS values 1–4; 1 = unburned/low, 2 = low, 3 = moderate, and 4 = high). See Figure 1 for large fire perimeters used to compare historical and observed fire severities within the study area.

mortality of older trees with time and/or subsequent disturbance. Post-fire tree establishment is not a direct measure of fire severity, and can vary significantly among species and in relation to seed availability and climate conditions, but it provides corroborative evidence of fire severity in combination with remnant trees, which provide a more direct measure of fire severity. Both metrics, pre-fire remnant trees and post-fire tree establishment, have uncertainties for evaluating fire severity and are not a perfect complement, but used in conjunction with fire dates and integrated into measureable metrics represent a robust way to reconstruct historical fire severity in montane forests (see [53] for more on methodological limitations and recommendations).

Based on the fire severity classifications, we assigned a predominant fire regime at each (stand- and plot-level) site based on the following criteria of cumulative fire effects over time. Sites characterized by frequent, non-stand replacing fires (MFI <30 years) that left multiple fire scars on individual trees throughout the site and without evidence of higher-severity fire were classified as having only low-severity fire regimes. Low-severity sites identified without historical evidence of moderate- or high-severity fire are assumed to have had only low-intensity (surface) fire and low to no canopy mortality. Sites characterized by varied fire effects that included canopy-replacing fire within sites were classified as having mixed-severity regimes (see [51–53] for within site sampling protocols). Mixed-severity fire regimes include low-, moderate- and high-severity fires where sites may illustrate different severities over time or space depending on weather, climate and stand conditions. Individual sites were differentiated based on evidence of low- and moderate-severity fire effects (no evidence of high-severity fire with <20% the stand age preceding a single fire) or low- to high-severity fire effects. Sites identified with either (low- and moderate-severity fire effects or low- to high-severity fire effects) were classified as a mixed-severity fire regime. Mixed-severity sites identified with moderate- or high-severity fire are assumed to have had components of intense fire behavior and canopy mortality, but varied in the extent of canopy mortality.

Landscape-level modeling of historical fire severity. Based on the fire severity classifications at the 232 field sites, we modeled historical fire-severity regime across the study area using potential predictor variables derived from two sources: 1) terrain variables from a 30-m digital elevation model (DEM); and 2) LANDFIRE biophysical variables. Environmental conditions at each of the 232 sites were classified in terms of the median elevation, slope steepness, aspect, cover type, distance to grassland, distance to ravine and the associated fire regime type. Elevation, slope steepness, aspect (arcsine and sine transformations) and slope curvature were all derived from the 30-m DEM. Slope curvature (concavity to convexity) and hillslope position are related to soil depth, texture and potential soil moisture [67]. Ravine drainages were delineated from the 30-m DEM using Arc GRID (ESRI 2002) hydrologic terrain modeling [68] and distance to ravine drainage was calculated. Distance to grassland was calculated as the distance to the edge of grassland areas of at least 0.1 ha.

Classification and regression trees (CART; $n = 5000$ trees) were used to identify explanatory variables with the most power to predict fire severity type (RandomForests Software, Salford Systems). CART has been increasingly used in ecological studies that require nonparametric techniques to explore complex, hierarchical interactions among variables [69–71]. We evaluated predictor variables for three response variables (historical low-, moderate- and high-severity fire) and a binary response variable (historical low- or mixed-severity fire) using all 232 sites to identify complex and non-linear interactions between variables that could be mapped across the study area (e.g. [71]). A 10-fold cross-validation process was used to avoid over fitting. The regression tree size with the smallest error was recorded for each run. The average tree size was used to prune the original tree (from the complete dataset), which tends to minimize complexity and results in a more generalizable tree [72]. There was strong overlap in the range of environmental conditions at sites with evidence of historical moderate and high fire severities. For this reason, we used the results evaluating the binary response variable (historical low- or mixed-severity fire) across all 232 sites for this study to produce a predictive map of the historical fire severity across the CFR study area in a GIS (ESRI ArcGIS 10.1). We identified the optimal model as the one with the fewest variables that best predicted the occurrence of sites with historical low and mixed severities. Model performance was evaluated using the overall percentage correctly classified (PCC) and area under the receiver operating characteristic curve (AUC) statistic. An AUC value of 0.5 indicates the prediction accuracy of sites picked at random, whereas 1.0 indicates perfect classification accuracy. AUC values of 0.7, 0.8, and 0.9 indicate fair, good, or excellent accuracy, respectively.

Comparison of historical fire severity to observed fire severity. Observed fire severities from nine large fires that occurred since 2000 (Table 2; Figure 1), including a range of fire severity, were compared with the spatially-explicit map of historical fire severity (Table 3). The 'observed' severity data came from the Monitoring Trends in Burn Severity (MTBS) program, which provides maps of the perimeters and severity of all fires larger than 1000 ha in the western United States from 1984-present [73]. MTBS thematic burn severity classification maps for each fire were used to evaluate the observed fire severity (unburned/low, low, moderate, and severe) and compared with the spatially-explicit map of historical fire severity within the study area (ESRI ArcGIS 10.1). We recognize these two datasets (observed fire severity and historical fire severity) are derived from different sources of information, each with their own limitations. Thus, our analysis focused on a summary comparison of the proportion of area burned by observed fire severity type compared to historical fire severity of the same area as a broad index of departure from historical fire regime reflecting landscape-level changes rather than a fine-scale (pixel-by-pixel) comparison. For each of the nine fires, we tested for differences (in the number of pixels) between the observed amount of low-severity fire (MTBS unburned/low and low severity classes) compared to the expected (historical low-severity) using χ^2 tests (IBM SPSS 20).

Comparison of historical fire severity to modeled fire severity. The spatially-explicit model of historical fire severity was overlaid with a model of fire potential under extreme (99th percentile) weather conditions to identify possible landscape-scale changes of historical low- and mixed-fire severity in the entire montane study area (564,413 ha; Table 3). We modeled potential fire behavior using FlamMap 3.0, a fire modeling system that predicts instantaneous fire behavior under fixed weather and fuel conditions (i.e. it is a static rather than dynamic model). FlamMap requires the following inputs: elevation, slope, aspect (derived from a 30 m USGS DEM); and percent canopy cover, stand height, canopy base height, canopy bulk density, and Scott and Burgan [74] fuel models (all from the LANDFIRE project [75]). Canopy cover and canopy base height values were adjusted downward on the recommendation of the LANDFIRE project [75] and a comparison of field-derived fuel layers in the study area [76]. The modeling procedure takes place in two steps (described in detail in [77–78] and summarized here). First, fuels are conditioned based on specified weather conditions. We used wind/weather conditions during the primary fire season (June–September) derived from four Remote Automated Weather Stations (RAWS) that span the study area and have the longest record: Redfeather (station 050505; 1964–2007), Estes Park (station 050507; 1964–2007), Corral Creek (station 051804; 1968–2007) and Bailey (station 052001; 1970–2007). In the second step, FlamMap calculates fireline intensity and crown fire activity using a suite of surface and crown fire models [79–82]. Fireline intensity is a measure of energy released per unit length along the flaming front of a fire

(kW/m). Crown activity (a binary variable representing surface fire or active/passive crown fire) was calculated using the Scott and Reinhardt [83] method. To address how historical fire severity compares with modeled potential wildfire behavior, we overlaid the areas modeled as historical low-severity with potential wildfire behavior (fireline intensity and crown fire) assuming that areas of higher fireline intensity (e.g. >33,000 kW/m) and crown fire today indicate a shift from the historical fire regime to one of higher potential for crown fire and canopy mortality (>33,000 kW/m and >40,000 kW/m are the approximate median fireline intensity values for moderate and high severity fire, respectively, for all verification fires with variable severity described in the next section; Table S1). A crown fire may or may not be lethal to all dominant vegetation, and a crown fire may be continuous or may occur in patches within a mixed-severity burn. Mixed-severity fires, by definition, have components of intense fire behavior and canopy mortality, and high fireline intensity and active crown fire would be within the HRV. Our focus is on the areas of greatest change that may have shifted from low severity to potential high fire intensity or crown fire today. These fire behavior outputs were chosen because they are commonly used, interpretable, and can be generally compared to historical fire severity.

The wildfire modeling utilized the 99th percentile monthly weather conditions from ca. 1964–2007 during the June-September fire season derived from four RAWS stations (listed above). Starting with a fixed southwest wind of 74 km/h, which is the 99.9th percentile daily wind speed during the primary fire season (June–September, ca. 1964–2007), we used the WindNinja program to refine wind speeds based on topography. FlamMap's input variables and modeling procedure follow Platt et al. [78] and were verified for a level of realism by comparing observed fire severity from seven major fires that occurred since 2002 (see the next section; Table 2).

To place historical 99th percentile conditions into context, we first compared them to the conditions of past fires and then to predicted future climate. During the Hayman Fire (06/08/2002), temperature and relative humidity reached up to the 98th and 99th percentiles, respectively (Cheesman RAWS station). Similarly, during the High Park Fire (06/09/2012), both temperature and humidity were up to the 99th percentile (Redstone RAWS station). The RAWS stations recorded wind gusts over 56 kph during the High Park Fire and over 80 kph during the Hayman Fire. We make the assumption that the 99th percentiles are of comparable weather conditions between historical and contemporary time periods.

We then compared average monthly temperature in the study area during the June-September fire season to two climate model runs used in the IPCC Fourth Assessment Report [84–85]: (1) climate conditions in 1964–1999 from the Climate of the 20th Century (20C3M) model run and (2) climate conditions in 2064–2099 from the SRESA2 scenario. During the June–September fire season in 1964–1999 (20C3M model run), the 99th percentile

Table 3. Comparisons of historical fire severity, observed fire severity, and modeled potential fire behavior.

Historical	Observed (MTBS thematic burn severity classifications)	Modeled (FlamMap output)
Low-severity fire only	Unburned to low severity, Low severity	Surface fire, e.g. median fireline intensity <33,000 kW/m
Mixed-severity fire: evidence of varied severities from low-, moderate- to high-severity fire	Moderate severity, High severity	Crown, torch, e.g., median fireline intensity > 33,000 kW/m

monthly temperatures were 24°C, and the 50th percentile monthly temperatures were 19°C. During the June–September fire season from 2064–2099 (SRESA2 scenario), the 99th percentile temperatures are predicted to increase to 30°C, and the 50th percentile temperatures are predicted to increase to 26°C. The 99th percentile temperatures from 1964–1999 are predicted to become 39th percentile temperatures under the SRESA2 scenario. It is thus reasonable to say that the 99th percentile conditions described in this study may become average conditions in 50 or more years from now.

Verification of modeled fire behavior. To assess the realism of the FlamMap models described above, we compared modeled fireline intensity and crown fire class to observed fire severity of the seven regional fires that occurred since 2002, the year for which the LANDFIRE (fuels) data is valid (Table 2). Within the perimeter of each fire, we generated randomly placed points, and at each point calculated median fireline intensity and percentage active and passive crown fire within four MTBS classes (unburned to low, low, moderate, and high severity; Table S1; ESRI ArcGIS 10.1). There are limitations to such a comparison. In an actual fire, intensity varies minute by minute depending on factors such as wind, relative humidity, fuel, and fire suppression. In contrast, FlamMap estimates intensity for an instant in time across an entire landscape for the generic "extreme conditions" described in the paper. Thus, while we cannot comprehensively validate a fire we can check whether, overall, areas identified as severely burned by MTBS correspond to higher modeled fireline intensity and crown fire. Kruskal-Wallis tests (IBM SPSS 20) were used to determine if the distribution of fireline intensity and percentage crown fire were significantly different between MTBS classes at the $p = 0.01$ level.

Results

Across all 232 field sites, cover type was dominated by a single species ($\geq 80\%$ of the canopy trees >4 cm dbh) of ponderosa pine (42.2%–98 sites) and lodgepole pine (8.6%–20 sites), and co-dominated ($<80\%$ of the canopy trees a single species) by mixed-conifer types (49.2%–114 sites). Considering pure ponderosa pine together with mixed-conifer sites, 61.2% (142) of the 232 sites were dominated by ponderosa pine and the remaining sites were dominated by Douglas-fir (20.6% and 48 sites), lodgepole pine (15.5% and 36 sites), and aspen (2.6% and 6 sites). The overall accuracy of the LANDFIRE EVT classification was relatively low with 39.2% of the 232 sites misclassified: 31 pure ponderosa pine, 5 lodgepole pine and 55 co-dominated mixed-conifer sites. Thus, the proportional values of EVT cover types across the study area, and our results defined by EVT cover types (the best available cover type at the landscape scale), should be interpreted cautiously (i.e. Figure 1 pie chart). Nevertheless, field observations support the overall LANDFIRE cover type trends illustrating ponderosa pine throughout the entire study area and an increase of lodgepole pine and mixed-conifer stands in the higher elevation portions of the study area.

Site-level fire-severity

A total of 7680 tree cores and 1262 fire-scar samples were used to delineate fire dates and fire severity across all sites. Across 120 stand-level sites, there was evidence of 322 spreading fires (fires with ≥ 2 trees scarred per site) between 1597 and 1995 with only 71 (22%) of the spreading fire dates unique to one site; all other fire dates were recorded at two or more sites. Fire years of 1654, 1786, and 1859–60 were particularly extensive with 36%, 43% and 48% of the available recorder sites (with ≥ 2 fire-scarred trees)

recording each fire year, respectively (Figure 2). These fires were recorded at multiple sites that extended over 9 km (1654), 7.5 km (1786) and up to 30 km (1859–60) distance away from one another (Figure 3). Other widespread fire dates recorded at four or more sites (recording on average 22% of the available recorder sites per year; ≥ 2 trees scarred per site) were 1624, 1712, 1793, 1809, 1813, 1842, 1850–52, 1857, 1863, 1866, 1871, 1880–81, 1886, and 1910 (Figure 2; multiple years indicate difficulty differentiating the exact fire year due to dormant season fires and/or missing rings before or after scarring).

Historical fire severity regime was delineated at individual sites using 297 fires that occurred prior to 1920 (pre-fire exclusion) and tree age data (remnant and establishment ages). Eighteen (7.8%) sites were characterized by only low-severity fires and 214 (92.2%) were characterized by mixed-severity fires. Across the montane region, the majority of trees established between 1820 and 1920 (66% of tree ages; Figure 4) primarily because of past mixed-severity fires during this period (this study; and also see [44–45,48,50–51,53,55]). Low-severity sites tended to have younger trees than mixed-severity sites (average date of tree establishment of 1907 compared to 1873, respectively; Figure 4). The low-severity fire regime was most common in pure ponderosa pine sites (12 sites; $\geq 80\%$ of the canopy trees as ponderosa pine), but also occurred in mixed-conifer sites (6 sites with co-dominance of ponderosa pine and Douglas-fir). Sites with the historical low-severity regime tended to occur at lower elevations than sites with mixed-severity fires with median elevations of 2101 m and 2445 m, respectively.

Of the 214 sites with evidence of mixed-severity fire (Figure 5), 108 sites had evidence of low and moderate-severity fires (no evidence of high-severity fire) and 106 sites had evidence of high-severity fires (evidence of up to ~50 ha patches of high-severity in some stand-level sites). Interpretation of the spatial extent of historical fire severity is limited by our sampling unit size (sites ranged between ~3–232 ha in size). Nevertheless, there is strong evidence that in most years before 1915 fires were much more extensive than the size of a single site (only 15.5% of the fire years were unique to one site; 46 of 297 fire dates; Figure 2). Many sites burned at moderate or high severity during the same year with an average of 1.1 km distance between sites (and up to 7 km distance between nearest sites burning in the same year with evidence of higher-severity fire). Evidence of mixed-severity fires occurred in all dominant cover types sampled, including sites of pure ($\geq 80\%$ of the canopy trees) lodgepole pine (20 sites) and ponderosa pine (89 sites), as well as mixed-conifer types with dominance of aspen (6 sites), lodgepole pine (16 sites), Douglas-fir (40 sites), and ponderosa pine (43 sites). Across sites classified with mixed-severity fire regimes, moderate- and high-severity fire effects were evident in all dominant cover types including pure lodgepole pine (8 and 12 sites) and ponderosa pine (48 and 41 sites), as well as mixed-conifer types with dominance of aspen (2 and 4 sites), lodgepole pine (8 and 8 sites), Douglas-fir (21 and 19 sites), and ponderosa pine (18 and 25 sites), respectively.

Landscape-level historical fire severity

Elevation and slope steepness were the most important predictive variables delineating locations of only low-severity versus mixed-severity sites. The CART model with the best predicted occurrence of sites with historical low- and mixed-severity fire had a PCC of 80.6% (187 of 232 sites predicted correctly) and an AUC value of 0.77 for the cross-validation test. The model indicated low-severity sites tended to occur at or below 2263 m, or on slopes equal or less than 4 degrees above 2263 m (17 of 18 low-severity sites fit this model, 94.4%). Mixed-severity

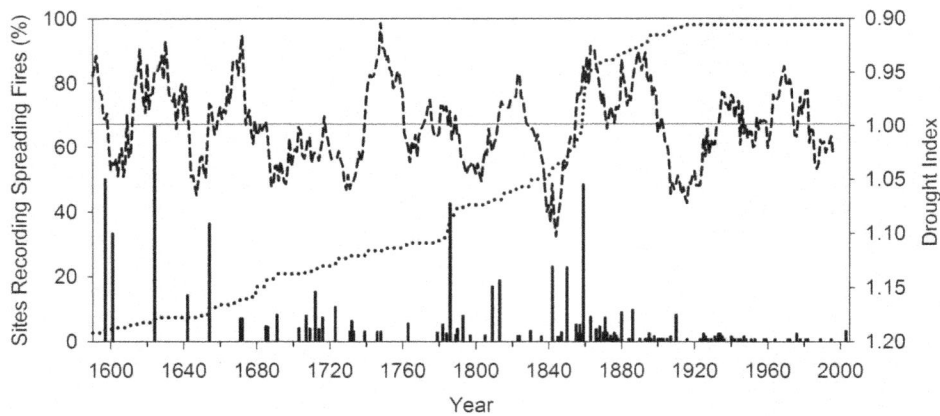

Figure 2. Percentage of sites recording spreading fires. Percentage of sites with spreading fires (minimum of 2 scarred trees per year at each stand-level site represented as histogram bars, n = 120 sites; left axis), sample depth of the percentage of sites available to record fires (minimum of 2 scarred trees available to record fires at each site; dotted line; left axis), and a 20-year smoothing average of a regional tree-ring drought index [60] with a mean value of 0.99 (dashed line with an inverted axis where smaller values are towards the top and indicate dry conditions; right axis).

sites occurred across a broad topographic gradient of the montane zone, but tended to occur on slopes greater than 4 degrees above 2263 m (170 of 214 mixed-severity sites fit this model, 79.4%). Mixed-severity sites with evidence of high-severity fire tended to occur at higher elevation, on greater slope steepness, and at greater distances from streams compared to sites with evidence of only low-severity fires. There were no strong generalizable differences between locations of mixed-severity sites with evidence of low- to high-severity fire and those without evidence of high-severity fire (only low- and moderate-severity fires). Cover type was not a significant predictor of fire severity.

The best-fit model from the CART analysis (described above) was used to map historical fire severity across the montane study area (564,413 ha). Across all cover types, 27.8% (156,198 ha) of the area was mapped with an historical regime of low-severity fire and 72.2% (406,173 ha) was mapped as mixed-severity fire (Figure 6a). Within the low-severity fire regime (27.8% of the study area) almost equal areas were classified by LANDFIRE EVT as ponderosa pine (45.4%) and mixed conifer (48.8%), intermixed with grassland/shrubland (5.7%). In comparison, areas mapped as the historical mixed-severity fire regime (72.2% of the study area) were classified with a lower percentage of ponderosa pine (16%) and grassland/shrubland (2.7%), and a higher percentage of mixed conifer (81.3%) cover types. Of areas mapped as pure ponderosa pine, 52.4% was classified as low-severity and 47.6% as mixed severity fire (Figure 6b). Of areas mapped as mixed conifer types, 18.8% was classified as low-severity and 81.2% as mixed severity fire (Figure 6c). The decrease in pure ponderosa pine dominance above 2263 m is consistent with the elevation shift in co-dominance with other species, primarily Douglas-fir, in the upper montane zone.

Comparison of historical fire severity to observed fire severity

For the nine recent fires in the study area since 2000 (Table 2; Figure 1), there was a 4% difference between the average observed (47%) and historical (51%) areas of low-severity fire severity, and a 7% difference between the total observed (39%) and historical (46%) areas of mixed-severity fire (Figure 7). The greatest differences between observed (MTBS) and expected (historical) low-severity fire was at the lowest elevations (median elevations <2155 m; Figure 7). Three fires (Picnic Rock Fire, Bobcat Fire,

Crystal Fire) had lower than expected observed low-severity fire compared to the historical fire severity map (X^2 value of 6.39 for Bobcat Fire and 5.49 for Crystal Fire, $p<0.05$; the X^2 test was invalid for the Picnic Rock Fire with no pixels classified as historical mixed-severity fire). For the remaining six fires there were either no significant differences in low-severity fire (High Park Fire, Four Mile Fire, Overland Fire, Hayman Fire, X^2 value of 1.04, 2.84, 1.56, and 2.43, respectively, $p>0.05$) or higher than expected low-severity fire observed compared to the expected historical fire severity map (High Meadows Fire and Big Elk Fire, X^2 values of 5.14, and 8.25, respectively, $p<0.05$).

Verification of modeled fire behavior

Overall, we found a positive association between modeled wildfire behavior (fireline intensity and percentage crown fire) and MTBS severity class (Table S1; Figures S1 and S2). The exception was the Picnic Rock Fire, for which the modeled percentage crown fire was consistently very low across observed severity classes. This is not surprising, as the Picnic Rock Fire occurred at the lower elevation grassland ecotone with over 50% of the fire classified as unburned/low or low severity and only 21% as high severity that occurred primarily in grassland fuel types (73% of high-severity area). Overall, severely burned areas in observed fires of variable fire severities (Table 2) corresponded to higher modeled fireline intensity and proportion of crown fire (Table S1), providing confidence in the model of potential wildfire behavior and its comparison with historical fire severity.

Comparison of historical fire severity to modeled current fire behavior

Across the entire montane study area present-day potential fire behavior was 36.1% surface and 73.9% crown (1.3% torch and 62.6% crown) fire occurrence under extreme (99th percentile) weather conditions (Figure 6a). Historical fire severity for the entire study area was mapped as 27.8% low-severity only and 72.2% mixed-severity fire regimes. For the pure ponderosa pine cover type, under the extreme fire weather scenario (99th percentile), present-day fuels predicted 40.2% of the area would burn as surface fire and 59.8% as crown (3.1% torch and 56.7% crown) fire (Figure 6b). For the mixed conifer cover type, the extreme fire weather scenario and present-day fuels predicted

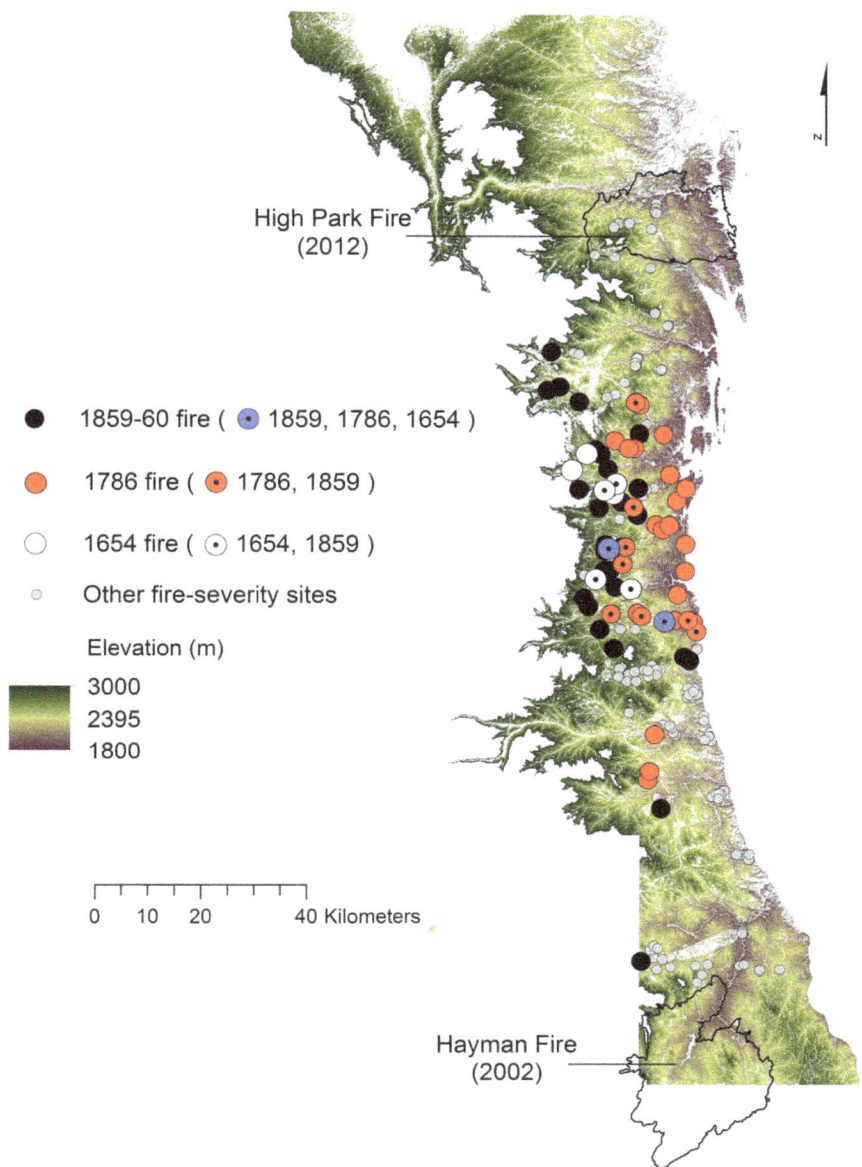

Figure 3. Historical fire-severity sites recording widespread fire dates in 1654, 1786, and 1859–60. Sites are shown across the elevation gradient of the study area. Symbols with centered dots indicate two or three of those fire years listed were recorded at that site. The two largest fires in recorded history for the region (Hayman Fire and High Park Fire) are shown for spatial comparison.

31.8% as surface and 68.2% as crown (0.7% torch and 67.5% crown) fire (Figure 6c).

Almost 12% (66,394 ha) of the entire study area showed little change in the low-severity fire regime (Figure 8 – green), based on a high spatial coincidence of historical low-severity and present-day potential for surface fire under extreme (99th percentile) weather conditions. These areas were predominantly in the lower montane zone (mean 2136 m elevation and 12 degree slope steepness), and had an average modeled fireline intensity of 14,020 kW/m. About 16% of the entire montane study area (Figure 8 – red) is now susceptible to crown fire but historically had lower-severity fire. These areas were also concentrated in the lower montane zone (mean 2177 m elevation and 16 degree slope steepness), had an average modeled fireline intensity of 43,620 kW/m, and were classified by LANDFIRE EVT as

grassland or shrubland (0.5%), lodgepole pine (5.2%), ponderosa pine (43.7%), and mixed conifer (50.7%). Of areas mapped only as the historical low-severity fire regime (27.8% and 156,198 ha of the study area; Figure 8 – green and red combined), 42.5% of the area showed little change in the fire regime (Figure 8 – green), whereas 57.5% had the potential for higher-severity fire (Figure 8 – red).

Above the lower montane zone (>2263 m), the historical and modeled fire behavior under extreme conditions showed high variability from surface to crown fire, as expected in a mixed-severity fire regime. Areas above 2263 m that were mapped as historical mixed severity fire, but today have potential for surface fire and crown fire were 24.8% and 47.4% of the entire montane study area, respectively (Figure 8 – yellow and orange). Of areas mapped only with the historical mixed-severity fire regime (72.2%

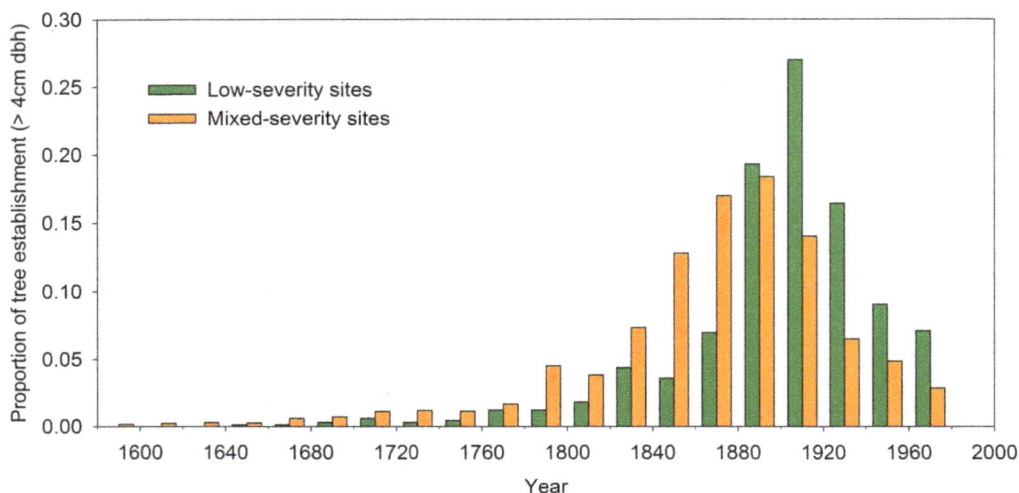

Figure 4. Proportion of tree establishment dates. Tree establishment (≥4 cm diameter) in 20-year bins for low-severity (green) and mixed-severity (orange) sites sampled at the stand-scale (120 sites). Total number of trees included are 663 and 5703, respectively.

and 406,173 ha of the study area; Figure 8 – yellow and orange), 34.4% has a low probability of crown fire today (mapped as potential for surface fire in Figure 8 (yellow); average fireline intensity of 15,861 kW/m), whereas 65.6% of the area has a high probability of crown fire activity today (mapped as potential for torch or crown fire in Figure 8 (orange); average fireline intensity of 42,290 kW/m).

Discussion

A key finding of our study of fire regime changes in the montane forest zone of the Colorado Front Range is that only 16% of the total study area recorded a shift from historical low-severity to a higher potential for crown fire today. This area of increased fire severity occurs in over half (57.5%) of the area mapped with the historical low-severity fire regime and is concentrated in the lower montane zone. A substantial portion (42.5%) of the area mapped as the historical low-severity fire regime (11.8% of the study area) shows little change in the fire regime (Figure 8), and is expected to support only surface fire even under extreme (99th percentile) weather conditions based on our modeling of present day fuels. Both areas occur primarily in the lower montane zone below 2263 m, but the areas with little change in the low-severity fire regime are on average slightly lower in elevation and slope steepness and closer to grasslands and ravine drainages than areas with higher potential for crown fire today. Observations from recent large wildfires and modeling of potential fire behavior under extreme weather conditions are consistent with the historical evidence of a varied fire regime of primarily low-severity fires at the lowest elevations to a mixed-severity fire regime at higher elevations in montane forests. Three regional fires predominantly in the lower montane forests showed the greatest differences between the observed (MTBS of recent large fires) and expected (historical fire severity) areas burned by low-severity fire. However, the differences between the observed and expected proportions of low-severity fire for the six other fires were non-significant, or showed higher than expected low-severity fire.

A decline in fire frequency over the past 100 years leading to substantial increases in stand density is supported only for the lowest elevations of forest below approximately 2200 m in the Colorado Front Range [13–14,45,51,60,86]. These areas were

characterized mainly by frequent (average return intervals <30 years) low-severity fires that maintained open forests by killing mostly juvenile trees, resulting in low densities of mature trees. Greater proximity to grassland in lower elevation areas probably promoted more frequent fire due to more abundant fine, herbaceous fuels [62]. The cessation of formerly frequent fires coincides with increased stand densities broadly throughout the lower montane zone. This pattern is especially evident below 2200 m, but also occurred at some sites at higher elevations on less steep slopes most likely where montane grasslands occurred. However, overall this represents a relatively small proportion of the montane forest of the northern Colorado Front Range (27.8% of the study area is mapped with the historical low-severity fire regime).

The dominant (72.2% of the 564,413 ha study area) historical fire regime of the northern Colorado Front Range consisted of a mixed-severity regime in which stand structures were shaped primarily by moderate-severity (46.5% of sites) and high-severity (45.7% of sites) fires; only 7.8% of the sites recorded predominantly low-severity fires. At higher elevation (>2263 m), spreading fires were typically less frequent (>30 year fire intervals), and had varied fire effects (mixed-severity) that included non-stand replacing fire to canopy-replacement fire both within sites and across broad landscapes, often in relation to topographic variability. Many sites experienced intervals between successive widespread fires that would have been sufficient for conifer seedlings to reach sizes that would survive low-severity surface fires. Specifically, evidence of high-severity fire tended to occur further from grasslands and on steeper sites than those with evidence of only low-severity fire, although there is overlap between sites that show moderate- or high-severity effects particularly above the lower montane zone. Evidence of mixed-severity fires occurred in all dominant cover types sampled including sites of pure (≥80% of the canopy trees) lodgepole pine and ponderosa pine, as well as mixed-conifer types with dominance of aspen, lodgepole pine, Douglas-fir, and ponderosa pine. Consistent site- to landscape-scale evidence indicates an historical mixed-severity fire regime in which moderate- and high-severity fire effects shaped current forest age structures in the mid- and upper montane zone [13,46,51,53] (and this study).

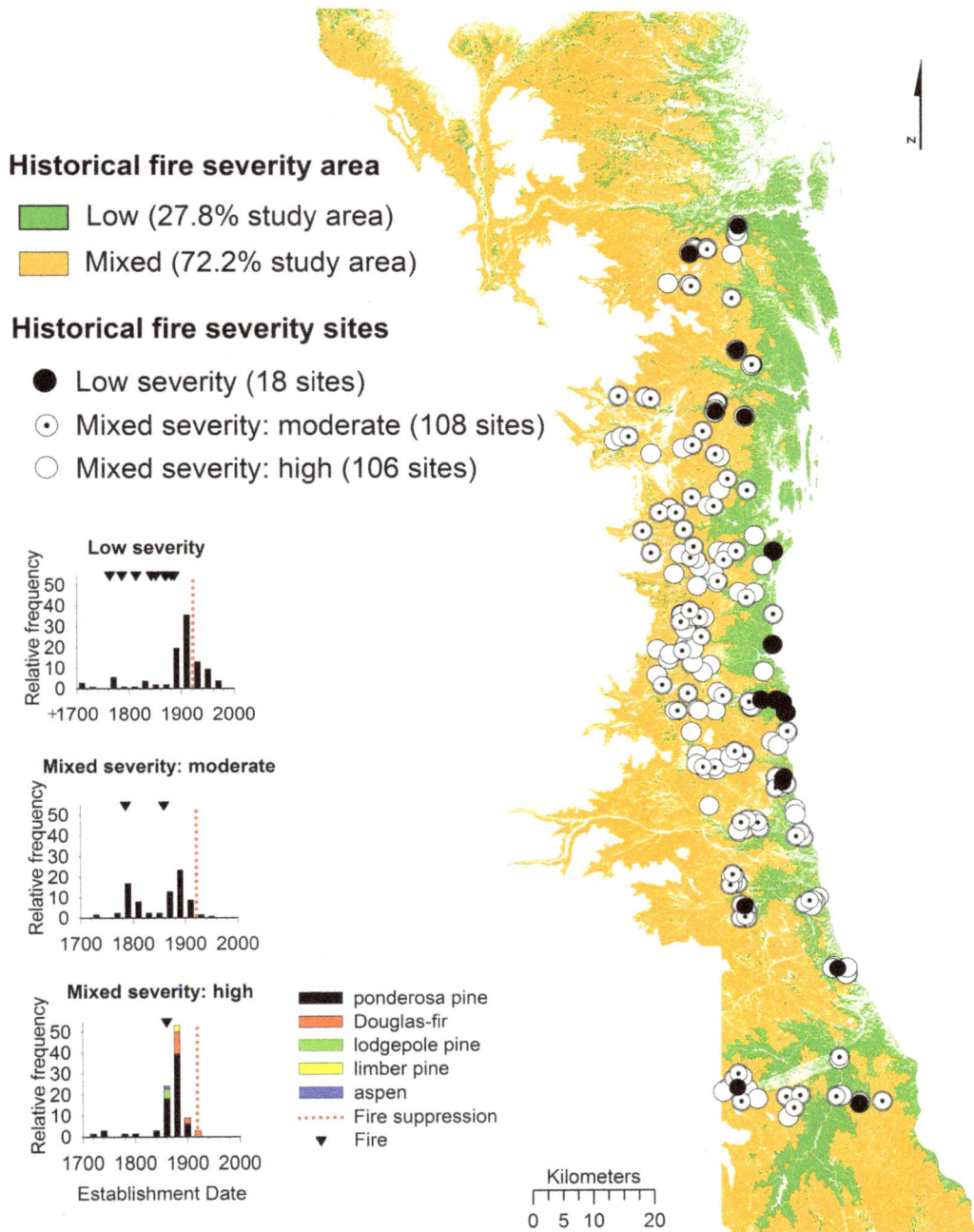

Figure 5. Distribution of 232 sites with historical (pre-1920) evidence of low-severity and mixed-severity fires. Sites with historical evidence of mixed-severity fires are differentiated with evidence of low- and moderate-severity fire (no evidence of high-severity fire; 108 sites) and low- to high-severity fire (106 sites). Areas mapped as historical low-severity fire only (27.8%) and mixed-severity fire (72.2%) are shown in green and orange, respectively. Three example graphs are also shown to illustrate the evidence (remnant trees, tree establishment and spreading fires) used to classify sites with evidence of low and mixed-severity fire effects at individual sites.

Most spreading fires recorded at our sites extended beyond a single site (88% of 322 fires with ≥2 trees scarred per site from 1597–1995). Moderate- and high-severity fires prior to the 20th century were documented at many sites, as illustrated by fires in 1654, 1786 and 1859–60 that extended at least 7–30 km between nearest sampling sites, are within the HRV for montane forests at a regional scale (Figure 3). For example, the spatial extent of sites recording fires in the same year is within the spatial scale of the largest modern wildfires on record in the study area (e.g. High Park Fire and Hayman Fire). Our ability to interpret the spatial extent of historical fire severity patches from individual fires is limited by our sampling unit size (3 ha–232 ha), subsequent fire events, and sample depth (although >20% of our sites record fires since 1700), but the evidence indicates mixed-severity fires occurred in patches up to at least 200 ha, with evidence of approximately 50 ha of high-severity in some sites. Direct

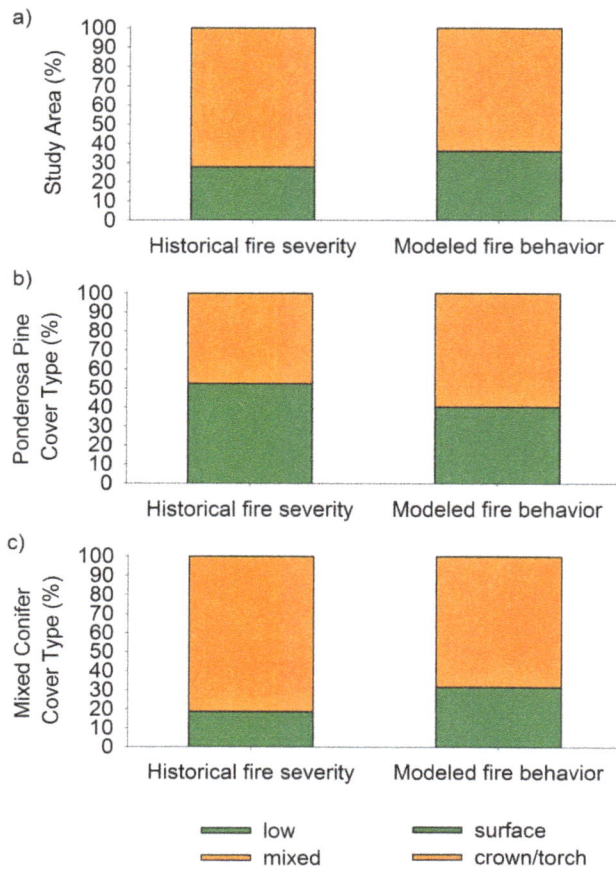

Figure 6. Percent study area and cover type classified with historical fire type and modeled fire behavior. A comparison of the percentage of the a) study area, b) ponderosa pine cover type, and c) mixed conifer cover type classified as historical fire (low-severity only or mixed-severity) and modeled current fire behavior (surface or crown/torch) under extreme weather conditions (99th percentile for 1964–2007).

comparison of patch sizes of fires is difficult given the presence of major highways and other land uses that have fragmented fuel continuity at a landscape scale. It is highly likely that variance in patch size of historical fires was great, similar to modern fire observations, rendering quantitative comparisons of rather limited value. Nevertheless, high-severity fire and high-stand densities are within the HRV for the mid- and upper montane forests (e.g. [44,51,53–54], this study), which needs to be taken into account when forest management goals consider restoring forests to pre-fire-exclusion conditions.

In the Colorado Front Range, at higher elevations (>2263 m) mixed-severity fire regimes appear to have been the predominant fire regime historically and today. The modeled potential wildfire behavior under current fuel conditions and extreme fire weather shows a mixed-severity fire regime throughout the montane forests of the Colorado Front Range. High fireline intensity and active crown fire are within the HRV of a mixed-severity fire regime particularly under extreme conditions, but would vary in extent of intensity and canopy mortality depending on existing conditions. Site-level evidence indicates high overlap in biophysical conditions that support moderate and high-severity fires (this study and [52]). These results provide a snapshot of the expectations into the near future for fire behavior in the Colorado Front Range. The

advantage of comparing historical and observed fire severity with fire behavior modeling across the study area under realistic, but also extreme, weather conditions is that it provides an important present-day comparison to the past. Recent large wildfire years in the Colorado Front Range are not without precedent and similar events occurred historically (i.e. 1654, 1786, 1859–60), under similar exceptional (interannual to multi-decadal scale) climate conditions [61]. The largest fires since the year 2000 – the Hayman and High Park Fires – were both characterized by up to 98–99th percentile conditions for both temperature and relative humidity. The situation will likely be exacerbated under climate change; under the 2007 IPCC A2 scenario for the study area, the 99th percentile temperatures of the past are expected to become the 39th percentile temperatures by 2064–2099 (also see [87]). Thus, the range of weather conditions during local to more widespread fire years in the last 300+ years likely represents a similar range of potential fire-climate conditions presently, and informs expectations of potential fire behavior that may occur under severe, yet increasingly more common weather conditions.

We recognize that each of our spatial datasets (mapping of historical, observed, and potential fire) have their own set of limitations. For example, distance to grassland was not a predictor in the best-fit model used to map the historical fire severity landscape. However, our prior research indicates areas adjacent to grasslands experience more frequent fires than sites farther from grasslands, owing to the proximity to prevalent fine fuels that increase the sensitivity of fire activity to interannual climate variability [62]. Thus, the historical model likely under-estimates the amount of low-severity fire in some areas at higher elevation within or adjacent to grasslands where we have limited or no sample sites (e.g. high elevation grassland areas in the northern portion of the study area - Figures 1 and 8). This could explain why some of the study area (Figure 8) shows a shift from historical mixed-severity fire to potential low-severity fire in the contemporary forest even under the extreme weather scenario. The observed fire behavior dataset (MTBR Thematic Burn Severity) has its own limitations. To classify fire severity into 'high' 'moderate', and 'low' severity classes, image analysts determine thresholds in dNBR values based on visual interpretation of imagery, field plot data, and expert knowledge. Though dNBR is sensitive to initial vegetation conditions [88], alternatives such as RdNBR are less parsimonious and in practice may not capture burn severity with greater accuracy than dNBR [89]. Also, while the fire behavior modeling from FlamMap operates on a fixed set of fuels and weather conditions, the historical severity model is based on the fire regime over a long period of time during which fires of different severity have the opportunity to burn. Because of these limitations, we report proportional area of landscape-level changes rather than finer-scale interpretations, and emphasize broad trends evident from multiple lines of inquiry. Additionally, while our model was based on the most recently available fuels data from LANDFIRE, characterizing vegetation changes across the landscape due to wildfire and management through 2010, mortality from recent insect outbreaks, for example, are not reflected. Inferring the effect of fuel and fire behavior changes in our model due to tree mortality from insects is challenging as such fuel changes are: 1) highly dependent on the timing and severity of the outbreak, 2) not characterized well in standard fuel models, and 3) not validated for fire behavior as few empirical studies have examined how the timing and extent of insect outbreaks in montane forests affects observed fire behavior.

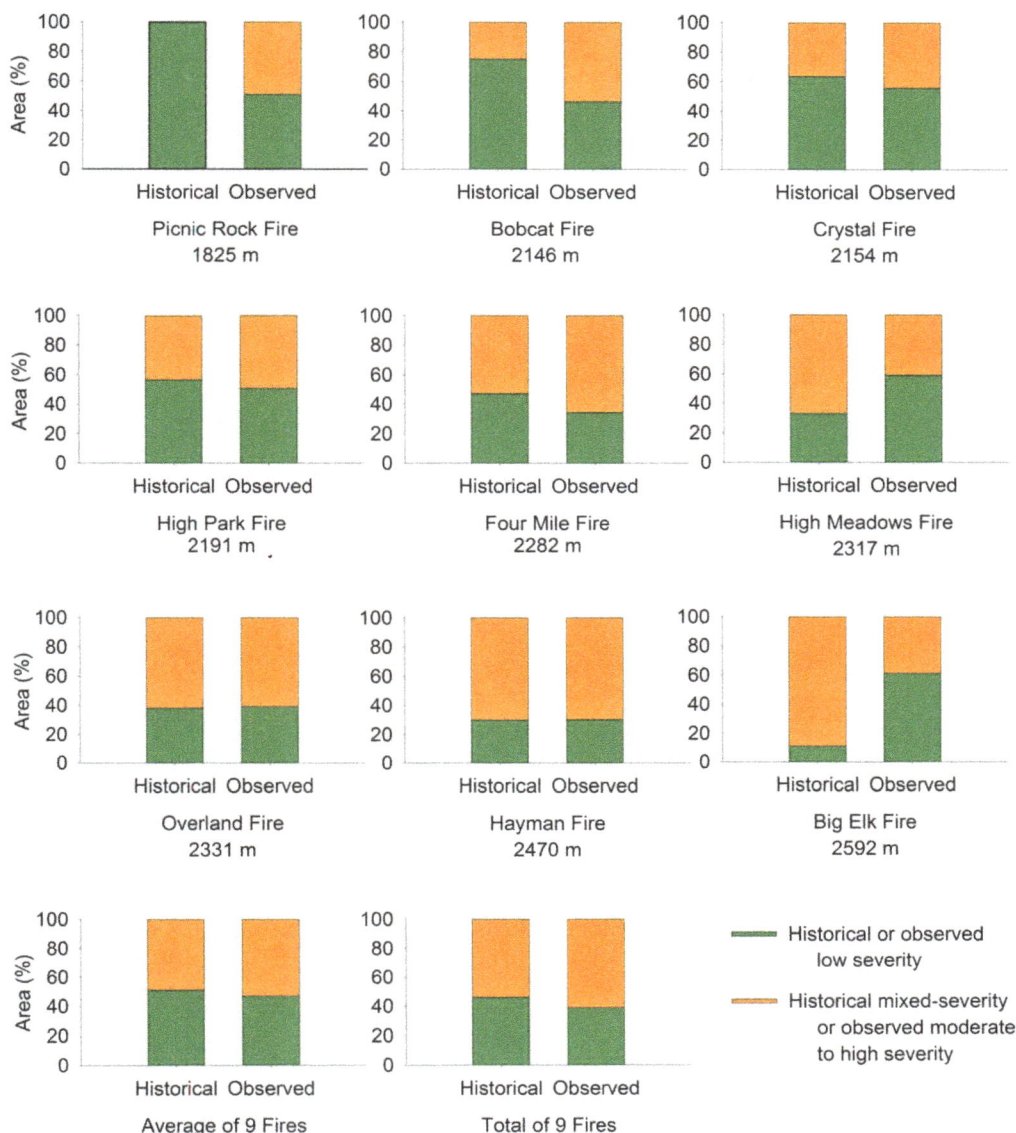

Figure 7. Comparisons of the historical and observed fire severities for nine recent fires (2000–2012). The comparison shows the proportion of low- (unburned/low and low) and mixed- (moderate and high) severity fire within the perimeter of nine recent fires (2000–2012), and the average and total (all pixels) proportions of fire severities across all nine fires in the study area. The median elevation for each fire is given under the fire name.

Implications for fuels management and ecological restoration

A clear delineation of the spatial extent of past fire regime types is a major concern for ecosystem managers in the context of wildfire risk and ecological restoration [90–92]. Evidence from previous studies in ponderosa pine and other montane forests in the Colorado Front Range have shown that the historical fire regime was variable over time and space, represented by a mixed-severity regime [44,46–48,50–54]. Goals of ecological restoration and wildland fire hazard mitigation are both compatible with management practices, like prescribed fire and thinning to reduce fuels, below approximately 2200 m in our study area, which experienced the greatest increase in fire severity, and likely fuels, since fire exclusion [93]. Disturbance from grazing and logging as well as periods of favorable climate probably also contributed to

increased tree establishment in the late 19th and early 20th century, but seedling survival clearly depended on the long fire-free periods. Even at low elevations, however, some sites had historical fire regimes dominated by infrequent rather than frequent fires (e.g. steep and north-facing slopes [86]). These infrequent fires, inferred to be high-severity fires from age structure data, killed high percentages of trees within a fire perimeter, and promoted the establishment of naturally dense forest patches [44,51,86]. This suggests that some of the areas (e.g. north-facing and steep slopes) with potential for crown fire today may support characteristic fire behavior and are not necessarily out of the HRV even in the lower montane zone.

Mixed-severity fire regimes where spreading fires occurred at lower frequencies (>30 year fire intervals) are less clearly candidates for thinning than are low-severity fire regimes and a

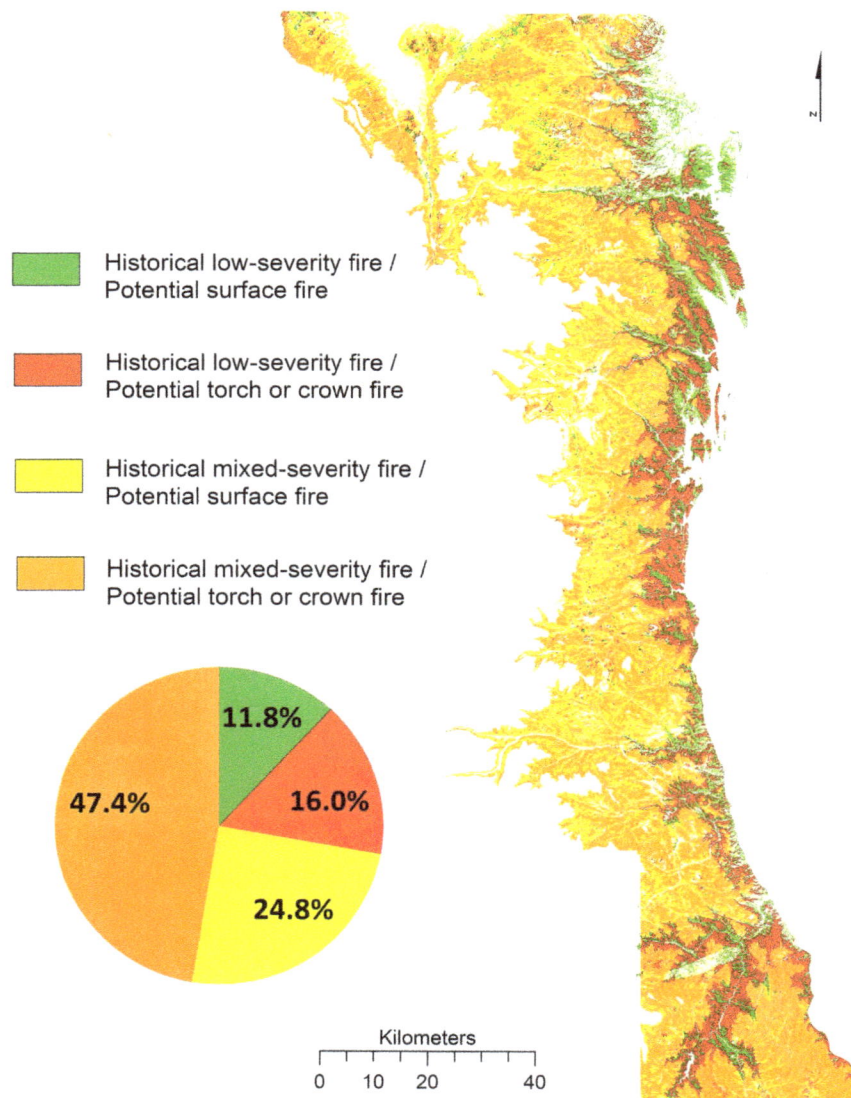

Figure 8. Historical fire severity overlaid with a model of fire potential under extreme (99th percentile weather). The areas and proportion of the study area mapped as historical low-severity fire with current potential for surface fire (green), historical low-severity fire with current potential for torch or crown fire (red), historical mixed-severity with current potential for surface fire (yellow), and historical mixed-severity with current potential for torch or crown fire (orange). Current potential fire behavior is modeled under extreme (99th percentile) weather (1964–2007).

cautious approach to restoration efforts has been recommended [15,40]. In areas naturally characterized by lower frequencies of moderate- to high-severity fires, fuel reduction through prescribed fire and thinning will likely not achieve both ecological restoration and fire hazard mitigation goals. Restoration thinning treatments will not return the fire regime to one of low severity across the Front Range montane zone, which was historically predisposed to periodic fires of varying severities, and are of questionable effectiveness in preventing severe wildfires [2]. Where extreme fire behavior appears within HRV, high-severity fire may largely be explained by extreme weather conditions (for example, high winds and low humidity during severe drought) rather than quantity of woody fuels [38]. This is illustrated by the 2002 Hayman Fire in Colorado in which more than 24,000 ha burned at high severity throughout ponderosa pine and mixed-conifer forests of variable stand structures in a single day [94]. The fire-

weather conditions presented here (99th percentile for the late 20th century and early 21st century) represent only the projected average expected conditions by the mid-21st century, which is a pressing issue for existing and future WUI development [95] and fire management [2]. Learning from recent experiences where wildfire damage has been high in the WUI [2], along with considering the costs of suppressing future wildland fires (i.e. reducing forest resiliency, unsustainable federal costs to society [1]) and the sociopolitical expectations of wildland fire management are critical for managing future fire risks and forest integrity.

For the ponderosa pine cover type, an increasing number of studies in the Pacific Northwest [15,96–100] have also document-ed fire regimes of variable severity. These studies and ours illustrate the importance of collecting evidence on the severity of past fires (i.e. remnant tree and cohort ages, growth releases, tree mortality) because fire severity cannot be reliably interpreted from fire interval data alone, which has often the approach applied to the study of fire history in ponderosa pine forests. Although current implementation of national programs (Collaborative Forest Landscape Restoration Program) and databases (LAND-FIRE) recognize variation in fire regimes within the same cover type across different geographical regions, fire regimes of a cover type within a region can still be considered relatively uniform. This is problematic in the Front Range and potentially elsewhere where the historical fire regime varied from low-severity at the lowest elevations to mixed-severity over most of the montane zone regardless of forest type. Additionally, the accuracy of the current cover type (EVT) layer for the region is low. Thus, the spatial and proportional reliability of the LANDFIRE classifications of fire regimes and vegetation do not accurately reflect the historical or current landscape, and therefore will not be highly effective in prioritizing locations for fire management and forest restoration objectives. Further corroboration of these issues are also documented at a national scale (see p. 19 in [92]) "The accuracy of various aspects of the LANDFIRE data is questionable, even when used at intended scale…Without accurate data, many assumptions and actions based on this data will be compromised. There is a need for more realistic and accurate depiction of where wildland fire hazard/risk actually occurs across the country, which can be used to base decisions upon."

Our study, along with prior studies, shows that an assumption of fire regime uniformity is not valid for the ponderosa pine and other montane cover types in the Colorado Front Range and elsewhere (i.e. [15]). These findings challenge ecologists and managers to re-consider the degree of variation in fire regimes within broadly distributed forest types. Thus, management efforts to create large areas of open woodlands in the higher elevation areas of the montane zone of the Colorado Front Range would not be consistent with historical fire regimes and stand structures.

Supporting Information

Figure S1 MTBS fire severity classification (observed) and median fire line intensity (modeled). The comparison was used as a measure of verification of the modeled fire behavior.

Figure S2 MTBS fire severity class (observed) and percentage of crown fire (modeled). The comparison was used as a measure of verification of the modeled fire behavior.

Table S1 Summary table of biophysical factors for each observed fire and potential fire model in analysis.

Table S2 List of plot-level sites. The fire severity classification for each site was documented from forest structure and fire scars in the field using an existing protocol [52].

Acknowledgments

We thank student assistants from Humboldt State University (Alicia Iverson, Kelly Muth, Madelinn Schriver) and Gettysburg College (Erin Palmer) for help with compiling datasets. We also thank Cameron Naficy, Ben Bond-Lamberty, and three anonymous reviewers for insightful comments that improved the manuscript.

Author Contributions

Conceived and designed the experiments: RLS RVP TTV TLS. Performed the experiments: RLS RVP TTV TLS MHG. Analyzed the data: RLS RVP. Wrote the paper: RLS RVP TTV TLS MHG.

References

1. Gorte R (2013) The rising cost of wildfire protection. Headwaters Economic Research Report. http://headwaterseconomics.org/wildfire/fire-cost-background. Accessed 2014 August 23.
2. Calkin DE, Cohen JD, Finney MA, Thompson MP (2013) How risk management can prevent future wildfire disasters in the wildland-urban interface. Proceedings of the National Academy of Science 756–751, doi: 10.1073/pnas.1315088111.
3. Mackun P, Wilson S (2011) Population distribution and change: 2000 to 2010. U.S. Department of Commerce Economics and Statistics Administration. U.S. Census Bureau. http://www.census.gov/prod/cen2010/briefs/c2010br-01.pdf. Accessed 2014 Aug 23.
4. Williams J (2013) Exploring the onset of high-impact mega-fires through a forest land management prism. Forest Ecology and Management 294: 4–10.
5. Stephens SL, Agee JK, Fulé PZ, North MP, Romme WH, et al. (2013) Managing forests and fire in changing climates. Science 342 (6154): 41–42, doi 10.1126/science.12402994.
6. Dillon GK, Holden ZA, Morgan P, Crimmins MA, Heyerdahl EK, et al. (2011) Both topography and climate affected forest and woodland burn severity in two regions of the western US, 1984–2006. Ecosphere 2: Article 130. http://www.esajournals.org/doi/pdf/10.1890/ES11-00271.1.
7. Odion DC, Hanson CT (2008) Fire severity in the Sierra Nevada revisited: conclusions robust to further analysis. Ecosystems 11: 12–15.
8. Miller JD, Safford HD, Crimmins M, Thode AE (2009) Quantitative evidence for increasing forest fire severity in the Sierra Nevada and southern Cascade Mountains, California and Nevada. Ecosystems 12: 16–32.
9. Miller JD, Skinner CN, Safford HD, Knapp EE, Ramirez CM (2012) Trends and causes of severity, size, and number of fires in northwestern California, USA. Ecological Applications 22: 184–203.
10. Miller JD, Collins BM, Lutz JA, Stephens SL, van Wagtendonk JW, et al. (2012) Differences in wildfires among ecoregions and land management agencies in the Sierra Nevada region, California, USA. Ecosphere 3: Article 80.
11. Miller JD, Safford HD (2012) Trends in wildfire severity: 1984 to 2010 in the Sierra Nevada, Modoc Plateau, and southern Cascades, California, USA. Fire Ecology 8: 41–57.
12. Hanson CT, Odion DC (2014) Is fire severity increasing in the Sierra Nevada mountains, California, USA? International Journal of Wildland Fire. 23: 1–8.
13. Williams MA, Baker WL (2012) Comparison of the Higher-Severity Fire Regime in Historical (A.D. 1800s) and Modern (A.D. 1984–2009) Montane Forests Across 624,156 ha of the Colorado Front Range. Ecosystems 15: 832–847.
14. Fulé PZ, Swetnam TW, Brown PM, Falk DA, Peterson DL, et al. (2013) Unsupported inferences of high-severity fire in historical dry forests of the western United States: response to Williams and Baker. Global Ecology and Biogeography. doi: 10.1111/geb.12136.
15. Odion DC, Hanson CT, Arsenault A, Baker WL, DellaSalla DA, et al. (2014) Examining historical and current mixed-severity fire regimes in ponderosa pine and mixed-conifer forests of western North America. PLoS ONE. doi: 10.1371/journal.pone.0087852.
16. Morgan P, Aplet GH, Haufler JB, Humphries HC, Moore MM, et al. (1993) Historical Range Of Variability: A Useful Tool for Evaluating Ecosystem Change. In: Sampson RN, Adams DL, editors. Assessing Forest Ecosystem Health in the Inland West. Sun Valley: Haworth Press. pp. 87–109.
17. Keane RE, Hessburg PF, Landres PB, Swanson FJ (2009) The use of historical range of variability (HRV) in landscape management. Forest Ecology and Management 258: 1025–1037.
18. Hayward GD, Veblen TT, Suring LH, Davis B (2012) Challenges in the application of historical range of variation to conservation and land management. In Historical Environmental Variation in Conservation and Natural Resource Management. Eds. Wiens JA, Hayward GD, Safford HD, Giffen CM. John Wiley & Sons, Ltd.
19. Millar CI, Stephenson NL, Stephens SL (2007) Climate change and forests of the future: managing in the face of uncertainty. Ecological Applications 17: 2145–2151.
20. Wiens JA, Hayward GD, Safford HD, Giffen C (2012) Historical environmental variation in conservation and natural resource management. John Wiley & Sons.
21. Moritz MA, Hurteau MD, Suding KN, D'Antonio CM (2013) Bounded ranges of variation as a framework for future conservation and fire management. Annals of the New York Academy of Sciences 1286: 92–107.
22. DeLeo GA, Levin S (1997) The Multifaceted Aspects of Ecosystem Integrity. Conservation Ecology 1: 3.
23. Parrish JD, Braun DP, Unnasch RS. (2003) Are we conserving what we say we are? Measuring ecological integrity within protected areas. Bioscience 53: 851–860.

24. Healthy Forests Restoration Act. HR 1904 (2003) http://www.forestsandrangelands.gov/resources/overview/hfra-implementation12-2004.shtml. Accessed 2014 August 23.

25. Peterson DL, Millar CI, Joyce LA, Furniss MJ, Halofsky JE, et al. (2011) Responding to climate change in National Forests: A guidebook for developing adaptation options. USDA Forest Service General Technical Report PNW-GTR-855.

26. Wildland Fire Executive Council (2014) The National Strategy: The Final Phase in the Development of the National Cohesive Wildland Fire Management Strategy. http://www.forestsandrangelands.gov/strategy/documents/strategy/CSPhaseIIINationalStrategyApr2014.pdf. Accessed 2014 August 23.

27. Brown JK, Oberheu RD, Johnston CM (1982) Handbook for inventorying surface fuels and biomass in the Interior West. USDA Forest Service General Technical Report. INT-129.

28. Covington WW, Moore M (1994) Southwestern Ponderosa Forest Structure: changes since Euro-American settlement. Journal of Forestry 92: 39–47.

29. Covington WW (2000) Helping western forests heal. Nature 408: 135–136.

30. Agee JK (2002) The Fallacy of Passive Management Managing for Firesafe Forest Reserves. Conservation in Practice 3: 18–26.

31. Allen CD, Savage M, Falk DA, Suckling KF, Swetnam TW, et al. (2002) Ecological restoration of southwestern ponderosa pine ecosystems: a broad perspective. Ecological Applications 12: 1418–1433.

32. Agee JK, Skinner CN (2005) Basic principles of forest fuel reduction treatments. Forest Ecology and Management 211: 83–96.

33. Stephens SL, Ruth LW (2005) Federal forest-fire policy in the United States. Ecological Applications 15: 532–542.

34. Roos CI, Swetnam TW (2012) A 1416-year reconstruction of annual, multidecadal, and centennial variability in area burned for ponderosa pine forests of the southern Colorado Plateau region, Southwest USA. The Holocene 22: 281–290.

35. United States Forest Service (2002) National Database of State and Local Wildfire Hazard Mitigation Programs, http://www.wildfireprograms.usda.gov. Accessed 2014 August 23.

36. Mell WE, Manzello SL, Maranghides A, Butry DT, Rehm RG (2010) The wildland–urban interface fire problem – current approaches and research needs. International Journal of Wildland Fire 19: 238–251.

37. Veblen TT (2003) An introduction to key issues in fire regime research for fuels management and ecological restoration. In: Omi P, Joyce L, Eds. Fire, fuel treatments and ecological restoration: conference proceedings 2002, 16–18 April, Fort Collins, CO. USDA, Forest Service, RMRS-P-29.

38. Schoennagel T, Veblen TT, Romme WH (2004) The Interaction of Fire, Fuels, and Climate across Rocky Mountain Forests. BioScience 54: 661–676.

39. Nagel TA, Taylor AH (2005) Fire and persistence of montane chaparral in mixed conifer forest landscapes in the northern Sierra Nevada, Lake Tahoe Basin, California, USA. Journal of the Torrey Botanical Society 132: 442–457.

40. Baker WL, Veblen TT, Sherriff RS (2007) Fire, fuels and restoration of ponderosa pine–Douglas fir forests in the Rocky Mountains, USA. Journal of Biogeography 34: 251–269.

41. Klenner W, Walton R, Arsenault A, Kremsater L (2008) Dry forests in the Southern Interior of British Columbia: Historic disturbances and implications for restoration and management. Forest Ecology and Management 256: 1711–1722.

42. Hutto RL (2008) The ecological importance of severe wildfires: some like it hot. Ecological Applications 18: 1827–1834.

43. Colombaroli D, Gavin DG (2010) Highly episodic fire and erosion regime over the past 2,000 y in the Siskiyou Mountains, Oregon. Proceedings of the National Academy of Sciences 107: 18909–18915.

44. Veblen TT, Lorenz DC (1986) Anthropogenic disturbance and recovery patterns in montane forests, Colorado Front Range. Physical Geography 7: 1–24.

45. Veblen TT, Lorenz DC (1991) The Colorado Front Range: A Century of Ecological Change. Salt Lake: University of Utah Press. 186 p.

46. Mast JN, Veblen TT, Linhart YB (1998) Disturbance and climatic influences on age structure of ponderosa pine at the pine/grassland ecotone, Colorado Front Range. Journal of Biogeography 25: 743–755.

47. Brown P, Kaufmann M, Shepperd W (1999) Long-term, landscape patterns of past fire events in a montane ponderosa pine forest of central Colorado. Landscape Ecology 14: 513–532.

48. Kaufmann MR, Regan CM, Brown PM (2000) Heterogeneity in ponderosa pine/Douglas-fir forests: age and size structure in unlogged and logged landscapes of central Colorado. Canadian Journal of Forest Research 30: 698–711.

49. Huckaby LS, Kaufmann MR, Stoker JM, Fornwalt PJ (2001) Landscape Patterns of Montane Forest Age Structure Relative to Fire History at Cheesman Lake in the Colorado Front Range. U.S.D.A. Forest Service Proceedings RMRS-P-22.

50. Ehle DS, Baker WL (2003) Disturbance and stand dynamics in ponderosa pine forests in Rocky Mountain National Park, USA. Ecological Monographs 73: 543–566.

51. Sherriff RL, Veblen TT (2006) Ecological effects of changes in fire regimes in Pinus ponderosa ecosystems in the Colorado Front Range. Journal of Vegetation Science 17: 705–718.

52. Sherriff RL, Veblen TT (2007) A spatially-explicit reconstruction of historical fire occurrence in the ponderosa pine zone of the Colorado Front Range. Ecosystems 10: 311–23.

53. Schoennagel T, Sherriff RL, Veblen TT (2011) Fire history and tree recruitment in the Colorado Front Range upper montane zone: implications for forest restoration. Ecological Applications 21: 2210–2222.

54. Williams MA, Baker WL (2012) Spatially extensive reconstructions show variable-severity fire and heterogeneous structure in historical western United States dry forests. Global Ecology and Biogeography 21: 1042–1052.

55. Romme WH, Veblen TT, Kaufmann MR, Sherriff RL, Regan CM (2003) Ecological effects of the Hayman fire, Part I: Historical (pre-1860) and current (1860–2002) fire regimes. In: Graham RT, editor. Hayman Fire Case Study. Fort Collins CO, Rocky Mountain Research Station: USDA Forest Service, General Technical Report RMRS-GTR-114. pp. 181–195.

56. Kaufmann MR, Veblen TT, Romme WH (2006) Historical fire regimes in ponderosa pine forests of the Colorado Front Range, and recommendations for ecological restoration and fuels management. Front Range Fuels Treatment Partnership Roundtable, findings of the Ecology Workgroup. www.frftp.org/roundtable/pipo.pdf.

57. Colorado Climate Center (2013) http://climate.atmos.colostate.edu/. Accessed 2014 August 23.

58. Graham RT (2003) Hayman Fire Case Study. Gen. Tech. Rep. RMRS-GTR-114. Ogden, UT: U.S. Department of Agriculture, Forest Service, Rocky Mountain Research Station. 396 p.

59. National Fire Protection Association Codes and Standards (1989) www.nfpa.org. Accessed 2014 August 23.

60. Veblen TT, Kitzberger T, Donnegan J (2000) Climatic and Human Influence on Fire Regimes in Ponderosa Pine Forests in the Colorado Front Range. Ecological Applications 10: 1178–1195.

61. Sherriff RL, Veblen TT (2008) Variability in fire–climate relationships in ponderosa pine forests in the Colorado Front Range. International Journal of Wildland Fire 17: 50–59.

62. Gartner MH, Veblen TT, Sherriff RL, Schoennagel TL (2012) Proximity to grasslands influences fire frequency and sensitivity to climate variability in ponderosa pine forests of the Colorado Front Range. International Journal of Wildland Fire 21: 562–571.

63. Duncan (1989) An evaluation of errors in tree age estimates based on increment cores in Kahikatea (Dacrycarpus dacrydioides). New Zealand Natural Sciences Journal 16: 31–37.

64. Barrett SW, Arno SE (1998) Increment-borer methods for determining history in coniferous forests. USDA Forest Service, General Technical Report INT-244.

65. Agee JK (1993) Fire Ecology of the Pacific Northwest Forest. Island Press, Washington DC. pp. 493.

66. Hessburg P, Salter RB, James K (2007) Re-examining fire severity relations in pre-management era mixed conifer forests: inferences from landscape patterns of forest structure. Landscape Ecology 22: 5–24.

67. Wilson JP, Gallant JC (2000) Terrain analysis: principles and applications. New York: Wiley.

68. Jenson SK, Domingue JO (1988) Extracting topographic structure from digital elevation data for geographic information system analysis. Photogramm Eng Remote Sens 54: 1593–1600.

69. Breiman L, Friedman JH, Olshen RA, Stone CJ (1984) Classification and Regression Trees. Monterrey: Wadsworth and Brooks/Cole. 358 p.

70. Breiman L (2001) Random Forests. Machine Learning 45: 5–32.

71. Moisen GG, Frescino TS (2002) Comparing five modelling techniques for predicting forest characteristics. Ecological Modelling 157: 209–225.

72. Qian SS (2010) Environmental and Ecological Statistics. Boca Raton: RC Press.

73. Eidenshink J, Schwind B, Brewer K, Zhu Z, Quayle B, et al. (2007) A project for monitoring trends in burn severity. Fire Ecology 3: 3–21.

74. Scott JH, Burgan RE (2005) Standard fire behavior fuel models: A comprehensive set for use with Rothermel's surface fire spread model. In: Fort Collins, CO: USDA Forest Service Rocky Mountain Research Station, General Technical Report RMRS-GTR-153. 72 p.

75. LANDFIRE (2010) U.S. Department of Agriculture, Forest Service, U.S. Department of Interior. http://www.landfire.gov/index.php. Accessed 2014 August 23.

76. Krasnow K, Schoennagel T, Veblen TT (2009) Forest fuel mapping and evaluation of LANDFIRE fuel maps in Boulder County, Colorado, USA. Forest Ecology and Management 257: 1603–1612.

77. Finney MA (2006) An overview of FlamMap fire modeling capabilities. In: Fuels management-how to measure success: conference proceedings. 2006 March 28–30; Portland, Oregon. Proceedings RMRS-P-41. Fort Collins, CO: U.S. Department of Agriculture, Forest Service, Rocky Mountain Research Station: 213–220.

78. Platt RV, Schoennagel T, Veblen TT, Sherriff RL (2011) Modeling wildfire potential in residential parcels: A case study of the north-central Colorado Front Range. Landscape and Urban Planning 102: 117–126.

79. Rothermel RC (1972) A mathematical model for predicting fire spread in wildland fuels. Ogden, UT: USDA Forest Service Intermountain Forest and Range Experiment Station, General technical report INT-115.

80. Van Wagner CE (1977) Conditions for the start and spread of crown fire. Canadian Journal of Forest Research 7: 23–34.

81. Rothermel RC (1991) Predicting behavior and size of crown fires in the Northern Rocky Mountains. Ogden, UT: USDA Forest Service Intermountain Forest and Range Experiment Station, Research paper INT-438. 46 p.

82. Nelson RM (2000) Prediction of diurnal change in 10-hour fuel moisture content. Canadian Journal of Forest Research 31: 1071–1087.

83. Scott JH, Reinhardt ED (2001) Assessing crown fire potential by linking models of surface and crown fire behavior. In: Fort Collins CO: U.S. Department of Agriculture, Forest Service, Rocky Mountain Research Station, Research Paper RMRS-RP-29. 59 p.

84. Intergovernmental Panel on Climate Change (2007) https://www.ipcc.ch/publications_and_data/publications_and_data_reports.shtml. Accessed 2014 August 23.

85. NCAR Science and Research Data for IPCC AR4 Data Guide (2012) http://eos-earthdata.sr.unh.edu/data/dataGuides/ci_science_dg.pdf. Accessed 2014 August 23.

86. Sherriff RL (2004) The historic range of variability of ponderosa pine in the northern Colorado Front Range: past fire types and fire effects. Ph.D. Dissertation. University of Colorado, Boulder, Colorado.

87. Rocca ME, Brown PM, MacDonald LH, Carrico CM (2014) Climate change impacts on fire regimes and key ecosystem services in Rocky Mountain forests. Forest Ecology and Management 327: 290–305.

88. Miller JD, Thode AE (2007) Quantifying burn severity in a heterogeneous landscape with a relative version of the delta Normalized Burn Ratio (dNBR). Remote Sensing of Environment 109: 66–80.

89. Soverel NO, Perrakis DDB, Coops NC (2010) Estimating burn severity from Landsat dNBR and RdNBR indices across western Canada. Remote Sensing of Environment 114: 1896–1909.

90. Front Range Fuels Treatment Partnership (2006) Living with Fire: Protecting Communities and Restoring Forests – Findings and Recommendations of the Front Range Fuels Treatment Partnership Roundtable. http://frftp.org/roundtable/report.pdf. Accessed 2014 August 23.

91. Schoennagel T, Nelson CR (2011) Restoration relevance of recent National Fire Plan treatments in forests of the western United States. Frontiers in Ecology and the Environment 9: 271–277.

92. The National Cohesive Wildland Fire Management Strategy and Risk Analysis – Phase III Report (2013) http://www.forestsandrangelands.gov/. Accessed 2014 August 23.

93. Platt RV, Veblen TT, Sherriff RL (2006) Are Wildfire Mitigation and Restoration of Historic Forest Structure Compatible? A Spatial Modeling Assessment. Annals of the Association of American Geographers 96: 455–470.

94. Finney MA, Bartlette R, Bradshaw L, Close K, Collins BM, et al. (2003) Fire behavior, fuel treatments, and fire suppression on the Hayman Fire. In Hayman Fire case study. Ed. Graham RT. USDA Forest Service, Rocky Mountain Research Station, General Technical Report RMRS-GTR-114, pp. 33–179.

95. Theobald DM, Romme WH (2007) Expansion of the US wildland-urban interface. Landscape and Urban Planning 83: 340–354.

96. Arno SF, Scott JH, Hartwell MG (1995) Age-class structure of old growth ponderosa pine/Douglas-fir stands and its relationship to fire history. USDA Forest Service, Research Paper INT-481.

97. Wright CS, Agee JK (2004) Fire and vegetation history in the eastern Cascade mountains. Ecological Applications 14: 443–459.

98. Everett RL, Schellhaas R, Keenum D, Spurbeck D, Ohlson P (2000) Fire history in the ponderosa pine/Douglas-fir forests on the east slope of Washington Cascades. Forest Ecology and Management 129: 207–225.

99. Hessburg PF, Salter RB, James KM (2007) Re-examining fire severity relations in pre-management era mixed conifer forests: inferences from landscape patterns of forest structure. Landscape Ecology 22: 5–24.

100. Perry DA, Hessburg PF, Skinner CN, Spies TA, Stephens SL, et al. (2011) The ecology of mixed severity fire regimes in Washington, Oregon, and northern California. Forest Ecology and Management 262: 703–717.

Vertical Profiles of Soil Water Content as Influenced by Environmental Factors in a Small Catchment on the Hilly-Gully Loess Plateau

Bing Wang[1], Fenxiang Wen[1], Jiangtao Wu[1], Xiaojun Wang[1]*, Yani Hu[2]

1 College of Environmental Science and Resources, Shanxi University, Taiyuan, China, **2** Library, Hebei University of Science and Technology, Shijiazhuang, China

Abstract

Characterization of soil water content (SWC) profiles at catchment scale has profound implications for understanding hydrological processes of the terrestrial water cycle, thereby contributing to sustainable water management and ecological restoration in arid and semi-arid regions. This study described the vertical profiles of SWC at the small catchment scale on the hilly and gully Loess Plateau in Northeast China, and evaluated the influences of selected environmental factors (land-use type, topography and landform) on average SWC within 300 cm depth. Soils were sampled from 101 points across a small catchment before and after the rainy season. Cluster analysis showed that soil profiles with high-level SWC in a stable trend (from top to bottom) were most commonly present in the catchment, especially in the gully related to terrace. Woodland soil profiles had low-level SWC with vertical variations in a descending or stable trend. Most abandoned farmland and grassland soil profiles had medium-level SWC with vertical variations in varying trends. No soil profiles had low-level SWC with vertical variations in an ascending trend. Multi-regression analysis showed that average SWC was significantly affected by land-use type in different soil layers (0–20, 20–160, and 160–300 cm), generally in descending order of terrace, abandoned farmland, grassland, and woodland. There was a significant negative correlation between average SWC and gradient along the whole profile ($P < 0.05$). Landform significantly affected SWC in the surface soil layer (0–20 cm) before the rainy season but throughout the whole profile after the rainy season, with lower levels on the ridge than in the gully. Altitude only strongly affected SWC after the rainy season. The results indicated that land-use type, gradient, landform, and altitude should be considered in spatial SWC estimation and sustainable water management in these small catchments on the Loess Plateau as well as in other complex terrains with similar settings.

Editor: Andrew C. Singer, NERC Centre for Ecology & Hydrology, United Kingdom

Funding: This work was supported by the Natural Science Foundation of China (No. 41201277 and 41101025) and the Natural Science Foundation of Shanxi Province of China (No. 2014011034-2). The funders had no role in study design, data collection and analysis, decision to publish, or preparation of the manuscript.

Competing Interests: The authors have declared that no competing interests exist.

* Email: xjwang@sxu.edu.cn

Introduction

Soil water content (SWC) is a critical factor for plant growth and a determinant of plant distribution in arid and semiarid areas such as China's Loess Plateau [1, 2]. Vertical distribution of SWC can greatly affect soil water movement [3], thereby greatly affecting the biomass production and water use efficiency of plants (e.g., switchgrass) under water stress [4]. Plant-available water stored in the soil profile has a buffering capacity, which, in deep layers, prolongs or alleviates the effects of seasonal or inter-annual drought on plant growth and soil water flux to the atmosphere [5–7]. Research has provided strong evidence that deep soil water depletion plays a key role in sustainable agriculture, ecological restoration, and terrestrial water cycling on the Loess Plateau [8–10]. However, measurement of SWC profiles has been frequently conducted at different spatial scales. The results thus need to be converted before comparison analysis or practical uses. The SWC profile in small catchment is considered to be at a moderate scale for data exchanging. In particular, small catchment is thought to be the basic unit for integrated soil and water loss management in

complicated terrain of the Loess Plateau [11, 12]. Characterization of SWC profiles and evaluation of relevant influencing factors at the small catchment scale have implications for hydrological modeling of soil water dynamics and sustainable management of soil water resources in similar areas.

Classical statistics is frequently used to analyze the variability of SWC profiles at the small catchment scale, which involves the estimation of descriptive parameters such as average (mean), variance, standard deviation (STD), and coefficient of variation (CV). Average SWC at individual soil depth intervals or across the whole soil profile is extensively determined. The CV of SWC is also routinely calculated as the temporal variable in a certain period of time or the spatial variable across a specific area. The SWC profile can be divided into distinct intervals by considering its average and CV which exhibit complex spatial-temporal relationships in several plots or watersheds [13–16]. Additionally, ranking method, clustering method, and semivariogram model have been applied for the division of SWC profile [3, 17–20]. However, the above-mentioned methods cannot clearly reflect the variation trend in SWC profiles. Thus, great effort has been made

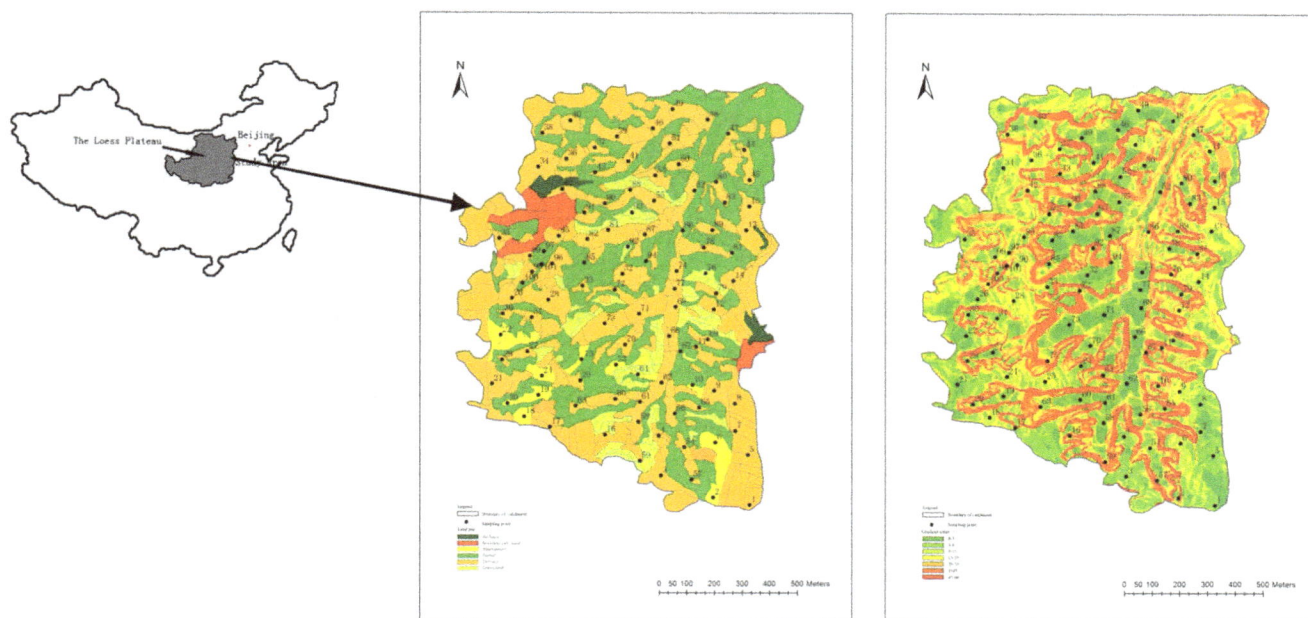

Figure 1. Location of Sanyanjing catchment and distribution of 101 sampling points in the Sanyanjing catchment.

to describe the vertical profiles of SWC through comparing variation curves or variation ranges, in small watersheds related to different land-use types, vegetation species, and/or terrain factors [3, 17, 18, 20–22]. If a massive sample size is involved, however, it becomes difficult to distinguish the vertical profiles and major influencing factors of SWC by direct comparisons.

In recent decades, a great number of studies have been conducted on the spatiotemporal variability of SWC and related influencing factors worldwide. Canton [23] pointed out that wasteland-scale spatial variability of SWC is mainly controlled by surface cover and soil properties in a semi-arid region of Spain, where surface cover counteracts the influence of terrain factors (including gradient, aspect, topographic wetness index, and distance from the river) on SWC distribution. Burnt [24] reported a topographic index which can simulate changes in high-level SWC in a humid climate zone of Devin County, UK. O'loughlin [25] estimated the spatial pattern of SWC distribution in a small catchment using humidity index model based on digital terrain dataset. Hawley [15] discovered that topography is the major factor responsible for the spatial distribution of SWC in an agricultural region, where resultant SWC variation is diminished by vegetation in a moist climate zone in Chickasha, USA. In arid and semi-arid areas, catchment-scale distribution of SWC is strongly affected by land-use/vegetation and topographic indices, e.g., land-use type, soil organic matter content, tillage, soil physical properties, gradient, and aspects [6, 17–19, 21, 26].

Many researchers have focused on the quantification of environmental parameters such as topographic factor, vegetation type, soil texture, and land-use type, in attempt to evaluate their impacts on the variability of SWC. At the small catchment scale, little information is available on the major factors affecting vertical profiles of SWC in cinnamon soil (Haplic Lixisols, FAO) zone on the hilly and gully Loess Plateau [3, 19]. As a regional water reservoir experiencing depletion, the plateau region requires measurements and characterization of deep SWC profiles for the thick soil layer. However, soil sampling for SWC profile analysis at the catchment scale has been commonly conducted at <200 cm

depth [18–20, 23]. Wang et al. [21] exceptionally examined SWC along the 0–21 m soil profile on the Loess Plateau, but the reliability of their tests might be affected by a small sample size (11 sites). Deep soil sampling at a larger number of sites and statistical analysis of parameters involving soil depth information will contribute to better understanding of the vertical profiles and influencing factors of SWC.

In the present study, we characterized the vertical profiles of SWC in a small catchment on the Loess Plateau by cluster analysis of two descriptive parameters (mean and regression gradient). Sampling was carried out in the 0–300 cm profile at 101 points throughout the catchment before and after the rainy season, to meet the demand for deep depth, spatial representativeness and temporal comparability. The influences of selected environmental factors (land-use type, topographic factors, and landform) on average SWC were examined by multi-linear regression [6, 26–32]. The results were discussed in order to provide new insights to the vertical profiles and influencing factors of SWC on the Loess Plateau, further providing reference data for sustainable management of water resource in small catchment areas in the semi-arid region with complex terrain.

Material and Methods

1 Site description

This study was conducted in Sanyanjing catchment ($112°2'13''$, $37°46'23''$), which is located on the east margin of the Loess Plateau in Shouyang county, mid-east Shanxi province, China (Figure 1). The catchment has a total area of 1.32 km^2 and the elevation ranges from 1001 to 1160 m. It is a hilly and gully area with mostly deep gully erosion slopes. The landform consists of ridge and gully.

The catchment area has a semi-arid continental climate (Cwa by Koppen Climate Classification) with an average annual precipitation of 474.2 mm (1967–1999). Snow in the winter accounts for ~8% and rainfall in July to September for ~73% of annual precipitation. Monthly average precipitation, potential evapotranspiration and precipitation in 2013 are shown in Figure 2. Annual mean temperature in this area is 8.1°C, with a maximum of 34.7°C and a minimum −20.6°C. The soil type is cinnamon soil (Haplic lixisols, FAO), which consists of 54–62% silt and 10.95–30.15% sand with the bulk density of 1.3–1.4 g/cm³. Soil texture was measured using a particle size analyzer (SEDIMAT 4–12, UGT, Germany). Soil bulk density was determined through sampling with cutting rings (inner diameter 5.0 cm, volume 100 cm³) and drying in an oven (105°C, 24 h). The profile of soil texture related to different landforms is listed in Table 1. Maximum soil depth is mostly down to 300 cm on the ridge and bare rock could be rarely seen only on the northern margin of the gully area.

The distribution of land-use types across the catchment is shown in Figure 1. Terrace is the dominant land-use type, accounting for about 60% of the study area. Few terraces had been abandoned for natural restoration of vegetation because of the Grain for Green project since 2000. Grassland is mainly covered with herbs and semi-shrubs, which had never been reclaimed for several decades. About 80% of the woodland is covered with semi-shrubs at steep slopes unsuitable for sampling.

2 Soil sampling

Ethics statement. Sampling activities at the farmland were allowed by the owners. No specific permissions were required at other locations because they were not privately-owned or protected in any way and the field activities did not involve any endangered or protected species.

A total of 101 sampling points were designed in a 150 m×150 m grid throughout the catchment area by considering major land-use types, including terrace (83), abandoned farmland (9), grassland (3), and woodland (6). Soil sampling was carried out during two periods in 2013, from April 29 to May 4 (before the rainy season) and from October 28 to November 1 (after the rainy season). No precipitation occurred during the two sampling periods or a week before sampling. Each sample was taken at 20 cm intervals along the 0–300 cm soil profile using an auger (inner diameter 5.0 cm). The samples were kept in capped aluminum boxes for transportation. Measurement of SWC was conducted using an oven-drying method (105°C, 24 h). At the majority of the sampling points, soils were collected along a vertical profile over 300 cm. A few exceptions were in the north of the gully at the lowest altitude where weathered rock was occasionally encountered. Background information of the 101 sampling points is summarized in Table 2.

3 Data analysis

To identify the variability of SWC profiles at the catchment scale, descriptive parameters were calculated for each profile. Further, we calculated the linear regression coefficient (K value) between SWC and soil depth to represent the variation trend of SWC vertical profiles and the mean value to describe the average level of SWC along the 0–300 cm profile.

The SWC profiles were classified using a combined cluster analysis of the K and mean values. Cluster analysis is the process of grouping a set to data objects into multiple groups (or clusters), so that objects within a cluster share high similarity but are dissimilar to those in other clusters. In this approach, dissimilarities and similarities are assessed based on the attribute values describing the objects and often involve distance measures. Cluster analysis is a statistical classification method for discovering whether the individuals of a population gall into different groups by making quantitative comparisons of multiple characteristics [34]. Here a combined cluster analysis was conducted in three steps: 1) cluster of the mean to three groups, which present the average level of SWC along the vertical soil profile, 2) cluster of K to three groups, which reflect the variation trend of SWC profiles (top to bottom), and 3) combination of the two sets of groups into nine new groups using the between-groups linkage method with squared Euclidean distance criteria [34].

For group comparisons, SWC profiles (0–300 cm) of the same group were averaged and re-plotted. The average curves of SWC were compared between groups to identify the major factors influencing SWC in individual soil layers (0–20, 20–160, and 160–300 cm). On the basis of cluster analysis, the influences of land-use type, topography and landform on average SWC in individual soil layers were examined by multi-regression analysis. The independent variables were land-use type, landform type, Sin(gradient), Sin(aspect), flow accumulation (calculated cell numbers to a grid cell from surrounding cells with the ArcGIS hydrology analysis

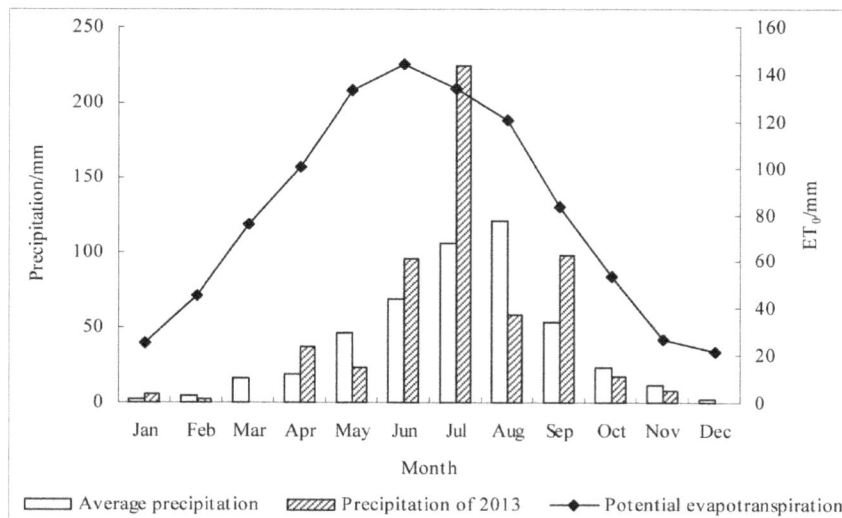

Figure 2. Average annual precipitation and potential evapotranspiration of 1967–1999 and precipitation in 2013 in the Sanyanjing catchment, Shanxi province, China.

Table 1. Soil texture in vertical profiles related to different landforms in the Sanyanjing catchment in Shanxi province, China.

Point description	Soil depth (cm)	Sand (%)		Silt (%)		Clay (%)
		>0.05 mm (%)	0.05–0.02 mm (%)	0.02–0.0063 mm (%)	0.0063–0.002 mm (%)	<0.002 mm (%)
	0–20	26.05	30.40	20.60	7.00	15.95
	20–40	21.35	34.80	20.80	6.90	16.15
	40–60	30.25	25.90	21.60	4.50	17.75
	60–80	28.25	28.80	21.70	2.80	18.45
	80–100	18.45	38.00	20.80	6.70	16.05
	100–120	26.45	30.10	20.40	5.20	17.85
Terrace at	120–140	26.95	27.90	22.00	4.40	18.75
ridge	140–160	16.35	32.60	24.20	8.10	18.75
(Point 33)	160–180	27.75	22.40	22.70	7.90	19.25
	180–200	36.55	17.80	20.60	8.00	17.05
	200–220	18.45	29.00	24.30	9.00	19.25
	220–240	20.45	28.50	23.20	8.60	19.25
	240–260	22.05	27.30	23.90	8.80	17.95
	260–280	31.35	20.30	22.20	9.40	16.75
	280–300	24.65	29.80	20.70	5.90	18.95
	0–20	30.15	20.00	23.20	6.40	20.25
	20–40	27.75	21.80	21.90	6.10	22.45
	40–60	25.95	26.80	20.50	4.50	22.25
	60–80	23.85	27.09	21.20	1.20	26.65
	80–100	0.95	31.00	20.10	20.10	27.85
	100–120	10.95	23.40	18.40	18.40	28.85
Terrace at	120–140	7.45	27.30	19.20	19.20	26.85
ridge	140–160	17.45	19.50	18.90	18.90	25.25
(Point 39)	160–180	10.45	25.80	18.20	18.20	27.35
	180–200	15.75	24.70	17.00	17.00	25.55
	200–220	13.35	21.40	18.60	18.60	28.05
	220–240	20.85	8.80	22.00	22.00	26.35
	240–260	20.25	12.40	20.00	20.00	27.35
	260–280	12.15	23.70	19.10	19.10	25.95
	280–300	24.95	29.00	20.50	0.50	25.05

module), and elevation. The former two factors were categorical variables converted into dummy variables before introduced into the regression analysis; and the latter four factors were continuous variables produced using digital elevation model at 1-m resolution.

SWC data were statistically analyzed in SPSS13.0 (SPSS Inc., Chicago, IL, USA), and topographic features were analyzed in ArcGIS 10.0 (ESRI, Redlands, CA, USA). SWC profiles were drawn in Microsoft Excel 2010 (Microsoft Corp., Redmond, WA, USA) and then clustered in SPSS 13.0 by considering descriptive parameters (maximum, minimum, mean, CV, STD, and K). Multi-regression analysis was performed in SPSS 13.0, with a P-value less than 0.05 considered statistically significant.

Results

1 Vertical profiles and descriptive parameters of SWC

The vertical profiles (0–300 cm) of SWC at 101 sampling points before the rainy season were drawn (Figure 3). These SWC profiles showed dynamic variations across the catchment study area, with substantial differences in the soil layers below 100 cm. At a few sampling points, there were obvious soil water depletion (e.g., 67, 75, 85, and 94) and an increasing trend (top to bottom) of SWC (e.g., 12 and 36). High degrees of soil desiccation were rarely detected in the lower soil layers, and low SWC was mainly found in the lower soil layers of woodland.

Descriptive parameters such as maximum, minimum, mean, and CV, STD are commonly used to reveal the spatial-temporal variability of SWC. However, these parameters cannot reflect the variation trend of SWC vertical profiles. To this end, the K value of SWC to profile depth was introduced for quantification of variation trend of SWC vertical profiles (Figure 4). Results showed that before the rainy season, SWC substantially varied between 5.87% and 34.72%, whereas the mean, STD, CV, and K values respectively ranged from 10.57% to 21.76%, 0.47 to 4.53, 3% to 24%, and −0.0405 to 0.0274 along the vertical soil profile (0–

Table 2. Background information of 101 soil sampling points in the Sanyanjing catchment study area in Shanxi province, China.

Land-use type	Vegetation	Landform type	Soil profile/cm	Sampling points
Terrace (n = 83)	Maize	Ridge	300	1, 3–5, 7–9, 12–18, 21–22, 25–26, 28, 32–35, 37, 51, 53, 91, 95, 96, 98
			260	99
		Gully	300	29, 30, 38–46, 48–50, 54–60, 62, 63, 65, 66, 68–74, 77–84, 87–90
			280	47
			260	31, 61
			220	52
			160	86
	Millet	Ridge	300	11
	Maize +five-year-walnut	Gully	300	27
Abandoned farmland	Subshrubs + herbs	Ridge	300	6, 10, 23
(n = 9)			240	2
	Subshrubs + herbs + few ulmus pumila	Gully	220	19
	Robinia peseudoacacia + subshrubs + herbs	Ridge	300	24
	Robinia peseudoacacia+ subshrubs + herbs	Gully	300	64
	Herbs + few almond-apricot	Gully	300	20
	Poplar + subshrubs +herbs	Gully	300	75
Grassland	Subshrubs +herbs	Ridge	300	92, 93
(n = 3)		Gully	300	76
Woodland	Poplar	Gully	300	97
(n = 6)			280	85
	Poplar + herbs	Gully	140	101
	Poplar + subshrubs	Gully	280	100
	Poplar + subshrubs + herbs	Gully	300	67, 94

doi:10.1371/journal.pone.0109546.t002

300 cm). The ranges of the parameters after the rainy season were generally similar with those before the rainy season.

According to the division criteria of Nielson [24], CV in the range of 10–100% indicates moderate variability. Thus, the vertical variability of SWC at all sampling points in Sanyanjing catchment (Figure 4) can be classified to the medium degree. K is the linear regression coefficient between SWC and soil depth. A positive value of K indicates that SWC increases with increasing soil depth. Inversely, a negative value of K indicates that SWC decreases with increasing soil depth. The positive and negative K values of SWC data (Figure 4) are indicative of different variation trends of SWC vertical profiles in the catchment.

2 Clustering of SWC profiles

The vertical profile of SWC across the catchment can be described more clearly using cluster analysis. The 101 SWC profiles before the rainy season were classified into the first three groups by considering the mean value of SWC (Figure 5a), and the second three groups by considering the K value of SWC to soil depth (Figure 5b). The mean value of SWC ranged from 10.57% to 13.13% (low level), 14.15% to 16.86% (medium level), and 17.13% to 21.76% (high level) in the first three groups, whereas the K value of SWC to soil depth ranged from −0.0405 to 0.0106 (decreasing trend), 0.0144 to 0.0163 (stable trend), and 0.0194 to 0.0274 (increasing trend) in the second three groups. By

combining the two cluster series, we obtained nine groups of SWC profiles (Table 3).

Before the rainy season, vertical profiles of SWC in groups 1–3 featured low-level SWC (Table 3). In group 1, SWC decreased along the vertical profile (0–300 cm) in woodland (2) and abandoned farmland (1) located in the gully area. In group 2, SWC remained stable along the vertical profile in terrace (2) and grassland (1) located on the ridge as well as woodland (1) located in the gully (Table 4). No sampling points were classified into group 3 with increasing SWC along the vertical profile.

Vertical profiles of SWC in groups 4–6 featured medium-level SWC (Table 3). In group 4, SWC decreased along the vertical profile on the ridge related to terrace (4) as well as in the gully related to abandoned farmland (1) and woodland (1) located in the gully. In group 5, SWC remained stable along the vertical profile in terrace (7) mostly located in the gully, few terrace (3) and abandoned farmland (1) located on the ridge, and grassland (1) and woodland (1) located in the gully. In group 6, SWC increased along the vertical profile in terrace (2) located on the ridge (Table 4).

Vertical profiles of SWC in groups 7–9 featured high-level SWC (Table 3). In group 7, SWC decreased along the vertical profile in terrace (5), grassland (1), and abandoned farmland (1) located on the ridge, and terrace (2), abandoned farmland (1) and woodland (1) located in the gully. In group 8, SWC remained stable along

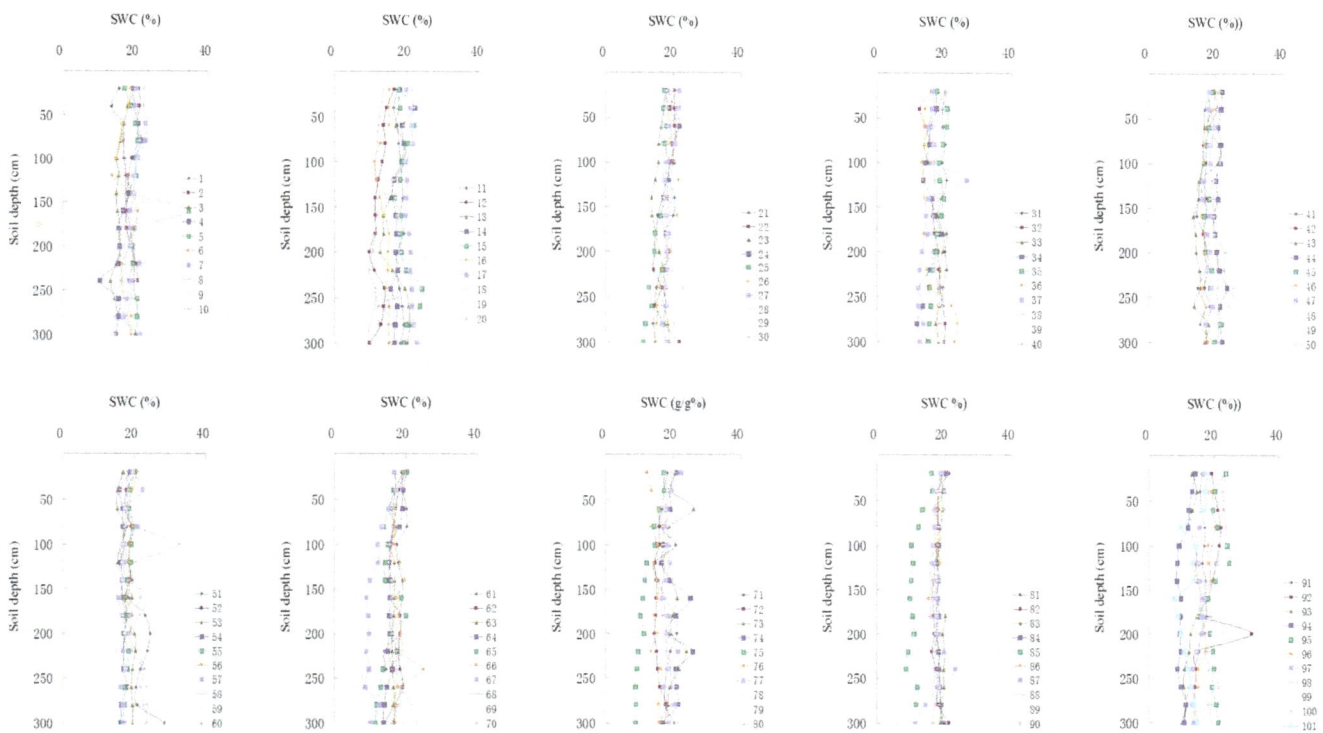

Figure 3. Vertical profiles of soil water at 101 sampling points in the Sanyanjing catchment.

the vertical profile at up to 58 of 101 sampling points, far more than other groups. Most sampling points of group 2 were located in the gully related to terrace (41), and few were on the ridge related to terrace (14) and abandoned farmland (3). In group 9, SWC increased along the vertical profile in terrace located on the ridge (3) and in the gully (1) (Table 4).

Similar grouping of SWC profiles was obtained with data collected after the rainy season. Overall, soil profiles of group 8 with high-level SWC in a stable trend were most commonly present in the catchment, more after the rainy season than before the rainy season. Group 3 of SWC profiles with low level and increasing trend was absent in the study area.

3 The relationships between average SWC and selected environmental factors

According to cluster analysis, there were nine combinations of SWC profiles in terms of average level and variation trend. We averaged SWC profiles of the same group and plotted the average curves (Figure 6), to examine differences of SWC profiles among various types. From Figure 4, we divided the whole soil profile (0–30 cm) into three layers (0–20, 20–160, and 160–300 cm) for multiple linear regression analysis. The results showed that selected environmental factors had significant linear correlations with average SWC at individual layers of 0–20 cm ($P<0.001$, $R^2 = 0.30$; $P<0.001$, $R^2 = 0.37$), 20–160 cm ($P = 0.01$, $R^2 = 0.19$; $P<0.001$, $R^2 = 0.39$), and 160–300 cm ($P<0.001$, $R^2 = 0.32$; $P<0.001$, $R^2 = 0.43$; Table 5).

Before the rainy season, average SWC in the lower soil layer (10–20 cm) was significantly lower in grassland and woodland than in terrace, with no significant difference between abandoned farmland and terrace ($P_{D51} = 0.109$, $P_{D52} = 0.003$, $P_{D53} = 0.047$, $P_{X1} = 0.013$, and $P_{D61} = 0.005$; Table 5). Additionally, average

SWC decreased with increasing gradient, with higher levels on the ridge than in the gully.

In the lower soil layer (20–160 cm), average SWC decreased significantly with increasing gradient ($P = 0.026$; Table 5), and was significantly lower in woodland than in the other three types of land-use types, with no significant differences among the latter three types. Other environmental factors had no significant linear correlation with average SWC ($P>0.05$).

In the deeper soil layer (160–300 cm), there also existed a significantly negative correlation between Sin(gradient) and average SWC ($P = 0.001$; Table 5). Average SWC obviously increased with increasing gradient and was significantly higher in terrace than in abandoned farmland, grassland, and woodland (in descending order).

Similar results can be seen in the data collected after the rainy season. That is, land-use type was the major environmental factor affecting average SWC, whereas landform and altitude strongly affected average SWC only in specific periods and soil layers (Table 5). In the whole vertical profile (0–300 cm), SWC occurred at high levels from upper to deeper layers in terrace, with the lowest level in woodland. Compared with data of terrace, average SWC was relatively low in grassland and woodland in the top (0–20 cm) and deeper soil layers (160–300 cm), with significantly low levels in abandoned farmland soils in the deeper layer only. In the deeper soil layer (160–300 cm), average SWC varied with different land-use types in descending order of terrace > abandoned farmland > grassland > woodland.

Discussion

1 Vertical profiles of SWC at the catchment scale

According to Wang [21], the variability of SWC (as indicated by the CV) varies notably across the whole Loess Plateau, i.e., 15% in

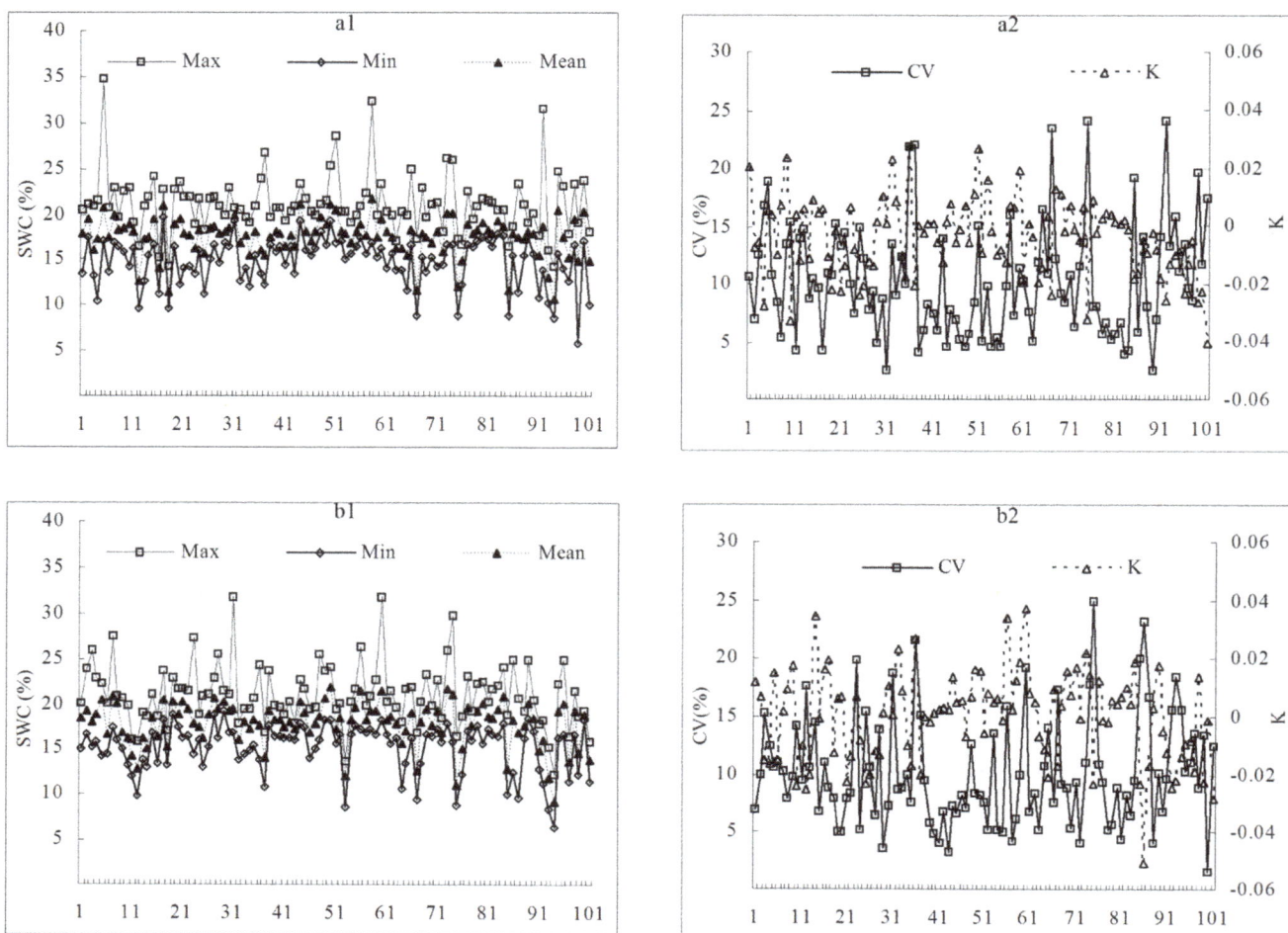

Figure 4. Statistical parameters of soil water content at 101 sampling points across the Sanyanjing catchment. (a. before the rainy season; and b. after the rainy season.)

Changwu and 55% in Shenmu. In the small catchment of Sanyanjing, SWC profiles exhibited weak and medium degrees of variability at 0–300 cm depth [33], with CV in the range of 3–

24% (Figure 2). The lower variability of SWC profiles in our study area may be related to the higher SWC levels across the catchment (Pearson correlation coefficient between average SWC and CV,

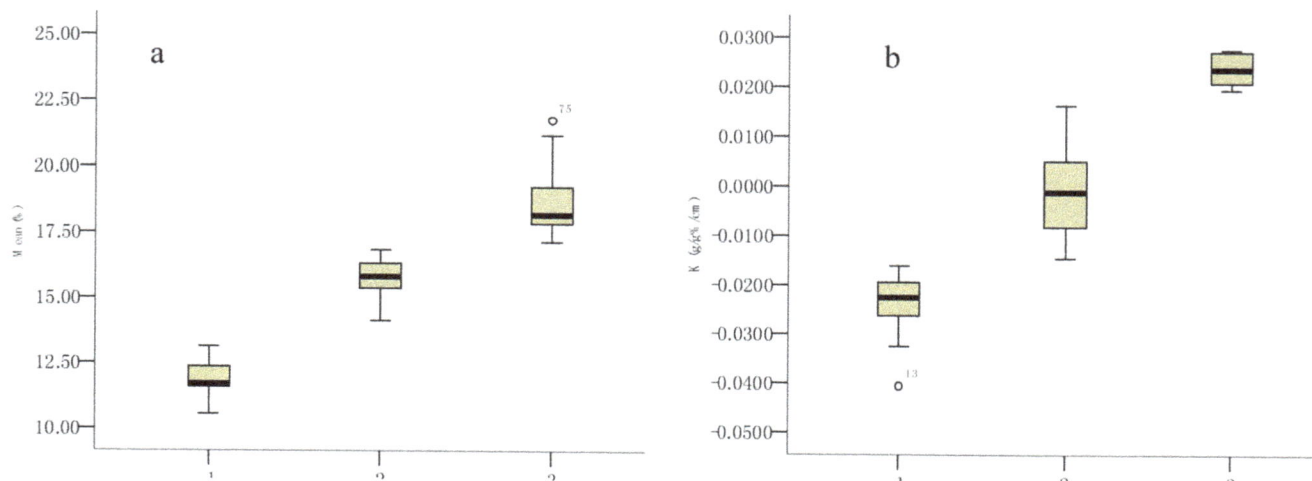

Figure 5. Grouping of 101 vertical soil water profiles in the Sanyanjing catchment before the rainy season by cluster analysis of the mean value (a) and regression gradient (K, b).

Table 3. Combined grouping of 101 vertical profiles of soil water content (0–300 cm) in Sanyanjing catchment by cluster analysis of the mean value and regression gradient.

Cluster by mean	Cluster by K	Combined grouping	Quantity of points Before the rainy season	Quantity of points After the rainy season	Point Nos. Before the rainy season	Point Nos. After the rainy season
1	1	1	3	0	65,67, 85	-
	2	2	4	1	12,18,93,94	94
	3	3	0	0	-	-
2	1	4	6	15	25,37,64,91,99,101	11,12,18,37,53,64,67,75,76,85,91,93,97,99,101
	2	5	13	0	3,16,23,34,41,43,56,63,65,70,72,76,97	3,16,23,34,41, 43,56,63,65,70,72,76,97
	3	6	2	1	32,36	14
3	1	7	11	1	4,10,19,21,26,35,61,86,92,96,100	86
	2	8	58	80	2,5–8,11,13–15,17, 20,22,24,27–31,33, 38–40,42,44–50,52, 53,55,56–59,62,66, 68,69,71,73,74, 77–84,87–90,95,98	2–10,13,15,16, 17,19–36,38–52,54,55,57–59, 61–63,65,66,68,69–74, 77–84, 87–90,92,95,96,98,100
	3	9	4	3	1,9,51,60	1, 51,60

Table 4. Grouping richness of 101 vertical profiles of SWC (0–300 cm) in relation to different land-use types in Sanyanjing catchment before the rainy season.

Land-use type	Grouping 1	2	3	4	5	6	7	8	9
Terrace	0	2	0	4	10	2	7	54	4
Abandoned	1	0	0	1	1	0	2	0	0
Grassland	0	1	0	0	1	0	1	0	0
Woodland	2	1	0	1	1	0	1	0	0

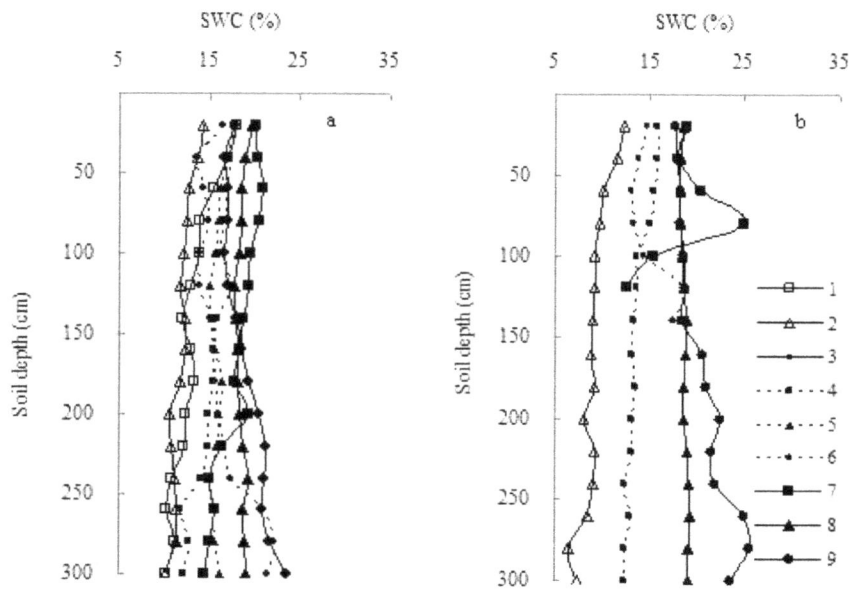

Figure 6. Vertical soil water profiles in relation to different groups in the Sanyanjing catchment study area before and after the rainy season (a. before the rainy season; and b. after the rainy season).

−0.40; $P<0.01$). Qiu [17] found that wetter soil with greater vertical variations in an increasing trend along the SWC profile (mean 13.03%; and STD, 2.3%) is representative in a dry year in Danangou catchment on the Loess Plateau, where the land-use pattern (including slope farmland, terrace, and orchard) differs from that in our study area.

Cluster analysis of the mean and K values provides a clear description for the overall variability of SWC in the vertical profile. Based on combined grouping, the 101 vertical SWC profiles were classified into nine groups with high, medium, and low levels associated with increasing, stable, and decreasing trends (Table 3). More than half of the SWC profiles were obtained from terrace soils in the gully and classified into group 8 (58/101 before the rainy season and 80/101 after the rainy season) with high-level SWC in a stable trend (Tables 3, 4). Despite that all sampling points of woodland were also located in the gully, their average SWC remained the lowest among different land-use types and mostly descended along the vertical profile (Table 4). The above differences can be attributed to the lower soil water consumption by maize crop in the terrace, which generally has shallower root distribution and less above-ground biomass than trees in the woodland. Our observations coincide with previous findings on the Loess Plateau that soil water conditions of terrace, gully farmland, and dam land are better than that of artificial woodland. The latter land-use type is associated with soil desiccation, especially in deep soil layers [2, 22, 35, 36].

Although the cluster analysis divided vertical SWC profiles into nine groups, only eight types were present in the Sanyanjing catchment and no sampling points were classified into group 3 (i.e., low-level SWC with an increasing trend from top to bottom). According to previous research in semi-arid regions, if SWC occurs at low level in the upper soil layer, deep-root crops, shrubs, and trees will consume more soil water in the deeper soil layers through root extraction [37–40]. Additionally, it is hard to achieve soil water recharge in the deeper soil layers by precipitation infiltration because the depth of soil water infiltration is shallow. Therefore, soil desiccation exists in the lower soil layer in case of

no groundwater recharge [8–10]. These mechanisms explain the absence of high-level SWC with an increasing trend along the vertical profile in the small catchment of Sanyanjing (Table 3).

2 Effects of environmental factors on average SWC at the catchment scale

Consistent with cluster analysis (Table 3), multiple regression analysis showed that land-use type had a significant effect on soil water status in the small catchment of Sanyanjing (Table 4, 5). This result coincides with the data previously reported in small catchments on the Loess Plateau [17, 18, 22, 23, 29]. For example, Zhang [23] concluded that average SWC (20–200 cm) descends with different land-use types (farmland > grassland > shrub land > and woodland, n = 80) in the small catchment of Zhifanggou. Bai [42] found that average SWC (0–500 cm) ranges from of 9% to 16% in orchard, gradient farmland, terrace, and grassland, but remains less than 10% in shrub land and most woodland (n = 91) in Nangou catchment in the central area of Loess Plateau, Ansai, Shaanxi. The consistency of the data demonstrates that cluster analysis is a reliable method for characterization of SWC profiles.

The effect of land-use type on SWC can be related to the differences existing in anthropogenic activity and vegetation type [22]. Average SWC was found significantly higher in terrace and abandoned farmland than in grassland and woodland along the 0–300 cm profile (Table 5). Abandoned farmland and terrace are associated with artificial tillage in the surface soil layer, which improves soil porosity and loosens soil structure, further enhancing soil water infiltration [22, 43]. Additionally, soil water consumption by crops is less than that in grassland and woodland due to lower leaf area index [21], contributing to the accumulation of SWC. The above mechanisms account for the greater average of SWC profiles with a stable trend to soil depth in terrace and abandoned farmland.

Difference in root distribution is another factor contributing the effect of land–use type on SWC [40]. In the Sanyanjing catchment, average SWC of woodland was higher in the 0–20 cm soil layer but lower in the 20–160 and 160–300 cm soil

Table 5. Multi-linear regression analysis of soil water content and selected environmental factors in three layers (0–20, 20–160, and 160–300 cm) of the vertical soil profile in the Sanyanjing catchment study area.

Before the rainy season

Model	Y$_1$ Unstandardized coefficients B	Y$_1$ Standardized coefficients (Beta)	Y$_1$ Sig.	Y$_2$ Unstandardized coefficients B	Y$_2$ Standardized coefficients (Beta)	Y$_2$ Sig.	Y$_3$ Unstandardized coefficients B	Y$_3$ Standardized coefficients (Beta)	Y$_3$ Sig.
Constant	14.714		0.066	6.031		0.498	18.514		0.069
X$_1$	−2.743	−0.23	0.013	−2.751	−0.22	0.026	−4.471	−0.296	0.001
X$_2$	−0.112	−0.037	0.68	0.042	0.013	0.889	−0.261	−0.068	0.449
X$_3$	4.86E−06	0.025	0.783	−7.80E−06	−0.038	0.694	1.20E−05	−0.05	0.578
X$_4$	0.005	0.081	0.507	0.011	0.177	0.175	0	0.005	0.97
D$_{51}$	1.09	0.146	0.109	0.019	0.002	0.98	−0.788	−0.084	0.354
D$_{52}$	−3.472	−0.278	0.003	−0.845	−0.064	0.503	−2.553	−0.162	0.073
D$_{53}$	−1.715	−0.191	0.047	−3.216	−0.342	0.001	−4.745	−0.384	0
D$_{61}$	−1.512	−0.348	0.005	−0.607	−0.134	0.307	−0.833	−0.151	0.212
	R^2 = 0.30	(P<0.001)		R^2 = 0.19	(P=0.001)		R^2 = 0.32	(P<0.001)	

After the rainy season

Model	Y$_1$ Unstandardized coefficients B	Y$_1$ Standardized coefficients (Beta)	Y$_1$ Sig.	Y$_2$ Unstandardized coefficients B	Y$_2$ Standardized coefficients (Beta)	Y$_2$ Sig.	Y$_3$ Unstandardized coefficients B	Y$_3$ Standardized coefficients (Beta)	Y$_3$ Sig.
Constant	11.426		0.111	−5.811		0.452	−9.045		0.41
X$_1$	−2.01	−0.178	0.041	−2.763	−0.221	0.01	−3.688	−0.206	0.015
X$_2$	0.313	0.11	0.201	−0.132	−0.042	0.618	0.294	0.063	0.44
X$_3$	6.30E−06	−0.034	0.692	1.80E−05	−0.089	0.285	1.60E−05	−0.054	0.508
X$_4$	0.007	0.125	0.278	0.023	0.359	0.002	0.027	0.291	0.009
D$_{51}$	−0.622	−0.088	0.307	−1.161	−0.149	0.08	−1.81	0.162	0.053
D$_{52}$	−2.374	−0.2	0.021	−2.372	−0.181	0.032	−4.166	−0.222	0.008
D$_{53}$	−2.599	−0.305	0.001	−4.807	−0.51	0	−6.862	−0.467	0
D$_{61}$	−2.082	−0.506	0	−1.512	−0.332	0.004	−3.14	−0.476	0
	R^2 = 0.37	(P<0.001)		R^2 = 0.39	(P=0.001)		R^2 = 0.43	(P<0.001)	

Dependent Variable: Y$_1$ (soil water content of 0-20 cm layer).
Y$_2$ (average soil water content of 20-160 cm layer).
Y$_3$ (average soil water content of 160-300 cm layer).
Independent Variables: X$_1$ = Sin(gradient), X$_2$ = Sin(aspect), X$_3$ = flowaccu, X$_4$ = elevation.
Dummy Variables: X$_5$ = terrace, (D$_{51}$, D$_{52}$, D$_{53}$) = (0,0,0); X$_5$ = abandoned farmland, (D$_{51}$, D$_{52}$, D$_{53}$) = (1,0,0).
X$_5$ = grassland, (D$_{51}$, D$_{52}$, D$_{53}$) = (0,1,0); X$_5$ = woodland, (D$_{51}$, D$_{52}$, D$_{53}$) = (0,0,1).
X$_6$ = ridge, (D$_{61}$) = 1; X$_6$ = gully, (D$_{61}$) = 0.
D represents sub-variable; binary variables 0 and 1 for the absence and presence of some land-use type or landform, respectively.

layers than data of grassland (Table 5). The varying trends of SWC profiles between grassland and woodland can be related to different distribution of root system in individual soil layers and stratified root extraction of soil water. The rooting depth of maize crop is reported to be approximately 100 cm and most maize roots are distributed in the soil layer of 0–20 cm, shallower than average rooting depths in grassland (20–60 cm) and woodland (20–100 cm) [40]. Diverse root distribution patterns can lead to different levels of soil water consumption by plants, contributing to great variability of SWC level.

In addition to land-use type, topographic factors strongly affected SWC in the study area (Table 4). This is because the distribution of wind and solar radiation varies with different topographic conditions, leading to different levels of soil evaporation, runoff on gradient, and soil water infiltration [41]. Gradient negatively affected SWC in the soil layers of 0–20, 20–160, and 160–300 cm (Table 5), possibly due to the increased runoff with increasing gradient and resultant reduction of precipitation infiltration [17, 27, 29, 41]. Other topographic factors including aspect and flow accumulation had no significant effects on average SWC in the three soil layers (Table 5). Similarly, Gómez [29] referred that aspect has no obvious influence on SWC in burned and unburned areas. Shi [19] suggested that aspect and catchment area significantly affect SWC during the wet period only, whereas elevation has a significant effect on SWC in arid and humid periods but not in semi-arid and semi-humid periods. In the present study, we found the effect of elevation on SWC of the three layers varying with the period of time and being significant after the rainy season only.

As for the landform type, location of sampling points significantly affected SWC only in the surface layer (0–20 cm) before the rainy season and throughout all the three layers (0–20, 20–160, and 160–300 cm) after the rainy season, with greater values in the gully than on the ridge (Table 5). The effect of landform type on SWC can be related to different levels of soil evaporation as affected by wind strength and solar radiation and soil physical properties. Similarly, Zhang [22] suggested that average SWC descends with different landforms as gully > terrace > slop land > hill top.

Overall, land-use type is the most significant factor affecting SWC while topographic factors and landform type are interacting jointly at the catchment-scale. Because the impact of environ-mental factors on SWC varies in different periods, it is necessary to increase the observation frequency, in order to better understand the spatiotemporal distribution and influencing factors of SWC in the small catchment. Such work will provide reference data for selecting reasonable environmental parameters in catchment scale SWC simulation over different periods of time.

Conclusions

In this study, cluster analysis enables catchment-scale characterization of soil water profiles in terms of average level and variation trend along the vertical profile, allowing for simple and clear interpretation of the results. A total of nine groups of soil water profiles are recognized but those with low-level soil water content and a decreasing trend are not present in the Sanyanjing catchment. Land-use type, gradient, landform type, and altitude are the major environmental factors significantly influencing average soil water content in the hilly and gully catchment with complex terrain. The former two factors strongly affect soil water content along the 0–300 cm soil profile, whereas effects exerted by the latter two factors vary by soil layer and season.

Understanding the vertical profile of soil water content and evaluation of related major influencing factors in individual soil layers can help with sustainable land use and water management in catchment areas on the hilly and gully Loess Plateau as well as in arid and semi-arid areas with complex terrain. For better estimation of soil water profiles in small catchments, other factors such as fertilization, coverage, and soil physical properties may be considered with respect to specific soil layers.

Acknowledgments

We give our thanks to Prof./Dr. Wenzhao Liu (Institute of Soil and Water Conservation, Northwest A & F University, Yangling) for suggestions on experimental arrangement.

Author Contributions

Conceived and designed the experiments: BW FW JW XW YH. Performed the experiments: BW FW JW XW YH. Analyzed the data: BW FW JW XW YH. Contributed reagents/materials/analysis tools: BW FW JW XW YH. Wrote the paper: BW FW JW XW YH.

References

1. Engelbrecht BMJ, Comita LS, Condit R, Kursar TA, Tyree MT, et al. (2007) Drought sensitivity shapes species distribution patterns in tropical forests. Nature 447: 80–82.

2. Yang L, Wei W, Mo BR, Chen LD (2011) Soil water under different artificial vegetation restoration in the semi-hilly region of the Loess Plateau. Acta Ecologica Sinica 31: 3060–3068 (in Chinese with English abstract).

3. Xing G, Zhang XM, Fei XL, Wu YX (2012) Study on soil moisture content under different land use types in Sunjiacha basin. Agricultural Research in the Arid Areas 30: 225–229 (in Chinese with English abstract).

4. Li JW, Zuo HT, Li QF, Fan XF, Hou XC (2011) Effect of soil water spatial distribution pattern on switchgrass during first growing season. Acta Agrista Sinica 19: 43–50 (in Chinese with English abstract).

5. Jipp PH, Nepstad DC, Cassel DK, Carvalho C (1998) Deep soil moisture storage and transpiration in forests and pastures of seasonally-dry Amazonia. Climatic Change 39: 395–412.

6. Grassini P, You JS, Hubbard KG, Cassman KG (2010) Soil water recharge in a semi-arid temperate climate in the central US Great Plains. Agricultural Water Management 97: 1063–1069.

7. Markewitz D, Devine S, Davidson EA, Brando P, Nepstad DC (2010) Soil moisture depletion under simulated drought in the Amazon: impacts on deep root uptake. New Phytologist 187: 592–607.

8. Li YS (2001) Fluctuation of yield on high-yield field and desiccation of the soil on dryland. Acta Pedologica Sinica 38: 353–356 (in Chinese with English abstract).

9. Huang MB, Dang TH, Gallichand J, Goulet M (2003) Effect of increased fertilizer applications to wheat crop on soil-water depletion in the Loess Plateau, China. Agricultural Water Management 58: 267–278.

10. Liu WZ, Zhang XC, Dang TH, Zhu OY, Li Z, et al. (2010) Soil water dynamics and deep soil recharge in a record wet year in the southern Loess Plateau of China. Agricultural Water Management 97: 1133–1138.

11. Beldring S, Gottschalk L, Seibert J, Tallaksen LM (1999) Distribution of soil moisture and groundwater levels at patch and catchment scales. Agricultural and Forest Meteorology 98–99: 305–324.

12. Li B, Rodell M (2013) Spatial variability and its scale dependency of observed and modeled soil moisture over different climate regions. Hydrology and Earth System Sciences 17: 1177–1188.

13. Henninger DL, Petersen GW, Engman ET (1976) Surface soil moisture within a watershed: Variations, factors influencing, and relationship to surface runoff. Soil Science Society of American Journal 40: 773–776.

14. Jones EB, Owe M, Schmugge TJ (1982) Soil moisture variation patterns observed in Hand county, South Dakota. Water Recources Bulletin 18: 949–954.

15. Hawley ME, Jackson TJ, Mccuen RH (1983) Surface soil moisture variation on small agricultural watersheds. Journal of Hydrology 62: 179–200.

16. Robinson M, Dean TJ (1993) Measurement of near surface soil water content using a capacitance probe. Hydrological Processes 7: 77–86.

17. Qiu Y, Fu BJ, Wang J, Chen LD (2000) Quantitative analysis of relationships between spatial and temporal variation of soil moisture content and

environmental factors at a gully catchment. Acta Ecologica Sinica 20: 741–747 (in Chinese with English abstract).

18. Zeng C, Shao MA, Wang QJ, Zhang J (2011) Effects of land use on temporal-spatial variability of soil water and soil-water conservation. Acta Agriculturae Scandinavica Section B-Soil and Plant Science 61: 1–13.

19. Shi ZH, Zhu HD, Chen J, Fang NF, Ai L (2012) Spatial heterogeneity of soil moisture and its relationships with environmental factors at small catchment level. Chinese Journal of Applied Ecology 23: 889–895 (in Chinese with English abstract).

20. Chen LD, Huang ZL, Gong J, Fu BJ, Huang YL (2007) The effect of land cover/vegetation on soil water dynamic in the hilly area of the loess plateau, China. Catena 70: 200–208.

21. Wang YQ, Shao MA, Liu ZP, Orton R (2013) Regional-scale variation and distribution patterns of soil saturated hydraulic conductivities in surface and subsurface layers in the loessial soils of China. Journal of Hydrology 487: 13–23.

22. Zhang R, Cao H, Wang YQ, Huang CQ, Tan WF (2012) spatial variability of soil moisture and its influence factors in watershed of gully region on the loess plateau. Research of Soil and Water Conservation 19: 52–58 (in Chinese with English abstract).

23. Canton Y, Sole-benet A, Domingo F (2004) Temporal and spatial patterns of soil moisture in semiarid badlands of SE Spain. Journal of Hydrology 285: 199–214.

24. Burnt TP, Butcher DP (1985) Topographic controls of soil moisture distributions. Journal of Soil Science 36: 469–486.

25. O'loughlin EM (1981) Saturation regions in catchments and their relations to soil and topographic properties. Journal of Hydrology 53: 229–246.

26. Huang J, WU P, Zhao XN (2012) Effects of rainfall intensity, underlying surface and slope gradient on soil infiltration under simulated rainfall experiments. Catena 104: 93–102

27. Qiu Y, FU BJ, Wang J, Chen LD (2003) Spatiotemporal prediction of soil moisture content using multiple-linear regression in a small catchment of the Loess Plateau, China. Catena 54: 173–195.

28. Qiu Y, Fu B, Wang J, Chen L, Meng Q, Zhang Y (2010) Spatial prediction of soil moisture content using multiple-linear regressions in a gully catchment of the Loess Plateau, China. Journal of Arid Environments 74: 208–220.

29. Gómez-Plaza A, Martínez-Mena M, Albaladejo J, Castillo VM (2001) Factors regulating spatial distribution of soil water content in small semiarid catchments. Journal of Hydrology 253: 211–226.

30. Dripps WR, Bradbury KR (2007) A simple daily soil-water balance model for estimating the spatial and temporal distribution of groundwater recharge in temperate humid areas. Hydrogeology Journal 15: 433–444.

31. Yao XL, Fu BJ, Lu YH, Sun FX, Wang S, et al. (2013) Comparison of four spatial interpolation methods for estimating soil moisture in a complex terrain catchment. PLoS One 8(1): e54660.

32. Wang MB, Li HJ (1995) Quantitative study on the soil water dynamics of various forest plantations in the loess plateau region in northwestern Shanxi. Acta Ecologica Sinica 15: 172–184 (in Chinese with English abstract).

33. Wang YQ, Zhang XC, Han FP (2008) Profile variability of soil properties in check dam on the Loess Plateau and its functions. Environmental Science 29: 1020–1026 (in Chinese with English abstract).

34. Jain AK (2010) Data clustering: 50 years beyond K-means. Pattern Recognition Letters 31: 651–666.

35. Huang YL, Chen LD, Fu BJ, Wang YL (2005) Spatial pattern of soil water and its influencing factors in gully catchment of the Loess Plateau. Journal of Natural Resources 20: 483–492 (in Chinese with English abstract).

36. Zou JL, Shao MA, Gong SH (2011) Effects of different vegetation and soil types on profile variability of soil moisture. Research of Soil and Water Conservation 18: 12–17 (in Chinese with English abstract).

37. Kizito F, Dragila M,Se'ne M, Lufafa A, Diedhiou I, et al. (2006) Seasonal soil water variation and root patterns between two semi-arid shrubs co-existing with Pearl millet in Senegal, West Africa. Journal of Arid Environments 67: 436–455.

38. Li J, Chen B, Li XF, Zhao YJ, Ciren YJ, et al. (2008) Effects of deep soil desiccation on artificial forestlands in different vegetation zones on the Loess Plateau, China. Acta Ecologica Sinica 28: 1429–1445 (in Chinese with English abstract).

39. Cheng LP, Liu WZ (2013) Long term effects of farming system on soil water content and dry soil layer in deep loess profile of Loess Tableland in China. Journal of Integrative Agriculture. 13(6): 1382–1392.

40. Wang XZ, Jiao F (2011) Partition of soil moisture profiles based on sequential clustering method. Journal of Northwest A &F University 39: 191–201,196 (in Chinese with English abstract).

41. Fu XL, Shao MA, Wei XR, Wang HM, Zeng C (2013) Effects of monovegetation restoration types on soil water distribution and balance on a hillslope in northern Loess Plateau of China. Journal of Hydrologic Engineering 18: 413–421.

42. Bai TL, Yang QK, Shen J (2009) Soil variability of soil moisture vertical distribution and related affecting factors in hilly and gully watershed region of Loess Plateau. Chinese Journal of Ecology 28: 2508–2514 (in Chinese with English abstract).

43. Lian G, Guo XD, Fu BJ, Hu CX (2006) Spatial variability of bulk density and soil water in a small catchment of the Loess Plateau. Acta Ecologica Sinica 26: 647–654 (in Chinese with English abstract).

Global Validation of a Process-Based Model on Vegetation Gross Primary Production Using Eddy Covariance Observations

Dan Liu[1], Wenwen Cai[1], Jiangzhou Xia[1], Wenjie Dong[1], Guangsheng Zhou[2,3], Yang Chen[1], Haicheng Zhang[1], Wenping Yuan[1]*

1 State Key Laboratory of Earth Surface Processes and Resource Ecology, Beijing Normal University, Beijing, China, **2** Chinese Academy of Sciences, Institute of Botany, State Key Laboratory of Vegetation and Environmental Change, Beijing, China, **3** Chinese Academy of Meteorological Sciences, Beijing, China

Abstract

Gross Primary Production (GPP) is the largest flux in the global carbon cycle. However, large uncertainties in current global estimations persist. In this study, we examined the performance of a process-based model (Integrated Biosphere Simulator, IBIS) at 62 eddy covariance sites around the world. Our results indicated that the IBIS model explained 60% of the observed variation in daily GPP at all validation sites. Comparison with a satellite-based vegetation model (Eddy Covariance-Light Use Efficiency, EC-LUE) revealed that the IBIS simulations yielded comparable GPP results as the EC-LUE model. Global mean GPP estimated by the IBIS model was 107.50 ± 1.37 Pg C year^{-1} (mean value ± standard deviation) across the vegetated area for the period 2000–2006, consistent with the results of the EC-LUE model (109.39 ± 1.48 Pg C year^{-1}). To evaluate the uncertainty introduced by the parameter V_{cmax}, which represents the maximum photosynthetic capacity, we inversed V_{cmax} using Markov Chain-Monte Carlo (MCMC) procedures. Using the inversed V_{cmax} values, the simulated global GPP increased by 16.5 Pg C year^{-1}, indicating that IBIS model is sensitive to V_{cmax}, and large uncertainty exists in model parameterization.

Editor: Christopher Carcaillet, Ecole Pratique des Hautes Etudes, France

Funding: This study was supported by the National Science Foundation for Excellent Young Scholars of China (41322005), the National High Technology Research and Development Program of China (863 Program) (2013AA122003), National Natural Science Foundation of China (41201078), Program for New Century Excellent Talents in University (NCET-322 12-0060) and the Fundamental Research Funds for the Central Universities. The funders had no role in study design, data collection and analysis, decision to publish, or preparation of the manuscript.

Competing Interests: The authors have declared that no competing interests exist.

* Email: wenpingyuancn@yahoo.com

Introduction

Terrestrial gross primary production (GPP) is the largest carbon flux in terrestrial ecosystems, and it is approximately 20 times larger than the amount of carbon introduced from anthropogenic sources [1]. Thus, even small fluctuations in GPP can cause large changes in the airborne fraction of carbon and subsequently influence future climate change [2]. Vegetation also contributes to human welfare by providing food, fiber and energy [3–4]. Therefore, regular monitoring and reliable estimation of global terrestrial GPP is important for improving our understanding of the global carbon cycle, accurately predicting future climate, and ensuring the long-term sustainability of terrestrial ecosystem services.

Ecosystem models serve as a backbone for evaluating large-scale and global GPP. Two categories of ecosystem models are widely used: process-based and satellite-based. Satellite-based models are driven by remotely sensed data and provide simple means of estimating GPP [5]; however, they are limited in their ability to model mechanisms. Process-based models typically exhibit detailed expressions of terrestrial processes, such as photosynthesis, respiration, phenology, and hydrological cycle. Therefore, process-based models play important roles in investigating the mechanisms underlying current biases in estimated ecosystem production [6],

predicting the future conditions of the terrestrial carbon cycle, and exploring its feedback to climate change [7].

Numerous attempts have been made to develop and improve process-based models. However, a recent study from the North American Carbon Project (NACP) showed that current models perform poorly and difference between observations and simulations far exceed the observational uncertainty [8]. Model parameter uncertainty is a key source limiting the accuracy of process-based models. Knorr and Heimann analyzed the uncertainties of process-based models [9]. They found that parameter uncertainties could explain much of the large variance among models and that the largest uncertainties arose from plant photosynthesis, respiration and soil water storage.

The maximum rate of carboxylation by the enzyme Rubisco (V_{cmax}) is fundamental in modeling photosynthesis [10]. Sensitivity analysis shows that the projections of ecosystem production are particularly sensitive to the fixed parameters associated with V_{cmax} [11]. Therefore, the parameterization scheme of V_{cmax} is essential for GPP simulation, and its impacts on model performance need to be tested.

Eddy covariance measurements recorded by the increasing number of eddy covariance (EC) towers provide a great opportunity for model validation and improvement. Concurrent measurements include carbon fluxes, latent heat, and sensible

Table 1. The values of V_{cmax} set in the IBIS model and the initial range of MCMC inversion.

Plant Functional Type (PFT)	V_{cmax} (10^{-6} mol CO_2 m^{-2} s^{-1})	V_{cmax} range (10^{-6} mol CO_2 m^{-2} s^{-1})
Tropical broadleaf trees	65	1–300
Warm-temperate broadleaf trees	40	1–300
Temperate broadleaf trees	30	1–300
Boreal broadleaf trees	30	1–300
Temperate conifer trees	30	1–300
Boreal conifer trees	20	1–300
Shrub	27.5	1–300
C3 herbaceous	25	1–300
C4 herbaceous	15	1–300

V_{cmax} set according to Kucharik et al. [47].

heat, as well as meteorological conditions such as air temperature and relative humidity, and provide unprecedented datasets for model validation and evaluations of parameter constraints [12]. The current network of EC sites covers a wide range of ecosystem types, thus it has the potential to significantly improve our understanding of the variation in GPP across time, space and biomes [13].

The goal of this study was to validate a process-based ecosystem model (Integrated BIosphere Simulator, IBIS) based on measurements from 62 EC sites [14]. The specific objectives were to (1) examine the performance of IBIS over several ecosystem types, (2) compare IBIS model performance with a satellite-based model (i.e., Eddy Covariance-Light Use Efficiency, EC-LUE), and (3) investigate the impacts of the parameter V_{cmax} on model performance.

Data and Methods

2.1 IBIS model and parameter inversion

The Integrated BIosphere Simulator (IBIS) is designed to integrate a variety of terrestrial ecosystem processes within a single, physically consistent modeling framework. It represents land surface processes, canopy physiology, vegetation phenology, long-term vegetation dynamics, and carbon and water cycling [14]. The photosynthesis module of the IBIS model is provided by the formulations of Farquhar [15].

C_3 photosynthesis and C_4 photosynthesis are expressed separately in the IBIS model. For C_3 plants, the gross photosynthesis rate per unit leaf area, A_g(mol CO_2 m^{-2} s^{-1}) is expressed as

$$A_g = \min(J_e, J_c, J_s) \qquad (1)$$

where J_e is light-limited rate of photosynthesis, J_c represents the Rubisco-limited rate of photosynthesis, and J_s is the photosynthesis limited by the inadequate rate of utilization of triose phosphate.

The light-limited rate of photosynthesis is given as

$$J_e = \alpha_3 Q_p \frac{C_i - \Gamma_*}{C_i + 2\Gamma_*} \qquad (2)$$

where α_3 is the intrinsic quantum efficiency of CO_2 uptake in C_3 plants (mol CO_2 mol^{-1} quanta), Q_p is the flux density of photosynthetically active radiation absorbed by leaf (mol quanta m^{-2} s^{-1}), C_i is the concentration of CO_2 in the intercellular air

spaces of the leaf (mol mol^{-1}), and Γ_* is the compensation point for gross photosynthesis (mol mol^{-1}).

The Rubisco-limited rate of photosynthesis is calculated as

$$J_c = \frac{V_m(C_i - \Gamma_*)}{C_i + K_C(1 + \frac{[O_2]}{K_O})} \qquad (3)$$

where V_m is the maximum carboxylase capacity of Rubisco (mol CO_2 m^{-2} s^{-1}) and K_C and K_O are the Michaelis-Menten coefficients (mol mol^{-1}) for CO_2 and O_2, respectively.

Under conditions of high intercellular CO_2 concentrations and high irradiance, photosynthesis is limited by the inadequate rate of utilization of triose phosphate. This limitation is expressed as

$$J_s = 3T(1 - \frac{\Gamma_*}{C_i}) + \frac{J_p\Gamma_*}{C_i} \qquad (4)$$

where T is the rate of triose phosphate utilization.

Photosynthesis in C_4 plants is similarly modeled as the minimum of three potential capacities to fix carbon [16]. The gross photosynthesis rate is given by

$$A_g = \min(J_i, J_e, J_c) \qquad (5)$$

where $J_i = \alpha_4 Q_p$ is the light-limited rate of photosynthesis, $J_e = V_m$ is the Rubisco-limited rate of photosynthesis and $J_c = k$ is the CO_2-limited rate of photosynthesis at low CO_2 concentrations.

The parameter V_{cmax} (mol CO_2 m^{-2} s^{-1}) is very important for simulating the photosynthesis process. In the IBIS model, it is established as a constant that differs among plant functional types (PFTs) (Table 1). To validate the IBIS model and investigate the impact of the V_{cmax} parameter scheme on model performance, we conducted two simulations (IBIS and IBIS-Type) for each site. IBIS simulation used the default values of V_{cmax} (Table 1). For the IBIS-Type simulation, the vegetation-specific V_{cmax} values were inversed for each PFT. The Markov Chain-Monte Carlo (MCMC) procedure was used for the parameter inversion, and the Metropolis-Hastings (M-H) algorithm was used as the MCMC sampler [17–18] (see Xu et al. [19] and Yuan et al. [20] for detailed descriptions of the MCMC procedure). We conducted 10000 samples for each site and assigned the V_{cmax} value with the highest frequency as the optimal value of V_{cmax}. Finally, we ran the model using the inversed V_{cmax} for each PFT as the IBIS-Type simulation.

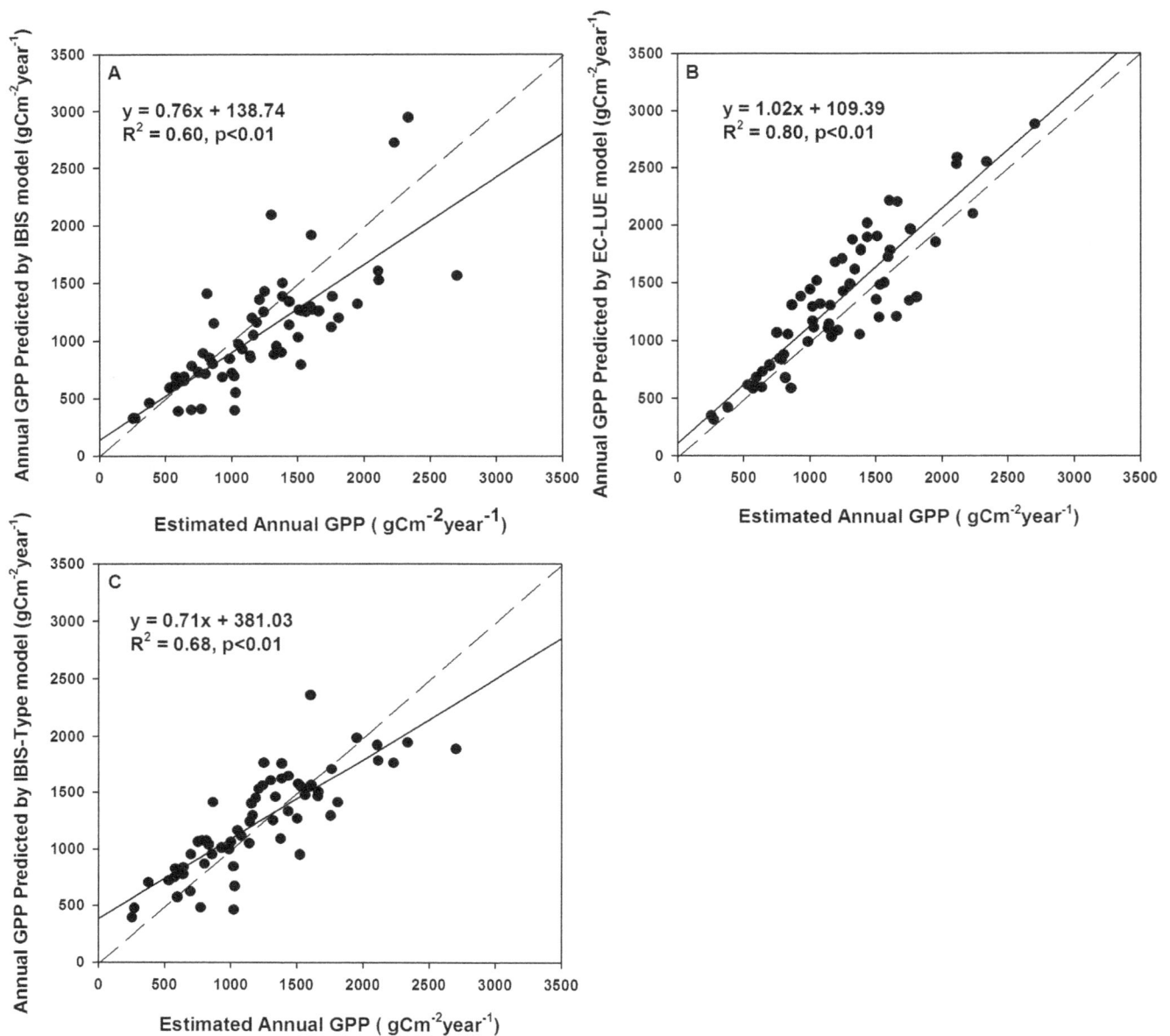

Figure 1. Comparison of predicted and measured GPP. Comparison between gross primary production (GPP) estimated from Eddy Covariance (EC) measurements and GPP predicted from different model simulations: (a) IBIS, (b) EC-LUE and (c) IBIS-Type. The solid lines are the linear regression lines and the short dashed lines are the 1:1 lines.

2.2 EC-LUE model

A satellite-based model (i.e. EC-LUE) [5,21–23] was used to compare the local and global GPP simulations with those of the IBIS model. The EC-LUE model is based on two assumptions: (1) ecosystem GPP has a direct relationship with the absorbed photosynthetically active radiation (APAR) via light use efficiency (LUE), where LUE is defined as the amount of carbon produced per unit of APAR; and (2) realized LUE may be reduced below its theoretical potential value by environmental stressors, such as low temperatures or water shortages. The EC-LUE model is driven by four variables: the normalized difference vegetation index (NDVI), photosynthetically active radiation (PAR), air temperature, and the ratio of sensible to latent heat flux (Bowen ratio).

2.3 Data

We used the eddy covariance (EC) data to validate the IBIS model. We used data obtained from the LaThuile dataset (http://www.fluxdata.org). The daily GPP values were estimated from the eddy covariance measurements using a community standard method [24]. Briefly, GPP was estimated from the equation:

$$GPP = R_{day} - NEE_{day} \qquad (6)$$

where NEE_{day} is daytime NEE. Daytime ecosystem respiration R_{day} was estimated using daytime temperature and an equation describing the temperature dependence of respiration, which was formulated from nighttime NEE measurements. For further details on the algorithm, see Reichstein et al. (2005) [24]. The gap filling

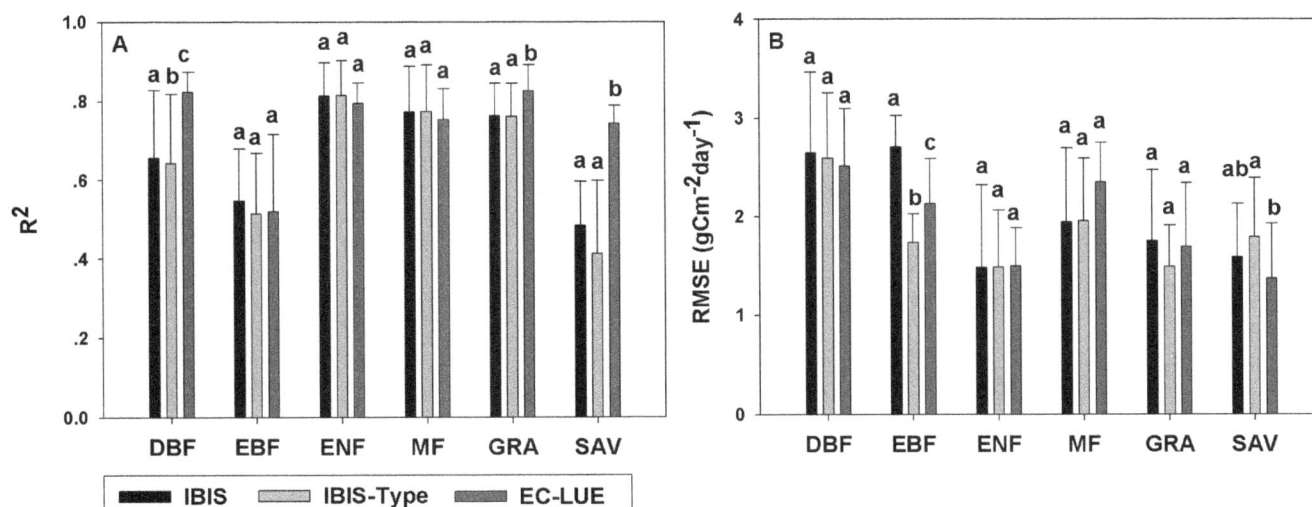

Figure 2. Comparison among models for different PFTs. Comparisons among IBIS, IBIS-Type and EC-LUE models for each plant functional type (PFT), where (a) and (b) are the results of R^2 and RMSE, respectively. Lowercase letters above the bars indicate significant differences among models.

and quality control of this dataset are conducted according to standard criteria [25–26], and the uncertainty in annual GPP can be controlled to some extent below $100 \text{ g C m}^{-2} \text{year}^{-1}$ [25]. This dataset is widely used in model studies. In the present study, we selected 62 EC sites for model validation (Table S1). The selected data covered six major terrestrial biomes: evergreen needleleaf forest (ENF), deciduous broadleaf forest (DBF), evergreen broadleaf forest (EBF), mixed forest (MF), grassland (GRA) and savanna (SAV). Additional information on the vegetation, climate and soil characteristics of each site was collected from the associated metadata of the LaThuile dataset. Daily average, maximum and minimum temperature, relative humidity, precipitation, cloud fraction, photosynthetically active radiation, latent heat and sensible heat were used to drive the IBIS, and GPP was used to evaluate its performance.

To decrease model uncertainty, we used the satellite-based leaf area index (LAI) from the Moderate Resolution Imaging Spectroradiometer (MODIS) as a model input. The Normalized Difference Vegetation Index (NDVI) data derived from MODIS were used to drive the EC-LUE model. The 8-day MODIS-NDVI data (MOD13) and the MODIS-LAI (MOD15) data with 1-km spatial resolution were used for model verification at the EC sites. Quality control (QC) flags, which signal cloud contamination in each pixel, were examined to filter out NDVI and LAI data of insufficient quality. We temporally filled missing or unreliable values for each 1-km MODIS pixel based on their corresponding quality assessment data fields, as proposed by Zhao et al. [27]. In addition, data on soil properties, including soil texture, organic carbon content and nitrogen content, were required to input soil information into the model, and were therefore collected from the sites where EC towers are established.

For global simulation, we used meteorological datasets from the Modern Era Retrospective Analysis for Research and Applications (MERRA) archive for 2000–2006 to drive the IBIS and EC-LUE models. MERRA is a NASA reanalysis dataset for the satellite era which uses a new version of the Goddard Earth Observing System Data Assimilation System Version 5 (GEO-5). We used climate conditions at 10 meters above the land surface and at a resolution of 0.5° latitude by 0.6° longitude. The Global Gridded Surfaces of Selected Soil Characteristics datasets were used to supply soil

properties for the IBIS model; detailed information is available from the website (http://ww.isric.org). The global distribution of plant functional types (PFTs) was derived by overlapping the MODIS land-cover type product with the Köppen-Geiger climate classification map, with the land cover classifications aggregated into nine PFTs (Table 1). The IBIS model was unable to simulate carbon cycle processes in cropland or wetland; therefore, these two vegetation types were replaced with C_3 grassland in this study.

2.4 Statistical analysis

Three metrics were used to evaluate model performance:

(1) the coefficient of determination, R^2, which represents how much variation in the observations is explained by the model simulations;

(2) the root mean square error (RMSE), which represents the total difference between the simulated and measured values;

(3) the relative predictive error (RPE), which represents the ratio of error to observation. It is computed as

$$RPE = \frac{\overline{P} - \overline{O}}{\overline{O}} \times 100 \qquad (7)$$

where $\overline{P}$ and $\overline{O}$ are the mean simulated and measured values, respectively.

One-way ANOVA was employed using SPSS software to test the significance of the differences in optimal V_{cmax} for each biome in the IBIS-Type scheme, and the paired-samples t-test was used to test the significance of the difference in statistical metrics between different models.

Results

3.1 Model validation at EC towers

The overall comparison of the estimated GPP with the EC measurements showed that the IBIS model performed well in capturing the variability in GPP. Across all study sites, the IBIS model explained approximately 60% of the variation in site-averaged GPP (Fig. 1). The coefficient of determination (R^2) varied from 0.11 at the ES-LMa site to 0.94 at the CA-Man site, with a mean value of 0.71 across all EC sites. The mean R^2 values

Table 2. The performance of the IBIS model and the EC-LUE model.

Site	Vegetation Type	IBIS			EC-LUE		
		R²	RMSE	RPE	R²	RMSE	RPE
CA-Oas	DBF	0.75	2.57	−32.34	0.87	2.26	26.75
DE-Hai	DBF	0.84	2.26	−17.92	0.88	1.62	−3.42
FR-Fon	DBF	0.68	2.98	−24.83	0.77	3.23	22.55
FR-Hes	DBF	0.74	3.07	−20.36	0.80	2.72	9.49
IT-Col	DBF	0.75	2.62	−18.94	0.75	2.88	20.36
IT-Ro1	DBF	0.61	2.03	6.49	0.77	2.33	32.39
IT-Ro2	DBF	0.24	4.05	−25.93	0.83	2.85	30.68
US-Bar	DBF	0.80	1.80	1.10	0.83	2.82	37.93
US-Bn2	DBF	0.72	1.12	10.49	0.86	1.04	14.97
US-Dk2	DBF	0.68	3.74	−23.88	0.81	2.86	19.87
US-Ha1	DBF	0.76	2.82	−19.57	0.82	2.38	11.84
US-MOz	DBF	0.32	3.62	−24.46	0.85	3.02	34.81
US-UMB	DBF	0.64	2.57	−2.44	0.92	2.47	41.20
US-Wi8	DBF	0.65	1.85	19.52	0.76	2.72	37.90
AU-Tum	EBF	0.69	2.80	22.60	0.75	2.27	−5.86
AU-Wac	EBF	0.43	2.92	24.27	0.39	2.66	5.82
FR-Pue	EBF	0.63	2.89	59.41	0.62	2.03	15.37
PT-Mi1	EBF	0.44	2.24	72.49	0.33	1.55	−17.23
CA-Ca3	ENF	0.74	1.38	14.26	0.76	1.87	14.11
CA-Man	ENF	0.94	0.61	6.83	0.80	1.19	13.54
CA-Obs	ENF	0.91	0.80	−11.11	0.86	1.16	9.49
CA-Ojp	ENF	0.88	0.77	17.75	0.78	0.99	5.13
CA-Qfo	ENF	0.84	0.91	4.25	0.80	1.19	9.92
CA-SJ3	ENF	0.79	1.24	33.48	0.75	2.12	51.35
FI-Hyy	ENF	0.87	1.43	−22.85	0.89	1.20	−2.87
FI-Sod	ENF	0.85	0.89	2.78	0.77	1.06	−7.73
IT-Ren	ENF	0.73	1.49	−7.29	0.69	1.74	−12.11
NL-Loo	ENF	0.81	2.28	−32.99	0.78	1.95	−21.84
RU-Fyo	ENF	0.87	2.30	−34.64	0.85	1.82	−23.40
RU-Zot	ENF	0.86	1.05	−14.51	0.85	1.00	0.63
SE-Fla	ENF	0.86	0.98	13.78	0.76	1.46	6.35
US-Bn1	ENF	0.75	0.99	6.62	0.76	1.10	1.76
US-Fmf	ENF	0.73	1.09	−14.31	0.73	1.50	21.17
US-Ha2	ENF	0.87	1.61	−21.33	0.85	1.96	11.45

Table 2. Cont.

Site	Vegetation Type	IBIS			EC-LUE		
		R^2	RMSE	RPE	R^2	RMSE	RPE
US-Ho1	ENF	0.92	2.11	−31.29	0.84	1.69	−9.90
US-Me2	ENF	0.61	2.73	−47.52	0.77	1.63	−20.61
US-Me3	ENF	0.75	0.89	−6.74	0.76	1.15	−31.66
US-NC2	ENF	0.71	4.05	−42.25	0.83	2.08	6.32
CA-Let	GRA	0.84	1.68	−34.70	0.90	1.10	13.64
CN-HaM	GRA	0.82	1.68	−42.69	0.83	1.47	12.36
CN-Xi2	GRA	0.67	0.77	21.88	0.70	0.82	10.69
DE-Meh	GRA	0.84	1.90	−25.53	0.85	1.49	0.20
HU-Bug	GRA	0.70	1.92	−30.21	0.85	2.30	46.43
IE-Dri	GRA	0.65	2.80	−33.36	0.75	2.25	−6.00
IT-MBo	GRA	0.84	2.32	−28.74	0.83	2.49	20.46
PT-Mi2	GRA	0.76	1.43	−24.61	0.77	2.14	52.77
RU-Ha1	GRA	0.79	0.86	−3.50	0.82	1.77	42.51
US-ARc	GRA	0.64	2.91	−33.45	0.90	2.21	41.67
US-Aud	GRA	0.84	1.03	16.20	0.89	0.57	10.06
BE-Bra	MF	0.81	1.31	12.90	0.71	1.91	0.95
BE-Jal	MF	0.90	1.48	−19.78	0.84	1.99	−4.06
BE-Vie	MF	0.82	1.88	−25.86	0.78	2.10	−27.43
CA-Gro	MF	0.75	1.72	−7.45	0.75	3.07	44.54
JP-Tef	MF	0.54	2.29	0.07	0.61	2.33	9.86
JP-Tom	MF	0.78	3.48	−34.21	0.83	2.55	−21.88
US-PFa	MF	0.81	1.44	−0.10	0.76	2.50	28.45
ES-LMa	SAV	0.59	1.55	−39.30	0.77	1.27	8.16
ZA-Kru	SAV	0.32	2.10	−51.83	0.74	1.35	22.69
BW-Ma1	SAV	0.42	1.58	−48.15	0.74	0.90	9.73
US-FR2	SAV	0.11	2.17	−46.41	0.77	1.27	8.16
US-Ton	SAV	0.50	1.48	−20.37	0.79	2.14	31.26
US-SRM	SAV	0.59	1.49	4.82	0.66	1.88	25.23

DBF: deciduous broadleaf forest; EBF: evergreen broadleaf forest; ENF: evergreen needleleaf forest; GRA: grassland; MF: mixed forest; SAV: savanna.

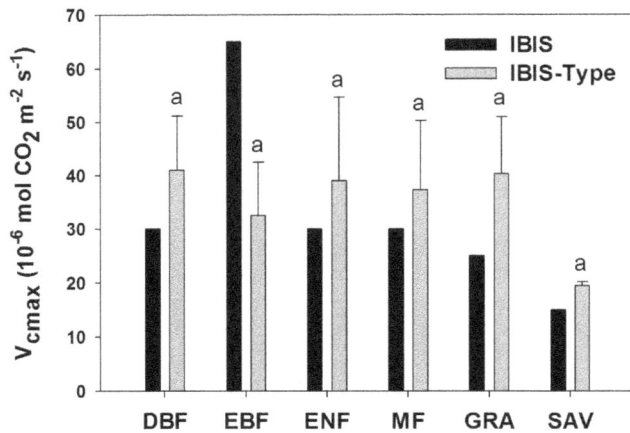

Figure 3. Comparison of V_{cmax} from the original scheme and the inversed V_{cmax} values. Comparison of model V_{cmax} values with the inversed values of the IBIS-Type. Same letter on the top of the bar indicates no significant difference among biomes.

were 0.66, 0.55, 0.81, 0.77, 0.76 and 0.48 for deciduous broadleaf forest, evergreen broadleaf forest, evergreen needleleaf forest, mixed forest, grassland and savanna, respectively (Fig. 2). The root mean square error (RMSE) varied from 1.48 to 2.71 g C m^{-2} day^{-1} among the six vegetation types (Fig. 2). On average, the mean relative predictive error (RPE) of all sites was -9.68%, and the RPE at most sites was less than 20% (Table 2).

Although the IBIS model explained most of the GPP variability at individual sites, large differences between the predicted and estimated GPP values from the EC measurements were apparent at some sites and for some vegetation types. The IBIS model underestimated GPP for the majority of PFTs and overestimated GPP in evergreen broadleaf forest. Specifically, over 40 of the 62 sites had negative RPE values, and the RPEs of 26 sites were below -20%. The largest underestimation occurred at two savanna sites, ZA-Kru and BW-Ma1, with PREs of -51.83% and -48.15%, respectively. The other 24 sites with RPEs below -20% were predominantly deciduous broadleaf forest (6 sites), evergreen needleleaf forest (7 sites) and grassland (8 sites). In addition, the model overestimated GPP at 6 sites with RPEs greater than 20%, four of which were evergreen broadleaf forest. Extreme overestimation occurred at two sites of evergreen broadleaf forest, FR-Pue and PT-Mi1, with RPEs of 59.41% and 72.49%, respectively.

3.2 Comparison of IBIS and EC-LUE

Compared to the satellite-based EC-LUE model, the IBIS model performed comparably at most sites, according to the R^2, RMSE and RPE values (Table 2). The IBIS mean R^2 values for evergreen broadleaf forest, evergreen needleleaf forest, mixed forest and grassland were similar to those of the EC-LUE model, which were 0.52, 0.79, 0.75, 0.83, respectively; no significant differences were found for most PFTs. Comparable results were also found for the RMSE, and no significant differences in RMSE were detected for any PFTs except for evergreen broadleaf forest.

However, the EC-LUE model had some advantages for some PFTs (Figure 2). For broadleaf forest, grassland and savanna, the R^2 of the EC-LUE was significantly higher than that of the IBIS, particularly for savanna, where the mean was 76% higher than that of the IBIS (0.74 for EC-LUE and 0.42 for IBIS). In addition, the EC-LUE had significantly low model error for evergreen

broadleaf forest, with a mean RMSE 22% lower than that of the IBIS (2.12 g C m^{-2} day^{-1} for EC-LUE and 2.71 g C m^{-2} day^{-1} for IBIS).

3.3 IBIS Performance with inversed V_{cmax}

Model parameterization did not significantly improve the performance of the IBIS, as indicated by the R^2 and RMSE. The mean R^2 values of the IBIS-Type were 0.64, 0.51, 0.81, 0.77, 0.76, 0.41 for deciduous broadleaf forest, evergreen broadleaf forest, evergreen needleleaf forest, mixed forest, grassland and savanna, respectively. These values were very similar to those of the IBIS model; only the R^2 of the deciduous broadleaf forest differed significantly, being higher in the IBIS-Type model. The overall R^2 increased from 0.60 to 0.68 after revising V_{cmax} (Fig. 1). The RMSEs of the IBIS and the IBIS-Type models were also comparable. The mean RMSE of the IBIS-Type varied from 1.49 g C m^{-2} day^{-1} for evergreen needleleaf forest to 2.60 g C m^{-2} day^{-1} for deciduous broadleaf forest. Evergreen broadleaf forest was the only vegetation type yielding a significant difference in RMSE between the two models.

The IBIS employs a set of parameter values for a given PFT (Table 1). In the IBIS-Type scheme, the V_{cmax} value, which was set to as the mean value of V_{cmax} inversed from each site within a given PFT, was largely differentiated from the original parameter values. V_{cmax} of the IBIS-Type was 36.67%, 30.00%, 24.29%, 61.09% and 30.00% higher than that of the IBIS for deciduous broadleaf forest, evergreen needleleaf forest, mixed forest, grassland and savanna, respectively; it was 50% lower for evergreen broadleaf forest. In addition, the V_{cmax} value inversed at each site did not indicate significant differences among different PFTs (Fig. 3).

3.4 Temporal and spatial patterns in global averaged GPP

The spatial pattern in average annual GPP estimated using the original IBIS, the IBIS-Type and the EC-LUE models from 2000 to 2006 were generally consistent (Fig. 4). The highest value was from the humid tropics (Amazonia, Central Africa and Southeast Asia), with an annual GPP over 2000 g C m^{-2} year^{-1}. Temperate regions had intermediate levels of GPP, and the lowest GPP was found in both cold and arid regions.

The magnitude of GPP estimated by the IBIS and EC-LUE models were comparable, reaching 107.50±1.37 Pg C year^{-1} and 109.39±1.48 Pg C year^{-1} (mean value ± standard deviation) globally, respectively (Fig. 5). Two-model comparisons revealed consistent GPP estimations for the various PFTs (Table 3) with the exception of savanna, for which the IBIS model greatly underestimated GPP. The GPP estimate of the IBIS-Type scheme was much higher than those of the other two models, with a global value of 123.97±1.76 Pg C year^{-1} (Fig. 5). Larger GPP estimations using the IBIS-Type simulation were found for most biomes (Table 3).

Discussion

Process-based ecosystem models are one of the most important components of earth system models used to predict future climate change [1]. The IPCC AR5 requested that all earth system models integrate the global carbon cycle module [7]. Previous studies have shown that the uncertainty in carbon cycle models can produce 40% differences in the predicted temperature by 2100 [28]. The GPP is the total photosynthetic uptake or carbon assimilation by plants, and it is a key component of terrestrial carbon balance. Any errors in the GPP simulations will propagate through the model, introducing errors into the simulated biomass and net ecosystem

A. IBIS GPP

| | NoData | | 500 - 1,000 | | 1,500 - 2,000 | | 2,500 - 3,000 |
| | 0 - 500 | | 1,000 - 1,500 | | 2,000 - 2,500 | | 3,000 - 3,355 |

B. EC-LUE GPP

| | NoData | | 500 - 1,000 | | 1,500 - 2,000 | | 2,500 - 3,000 |
| | 0 - 500 | | 1,000 - 1,500 | | 2,000 - 2,500 | | 3,000 - 3,122 |

C. IBIS-Type GPP

| | NoData | | 500 - 1,000 | | 1,500 - 2,000 | | 2,500 - 2,623 |
| | 0 - 500 | | 1,000 - 1,500 | | 2,000 - 2,500 | |

D. IBIS - EC-LUE

| | NoData | | -1000 - -500 | | -200 - 0 | | 200 - 500 | | 1,000 - 2,496 |
| | -2195 - -1,000 | | -500 - -200 | | 0 - 200 | | 500 - 1,000 | |

E. Plant Functional Type

Unit: $gCm^{-2}year^{-1}$

- Tropical Broadleaf Evergreen
- Tropical Broadleaf Drought-Deciduous
- Warm-Temperate Broadleaf Evergreen
- Temperate Conifer Evergreen
- Temperate Broadleaf Cold-Deciduous
- Boreal Conifer Evergreen
- Boreal Broadleaf Cold-Deciduous
- Boreal Conifer Cold-Deciduous
- Evergreen Shubs
- Cold-Deciduous Shrubs
- C_4 Herbaceous
- C_3 Herbaceous
- Non-vegertated

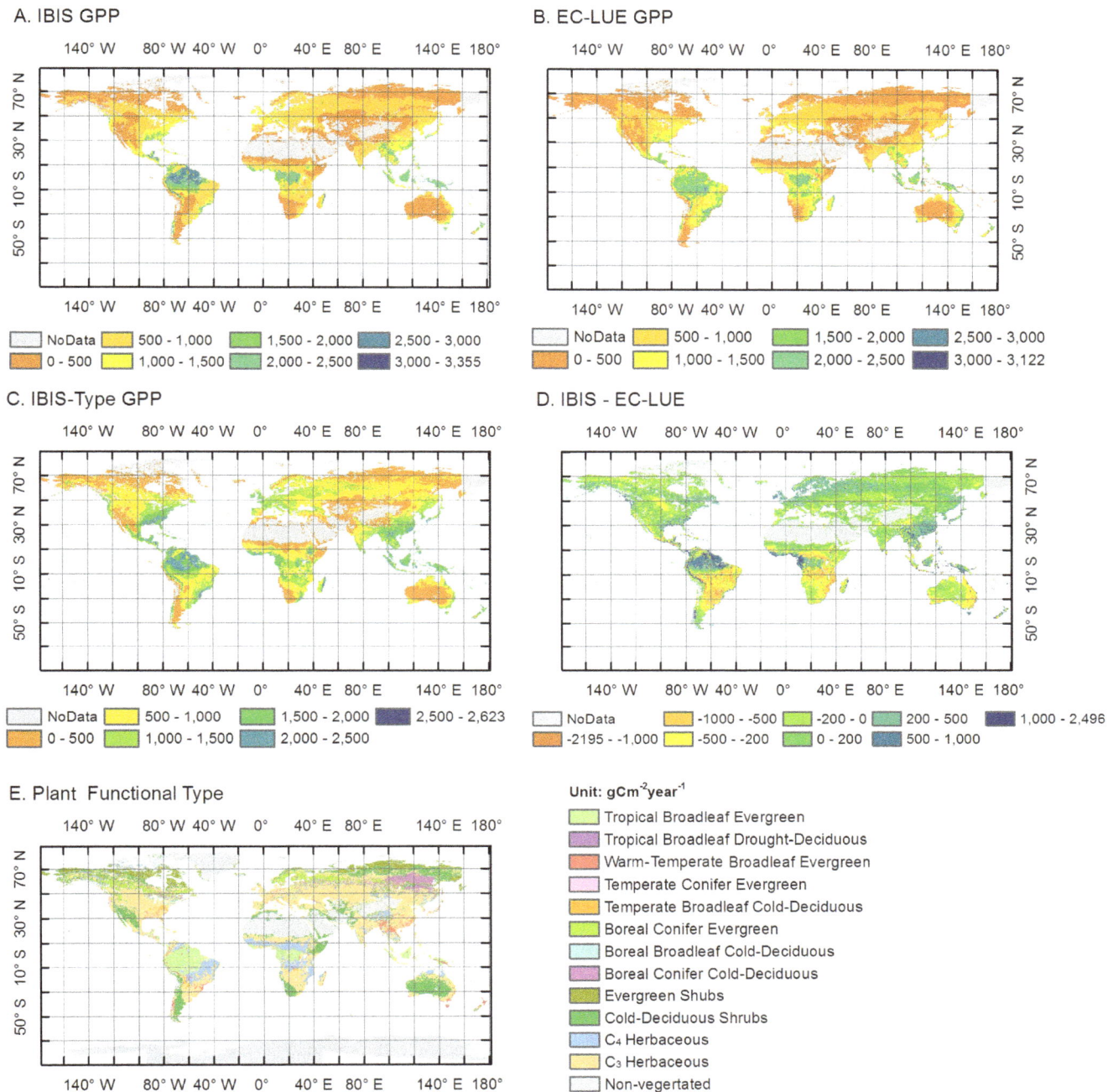

Figure 4. The global pattern of GPP. The global pattern of annual vegetation gross primary production (GPP) from 2000 to 2006. (a) estimated GPP using the IBIS model, (b) estimated GPP using the EC-LUE model, (c) estimated GPP using the IBIS-Type model, (d) the difference between the IBIS and EC-LUE models and (e) the spatial distribution of plant functional types (PFTs).

fluxes. If the simulated GPP is too low or too high, predicted leaf area index, wood biomass, crop yield, and soil biomass may also be too low or too high [29].

In this study, we examined the performance of the IBIS, which has been widely used to evaluate the regional and global terrestrial ecosystem carbon balance and has been integrated into earth system models (e.g., Brazilian Earth System) [30]. Our results indicate that the IBIS model is a good candidate for simulating GPP at regional-to-global scales, and its performance was comparable to that of the satellite-based EC-LUE model based

on EC site validation and comparison. The magnitude of global GPP estimated by the IBIS is also consistent with results of previous studies. Our estimate of annual global mean GPP was 107.50 ± 1.37 Pg C $year^{-1}$ (mean value $\pm$ standard deviation). Beer et al. estimated global GPP as 123.8 Pg C $year^{-1}$ [31]. Two satellite-based light-use efficiency models revealed similar estimates of global GPP: 109.29 Pg C $year^{-1}$ by the MODIS algorithm [27] and 111 Pg C $year^{-1}$ by the EC-LUE [21]. Interestingly, we found that the IBIS was consistent with the EC-LUE model across different PFTs (Table 3), despite the large

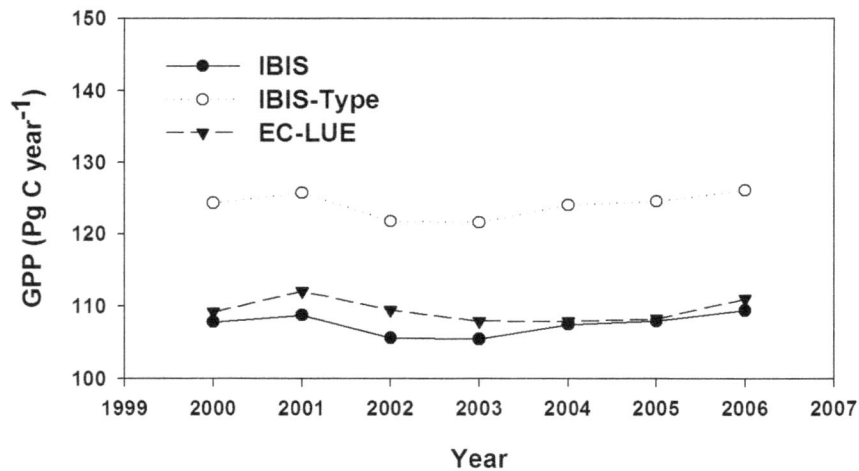

Figure 5. Interannual variability in GPP derived from different models. Interannual variability in global mean gross primary production (GPP) derived from the IBIS, IBIS-Type and EC-LUE models.

differences between the two approaches (i.e., satellite-based vs. process-based).

However, the IBIS did not perform well for two PFTs: evergreen broadleaf forest and savanna. Among the four evergreen broadleaf forest sites, two (PT-Mi1 and FR-Pue) have subtropical Mediterranean climate with dry summers and wet winters [32−33], and the AU-Tum site also suffers drought in summer [34]. At these sites, rainfall is the key driver of water and carbon fluxes [35]. Leuning analyzed the CO_2 and H_2O fluxes at the AU-Tum site [34], and found that carbon uptake was more strongly constrained by water stress than by temperature, and strongly affected by soil water availability. Moreover, savannas are characterized by climate with distinct wet and dry seasons, and this climate forcing causes savanna to form open, heterogeneous woodland canopies with grass understories [36].

These particular ecosystem properties result in greater complexity of modeling fluxes [37]. Process-based models need to simulate variation in soil moisture and plant phenology. However, previous studies have identified significant biases when simulating soil moisture [38]. We examined the performance of the IBIS on soil water at the savanna sites and also found obvious differences between the simulated values and the EC measurements (Fig. 6). Moreover, the IBIS integrated temperature-dominated phenology algorithms developed by Botta et al. [39]. However, field studies suggest that for many drought-deciduous species, the first large precipitation event at the start of the rainy season initiates rapid leaf flush [40−41], and leaf senescence is closely related to soil water availability in the dry season [41−42]. This relationship may explain why the IBIS model did not effectively capture the variance in GPP at savanna sites.

Parameterization is another large source of uncertainty in process-based models. V_{cmax} is the key parameter of the photosynthesis process [11], and a study by Bonan et al. [43] suggests that uncertainty in this parameter could account for 30 Pg C year^{-1} variations in model estimation of global GPP. In the present study, the differences in setting V_{cmax} values between the IBIS and IBIS-Type schemes caused an increase in global GPP of over 16.5 Pg C year^{-1}, which also indicated that V_{cmax} is a salient parameter for simulating GPP. Unfortunately, the determination of V_{cmax} in current models contains large uncertainty. Rogers surveyed V_{cmax} in current state-of-the-art models [44] and found that V_{cmax} varied within a wide range of −46 to +77% of the PFT mean. Thus, the determination of V_{cmax} and the reduction of uncertainty in this parameter are important issues for model development.

Many parameters in process-based models are established by PFTs, which are based on the assumption that the same type of vegetation responds similarly to the environment. However, a current study found that model parameters were more variable than previously assumed within the given PFTs [45] and that categorization of vegetation into less than eight PFTs may result in artificial multiple steady-states in a model of the Earth's climate-vegetation system depending on the number of PFTs used [46]. In the present study, the variation analysis showed that V_{cmax} did not significantly differ among PFTs. The predetermined parameterization scheme that sets the V_{cmax} constant values for each PFT in the IBIS model may cause systematic error.

We attempted to test the impact of V_{cmax} on model performance. However, variations in V_{cmax} cannot explain the overall uncertainty of the IBIS model. Validation of the IBIS on three flux sites demonstrated that parameterization and formula-

Table 3. The magnitude of gross primary production (GPP) in each plant functional type (PFT).

GPP (kgCm^{-2}year^{-1})	DBF	EBF	ENF	MF	GRA	SAV
IBIS	0.89	2.00	1.22	1.19	0.68	0.63
IBIS-TYPE	1.05	1.68	1.51	1.53	0.89	0.74
EC-LUE	0.78	1.76	0.96	1.16	0.73	0.92

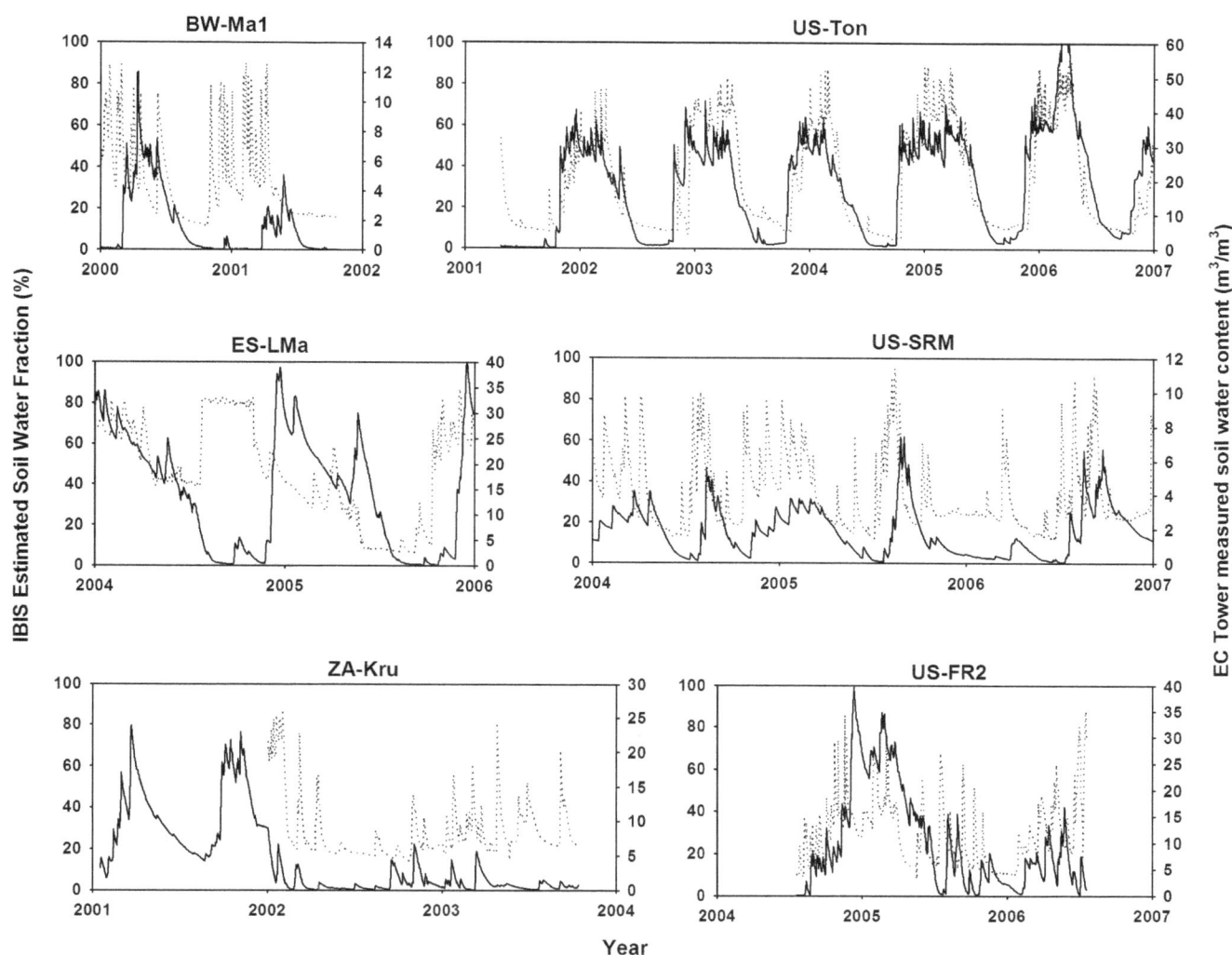

Figure 6. Comparison of estimated and observed soil water. Daily variation in the estimated soil water fraction of the IBIS model (i.e., fraction of soil pore space containing liquid water) and in the observed soil water content at EC sites. The solid lines represent simulation data, and the dotted lines represent the observed data.

tions of phenology also limit the model's ability to capture seasonal fluctuations in carbon and water exchange [38]. Particularly in regions with summer drought, phenology is primarily controlled by water supply rather than temperature [34]. This relationship hinders model simulation because biases in phenology and the dynamics of the leaf area index can affect the simulation of evapotranspiration. Such errors can pass to simulations of soil water content and other variables associated with the water and carbon cycles [38]. Therefore, revisions of not only the parameterization of V_{cmax} but also other parameters and formulations are needed for model improvement.

Summary

Process-based models are important tools for carbon cycle research, but current models incorporate substantial uncertainty. This study examines the performance of the IBIS model at global EC sites. Our results showed that the IBIS model explained 60% of the variation in GPP at all EC sites and performed comparably to the EC-LUE model, which explained 80% of the variation in observed GPP. At the global scale, the magnitudes of GPP

estimated by the IBIS and EC-LUE models were comparable, being 107.50 ± 1.37 Pg C year^{-1} and 109.39 ± 1.48 Pg C year^{-1}, respectively. The parameter V_{cmax} is a key parameter in the photosynthesis model. In the IBIS model, V_{cmax} was set as a constant for each PFT. The inversed V_{cmax} value was largely differentiated from the original setting, and no significant differences were detected among PFTs.

Acknowledgments

This work used eddy covariance data acquired by the FLUXNET community; in particular, by the following networks: AmeriFlux (U.S. Department of Energy, Biological and Environmental Research, Terrestrial Carbon Program (DE-FG02-04ER63917 and DE-FG02-04ER63911), AfriFlux, AsiaFlux, CarboAfrica, CarboEuropeIP, CarboItaly, Carbo-Mont, ChinaFlux, Fluxnet-Canada (supported by CFCAS, NSERC, BIOCAP, Environment Canada, and NRCan), GreenGrass, KoFlux,

LBA, NECC, OzFlux, TCOS-Siberia, USCCC. We acknowledge the financial support of the eddy covariance data harmonization provided by CarboEuropeIP, FAO-GTOS-TCO, iLEAPS, Max Planck Institute for Biogeochemistry, National Science Foundation, University of Tuscia, Université Laval, Environment Canada and the US Department of Energy. We also acknowledge the database development and technical support from Berkeley Water Center, Lawrence Berkeley National Laboratory, Microsoft Research Science, Oak Ridge National Laboratory, University of California–Berkeley and the University of Virginia.

Author Contributions

Conceived and designed the experiments: WY. Performed the experiments: DL. Analyzed the data: DL. Wrote the paper: DL WC JX WD GZ YC HZ WY.

References

1. IPCC (2007) Climate Change 2007: The Physical Science Basis. Contribution of Working Group I to the Fourth Assessment Report of the Intergovernmental Panel on Climate Change [Solomon S, Qin D, Manning M, Chen Z, Marquis M, et al. (eds).]. Cambridge University Press, Cambridge, United Kingdom and New York, NY, USA.

2. Raupach MR, Canadell JG, Quere CL (2008) Anthropogenic and biophysical contributions to increasing atmospheric CO_2 growth rate and airborne fraction. Biogeosciences 5(6): 1601–1613.

3. Rojstaczer S, Sterling SM, Moore NJ (2001) Human appropriation of photosynthesis products. Science 294(5551): 2549–2552.

4. Imhoff ML, Bounoua L, Ricketts T, Loucks C, Harriss R, et al. (2004) Global patterns in human consumption of net primary production. Nature 429(6994): 870–873.

5. Yuan WP, Liu SG, Zhou GS, Zhou GY, Tieszen LL, et al. (2007) Deriving a light use efficiency model from eddy covariance flux data for predicting daily gross primary production across biomes. Agricultural and Forest Meteorology 143(3–4): 189–207.

6. Zhang FM, Chen JM, Chen JQ, Gough CM, Martin TA, et al. (2012) Evaluating spatial and temporal patterns of MODIS GPP over the conterminous US against flux measurements and a process model. Remote Sensing of Environment 124: 717–729.

7. Friedlingstein P, Cox P, Betts R, Bopp L, von Bloh W, et al. (2006) Climate-carbon cycle feedback analysis: Results from the C4MIP model intercomparison. Journal of Climate 19(14): 3337–3353.

8. Schwalm CR, Williams CA, Schaefer K, Anderson R, Arain MA, et al. (2010) A model-data intercomparison of CO_2 exchange across North America: Results from the North American Carbon Program site synthesis. Journal of Geophysical Research: Biogeosciences 153(G3): G00H05.

9. Knorr W, Heimann M (2001) Uncertainties in global terrestrial biosphere modeling 1. A comprehensive sensitivity analysis with a new photosynthesis and energy balance scheme. Global Biogeochemical Cycles 15(1): 207–225.

10. Wolf A, Akshalov K, Saliendra N, Johnson DA, Laca EA (2006) Inverse estimation of V_{cmax}, leaf area index, and the Ball-Berry parameter from carbon and energy fluxes. Journal of Geophysical Research: Atmosphere 111(D8) DOI: 10.1029/2005JD005927.

11. Friend AD (2010) Terrestrial plant production and climate change. Journal of Experimental Botany 61(5): 1293–1309.

12. Friend AD, Arneth A, Kiang NY, Lomas M, Ogee J, et al. (2007) FLUXNET and modeling the global carbon cycle. Global Change Biology 13(3): 610–633.

13. Baldocchi D, Falge E, Gu LH, Olson R, Hollinger D, et al. (2001) FLUXNET: A new tool to study the temporal and spatial variability of ecosystem-scale carbon dioxide, water vapor, and energy flux densities. Bulletin of the American Meteorological Society 82(11): 2415–2434.

14. Foley JA, Prentice IC, Ramankutty N, Levis S, Pollard D, et al. (1996) An integrated biosphere model of land surface processes, terrestrial carbon balance, and vegetation dynamics. Global Biogeochemical Cycles 10(4): 603–628.

15. Farquhar GD, Caemmerer SV, Berry JA (1980) A biochemical-model of photosynthetic CO_2 assimilation in leave of C-3 species. Planta 149(1): 79–90.

16. Collatz GJ, Ball JT, Grivet C, Berry JA (1991) Physiological and environmental-regulation of stomata conductance, photosynthesis and transpiration - a model that includes a laminar boundary-layer. Agricultural and Forest Meteorology 54(2–4): 107–136.

17. Metropolis N, Rosenbluth AW, Rosenbluth MN, Teller AH, Teller E (1953) Equation of state calculations by fast computing machimes. Journal of Chemical Physics 21(6): 1087–1092.

18. Hastings WK (1970) Monte-Carlo sampling methods using Markov Chains and their applications. Biometrika 57(1): 97.

19. Xu T, White L, Hui DF, Luo YQ (2006) Probabilistic inversion of a terrestrial ecosystem model: Analysis of uncertainty in parameter estimation and model prediction. Global Biogeochemical Cycles 20(2) DOI: 10.1029/2005GB002468.

20. Yuan WP, Liang SL, Liu SG, Weng ES, Luo YQ, et al. (2012) Improving model parameter estimation using coupling relationships between vegetation production and ecosystem respiration. Ecological Modelling 240: 29–40.

21. Yuan WP, Liu SG, Yu GR, Bonnefond JM, Chen JQ, et al. (2010) Global estimates of evapotranspiration and gross primary production based on MODIS and global meteorology data. Remote Sensing of Environment 114(7): 1416–1431.

22. Yuan W, Luo Y, Li X, Liu S, Yu G, et al. (2011) Redefinition and global estimation of basal ecosystem respiration rate. Global Biogeochemical Cycles 25(4) DOI: 10.1029/2011GB004150.

23. Li X, Liang S, Yu G, Yuan W, Cheng X, et al. (2013) Estimation of gross primary production over the terrestrial ecosystems in China. Ecological Modelling 261–262(0): 80–92.

24. Reichstein M, Falge E, Baldocchi D, Papale D, Aubinet M, et al. (2005) On the separation of net 20 ecosystem exchange into assimilation and ecosystem respiration: review and improved algorithm. Global Change Biology 11: 1424–1439.

25. Papale D, Reichstein M, Aubinet M, Canfora E, Bernhofer C, et al. (2006) Towards a standardized processing of Net Ecosystem Exchange measured with eddy covariance technique: algorithms and uncertainty estimation. Biogeosciences 3: 571–583.

26. Moffat AM, Papale D, Reichstein M, Hollinger DY, Richardson AD, et al. (2007) Comprehensive comparison of gap-filling techniques for eddy covariance net carbon fluxes. Agricultural and Forest Meteorology 147(3–4): 209–232.

27. Zhao M, Heinsch FA, Nemani RR, Runing SW (2005) Improvements of the MODIS terrestrial gross and net primary production global data set. Remote Sensing of Environment 95(2): 164–176.

28. Huntingford C, Lowe JA, Booth BBB, Jones CD, Harris GR, et al. (2009) Contributions of carbon cycle uncertainty to future climate projection spread. Tellus B 61(2): 355–360.

29. Schaefer K, Collatz JG, Tans P, Denning SA, Baker I, et al. (2008) Combined Simple Biosphere/Carnegie-Ames-Stanford Approach terrestrial carbon cycle model. Journal of Geophysical Research-Biogeosciences 113(G3) DOI: 10.1029/2007JG000603.

30. Nobre P, Siqueira LSP, de Almeida RAF, Malaguitti M, Giarolla E, et al. (2013) Climate simulation and change in the Brazilian Climate Model. Journal of Climate 26(17): 6716–6732.

31. Beer C, Reichstein M, Tomelleri E, Ciais P, Jung M, et al. (2010) Terrestrial gross carbon dioxide uptake: Global distribution and covariation with climate. Science 329(5993): 834–838.

32. Reichstein M, Tenhunen J, Roupsard O, Ourcival JM, Rambal S, et al. (2003) Inverse modeling of seasonal drought effects on canopy CO_2/H_2O exchange in three Mediterranean ecosystems. Journal of Geophysical Research Atmospheres 108(D23): 4726.

33. David TS, Ferreira MI, Cohen S, Pereira JS, David JS (2004) Constraints on transpiration from an evergreen oak tree in southern Portugal. Agricultural and Forest Meteorology 122(3–4): 193–205.

34. Leuning R, Cleugh HA, Zegelin SJ, Hughes D (2005) Carbon and water fluxes over a temperate Eucalyptus forest and a tropical wet/dry savanna in Australia: measurements and comparison with MODIS remote sensing estimates. Agricultural and Forest Meteorology 129(3–4): 151–173.

35. Yuan W, Luo Y, Richardson AD, Oren R, Luyssaert S, et al. (2009) Latitudinal patterns of magnitude and interannual variability in net ecosystem exchange regulated by biological and environmental variables. Global Change Biology 15(12): 2905–2920.

36. Eamus D, Prior L (2001) Ecophysiology of trees of seasonally dry tropics: Comparisons among phenologies. Advances in Ecological Research 32: 113–197.

37. Baldocchi D, Xu L, Kiang N (2004) How plant functional-type, weather, seasonal drought, and soil physical properties alter water energy fluxes for an oak-grass savanna and an annual grassland. Agricultural and Forest Meteorology 123(1–2): 13–39.

38. Kucharik CJ, Barford CC, Maayar ME, Wofsy SC, Monson RK, et al. (2006) A multiyear evaluation of a Dynamic Global Vegetation Model at three AmeriFlux forest sites: Vegetation structure, phenology, soil temperature, and CO_2 and H_2O vapor exchange. Ecological Modelling 196(1–2): 1–31.

39. Botta A, Viovy N, Ciais P, Friedlingstein P, Monfray P (2000) A global prognostic scheme of leaf onset using satellite data. Global Change Biology 6(7): 709–725.

40. Monasterio M, Sarmiento G (1976) Phenological strategies of plant species in the tropical savanna and the semi-deciduous forest of the Venezuelan Llanos. Journal of Biogeography 3(4): 325–355.

41. Borchert R (1994) Soil and stem water storage determine phenology and distribution of tropical dry forest trees. Ecology 75(5): 1437–1449.

42. Childes SL (1988) Phenology of nine common woody species in semi-arid, deciduous Kalahari Sand vegetation. Vegetatio 79(3): 151–163.

43. Bonan GB, Lawrence PJ, Oleson KW, Levis S, Jung M, Reichstein M, et al. (2011) Improving canopy processes in the Community Land Model version 4 (CLM4) using global flux fields empirically inferred from FLUXNET data. Journal of Geophysical Research-Biogeosciences 116(G2) DOI: 10.1029/2010JG001593.

44. Rogers A (2013) The use and misuse of V_{cmax} in Earth System Models. Photosynthesis Research 1–15.

45. Groenendijk M, Dolman AJ, van der Molen MK, Leuning R, Arneth A, et al. (2011) Assessing parameter variability in a photosynthesis model within and between plant functional types using global Fluxnet eddy covariance data. Agricultural and Forest Meteorology 151(1): 22–38.

46. Kleidon A, Fraedrich K, Low C (2007) Multiple steady-states in the terrestrial atmosphere-biosphere system: a result of a discrete vegetation classification. Biogeosciences 4(5): 707–714.

47. Kucharik CJ, Foley JA, Delire C, Fisher VA, Coe MT, et al. (2000) Testing the performance of a Dynamic Global Ecosystem Model: Water balance, carbon balance and vegetation structure. Global Biogeochemical Cycles 14(3): 795–825.

Spatial and Temporal Variations of Ecosystem Service Values in Relation to Land Use Pattern in the Loess Plateau of China at Town Scale

Xuan Fang[1], Guoan Tang[1]*, Bicheng Li[2], Ruiming Han[3]

1 Key Laboratory of Virtual Geographic Environment, Ministry of Education, School of Geography Science, Nanjing Normal University, Nanjing, China, **2** Research Center of Soil and Water Conservation and Ecological Environment, Chinese Academy of Sciences, Yangling, Shaanxi, China, **3** School of Geography Science, Nanjing Normal University, Nanjing, China

Abstract

Understanding the relationship between land use change and ecosystem service values (ESVs) is the key for improving ecosystem health and sustainability. This study estimated the spatial and temporal variations of ESVs at town scale in relation to land use change in the Loess Plateau which is characterized by its environmental vulnerability, then analyzed and discussed the relationship between ESVs and land use pattern. The result showed that ESVs increased with land use change from 1982 to 2008. The total ESVs increased by 16.17% from US$ 6.315 million at 1982 to US$ 7.336 million at 2002 before the start of the Grain to Green project, while increased significantly thereafter by 67.61% to US$ 11.275 million at 2008 along with the project progressed. Areas with high ESVs appeared mainly in the center and the east where largely distributing orchard and forestland, while those with low ESVs occurred mainly in the north and the south where largely distributing cropland. Correlation and regression analysis showed that land use pattern was significantly positively related with ESVs. The proportion of forestland had a positive effect on ESVs, however, that of cropland had a negative effect. Diversification, fragmentation and interspersion of landscape positively affected ESVs, while land use intensity showed a negative effect. It is concluded that continuing the Grain to Green project and encouraging diversified agriculture benefit to improve the ecosystem service.

Editor: Ricardo Bomfim Machado, University of Brasilia, Brazil

Funding: This study was sponsored by the Jiangsu Planned Projects for Postdoctoral Research Funds (No. 1401033C), the National Natural Science Foundation of China (No. 41401441), and the Priority Academic Program Development of Jiangsu Higher Education Institutions (PAPD) (No. 164320H101). The funders had no role in study design, data collection and analysis, decision to publish, or preparation of the manuscript.

Competing Interests: The authors have declared that no competing interests exist.

* Email: tangguoan@njnu.edu.cn

Introduction

Ecosystem contributes to human welfare by providing goods and services directly and indirectly [1–2]. With widely spreading of environmental problems, ecosystem service received increasing attention. Many studies showed human factors, such as urban sprawl [3,4,5], socioeconomic changes [6], agricultural policies [7,8], could affect natural or artificial ecosystems. Land use, an original and foundational human activity and represents the most substantial human alteration to systems on the planet of earth for long-term study [9], plays an important role in providing ecosystem services, including biodiversity, water filtration, retention of soil, etc. [10] Inappropriate land use may lead to significant degradation of local and regional ecological services [11]. Moreover, there were studies showed that ecosystem service trade-offs could successful apply to land use planning [12,13]. Understanding the relationship between ecosystem services and land use change is essential for maintaining a healthy ecosystem and getting sustainable services.

The growing body of literatures focused on how ecosystem service changes in response to land use change of different regions [14,15,16,17,18]. However, these studied focus on the impact of land use type on ecosystem service, while the spatial pattern of land that reflects ecological processed and functions [19] get less attention. Monitoring the characteristic of landscape patterns including area, shape, diversity, etc., is helpful to deeply understand the relationship between ecosystem service and land use change and then to provide complete references for land use planning.

The Loess Plateau is the area suffered from the most severe soil erosion in the world, and it is also a major agricultural production region in China [20]. Long-term poor land use has resulted in vegetation destruction and accelerated soil erosion [21]. To control soil erosion and restore the ecosystem, the Grain for Green project converting slope cropland to grassland or forestland was implemented in 1999 by the Chinese Government [22]. The land use on the plateau under the project has changed significantly. Studying the ecosystem service in relation to land use change before and after the Grain to Green project was crucial for ecosystem protection and agricultural sustainability for the area. Researchers have analyzed ecosystem service at different scales within the Loess Plateau [17,18,23]. However, town is a basic administrative area in China. Exploring the characteristic of ecosystem services change at town scale is of practical significance to provide operable land use planning.

Ecosystem service values (ESVs) is monetary assessment of ecosystem services. This paper examined the characteristics of ESVs at Hechuan town, a typical town in the hilly and gully region of the Loess Plateau. The objectives of this study were: 1) to analyze the changes in land use pattern from 1982 to 2008; 2) to access the spatial and temporal variation in ESVs in response to land use during this period; 3) to quantitively analysis the relationship between ESVs and land use pattern; and 4) to discuss how land use management is favorable for ecosystem service supply and the ecological and economic sustainable development.

Data and Methods

2.1 Ethics statement

No specific permits were required for the described studies, and the work did not involve any endangered or protected species.

2.2 Study area

The study area, Hechuan town (106°18′43″~106°32′16″E, 35°54′59″~36°06′05″N), is located in Guyuan city of the Ningxia Hui Autonomous Region of northwest China (Fig. 1), consisting 12 villages with 16,524 people. The reasons that Hechuan Town was chosen as the study area were, on the one hand, Hechuan town has the typical characteristics of Loess Plateau including the

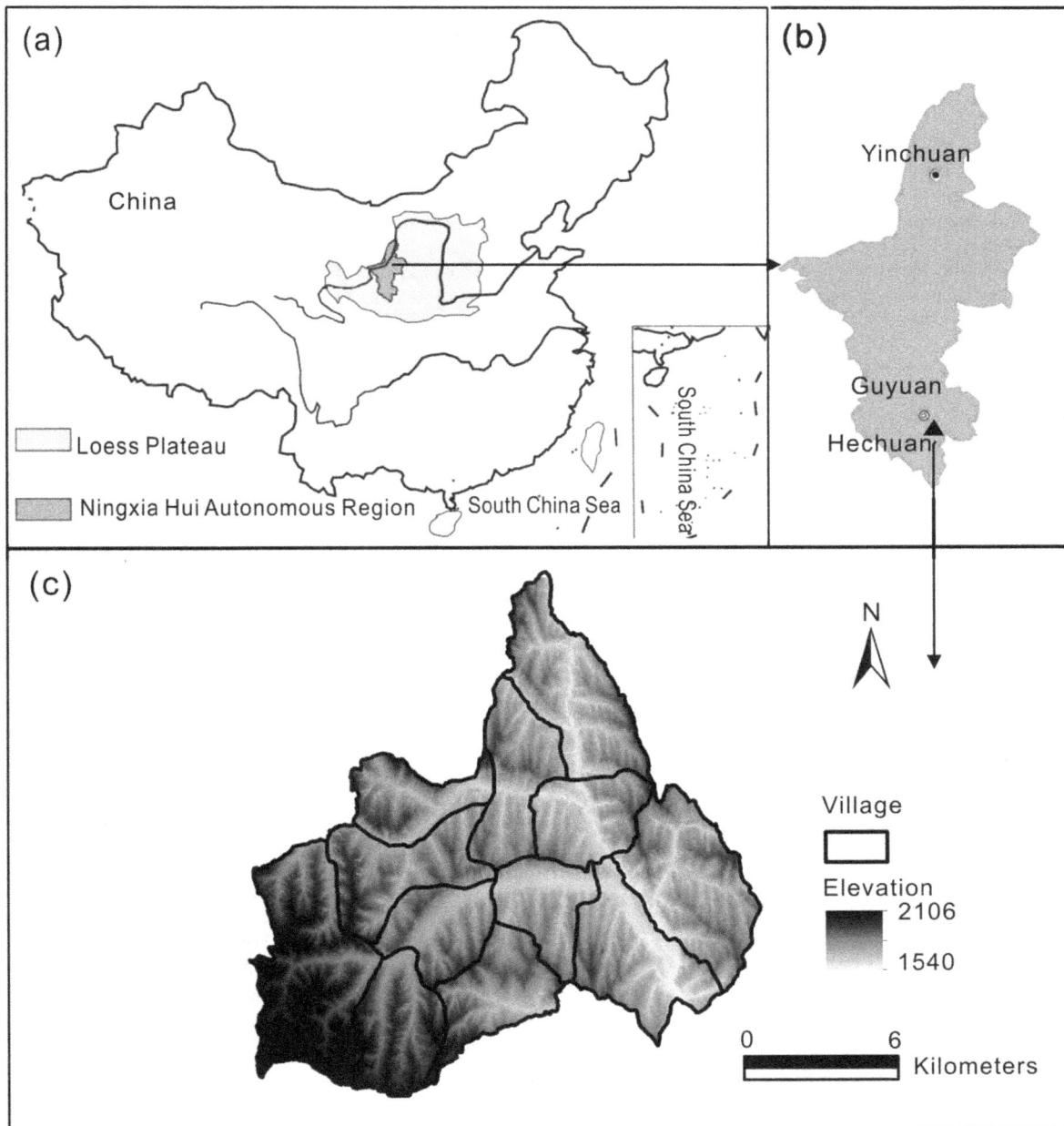

Figure 1. Location of the study area. Ningxia Province and the Loess Plateau, China (a), the location of Hechuan Town in the Loess area of Ningxia Province (b) and, the village boundary and the digital elevation model (DEM) map of Hechuan Town (c).

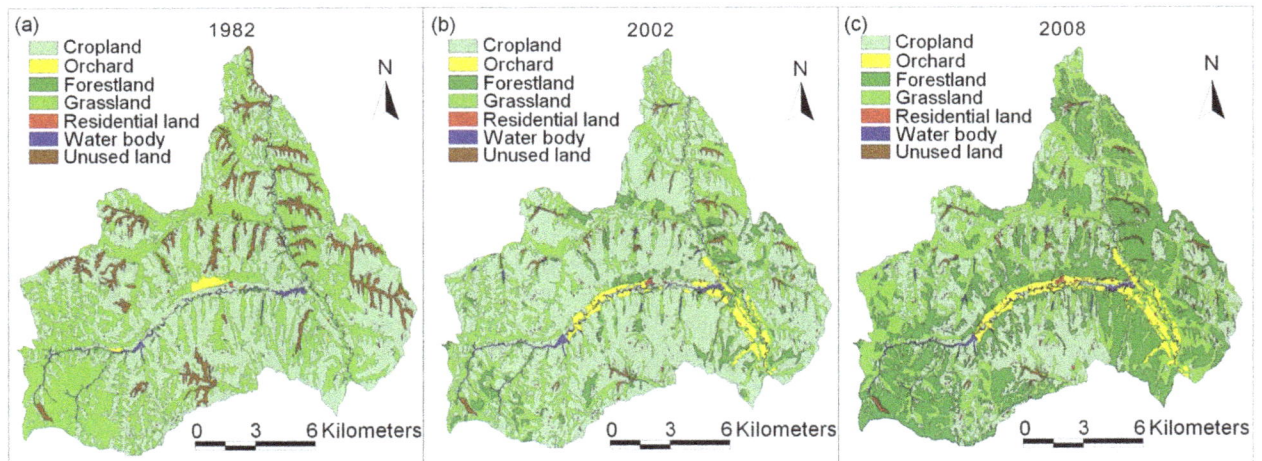

Figure 2. Land use maps of Hechuan town in 1982 (a), 2002 (b) and 2008 (c).

terrain of hill and gull, the fragile ecosystem and the backward economy; on the other hand, there was a long term ecological observation and experiment station in the study area, which facilitated the survey of land use and ecosystems. This town has an altitude ranging from 1540 to 2106 m, covering an area of 215.58 km^2. There exist the topographic differences in the town. The central area with river terrace stretches smoothly with a low elevation. The terrain in the northern area is fragmented while that of southern area is relatively simple. Hechuan town has a semi-arid continental temperate climate with the average annual temperature of 6.9°C and precipitation of 419 mm (1982–2002). Most of the annual precipitation is concentrated between June to September in the form of heavy storms that can cause severe soil erosion. The soil is composed of loessial soil and Dark loessial soils, which is erodible due to its weak cohesion and high infiltrability.

The ecosystem in Hechuan town is fragile with serious soil erosion and frequent natural disasters. Human disturbances of excessive land use, such as deforestation, overgrazing and over-reclamation further destructed the native natural grassland. Therefore, this area has long been in a vicious circle, endless cultivation and poverty. Since the early 1980s, a variety of comprehensive investigation of soil erosion was practiced by Chinese Academy of Sciences. Shanghuang watershed, located in the east of Hechuan town, was taken as a key test area. The

ecological restoration covering the whole town was started from implementing the Grain for Green project after 2002 (launched in 1999 by China government). Since then, abandoned cropland, shrubland (*Caragana korshinskii*, *Hippophae rhamnoides*) and artificial grassland (*Medicago sativa*) was generated, which made a significant change on landscape pattern and ecosystem components providing a variety of ecosystem services. Meanwhile, farming and grazing, the traditional way of living, had to be changed, and raising livestock, orchards, and migrant working diversified their incomes.

2.3 Data acquisition and preprocessing

Land use data was the key data for evaluating landscape pattern and ecosystem service. The land use data of 1982 was obtained by digitizing the land use patches from the 1:10,000 scale topographic maps of 1982, in which the information of land use types and its boundary are clearly shown. The 10 m resolution of remote sensing image could be considered to be corresponding with the scale of 1:50000 [24,25]. The land use data of 1982 acquired from 1:10000 topographic maps was therefore generalized to be at 1:50000 scale [26]. The land use data of 2002 and 2008 were respectively extracted from the 10 m resolution multispectral Spot-5 image of 2002 and 2008 by updating the land use patches of

Table 1. Equivalent weight factor of ecosystem service values (ESVs) per hectare of terrestrial ecosystem in China [30].

	Cropland	Forestland	Grass land	Water body	Barren land
Gas regulation	0.72	4.32	1.5	0.51	0.06
Climate regulation	0.97	4.07	1.56	2.06	0.13
Water supply	0.77	4.09	1.52	18.77	0.07
Soil formation and retention	1.47	4.02	2.24	0.41	0.17
Waste treatment	1.39	1.72	1.32	14.85	0.26
Biodiversity protection	1.02	4.51	1.87	3.43	0.40
Food production	1.00	0.33	0.43	0.53	0.02
Raw material	0.39	2.98	0.36	0.35	0.04
Recreation and culture	0.17	2.08	0.87	4.44	0.24
Total	7.90	28.12	11.67	45.35	1.39

Table 2. The ecosystem service values (ESVs) per hectare of different land use types in Hechuan town (US$·ha-1·yr-1).

	Cropland	Orchard	Forestland	Grass land	Water body	Unused land
Gas regulation	22.570	91.222	135.422	47.022	15.987	1.881
Climate regulation	30.407	88.244	127.585	48.902	64.576	4.075
Water supply	24.138	87.930	128.212	47.649	588.397	2.194
Soil formation and retention	46.081	98.118	126.018	70.219	12.853	5.329
Waste treatment	43.573	47.649	53.918	41.379	465.514	8.150
Biodiversity protection	31.975	99.999	141.378	58.620	107.523	12.539
Food production	31.348	11.912	10.345	13.480	16.614	0.627
Raw material	12.226	52.351	93.416	11.285	10.972	1.254
Recreation and culture	5.329	46.238	65.203	27.272	139.184	7.523
Total	247.647	623.663	881.498	365.828	1421.619	43.573

1982 one by one in visual interpretation method. The interpretation sign was established by understanding the Spot image characteristics and carrying out field surveys in order to further determine the relationship between the true ground and the image. The kappa accuracy index [27] was used to assess the accuracy of the interpretation. The stratified random sampling method was used to generate the reference points on the classified image for the accuracy test. These reference points were located in the field with a GPS with 5-m precision for ground truth. The total kappa indexes are all higher than 0.85, which are higher than the minimum acceptable (0.7) [28]. Considering the characteristic of the land use in study area and the interpretation level of the data and to facilitate the calculation of ESVs, the land use was classified into seven types: cropland, orchard, forestland, grassland, residential area, water area, and unused land (Fig. 2).

To acquire accurate area data of the land use for ESVs estimation and facilitate analyzing the spatial distribution of ESVs, the topographic maps and Spot images were transformed to the same projection and coordinate system (the Albers-Conical-Equal-Area projection system and Krasovsky 1940 coordinate system) before the extraction of land use data, and all acquired land use data were transformed to Arc-grid formats with the same grid size (10 m×10 m). The above data processing was completed using ERDAS and ArcGIS software.

2.4 Analysis on land use pattern

The transfer matrix analysis of land use was produced to understand how land use changed. Landscape metrics analysis was used for spatial pattern analysis of land use. Landscape metrics has been adopted widely; meanwhile, its abilities to indicate ecological process gained increasing attention [29,30,31]. Conceptual flaws in landscape pattern analysis, limitations inherent in landscape metrics and the improper use of pattern analysis may lead to the misuse of landscape metrics [32]. For better explanations and predictions of ecological phenomena from ecological pattern, the landscape metrics in this study was therefore selected by two steps. Firstly, the diversity, the fragmentation and the dominance of landscape were all considered, and then 34 metrics was selected, by understanding the knowledge of the landscape pattern and the ecological services indication of landscape metrics [33,34] and referring to the previous studies on landscape pattern [4,31,35,36]. Secondly, a correlation analysis for the 34 metrics was employed to ensure the low redundancy among landscape metrics. If the coefficient between two metrics was significant at 0.05 level, only one metric of them could be eventually selected.

Landscape-level metrics providing general landscape information and class-level metrics providing more specific information about variations at the local level and spatial patterns of land use classes [37] were used to monitor the characteristics of landscape pattern. The selected landscape-level metrics were patch density (PD), area-weighted mean shape index (SHAPE_AM), Interspersion and Justaposition Index (IJI), and Shannon's diversity index (SHDI). The selected class-level metrics were PD, the percentage of landscape (PLAND), SHAPE_AM and IJI. PD and SHAPE_AM could show the fragmentation of landscape. SHDI and PLAND reflect the dominance of some land use type and the diversity of landscape, respectively. IJI reflects whether the patches or classes are contiguous. Landscape metrics analysis was conducted with above metrics by FRAGSTATS 3.3, in which the eight-neighbor rule was used to derive the patch number. Besides these metrics, the land use intensity index (LUII) was also used to describe the landscape pattern. It was calculated by the following equation [31]:

$$LUII = \sum_{i=1}^{n} A_i \times C_i \qquad (1)$$

where $LUII$ is the land use intensity index, A_i is the percentage of for a give land use type i, and C_i is the coefficient value of intensity for a give land use type i, that is assigned 4 for build-ups, 3 for farmland and 2 for forest, orchard, grassland and water bodies, and 1 for unused land.

2.5 Estimation of ESVs

Costanza et al.'s model of ESVs estimation was adopted in this study [1,2]. The model classified ecosystem service into 17 types of service functions and estimated the ESVs by placing an economic value on different biomes [34]. For the defects of this model, such as overestimating the agriculture ESVs and underestimating the wetland ESVs, Xie et al. proposed refined coefficients for ESVs assessment both solving the above problem and making it apply to China [33,34]. Based on this model, the total ESVs in the study area was calculated using the following formulas:

$$ESV_k = \sum_f A_k VC_{kf} \qquad (2)$$

Table 3. Land use transition matrix from 1982 to 2002 and from 2002 to 2008 (%).

1982	Cropland	Orchard	Forestland	Grassland	Residential land	Water body	Unused land	Total	Loss
2002									
Cropland	44.42	2.33	1.18	2.50	0.35	0.04	0.01	50.83	6.41
Orchard	0.17	0.05	0.26	0.09	0.00	0.00	0.00	0.57	0.53
Forestland	0.10	0.01	0.15	0.02	0.00	0.00	0.00	0.28	0.13
Grass land	12.63	0.39	4.74	22.15	0.01	0.08	0.00	40.01	17.86
Residential land	0.05	0.01	0.00	0.01	0.14	0.00	0.00	0.22	0.07
Water body	0.01	0.00	0.01	0.00	0.00	0.80	0.00	0.81	0.02
Unused land	1.39	0.00	0.16	3.98	0.00	0.07	1.66	7.27	5.61
Total	58.76	2.78	6.51	28.77	0.51	0.99	1.68	100.00	
Gain	14.35	2.73	6.36	6.62	0.37	0.19	0.01		

2002	Cropland	Orchard	Forestland	Grass land	Residential land	Water body	Unused land	Total	Loss
2008									
Cropland	27.09	1.00	20.75	9.84	0.07	0.00	0.00	58.76	31.68
Orchard	0.00	2.75	0.01	0.00	0.01	0.00	0.00	2.78	0.03
Forestland	0.05	0.04	5.84	0.57	0.01	0.01	0.00	6.51	0.67
Grass land	0.12	0.00	7.17	21.42	0.02	0.04	0.00	28.77	7.35
Residential land	0.00	0.00	0.00	0.00	0.51	0.00	0.00	0.51	0.00
Water body	0.02	0.02	0.00	0.01	0.00	0.94	0.00	0.99	0.05
Unused land	0.00	0.00	0.27	0.08	0.00	0.00	1.33	1.68	0.35
Total	27.27	3.82	34.05	31.91	0.62	1.00	1.33	100.00	
Gain	0.18	1.07	28.21	10.50	0.11	0.05	0.00		

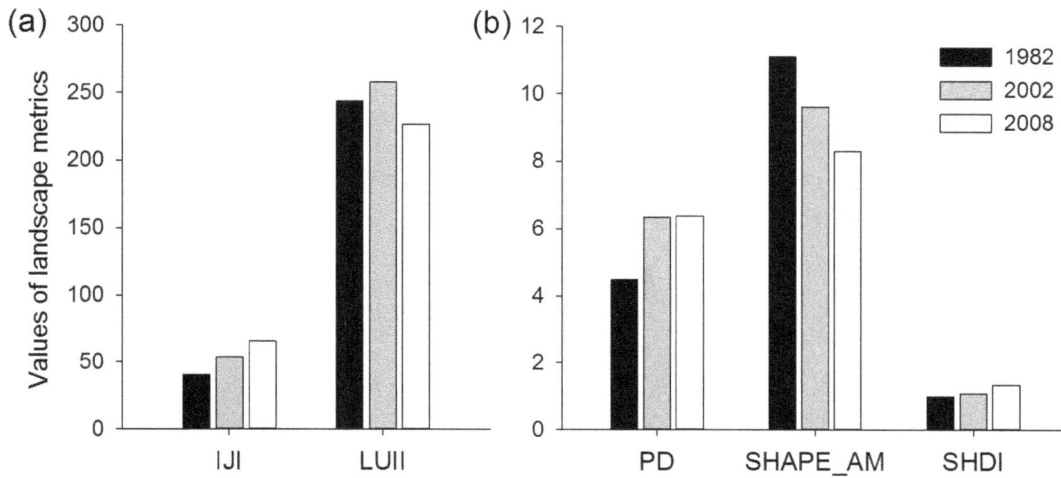

Figure 3. Landscape metrics at the landscape level in Hechuan Town in 1982, 2002 and 2008. IJI: Interspersion and Justaposition Index; LUII: land use intensity index; PD: patch density; SHAPE_AM: area-weighted mean shape index; SHDI: Shannon's diversity index.

$$ESV_f = \sum_k A_k VC_{kf} \qquad (3)$$

$$ESV = \sum_k \sum_f A_k VC_{kf} \qquad (4)$$

where ESV_k, ESV_f, ESV are the ESVs of land use type k, the ESVs of ecosystem service function type f, and the total ESVs respectively. A_k is the area (ha) for land use types. VC_{kf} is the value coefficient (US\$·ha-1·yr-1) for land use type k and ecosystem service function type f, which is the key for ESVs estimating. Xie et al.'s model was used to determine VC_{kf}, which can be expressed as follows:

$$VC_{kf} = R_{kf} \times V_f \qquad (5)$$

where R_{kf} is the equivalent weight factor of ecosystem service, V_f is food production values of agriculture land per area per year.

The equivalent weight factor was presented for customizing Chinese terrestrial ecosystem based on Costanza et al.'s model by surveying 500 Chinese ecologists (Table 1) [34]. It is the ratio of the ESVs to the economic value of average natural food production provided by agricultural land per hectare per year. The factors of land use types in our study were basically assigned based on the nearest ecosystems in Xie et al.'s model. However, minor adjustments were made. The equivalent weight factor of orchard which was not put forward clearly in Xie et al.'s model was determined by the mean of grassland and forestland by referring some researches [5,18]. The factor of unused land equates to that of barren land, and that of residential land was determined to zero.

The value of food production service of agriculture land per area per year was considered to be 1/7 of the actual price of food production in Xie et al.'s model. With the average actual food production of cropland in Hechuan town from 1982 to 2008 of 901.77 kg/ha which was get from *Statistic yearbook of the Yuanzhou District, Guyuan City, Ningxia Hui Autonomous Region* and the average grain price of US\$ 0.243 per kilogram (i.e. an equivalent of RMB Yuan 1.69 according to the average exchange

rate of 2008) in 2008, the value of food production service of cropland per area per year was calculated to be US\$ 31.348 (i.e. an equivalent of RMB Yuan 217.713 according to the average exchange rate of 2008). ESVs of one unit area of each land use types were then assigned as shown in Table 2.

After the ESVs were calculated by above processing, a sensitivity analysis was conducted to test the land use type's representative for ecosystem types and the certainty of the coefficients value for ecosystem service. A coefficient of sensitivity (CS) was used to indicate the degree of sensitivity of ESVs to a coefficients value, calculated by the following formula [5]:

$$CS = \left| \frac{(ESV_j - ESV_i)/ESV_i}{(VC_{jk} - VC_{ik})/VC_{ik}} \right| \qquad (6)$$

where ESV_j an ESV_i are the total ESVs of the initial status j and the adjusted status i, and VC_{jk} and VC_{ik} are the initial and adjusted coefficients. A 50% adjustment in the coefficients was made in the study. The greater the CS responded to the adjustment, the more critical is the use of an accurate coefficient [38]. A CS lower than 1 indicates the ESVs is inelastic to the coefficient and the estimation of ESVS is reliable. Otherwise, a CS greater than 1 indicates the estimation of ESVs is sensitive to the coefficient.

2.6 Correlation and regression analysis

The data of ESVs and landscape metrics was used to analysis the relationship between ecosystem service and land use pattern change. Because the spatial variation of landscape pattern exist among 12 villages in Hechuan town, the land use data of the three years (1982, 2002 and 2008) for the 12 villages can be considered as representing different landscape pattern on a time-for-space perspective [39]. Therefore, there were totally 36 sample data. Correlation and regression was employed for the relationship analysis, in which Multiple stepwise regression was specifically chosen considering the multicollinearity among landscape metrics. The dependents were the nine categories and total ESVs, while the corresponding independents were the landscape-level and class-level landscape metrics.

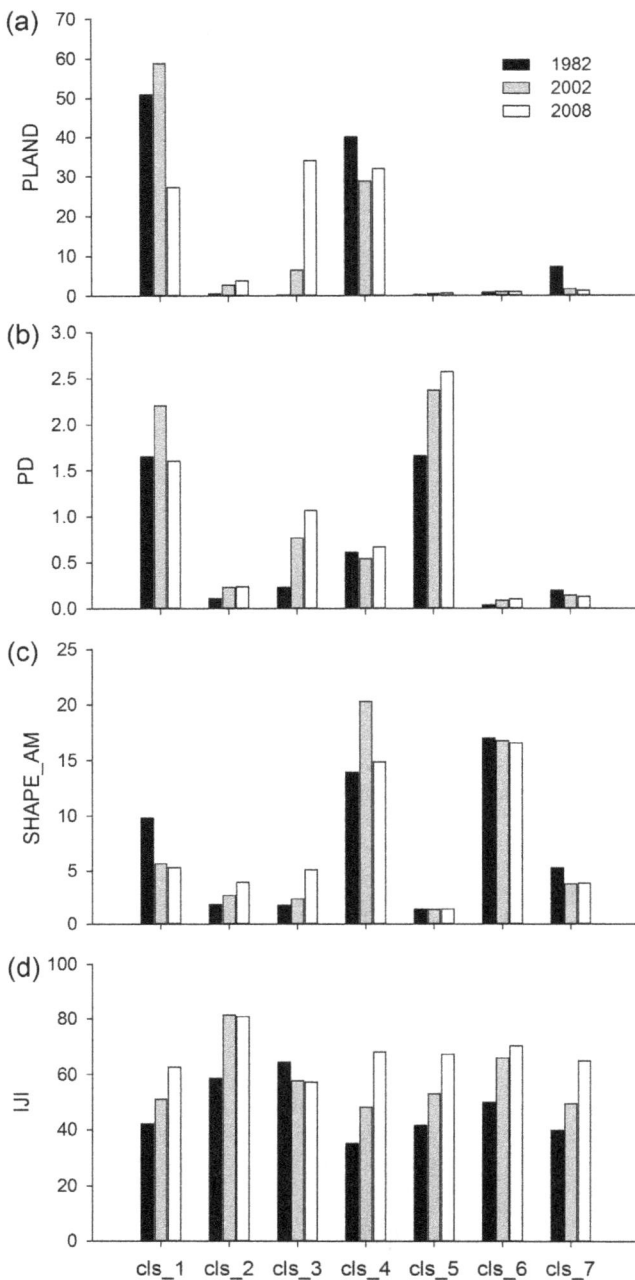

Figure 4. Landscape metrics at the class-level in Hechuan Town in 1982, 2002 and 2008. cls_1, cls_2, cls_3, cls_4, cls_5, cls_6, and cls_7 represent cropland, orchard, forestland, grassland, residential land, water body and unused land. PLAND: the percentage of landscape; PD: patch density; SHAPE_AM: area-weighted mean shape index; IJI: Interspersion and Justaposition Index.

Results

3.1 Changes of land use pattern

Table 3 showed the land use transition matrix. From 1982 to 2002, cropland as the dominant land use type increased from 50.83% to 58.76%. Grassland was the land use type with the largest change in area, decreasing from 40.01% to 28.77%. Orchard increased by 6.24% of total area, indicating the economic driving force of fruit trees on land use change. Forestland

Table 4. The change of ecosystem service values (ESVs) in Hechuan Town from 1982 to 2008.

		Cropland	Orchard	Forestland	Grass land	Water body	Unused land	Total
ESVs (10^6 US$ yr^{-1})	1982	2.714	0.077	0.051	3.155	0.249	0.068	6.315
	2002	3.137	0.374	1.237	2.269	0.304	0.016	7.336
	2008	1.456	0.514	6.470	2.517	0.305	0.013	11.275
Change of ESVs (10^6 US$ yr^{-1})	1982–2002	0.423	0.297	1.186	−0.887	0.054	−0.053	1.021
	2002–2008	−1.681	0.140	5.234	0.248	0.002	−0.003	3.939
	1982–2008	−1.258	0.437	6.419	−0.638	0.056	−0.056	4.960
Change of ESVs (%)	1982–2002	2.248	55.387	333.431	−4.045	3.145	−11.081	2.328
	2002–2008	−7.716	5.404	60.934	1.574	0.072	−2.992	7.731
	1982–2008	−6.674	81.578	1805.410	−2.913	3.232	−11.771	11.309
Average annual Change (%yr^{-1})	1982–2002	0.112	2.769	16.672	−0.202	0.157	−0.554	0.117
	2002–2008	−1.286	0.901	10.155	0.262	0.012	−0.498	1.289
	1982–2008	−0.256	3.137	69.439	−0.112	0.124	−0.452	0.435

Table 5. Values of different ecosystem service functions in 1982, 2002, and 2008.

	1982			2002			2008		
	ESVs (10^6 US$·yr^{-1})	%	Rank	ESVs (10^6 US$·yr^{-1})	%	Rank	ESVs (10^6 US$·yr^{-1})	%	Rank
Gas regulation	0.678	10.73	6	0.826	11.26	6	1.529	13.56	5
Climate regulation	0.791	12.53	5	0.936	12.75	5	1.540	13.65	4
Water supply	0.800	12.67	4	0.960	13.09	4	1.610	14.28	3
Soil formation and retention	1.141	18.06	1	1.260	17.17	1	1.764	15.65	1
Waste treatment	0.938	14.85	2	1.015	13.84	3	1.078	9.56	6
Biodiversity protection	0.915	14.49	3	1.054	14.37	2	1.738	15.42	2
Food production	0.466	7.38	7	0.506	6.90	7	0.367	3.25	9
Raw material	0.247	3.91	9	0.390	5.32	8	0.881	7.81	7
Recreation and culture	0.339	5.37	8	0.388	5.29	9	0.768	6.81	8
Total	6.315	100.00		7.336	100.00		11.275	100.00	

increased from 0.57% to 2.78%, reflecting that ecological restoration began to gain attention. From 2002 to 2008, cropland and forestland changed significantly, decreasing from 58.76% to 27.27% and increasing from 6.51% to 34.05% respectively. Land use structure was transferred from cropland dominated (58.76%) to cultivated land (27.27%), forestland (34.05%) and grassland (31.91%) relatively balanced distributed.

The most notable change of land use from 1982 to 2002 was the conversion from grassland to cropland and forestland with 12.63% and 4.74% of the total area respectively. The conversions from cropland (2.50%) and unused land (3.98%) to grassland were not adequate to compensate for the grass loss. From 2002 to 2008, the notable changes of land use were cropland to forestland, cropland to grassland, and grassland to forestland, with the rates of 20.75%, 9.84%, and 7.17% respectively. It was found that the conversion among land use types was more outstanding and concentrated than that before 2002, reflecting that the Grain for Green project as an ecological policy had great influence on land use change.

The results of landscape-level metric analysis were exhibited in Fig. 3. The significant increased PD from 1982 to 2002 reflected the landscape fragmentation. It was relative to the increase of patches on the land use types with intense human disturbance, such as cropland, residential land and artificial reservoir. Oppositely, the slight change of PD from 2002 to 2008 reflected that human disturbance became stable. The change of human disturbance was also demonstrated by the change of LUII which increased before 2002 and decreased after 2002. SHAPE_AM decreased in the study period, showing the landscape became more regular in shape. The increase of IJI suggested that the landscape became more contiguous and the ecological connectivity among land use types increased. SHDI increase obviously from 2002 to 2008, which related to that the land use structure became even.

Fig. 4 showed the change of class-level metrics. The PLAND of land use types indicated that cropland, forestland, and grassland had significantly influence on land use pattern. PD in orchard, forestland, and residential land increased obviously, attributing to the increasing area of these land use types and the fragmental terrain. SHAPE_AM showed that cropland and unused land became more regular in shape, while orchard and forestland more complicated. IJI increased generally in land use types. Orchard was the most contiguous with high IJI, which was relative to its concentrated distribution across the river terrace.

3.2 ESVs from 1982 to 2008

The ESVs of each land use type and the total ESVs was shown in Table 4. The total ESVs of Hechuan town was US$ 6.315, US$ 7.336 and US$ 11.275 million in 1982, 2002 and 2008, respectively. From 1982 to 2002, the decline of ESVs caused by the decrease of grassland was offset by the increase of forestland, orchard and cropland, resulting that the total ESVs increased by US$ 1.021 million. From 2002 to 2008, the total ESVs increased by US$ 3.939 million, mainly due to the increase of forestland. The average annual change rate of total ESVs before and after 2002 was quite different, that is 0.81% and 8.95% respectively. It indicated the Grain to Green project implemented since 2002 had a significant effect on the ecosystem service. It was also shown from the value of ESVs produced by forestland occupying 57.39% of the total ESVs. Overall, the total ESVs increased US$ 4.960 million during the study period, mainly due to the increase of ESVs by the increase of forestland and orchard far beyond the decrease of ESVs by the decrease of cropland and grassland. It was essentially because of the higher coefficient value of forestland and orchard than that of cropland and grassland.

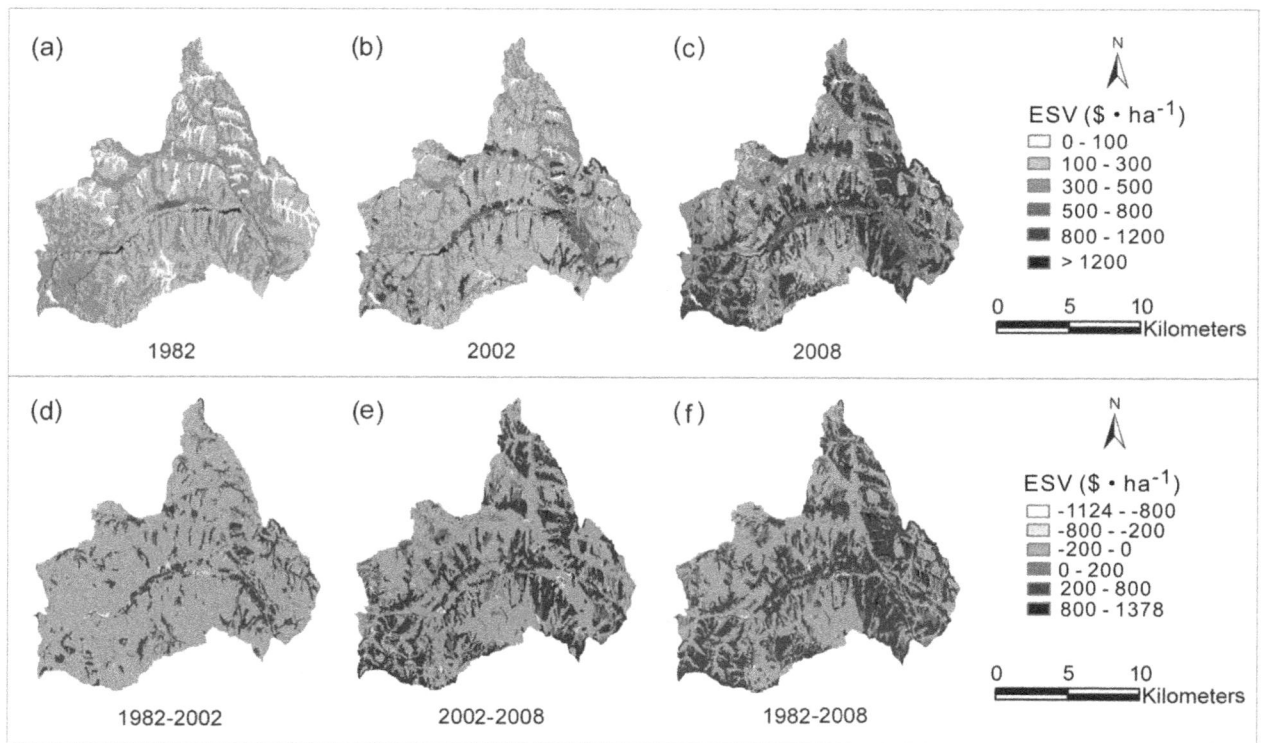

Figure 5. Spatial and temporal distribution of ecosystem service values (ESVs) in Hechuan Town from 1982 to 2008. The spatial distribution of ESVs in 1982 (a), 2002 (b) and 2008 (c), and the spatial-temporal changes of ESVs between time intervals from 1982 to 2002 (d), 2002 to 2008 (e) and 1982 to 2008 (f).

The ESVs of each ecosystem function type was shown in Table 5. Expect for food production, the values of ecosystem service functions increased especially after 2002. The decrease of food production was due to the great decline of cropland in the Grain to Green project. The ESVs proportion of each ecosystem function type to the total ESVs represented the contribution of each ecosystem function to the total ESVs. It was found that the functions of soil formation and retention, waste treatment, and food production were decline during 1982 to 2008, while other functions were improved. The rank of the contribution by each ecosystem service function was also estimated. It was basically stable except for relatively obvious decline in the rank of waste treatment and food production. In 2008, the rank order for each ecosystem service was as follows from high to low, soil formation and retention, biodiversity protection, water supply, climate regulation, gas regulation, waste treatment, raw material, recreation and culture, and food production. Soil formation and retention was the highest during the study period.

3.3 Spatial distribution of ESVs

Maps of ESVs in different periods (Fig. 5) showed the spatial distribution of ESVs of unit area in Hechuan town, directly reflecting the difference of ESVs among land use types. In 1982, the ESVs>4000 mostly appeared in the center of the town where river and river terrace located. It was because water body and orchard which intensely distributed in river terrace for its high water demand both had high ESVs. Therefore, due to the orchard increasing intensely and the forest increasing scatteredly, the increase of ESVs also mainly happened across the river terrace in 2002. Since 2008, the ESVs>4000 spread widely with the increase of forestland transformed from cropland. The lowest ESVs mostly

occurred in the gully where unused land was distributed in 1982. With vegetation recovery in the gully, the low ESVs happened from gully to terraced hillside where cropland with low ESVs was distributed in 2008. Fig. 5d–f showed the temporal change of ESVs spatial distribution. The change characteristic of 2002 to 2008 was adjacent to that during the total study period, reflecting that the change of ESVs mainly occurred after 2002, just after the Grain to Green project.

3.4 Relationship between ESVs and land use pattern

From the above analysis on the change of land use and ESVs in quantity and spatial distribution, we could infer there was some relationship between land use change and ecosystem service. To quantitively understand the relationship, the correlation analysis and regression analysis between ESVs and landscape pattern metrics was conducted.

Table 6 showed there existed significant correlations between ESVs and many landscape metrics (p<0.01), which explained that landscape pattern affected ESVs significantly. For example, the correlation coefficients between total ESVs and landscape metrics showed that there existed significantly positive relationship between SHDI (0.433), PLAND_3 (0.677), SHAPE_AM_3 (0.744), IJI_4 (0.513) and ESVs, and negative relationship between LUII (−0.634), PLAND_1 (−0.752) and ESVs. It reflected that the diversity and intensity of land use had important effects on total ESVs. It also reflected that cropland, forestland and grassland were the land use types which had significant effects on total ESVs. On quantity,the less the cropland and the more the forestland, the higher the total ESVs were. As to the landscape shape, the more regular the cropland and the more complex the forestland, the higher the total ESVs were. The higher the IJI of grassland, the

Table 6. Correlation coefficients between ecosystem service values (ESVs) and landscape pattern metrics.

	TESVs	ESVs_1	ESVs_2	ESVs_3	ESVs_4	ESVs_5	ESVs_6	ESVs_7	ESVs_8	ESVs_9
PD	0.035	0.497*	0.509*	0.539*	0.477*	0.516*	0.499*	-0.221	0.547	0.478*
SHAPE_AM	0.326	-0.216	-0.220	-0.188	-0.177	-0.088	-0.205	0.026	-0.290	-0.166
IJI	0.292	0.624*	0.635*	0.639*	0.597*	0.534*	0.621*	-0.293	0.687	0.586*
SHDI	0.433*	0.763*	0.764*	0.766*	0.741*	0.507*	0.765*	-0.636*	0.775*	0.765*
LUII	-0.634*	-0.681*	-0.658*	-0.618*	-0.675*	-0.113	-0.684*	0.977*	-0.599*	-0.734*
PLAND_1	-0.752*	-0.810*	-0.795*	-0.772*	-0.811*	-0.334	-0.815*	0.952*	-0.742*	-0.853*
PD_1	0.045	-0.055	-0.063	-0.056	-0.054	-0.113	-0.051	-0.134	-0.091	-0.022
SHAPE_AM_1	-0.476*	-0.369	-0.358	-0.330	-0.368	-0.063	-0.369	0.495	-0.328	-0.390
IJI_1	0.189	0.542*	0.552*	0.530*	0.510*	0.405	0.533*	-0.199	0.619	0.485*
PLAND_2	0.323	0.527*	0.541*	0.590*	0.520*	0.595*	0.534*	-0.246	0.558	0.525*
PD_2	0.420	0.457*	0.471*	0.515*	0.449	0.529*	0.463*	-0.193	0.491	0.452
SHAPE_AM_2	0.159	0.450	0.468*	0.541*	0.439	0.629*	0.460*	-0.149	0.493	0.452
IJI_2	0.207	0.392	0.409	0.483*	0.376	0.583*	0.402	-0.115	0.438	0.396
PLAND_3	0.677*	0.984*	0.983*	0.941*	0.975	0.558*	0.980*	-0.770*	0.988*	0.961*
PD_3	0.276	0.631*	0.637*	0.629	0.625	0.477*	0.629*	-0.383	0.653	0.606*
SHAPE_AM_3	0.744*	0.828*	0.827*	0.780*	0.820	0.449*	0.821*	-0.623*	0.836*	0.799*
IJI_3	0.040	0.231	0.241	0.291	0.208	0.356	0.236	-0.069	0.276	0.231
PLAND_4	0.224	-0.192	-0.212	-0.203	-0.159	-0.298	-0.181	-0.264	-0.311	-0.110
PD_4	-0.294	0.199	0.201	0.190	0.160	0.101	0.192	-0.086	0.257	0.171
SHAPE_AM_4	0.455	-0.061	-0.065	-0.049	-0.028	-0.029	-0.053	-0.086	-0.125	-0.022
IJI_4	0.513*	0.717*	0.719*	0.705*	0.697*	0.457*	0.715*	-0.539*	0.739*	0.701*
PLAND_5	-0.035	0.290	0.313	0.381	0.279	0.578*	0.297	0.081	0.357	0.271
PD_5	-0.047	0.244	0.269	0.322	0.244	0.548*	0.248	0.188	0.314	0.208
SHAPE_AM_5	0.118	0.081	0.082	0.053	0.072	-0.009	0.075	-0.004	0.105	0.055
IJI_5	0.307	0.461*	0.470*	0.525*	0.446*	0.512*	0.471*	-0.312	0.482	0.477*
PLAND_6	0.047	0.139	0.167	0.360	0.148	0.852	0.170	0.064	0.153	0.207
PD_6	-0.160	0.122	0.137	0.198	0.105	0.371	0.128	0.080	0.172	0.118
SHAPE_AM_6	0.378	0.088	0.086	0.140	0.080	0.141	0.099	-0.218	0.067	0.137
IJI_6	0.020	0.201	0.217	0.276	0.197	0.438	0.208	0.028	0.239	0.197
PLAND_7	-0.385	-0.447	-0.474*	-0.525*	-0.507*	-0.769*	-0.456*	-0.041	-0.442	-0.422
PD_7	-0.270	-0.105	-0.129	-0.182	-0.154	-0.494*	-0.114	-0.245	-0.108	-0.089
SHAPE_AM_7	-0.236	-0.313	-0.329	-0.434	-0.348	-0.650*	-0.333	0.150	-0.279	-0.358
IJI_7	0.210	0.416	0.408	0.287	0.402	-0.083	0.394	-0.254	0.443	0.341

TESVs: the total ecosystem service values (ESVs); ESVs_1: the ESVs of gas regulation; ESVs_2 climate regulation; ESVs_3: the ESVs of water supply; ESVs_4: the ESVs of soil formation and retention; ESVs_5: the ESVs of waste treatment; ESVs_6: the ESVs of biodiversity protection; ESVs_7 the ESVs of food production; ESVs_8: the ESVs of raw material; ESVs_9: the ESVs of recreation and culture. PD: patch density; SHAPE_AM: area-weighted mean shape index; IJI: Interspersion and Justaposition Index; LUII: land use intensity index; PLAND: percentage of landscape. The 1, 2, 3, 4, 5, 6, 7 after the above landscape metrics respects different landscape, that is cropland, orchard, forestland, grassland, residential land, water body and unused land, respectively.
*significant at 0.01 level.

Table 7. Regression analysis between ecosystem service values (ESVs) and landscape patterns (n = 36).

Dependent	Standardized coefficients regression	R²	Sig.
Gas regulation	0.878×PLAND_3+0.166×PLAND_2-0.099×PLAND_1-0.068×IJI_1	0.990	*
Climate regulation	0.790×PLAND_3-0.197×PLAND_7-0.190×LUII+0.081×PLAND_2	0.998	*
Water supply	0.665×PLAND_3-0.317×PLAND_7-0.254×LUII+0.106×PLAND_2	0.955	*
Soil formation and retention	0.684×PLAND_3-0.301×PLAND_7-0.284×LUII+0.066×PLAND_2	0.998	*
Waste treatment	0.672×PLAND_6+0.365×PLAND_3-0.352×PLAND_7+0.051×PLAND_2+0.049×PLAND_5	0.993	*
Biodiversity protection	0.059×SHDI +0.861×PLAND_3-0.033×SHAPE_AM_3+0.133×PLAND_1	0.967	*
Food production	0.742×LUII-0.173×PLAND_3-0.052×SHDI+0.106×PLAND_1	0.991	*
Raw material	0.964×PLAND_3+0.091×SHDI+0.068×LUII	0.981	*
Recreation and culture	−0.747×PLAND_1+0.380×IJI_1	0.853	*
Total	-0.588×PLAND_1+0.569× SHAPE_AM_3-0.303×SHDI	0.709	*

*significant at 0.01 level.
PLAND_1: the percentage of cropland; PLAND_2: the percentage of orchard; PLAND_3: the percentage of forestland; PLAND_5: the percentage of residential land; PLAND_6: the percentage of water body; PLAND_7: the percentage of unused land; SHAPE_AM_3: the area-weighted mean shape index of forestland; IJI_1: the Interspersion and Justaposition Index of cropland; LUII: land use intensity index; SHDI: Shannon's diversity index.

higher the total ESVs were. This indicated that the connectivity of grassland was important for ecosystem service.

Correlation also occurred between ESVs of all the functions and landscape metrics (Table 6). However, the relationships between ESVs of different functions and landscape pattern were different. For example, the correlation between food production and landscape pattern was almost opposite from that between other ecosystem functions and landscape pattern. For example, PLAND_1 had a positive effect on food production; SHDI, PLAND_3, SHAPE_3, and IJI_4 had a negative effect on food production. It could infer that there were contradictions between food production and other ecosystem functions.

As shown in Table 7, the result of regression analysis further explained that the ESVs was correlated significantly with landscape pattern. The total ESVs could be predicted by PLAND on cropland, SHAPE_AM on grassland, and SHDI. ESVs of all kinds of ecosystem functions also could be explained by landscape metrics. These regression equations indicated that landscape-level metrics (such as SHDI and LUII) and class_level metrics (such as PLAND of forestland, orchard, and cropland, unused land, SHAPE of forestland, IJI of cropland) acted as predictors for categories of ecosystem services. Specifically, the proportion of forest (PLAND_3) accounted for almost all of the categories of ecosystem services.

Discussion

4.1 Reliability of ESVs

This study estimated ESVs by multiplying the area for each land use types by the corresponding value coefficients. As discussed in the previous researches, estimations using this method was coarse

with high variation and uncertainty for the following reasons, limitations on the economic evaluation [1], problems of double counting and scales [40,41,42], the complex, dynamic and nonlinear ecosystems [43], the imperfect matches of land use categories as proxies [38] and the accuracy of the ecosystem value coefficients [5]. This study also existed such uncertainty on ESVs estimation. For example, the value coefficient of orchard, determined by the average of forest and grassland, was an approximate estimation and need a further exploration. However, the estimation of temporal variation on ESVs was considered to be more reliable than that of cross-sectional analysis [5]. In addition, the sensitivity analysis of the estimated ESVs with 50% adjustment in the value coefficients was conducted. The result showed that the sensitivity coefficients of all land use categories were lower than 1 (Table 8), which suggested that despite of the above limitations, the estimated ESVs are reliable and useful for subsequent study.

4.2 Relationship between ESVs and landscape pattern

It is usually assumed that land use can affect the ecosystem service. Moreover, a few studies showed that there was a correlation between landscape pattern and ESVs [41,44]. This study signified this statement at town scale on the Loess Plateau. Land use configuration, land use intensity, landscape diversity, fragmentation and connectivity all affected ecosystem service.

The correlation analysis between ESVs and PLAND implied land use structure had significant impact on ecosystem service. Especially, the increase of forestland and the decrease of cropland played an important part in improving the ESVs in the past twenty years. It is closely related to the Grain to Green project comprehensively started in study area since 2002. In the project,

Table 8. The coefficient of sensitivity (CS) resulting from adjustment of ecosystem valuation coefficients.

	Cropland	Orchard	Forestland	Grass land	Water body	Unused land
1982	−0.430	−0.012	−0.008	−0.500	−0.039	−0.011
2002	−0.428	−0.052	−0.169	−0.309	−0.041	−0.002
2008	−0.129	−0.048	−0.574	−0.223	−0.027	−0.001

measures for optimizing land use structure were implemented, including restoring slope cropland into forest and grassland, banning grazing, transforming slopes into terraces, and building reservoirs, etc. Forestland and grassland increased by 423.19% (27.54% of the study area) and 10.93% (3.15% of the study area), and cropland decreased by 53.59%(31.49% of the study area) (Table 3). The increase of ESVs due to the increase of forestland occupied 46.28% of the total ESVs in 2008 (Table 4). The result of the correlation analysis between ESVs and PLAND reflected that vegetation recovery could strongly enhance ecosystem service, and it was coincident with many other studies on the Loess Plateau [17,18,23,45]. LUII, which also related to the proportion of land use types, implied the intensity of human activities. This study showed land use intensity had a negative effect on ecosystem service with negative correlation coefficients (-0.634) (Table 6). It was coincident with some studies on ESVs change under urbanization [5,31]. These studies showed that urbanization which means the increase of land use intensity led to considerable declines in ESVs.

Landscape diversity always presents high positive relevance with biodiversity [46]. Our results were coincident to previous statements given the positive relationships between SHDI and biodiversity conservation. However, there were studies reporting the negative relationships between them, in which the increase of SHDI was the result of rapid urban sprawl [31]. In our study, the increase of SHDI was because land use structure became more balanced, which was the result of the increase of forestland. In addition, landscape diversity could also promote agricultural production [47]. Our study disagreed with this statement, and showed that food production was weakened with landscape diversification. It was because that the increase of SHDI was the result of a larger number of conversion from cropland to forestland. Therefore, the relationship between landscape diversity and biodiversity conservation as well as food production should not be treat as the same but be understood considering the driving force of SHDI change.

Fragmentation could lead to declining habitat quality, lower wildlife survival, and limited movement of soil microorganisms [48], and subsequently cause the decrease of ecosystem service [30]. Our study disagreed with this statement. For example, PD of the total landscape, PD_Forest, PD_orchard and shape_ Forest revealed significantly positive impacts on most categories of ESVs (Table 6–7). The increase of PD and the decrease of connectivity of landscape were usually simultaneous, which is disagreed in our study (Fig. 3 and Fig. 4). The landscape became more contiguous as IJI shown. Table 6 showed the IJI had significantly positive impacts on ESVs. Especially, the increase of IJI of grassland promoted the total ESVs and all categories of ESVs. This maybe because the connectivity of landscape has contribution to habitat corridors [49] and forest production [50].

Based on the relationship between ESVs and landscape pattern, we could improve the ecosystem service by the adjustment of land use policy. On the one hand, continuing to implement the Grain to Green project is helpful for improving ESVS, because it could increase the vegetation coverage, decline the intensity of land use, and make cropland become regular by canceling the slope cropland. On the other hand, diversified agriculture gathering planing fruit trees, planting crops and breeding, which could promote the diversification of land use, should be encouraged to increase both ESVs and farmer's incomes.

Conclusion

ESVs at town scale in the Loess Plateau were estimated in Hechuan town of Ningxia Hui Autonomous Region from 1982 to 2008. It was concluded that ESVs varied with land use change. ESVs in 1982, 2002, and 2008 were US\$ 6.315, US\$ 7.336 and US\$ 11.275 million respectively. Among all the land use types, forestland, grassland and cropland had important contribution (> 90%) on ESVs. The total ESVs increased slowly by 16.17% due to the decrease of grassland from 1982 to 2002, while the total ESVS increased significantly by 67.61% due to the increase of forestland from 2002 to 2008. Areas with high services level were mainly located in the center due to orchard and east due to forestland, while areas with low services level mainly located in the north and south sides due to cropland.

Land use pattern had a significant effect on ecosystem service in our study by analyzing and discussing the relationship between landscape pattern and ESVs. The proportion of forestland had a positive effect on ecosystem service while that of cropland had a negative effect on ESVs. The diversity and interspersion of landscape both had a positive effect on ESVs. Land use intensity which reflects the intensity of human activities had a negative effect on ESVs. Fragmentation had positive effect on ESVs, which was disagreed with the previous studies because the fragmentation in study area was related to the increased patch of such land use types as forestland, water body, orchard.

Based on the results of this study, it was conclude that land use pattern was important for ecosystem service. Therefore, we could improve the ecosystem service by the adjustment of land use policy. Continuing the Grain to Green project is reasonable and significant because it could increase the vegetation coverage and decline land use intensity. Diversified agriculture collecting planing fruit trees, growing food and breeding should be encouraged, because it could not only promote ecosystem service by increasing landscape diversification but also improve people's incomes.

Author Contributions

Conceived and designed the experiments: XF. Analyzed the data: XF. Contributed reagents/materials/analysis tools: GAT BCL. Contributed to the writing of the manuscript: XF GAT RMH.

References

1. Costanza R, Arge DR, Groot DR, Farber S, Grasso M, et al. (1997) The value of the world's ecosystem services and natural capital. Nature 387: 253–260.
2. Costanza R, Cumberland J, Daly H, Goodland R, Norgaard R (1997) An Introduction to ecological economics. Delray Beach Fla USA: St Lucie Press.
3. Kreuter UP, Harris HG, Matlock MD, Lacey RE (2001) Change in ESVs in the San Antonio area, Texas. Ecological Economics 39: 333–346.
4. Ronald CE, Yuji M (2013) Landscape pattern and ESV changes: Implications for environmental sustainability planning for the rapidly urbanizing summer capital of the Philippines. Landscape and Urban Planning 116: 60–72.
5. Li TH, Li WK, Qian ZH (2010) Variations in ESV in response to land use changes in Shenzhen. Ecological Economics 69: 1427–1435.
6. Cai YB, Zhang H, Pan WB, Chen YH, Wang XR (2013) Land use pattern, socio-economic development, and assessment of their impacts on ESV: study on

natural wetlands distribution area (NWDA) in Fuzhou city, southeastern China. Environ Monit Assess 185: 5111–5123.
7. Zaehle S, Bondeau A, Carter RT, Cramer W, Erhard M, et al. (2007) Projected changes in terrestrial carbon storage in europe under climate and land-use change, 1990–2100. Ecosystems 10: 380–401.
8. Eliska L, Jana F, Edward N, David V (2013) Past and future impacts of land use and climate change on agricultural ecosystem services in the Czech Republic. Land Use Policy, 33: 183–194.
9. Vitousek PM, Mooney HA, Lubchenco J, Melillo JM (1997) Human domination of earth's ecosystems. Science 277: 494–499.
10. Nasiri F, Huang GH (2007) Ecological viability assessment: A fuzzy multi-pleattribute analysis with respect to three classes of ordering techniques. Ecol Inform 2: 128–137.

11. Collin ML, Melloul AJ (2001) Combined land-use and environmental factors for sustainable groundwater management. Urban Water 3: 229–237.

12. Schmidta JP, Mooreb R, Alber M (2014) Integrating ecosystem services and local government finances into land use planning: A case study from coastal Georgia. Landscape and Urban Planning 122: 56–67.

13. Ernesto FV, Federico CF (2006) Land-use options for Del Plata Basin in South America: Tradeoffs analysis based on ecosystem service provision. Ecological Economics 57: 140–151.

14. Christine F, Susanne F, Anke W, Lars K, Franz M (2013) Assessment of the effects of forestland use strategies on the provision of ecosystem services at regional scale. Journal of Environmental Management 127: 96–116.

15. Ignacio P, Berta M, Pedro Z, David GDA, Carlos M (2014) Deliberative mapping of ecosystem services within and around Donana National Park (SW Spain) in relation to land use change. Reg Environ Change14: 237–251.

16. Mendoza-Gonzalez G, Martinez ML, Lithgow D, Perez-Maqueo O, Simonin P (2012) Land use change and its effects on the value of ecosystem services along the coast of the Gulf of Mexico. Ecological Economics 82: 23–32.

17. Su CH, Fu BJ (2013) Evolution of ecosystem services in the Chinese Loess Plateau under climatic and land use changes. Global and Planetary Change 101: 119–128.

18. Si J, Nasiri FZ, Han P, Li TH (2014) Variation in ESVs in response to land use changes in Zhifanggou watershed of Loess plateau: a comparative study. Environmental Systems Research 3: 2.

19. Turner MG, Gardner RH, O'Neill RV (2001) Landscape Ecology in theory and practice. New York: Springer-Verlag.

20. Ritsema CJ (2003) Introduction: soil erosion and participatory land use planning on the Loess Plateau in China. Catena 54: 1–5.

21. Fu BJ, Wang YF, Lu YH, He CS, Chen LD, et al. (2009) The effects of land-use combinations on soil erosion: a case study in the Loess Plateau of China. Progress in Physical Geography 33: 793–804.

22. Fu BJ, Chen DX, Qiu Y, Wang J, Meng QH (2002) Land Use Structure and Ecological Processes in the LoessHilly Area, China. Beijing: Commercial Press. (in Chinese).

23. Jing L, Zhiyuan R (2011) Variations in ESV in Response to Land use Changes in the Loess Plateau in Northern Shaanxi Province, China. Int. J. Environ. Res 5: 109–118.

24. Zhang TB, Tang JX, Liu DZ (2006) Feasibility of Satellite Remote Sensing Image About Spatial Resolution. Journal of Earth Sciences and Environment 28: 79–83.

25. Chu YF, Li ES, Lu J, Zhang KK (2007) The Adaptability Analysis to the Satellite Image Spatial Resolution and Mapping Scale. Hydrographic Surveying and Charting 27: 47–50.

26. Li Q, Liu C, Xi CY, Liu ML (2002) Cartographic Generalization of Digital Land Use Current Situation Map. Bulletin of Surveying and Mapping 9: 59–63.

27. Congalton RG (1991) A review of assessing the accuracy of classifications of remotely sensed data. Remote Sensing of Environment 37: 35–46.

28. Wang Y, Gao JX, Wang JS, Qiu J (2014) Value Assessment of Ecosystem Services in Nature Reserves in Ningxia, China: A Response to Ecological Restoration. PloS One 9: e89174. doi:10.1371/journal.Pone.0089174.

29. Ribeiro SC, Lovett A (2009) Associations between forest characteristics and socio-economic development: a case study from Portugal. Journal of Environmental Management 90: 2873–2881.

30. Su S, Jiang Z, Zhang Q, Zhang Y (2011) Transformation of agricultural landscapes under rapid urbanization: a threat to sustainability in Hang-Jia-Hu region, China. Applied Geography 31: 439–449.

31. Su SL, Xiao R, Jiang ZL, Zhang Y (2012) Characterizing landscape pattern and ESV changes for urbanization impacts at an eco-regional scale. Applied Geography, 34: 295–305.

32. Li H, Wu J (2004) Use and misuse of landscape indices. Landscape Ecology 19: 389–399.

33. Xie GD, Lu CX, Xiao Y, Zheng D (2003) The Economic Evaluation of Grassland Ecosystem Services in Qinghai Tibet Plateau, Journal of Mountain Science 21: 50–55. (in Chinese).

34. Xie GD, LU CX, Leng YF, Zheng D, Li SC (2008) Ecological assets valuation of Tibetan Plateau. Journal of Natural Resources 18: 190–196. (in Chinese).

35. Liu DL, Li BC, Liu Xianzhao Z, Warrington DN (2011) Monitoring land use change at a small watershed scale on the Loess Plateau, China: applications of landscape metrics, remote sensing and GIS. Environmental Earth Sciences 64: 2229–2239.

36. Pan WKY, Walsh SJ, Bilsborrow RE, Frizzelle BG, Erlien CM, et al. (2004) Farm-level models of spatial patterns of land use and land cover dynamics in the Ecuadorian Amazon. Agriculture, Ecosystems and Environment 101: 117–134.

37. de Groot RS, Wilson MA, Boumans RMJ (2002) A typology for the classification, description and valuation of ecosystem functions, goods and services. Ecological Economics 41: 393–408.

38. Kreuter UP, Harris HG, Matlock MD, Lacey RE (2001) Change in ESVs in the San Antonio area, Texas. Ecological Economics 39: 333–346.

39. Wu J, Jenerette GD, Buyantuyev A, Redman CL (2011) Quantifying spatiotemporal patterns of urbanization: the case of the two fastest growing metropolitan regions in the United States. Ecological Complexity 8: 1–8.

40. Turner RK, Paavola J, Coopera P, Farber S, Jessamya V, et al. (2003) Valuing nature: lessons learned and future research directions. Ecological Economics 46: 493–510.

41. Hein L, Koppen VK, de Groot RS, van Ierland EC (2006) Spatial scales, stakeholders and the valuation of ecosystem services. Ecological Economics 57: 209–228.

42. Konarska KM, Sutton PC, Castellon M (2002) Evaluating scale dependence of ecosystem service valuation: a comparison of NOAA-AVHRR and Landsat TM datasets. Ecological Economics 41: 491–507.

43. Limburg KE, O'Neill RV, Costanza R, Farber S (2002) Complex systems and valuation. Ecological Economics 41: 409–420.

44. Zhang MY, Wang KL, Liu HY, Zhang CH (2011) Responses of Spatial-temporal Variation of Karst Ecosystem Service Values to Landscape Pattern in Northwest of Guangxi, China. Chin. Geogra. Sci. 21: 446–453.

45. Deng L, Shangguan ZP, Li R (2012) Effects of the grain-for-green program on soil erosion in China. International Journal of Sediment Research 27: 120–127.

46. Nagendra H (2002) Opposite trends in response for the Shannon and Simpson indices of landscape diversity. Applied Geography, 22: 175–186.

47. Shrestha RP, Schmidt-Vogt D, Gnanavelrajah N (2010) Relating plant diversity to biomass and soil erosion in a cultivated landscape of the eastern seaboard region of Thailand. Applied Geography 30: 606–617.

48. Sherrouse BC, Clement JM, Semmens DJ (2011) A GIS application for assessing, mapping, and quantifying the social values of ecosystem services. Applied Geography 31: 748–760.

49. Li M, Zhu Z, Vogelmann JE, Xu D, Wen W, et al. (2011) Characterizing fragmentation of the collective forests in southern China from multitemporal Landsat imagery: a case study from Kecheng district of Zhejiang province.Applied Geography 31: 1026–1035.

50. Long JA, Nelson TA, Wulder MA (2010) Characterizing forest fragmentation: distinguishing change in composition from configuration. Applied Geography 30: 426–435.

Effects of Warming and Clipping on Ecosystem Carbon Fluxes across Two Hydrologically Contrasting Years in an Alpine Meadow of the Qinghai-Tibet Plateau

Fei Peng*, Quangang You, Manhou Xu, Jian Guo, Tao Wang, Xian Xue*

Key Laboratory of Desert and Desertification, Chinese Academy of Sciences, Cold and Arid Regions Environmental and Engineering Research Institute, Chinese Academy of Sciences, Lanzhou, China

Abstract

Responses of ecosystem carbon (C) fluxes to human disturbance and climatic warming will affect terrestrial ecosystem C storage and feedback to climate change. We conducted a manipulative experiment to investigate the effects of warming and clipping on soil respiration (Rs), ecosystem respiration (ER), net ecosystem exchange (NEE) and gross ecosystem production (GEP) in an alpine meadow in a permafrost region during two hydrologically contrasting years (2012, with 29.9% higher precipitation than the long-term mean, and 2013, with 18.9% lower precipitation than the long-tem mean). Our results showed that GEP was higher than ER, leading to a net C sink (measured by NEE) over the two growing seasons. Warming significantly stimulated ecosystem C fluxes in 2012 but did not significantly affect these fluxes in 2013. On average, the warming-induced increase in GEP (1.49 μ mol m^{-2}s^{-1}) was higher than in ER (0.80 μ mol m^{-2}s^{-1}), resulting in an increase in NEE (0.70 μ mol m^{-2}s^{-1}). Clipping and its interaction with warming had no significant effects on C fluxes, whereas clipping significantly reduced aboveground biomass (AGB) by 51.5 g m^{-2} in 2013. These results suggest the response of C fluxes to warming and clipping depends on hydrological variations. In the wet year, the warming treatment caused a reduction in water, but increases in soil temperature and AGB contributed to the positive response of ecosystem C fluxes to warming. In the dry year, the reduction in soil moisture, caused by warming, and the reduction in AGB, caused by clipping, were compensated by higher soil temperatures in warmed plots. Our findings highlight the importance of changes in soil moisture in mediating the responses of ecosystem C fluxes to climate warming in an alpine meadow ecosystem.

Editor: Ben Bond-Lamberty, DOE Pacific Northwest National Laboratory, United States of America

Funding: Financial support came from the Foundation for Excellent Youth Scholars of CAREERI, CAS (351191001), National Natural Science Foundation of China (41301210, 41201195 and 41301211), and Chinese Academy of Sciences (Hundred Talents Program). The funders had no role in study design, data collection and analysis, decision to publish, or preparation of the manuscript.

Competing Interests: The authors have declared that no competing interests exist.

* Email: pengguy02@yahoo.com (FP); xianxue@lzb.ac.cn (XX)

Introduction

Global mean temperature has increased by 0.76°C since the year 1850 and is predicted to rise an additional 1.8–4°C by the end of the 21st century [1]. Elevated global temperature can substantially impact the global carbon (C) budget, resulting in positive or negative feedbacks to global climate change [2,3]. The balance between C fixed by photosynthesis and C emitted to the atmosphere through plant and heterotrophic respiration determines the rate of terrestrial C storage [4].

Studies have shown that global warming could stimulate both ecosystem C uptake and emission across various terrestrial biomes [5]. However, the response of net C balance to warming is highly variable because of different temperature and soil moisture sensitivities in the processes that control C uptake and emission [6]. It is generally assumed that the terrestrial ecosystem might act as a net C source under a global warming scenario because the processes controlling ecosystem C emission are more sensitive to higher temperatures than the processes controlling C uptake [3,7,8]. However, some evidence indicates that warming could increase net C uptake, and global C models project enhanced terrestrial CO_2 uptake in response to warming through the middle of this century [9,10]. Current and completed experimental studies that have investigated warming effects have focused mostly on net primary productivity (NPP), biomass and soil respiration [5,11], from which the change in the C balance change was estimated. However, responses of gross primary production (GPP) and ecosystem respiration (ER), the major components net ecosystem exchange (NEE), to warming in field experiments [12] have received less attention in the alpine area [13].

Mowing (clipping) or grazing in grasslands, which account for 20% of the land use of the global terrestrial ice-free surface, may have substantial effects on ecosystem C fluxes, especially on a short-term basis [14]. Clipping would result in rapid changes in nutrient cycling [15], vegetation cover, plant community composition [16], and soil microclimate [17]. Collectively, these processes appear to stimulate the rate of ecosystem C cycling, however, their impacts on the net C balance are inconsistent [18,19].

Carbon stored in permafrost at high latitudes and in mountain areas is one of the major components of the terrestrial C pool. It is

Table 1. Results (F-values) of a three-way ANOVA on the effects of warming (W), clipping (C), measuring month (M), and their interactions on soil respiration (Rs), ecosystem respiration (ER), net ecosystem exchange (NEE) and gross ecosystem production (GEP).

	M	W	C	M×W	M×C	W×C	M×C×W
Rs	88.3**	3.9^	0.1	0.1	0.3	4.2*	1.1
ER	21.8**	8.3**	0.2	0.4	0.1	0.1	0.4
NEE	43.0**	4.8*	0.8	1.2	0.4	0.1	0.1
GEP	44.3**	9.5**	0.0	0.7	0.1	0.1	0.1
Rs/ER	1.5	0.3	0.1	2.5**	1.8	4.0**	0.1
ER/GEP	13.3**	0.4	0.5	2.1^	0.9	2.1	0.8
AGB	117.0**	3.8^	22.2**	1.2	2.3^	0.3	2.1^
RB	1.0	1.3	2.0	0.1	0.1	3.0^	0.2
AGB/RB	26.6**	10.3**	19.6**	6.6**	8.1**	13.0**	8.5**

Significance: ^, P<0.1; *, P<0.05; **, P<0.01.

estimated that soils in the permafrost regions store as much as 1672 Pg C (1 Pg = 10^{15} g), which is equivalent to double the atmospheric C pool [20,21]. Ecosystems in permafrost regions are C sinks because microbial decomposition of soil organic matter is inhibited under low annual mean temperature, and there is limited availability of organic C in frozen soil [22–24]. Altered growing season length, and changes in plant growth, ecosystem energy exchange and land use, together with the thawing of permafrost under a changing climate are projected to enhance the capability of ecosystem C uptake [21]. However, these altered dynamics do not appear to be able to compensate for the C released from thawing permafrost, resulting in ecosystems in permafrost regions acting as positive feedback to global change [21].

The Qinghai-Tibet Plateau (QTP) is experiencing a "much greater than average" increase in surface temperature, based on data observed at meteorological stations [25] and predictions from coupled climate-carbon models [1]. Grassland in the QTP is the largest vegetation unit of the Eurasian continent and covers an area of approximately 2.5 million km^2 [26]. Grazing is the most prevalent land use practice in the grassland. Results from eddy covariance measurements showed that the alpine meadow in the QTP is a weak C sink with annual variations [23,24]. Several studies have examined responses of ER and aboveground biomass to warming and clipping [17,27,28], in which the alpine meadow is thought to be a net C sink based on C balance calculations [28]. However, no field experiment has been conducted in the permafrost region of the QTP to measure the response of NEE, which provides a direct measure of the C balance. We conducted a two-year warming and clipping experiment to investigate how NEE and its components (GPP and ER) respond to warming and clipping, and how the associated changes in soil moisture, soil temperature, above- and belowground biomass affect the responses of ecosystem C fluxes in the permafrost region of the QTP.

Materials and Methods

Experimental site

The study site is situated in the region of the Yangtze River source, inland of the QTP near the Beilu River research station (34°49′N, 92°56′E, no specific permissions were required for activities in this location) at an altitude of 4635 m. This area has a typical alpine climate: mean annual temperature is −3.8°C and monthly air temperature ranges from −27.9°C in January to 19.2°C in July. Mean annual precipitation is 290.9 mm, of which over 95% falls during the warm growing season (May to October). Mean annual potential evaporation is 1316.9 mm, mean annual relative humidity is 57%, and mean annual wind velocity is 4.1 m s^{-1} [29]. The study site is a winter-grazed range, dominated by alpine meadow vegetation: *Kobresia capillifolia*, *K. pygmaea*, and *Carex moorcroftii*, with a mean plant height of 5 cm. Plant roots occur mainly within the 0–20 cm soil layer, and average soil organic C is 1.5%. The soil development is weak, and the soil belongs to alpine meadow soil (Chinese soil taxonomy), or is classified as a Cryosol according to World Reference Base, with a Mattic Epipedon at a depth of approximately 0–10 cm, and an organic-rich layer at a depth of 20–30 cm [30]. The parent soil material is of fluvioglacial origin and is composed of 99% sand. The Mattic Epipedon lowers the saturated soil water content, but increases soil water storage, and plant roots are dense and compressed within this layer. Permafrost thickness observed near the experimental site is 60–200 m and the depth of the active layer is 2.0–3.2 m [29,31]. However, because of climatic warming, the thickness of the active layer has been increasing at a rate of 3.1 cm y^{-1} since 1995 [32]. The experimental field is on a mountain slope

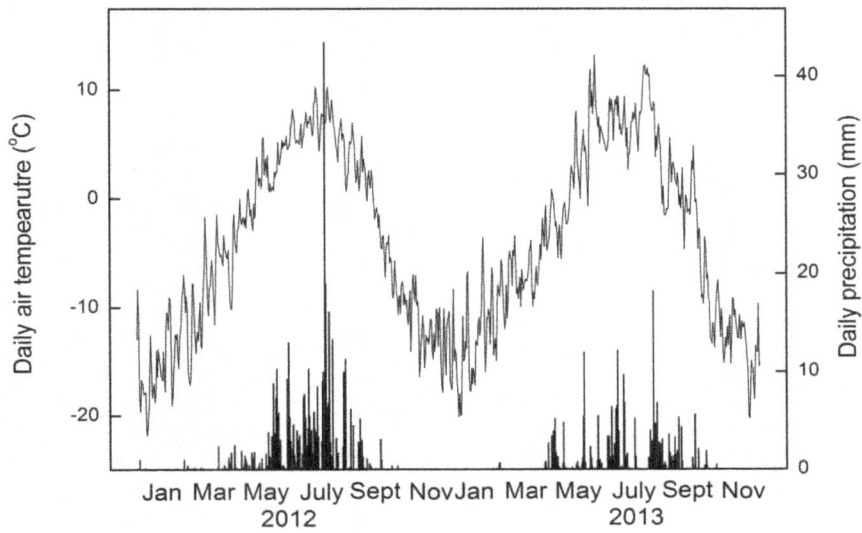

Figure 1. Daily precipitation (columns) and daily mean air temperature (line) in 2012 and 2013. Data are from the micro-meteorological station adjacent (approx. 100 m) to the experimental plots.

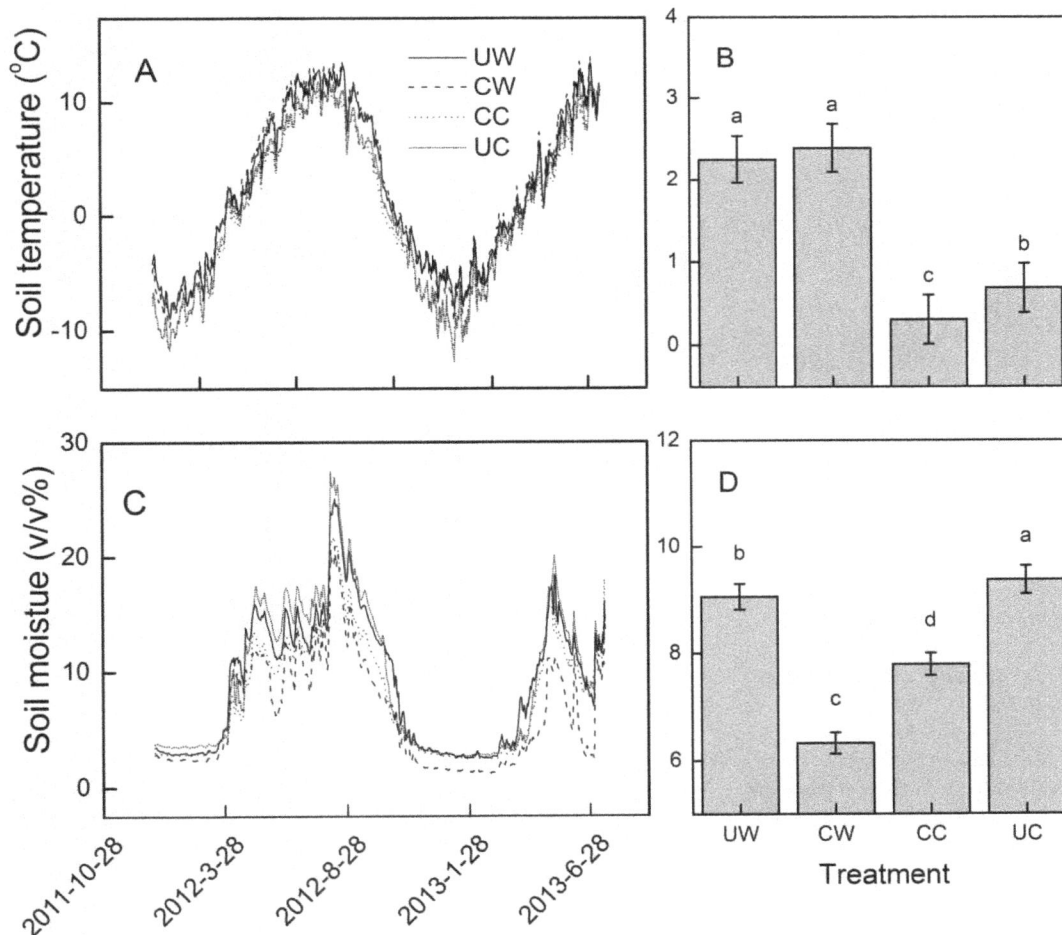

Figure 2. Daily soil temperature measured at a depth of 5 cm and volumetric soil moisture measured at a depth of 0–10 cm (A, C), and effects of warming and clipping on average (2012–2013) soil temperature (B) and soil moisture (d). UW, unclipped warming; CW, clipped warming; CC, clipped control; UC, unclipped control.

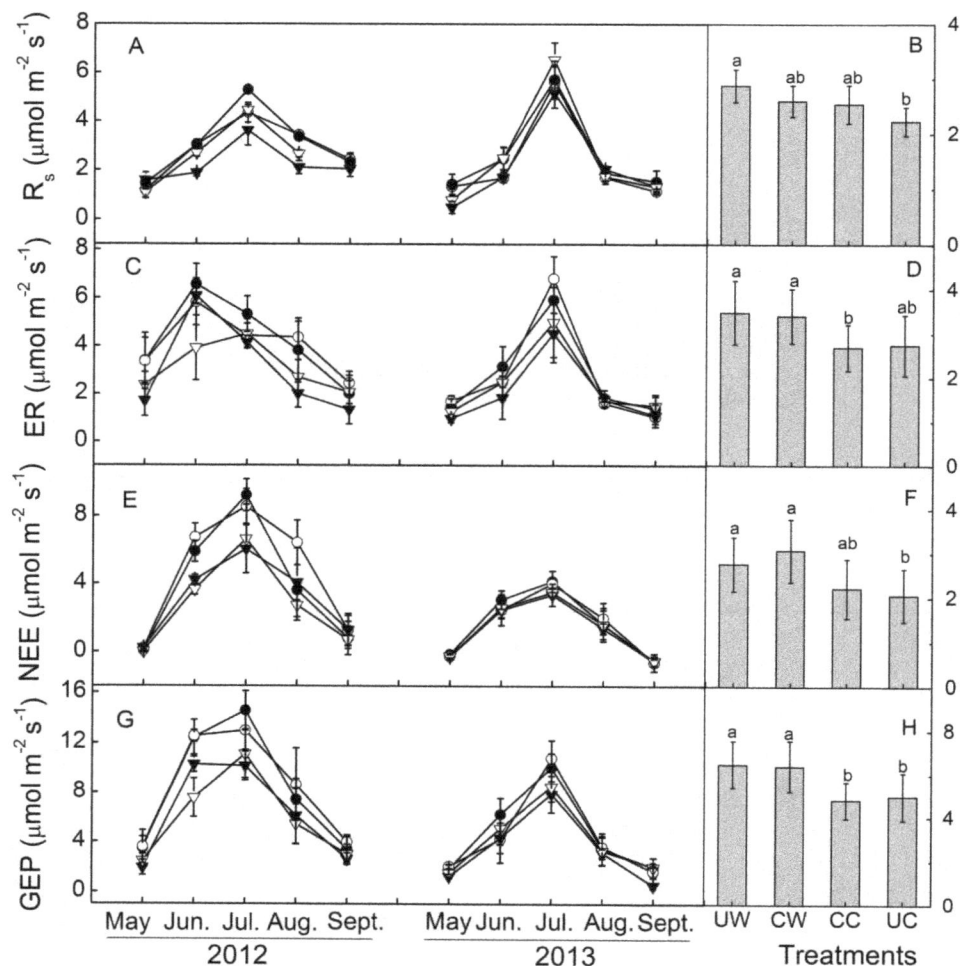

Figure 3. Mean monthly and overall means of soil respiration (Rs, A, B), ecosystem respiration (ER, C, D), net ecosystem exchange (NEE, E, F) and gross ecosystem production (GEP, G, H). Clipping was conducted in September 2011 and 2012. UW, unclipped warming; CW, clipped warming; CC, clipped control; UC, unclipped control.

with a mean incline of 5°. Detailed information about the soil properties is presented in Table 1.

Experimental design and measurement protocols

Experimental design. A two factorial experimental design (warming and clipping) was used with five replicates in each of the four treatments, i.e. unclipped control (UC), clipped control (CC), unclipped warming (UW) and clipped warming (CW). In total, 20 plots (2×2 m) were used in a complete randomized block distribution in the field. Plots were selected for homogeneity of topography, soil texture, aboveground biomass, and species composition. In each warmed plot, one 165 cm×15 cm infrared heater (MR-2420, Kalglo Electronics Inc., Bethlehem, PA, USA) was suspended in the middle of each plot at a height of 1.5 m above the ground with a radiation output of 150 watts m^{-2}. The heating has been operating continuously since July 1st 2010. To simulate the shading effects of the heaters, one "dummy" heater,

made of a metal sheet with the same shape and size as the heaters, was also installed in the control plot.

Plants in the clipped plots were clipped at the soil surface on an annual basis, usually in last September. The rotational grazing system is that two ranches for each family, one is for the summer grazing and another is for the winter grazing. The rotational use of ranches is implemented usually in September. We consulted to the owner and they ensured that grassland of our study site is a winter grazing ranch. In our study, the clipping treatment was conducted in last September. One sheep needs 1.46×10^6 g grass per year [33] and the carrying capacity for alpine meadow is 1.39 head of sheep per hectare. Based on those data, about 203 g m^{-2} grass per year would be grazed for the alpine meadow. The aboveground biomass in September 2013 was about 320 g m^{-2}. In the clipping treatment, biomass cut was <320 g m^{-2}. Therefore we believe that our clipping treatment provided a reasonable simulation of local grazing practices.

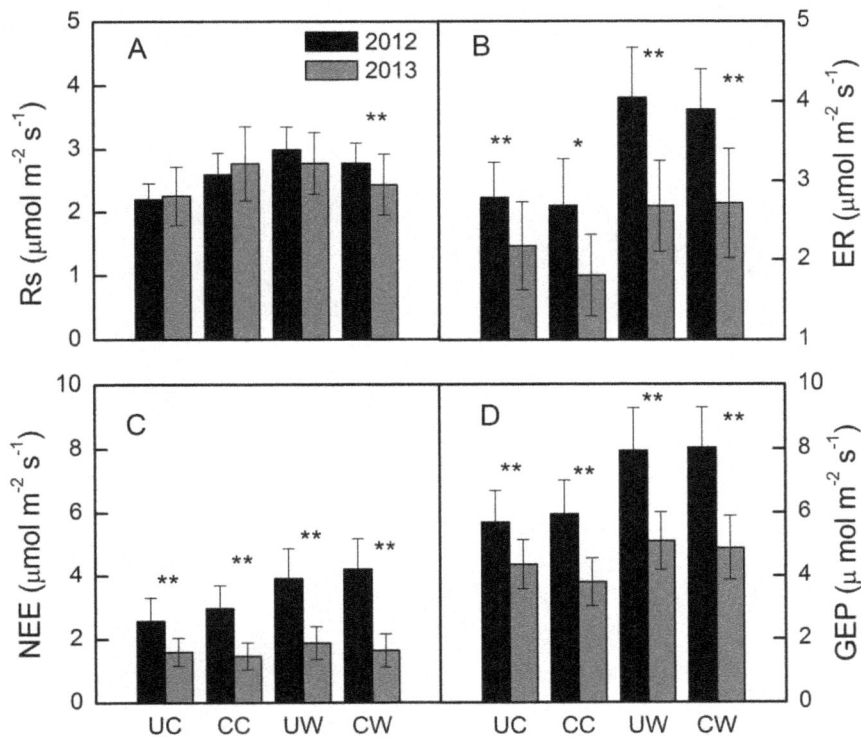

Figure 4. Mean annual soil respiration (Rs), ecosystem respiration (ER), net ecosystem exchange (NEE) and gross ecosystem production (GEP) under unclipped control (UC), clipped control (CC), unclipped warming (UW) and clipped warming (WC) treatments in 2012 and 2013. Symbols above the bars represent significant differences at p<0.05 (*) and p<0.01 (**).

Measurement protocol. Air temperature, water vapor pressure and relative humidity were monitored automatically at a height of 20 cm above the soil surface in the center of each plot using a Model HMP45C probe (Campbell Scientific Inc., Bethlehem, PA, USA). Nine thermistors were installed to monitor soil temperatures at depths of 5, 15, 30, 60, 100, 150, 200, 250 and 300 cm. All the probes were connected to a CR1000 datalogger (Campbell Scientific Inc.). Data recorded every 10 min were averaged and reported as daily values. Pavelka *et al.* (2007) stated that for grassland ecosystems, surface soil temperature is the most suitable depth for measuring soil temperature because of the optimized regression coefficient between surface soil temperature and soil respiration [34]. Therefore, we used soil temperature measured at a depth of 5 cm in the following analyses.

An EnviroSmart sensor (Sentek Pty Ltd., Stepney, Australia), which used frequency domain reflection, was used to monitor volumetric soil moisture at depths of 0–10, 10–20, 20–40, 40–60 and 60–100 cm. These soil moisture data were also recorded using a CR1000 datalogger. When analyzing the relationships between C fluxes and soil moisture, we used the daily average soil moisture data that were collected when ecosystem C flux measurements were conducted.

Soil respiration (Rs) was measured by using Licor-6400-09 (Lincoln, NE, USA) on PVC collars 5 cm in height and 10.5 cm in diameter, which were permanently inserted 2–3 cm into the soil in the center of each plot. Small living plants were cut at the soil surface at least one day before measurements to eliminate the effect of respiration from aboveground biomass [35]. ER and NEE were measured with a transparent chamber (0.5×0.5×0.5 m) attached to an infrared gas analyzer (IRGA, Licor-6400, Lincoln, NE, USA). The transparent chamber is a custom-designed

chamber made of Polytetrafluoroethene (4 mm in thickness) with light transmittance about 99%. During measurements, a foam gasket was placed the chamber and the soil surface to minimize leaks. One small fan ran continuously to mix the air inside the chamber during measurements. Nine consecutive recordings of CO_2 concentration were taken in each plot at 10 s intervals during a 90 s period. Following the measurement of NEE, the chamber was vented for several minutes and covered with an opaque cloth for measuring ER, as the opaque cloth eliminated light (and hence photosynthesis). CO_2 flux rates were determined from the time-course of the CO_2 concentrations used to calculate NEE and ER. The method used was similar to that reported by Steduto *et al.* (2002) [36] and Niu *et al.* (2008) [37]. Gross ecosystem productivity (GEP) was the calculated as the sum of NEE and ER. Rs, NEE and ER were measured in each plot on a monthly basis from May to September in 2012 and 2013.

Aboveground biomass (AGB) was obtained from a step-wise linear regression with AGB as the dependent variable, and coverage and plant height as independent variables. 100 small plots (30 cm×30 cm) were included in the regression analysis (AGB = 22.76×plant height +308.26×coverage −121.80, R^2 = 0.74, P<0.01). Coverage of each experimental plot was measured using a 10 cm×10 cm frame in four diagonally divided subplots replicated eight times. Plant height was measured 40 times by a ruler and averaged for each experimental plot. A biomass index was used as the ratio of the derived biomass on any given date to the maximum biomass during the entire study period [38]. Root biomass (RB) was obtained from soil samples that were air-dried for one week and passed through a 2- mm diameter sieve to remove large particles. Roots were separated from the soil by washing, and a 0.25-mm diameter sieve was used to retrieve fine

Table 2. Results (F-values) of a three-way ANOVA on the effects of warming (W), clipping (C), measuring month (M), and their interactions on soil respiration (Rs), ecosystem respiration (ER), net ecosystem exchange (NEE) and gross ecosystem production (GEP) in contrasting years.

	M	W	C	M×W	M×C	W×C	M×C×W
2012							
Rs	54.2**	11.6**	0.8	2.2^	0.3	5.1*	1.4
ER	13.9**	9.4**	0.1	0.6	0.4	0.01	1.2
NEE	46.9**	8.6**	2.6	2.0	1.3	0.0	0.04
GEP	40.3**	12.2**	0.2	0.9	0.6	0.1	0.2
ER/GEP	16.6**	0.05	0.2	0.6	0.1	0.8	0.8
AGB	33.9**	2.1	2.0	0.6	0.4	0.5	0.2
RB	7.2**	0.7	1.0	0.5	0.09	2.8	0.2
AGB/RB	2.4^	1.2	2.2	0.5	0.2	1.6	0.2
2013							
Rs	78.3**	0.1	0.02	0.8	0.6	3.3^	0.6
ER	31.0**	2.6	0.2	1.1	0.2	0.2	0.3
NEE	51.2**	1.1	0.3	0.3	0.1	0.01	0.1
GEP	44.2**	3.2^	0.9	0.7	0.4	0.2	0.3
ER/GEP	22.1**	9.0**	11.0**	13.9**	8.3**	8.8**	10.9**
AGB	89.9**	0.06	20.1**	1.3	0.8	0.5	1.1
RB	2.9*	0.7	1.2	0.2	0.3	0.8	0.2
AGB/RB	20.7**	1.5	6.2*	1.5	2.3^	2.6	2.7*

Significance: ^, $P<0.1$; *, $P<0.05$; **, $P<0.01$.

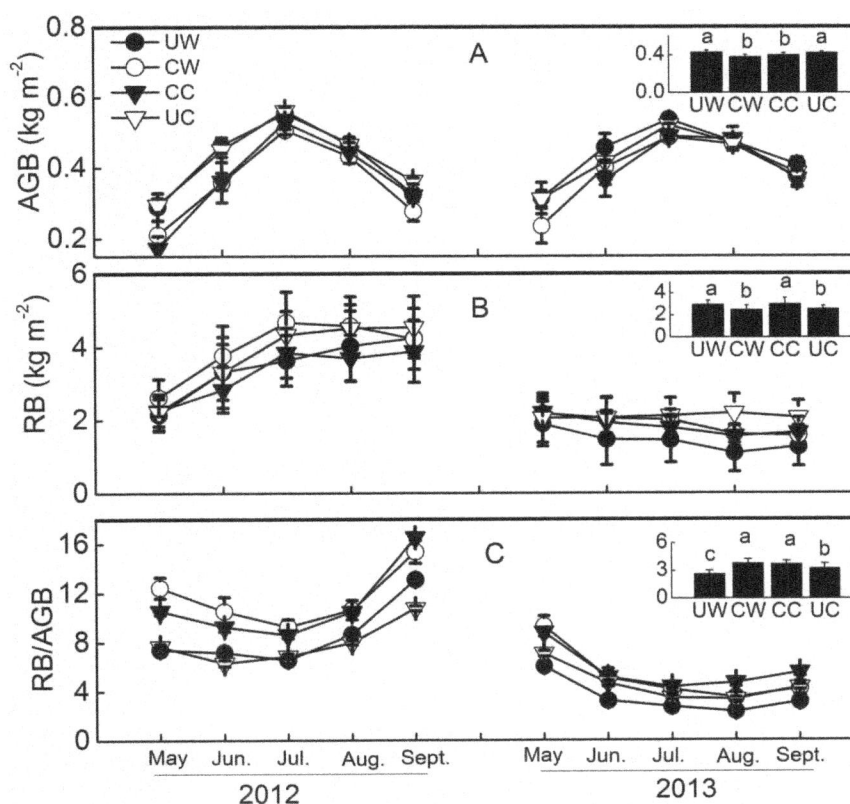

Figure 5. Temporal variations and overall means (inserted panels) of aboveground biomass (AGB, A), root biomass (RB, B) and the ratio of RB to AGB (RB/AGB, C). Clipping was conducted in September 2011 and 2012. See Figures 2 and 3 for notes and abbreviations.

roots. Living roots were separated from dead roots by their color and consistency [39]. Separated roots were dried at 75°C for 48 h.

Data analysis

Temperature and soil moisture data used in analyses were from January 1st 2012 to July 18th 2013 because a power failure prevented data from being collected from July 19th 2013 onwards. The effect of the warming and clipping treatments on soil temperature (5 cm), soil moisture (0–10 cm), Rs, ER, NEE, GEP, AGB and RB were determined with a three-way analysis of variance (ANOVA) using SPSS Version 18.0. (SPSS, Inc., Chicago, IL, USA).

Relationships between C fluxes and soil microclimate (soil temperature and soil moisture) were examined using daily soil microclimate data that were collected when ecosystem C fluxes

were measured. Linear regression analyses were used to examine the relationships of C fluxes with abiotic (soil moisture and soil temperature) and biotic factors (monthly AGB and RB).

Results

Microclimate

In comparison to the long-term average (1981–2008) mean annual air temperature (MAT, −5.1°C), higher MAT values were recorded in 2012 and 2013 (−3.5°C and −3.8°C, respectively). Annual precipitation in 2012 (420.1 mm, Fig. 1) was higher than the long-term mean annual precipitation (294.5 mm), but it was lower than the long-term mean in 2013 (238.8 mm) (Fig. 1).

Experimental warming significantly elevated annual mean soil temperature (Figs. 2A and 2B, $P<0.05$). In unwarmed plots, the average daily soil temperature at 5 cm depth was 0.65°C and

Table 3. Fitted quadratic models of the relationships between ecosystem respiration (ER), net ecosystem exchange (NEE), gross ecosystem production (GEP) and soil moisture (θ, v/v%, 10 cm). Max. F, θ represents the value of θ when ER, NEE and GEP are at their maximum.

	ER/μmol m^{-2}s^{-1}	NEE/μmol m^{-2}s^{-1}	GEP/μmol m^{-2}s^{-1}
Fitted model	$ER = -0.058\theta^2 + 1.71\theta - 7.72$	$NEE = -0.076\theta^2 + 2.37\theta - 12.53$	$GEP = -0.139\theta^2 + 4.21\theta - 21.02$
R^2	0.31	0.45	0.36
p	0.007	<0.001	<0.001
Max. F, θ/%	14.7%	15.6%	15.1%

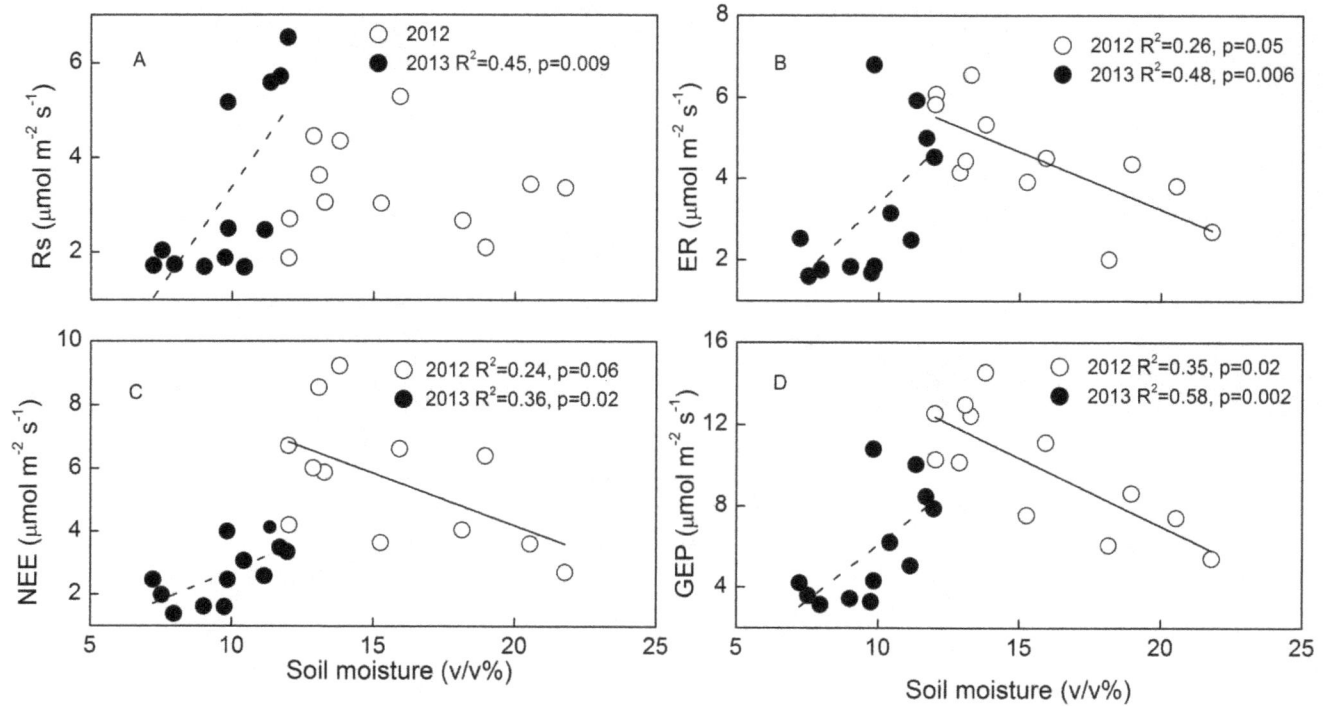

Figure 6. The relationships between soil moisture (0–10 cm) and ecosystem C fluxes: soil respiration (Rs), ecosystem respiration (ER), net ecosystem exchange (NEE) and gross ecosystem productivity (GEP) in 2012 (hollow circles) and 2013 (solid circles), respectively. Soil moisture and ecosystem C fluxes data were the average for all plots in each month. Data for both years were collected from June to August.

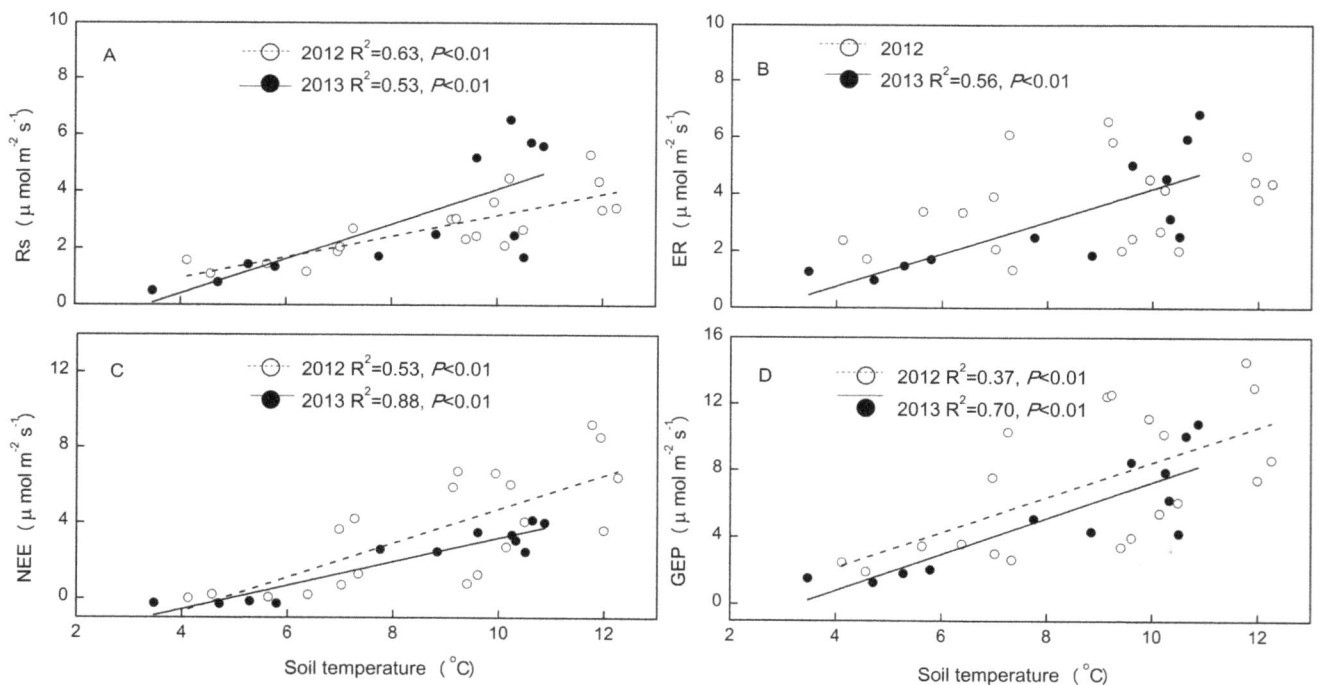

Figure 7. The relationships between soil temperature (5 cm) and ecosystem C fluxes: soil respiration (Rs), ecosystem respiration (ER), net ecosystem exchange (NEE) and gross ecosystem productivity (GEP). Soil temperature and ecosystem C fluxes data were the average of all plots in each month.

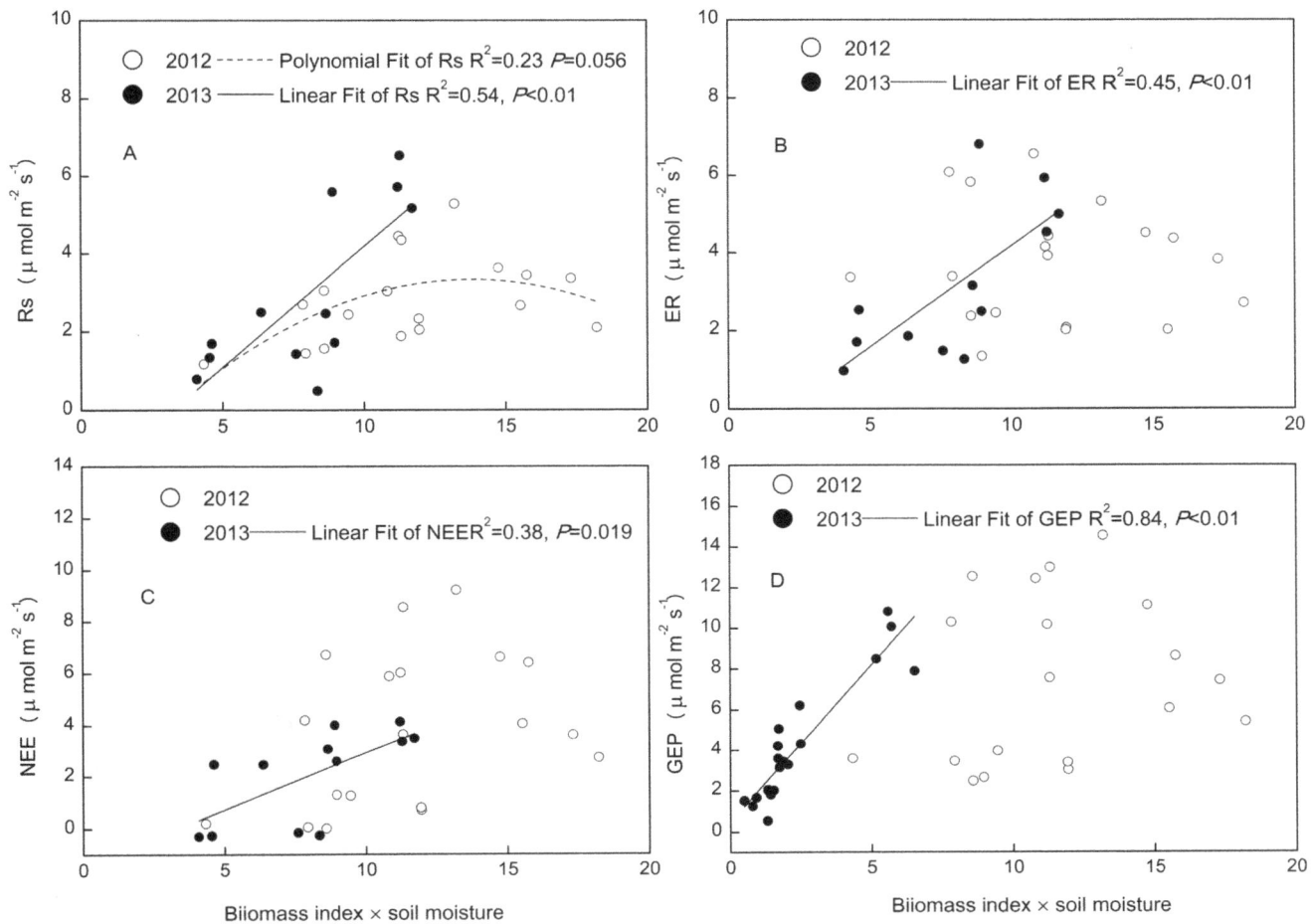

Figure 8. The relationships between changes in ecosystem C fluxes: soil respiration (Rs), ecosystem respiration (ER), net ecosystem exchange (NEE) and gross ecosystem productivity (GEP), and the changes in the product of abveground biomass index and soil moisture during 2012 and 2013.

0.14°C in 2012 and 2013, respectively. Warming significantly increased the soil temperature by 1.96°C (2012) and 1.59°C (2013) (Fig. 2B). The average daily soil temperature in unclipped plots was 1.69°C and 0.99°C in 2012 and 2013, respectively, but was unaffected by clipping. Volumetric soil moisture measured over 0–10 cm fluctuated greatly over the study period (Fig. 2C). The average daily soil moisture in unwarmed plots was 9.31 (v/v%) and 7.22 (v/v%), and warming significantly reduced soil moisture by 7.4% and 17.4% ($P<0.05$) in 2012 and 2013, respectively (Fig. 2D). The average daily soil moisture in unclipped plots was 10.08 (v/v%) and 7.61 (v/v%), and clipping significantly decreased it by 28.3% and 36.5% in 2012 and 2013, respectively (Fig. 2D).

Warming and clipping effects on C fluxes

The temporal dynamics of Rs, ER, NEE, and GEP followed the seasonal patterns of air and soil temperature in both years, which peaked in mid-growing season (Figs. 1 and 3). Substantial inter-annual variations in ecosystem C fluxes were observed in this study (Fig. 3). The annual average ER, NEE and GEP were all significantly higher in 2012 than in 2013 (Fig. 4) in all treatments, but higher Rs in 2012 was only observed in the CW treatment (Fig. 4A). On average, NEE, ER and GEP were 47%, 22% and 34% higher, respectively, in 2012 than in 2013.

Warming significantly increased NEE ($P = 0.03$), whereas no significant effects of clipping ($P = 0.37$) or its interaction with warming ($P = 0.83$) were detected (Table 1). When analyzed separately by year using a three-way ANOVA, warming only significantly increased NEE in 2012, by 28.5% (Table 2). Warming induced an enhanced growing season mean NEE in 2012, which was lower in clipped (17%) than in unclipped plots (30%). Measuring date had a significant effect on NEE, but the interaction of measuring date with warming or clipping had no effect on NEE in either year (Table 2).

Similar to NEE, average GEP was significantly increased by warming ($P = 0.003$) but not by clipping ($P = 0.97$) or by their interaction ($P = 0.87$, Table 1). When analyzed separately by year using a three-way ANOVA, warming significantly increased average GEP (Table 2) by 2.13 μ mol m^{-2}s^{-1} in 2012 and marginally enhanced it by 0.82 μ mol m^{-2}s^{-1} in 2013. The increased GEP caused by warming was lower in clipped (23%) than in unclipped plots (28.3%) in 2012, but GEP was higher in clipped (22.6%) than in unclipped plots (13.4%) in 2013.

Warming also significantly increased average Rs ($P = 0.052$) and ER ($P = 0.005$), but clipping had no significant effect on average Rs ($P = 0.73$) or ER ($P = 0.66$, Table 1). Similar to NEE, when analyzed separately by year using a three-way ANOVA, the effect of warming on ER and Rs was only significant in 2012 (Table 2),

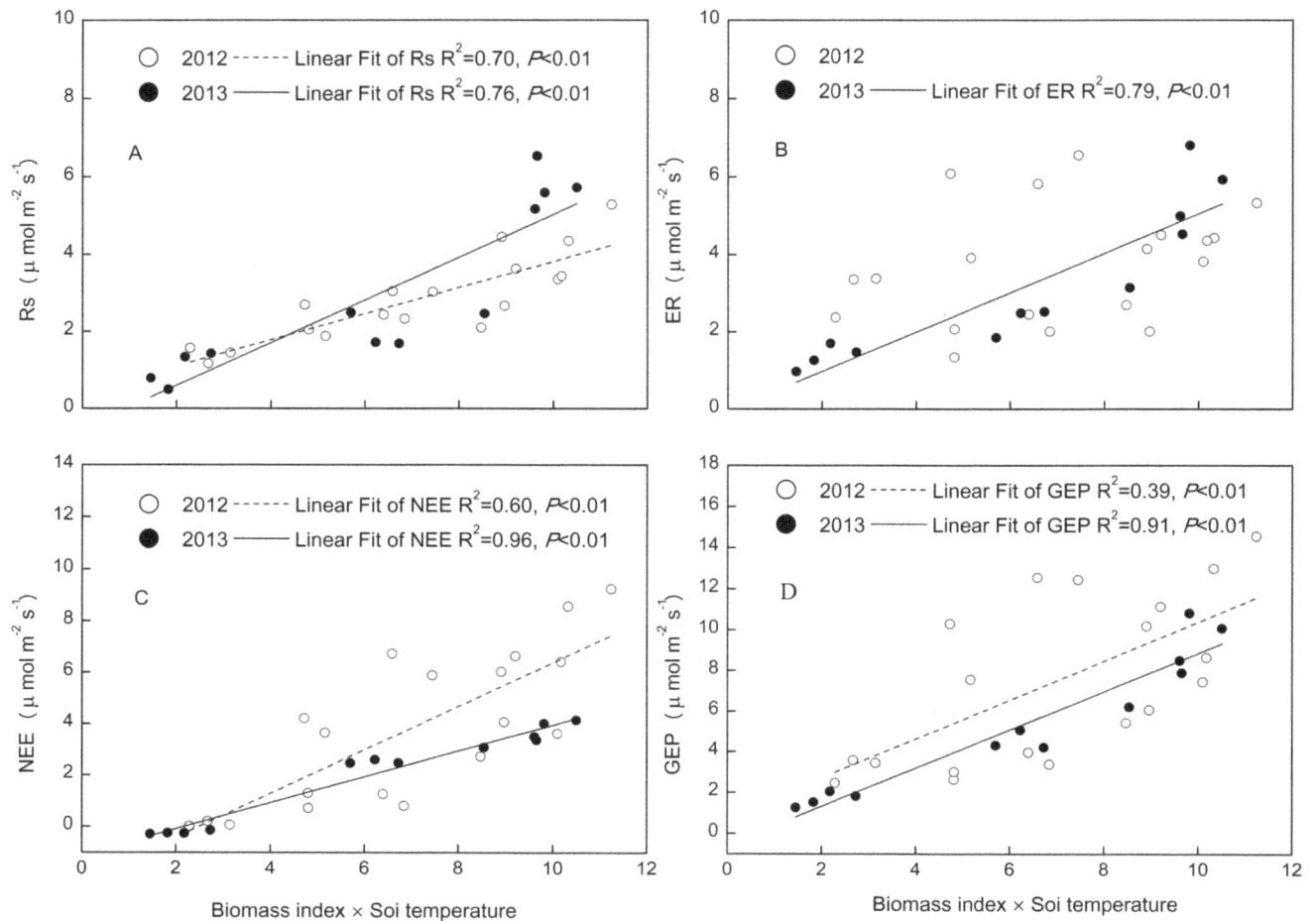

Figure 9. The relationships between changes in ecosystem C fluxes: soil respiration (Rs), ecosystem respiration (ER), net ecosystem exchange (NEE) and gross ecosystem productivity (GEP), and the changes in the product of aboveground biomass index and soil temperature during 2012 and 2013.

which increased by 25.8% and 17%, respectively. The interaction between warming and clipping had a significant effect on Rs but not on ER (Table 1). The average increase in Rs caused by warming was lower in clipped (5.9%) than in unclipped plots (27.4%).

There was no significant effect of warming or clipping on the ER to GEP ratio (ER/GEP, $P = 0.51$ and $P = 0.46$ for 2012 and 2013, respectively, Table 1). When analyzed separately by year using a three-way ANOVA, ER/GEP was significantly affected by warming, clipping, measurement date and their interactions in 2013 (Table 2).

Warming and clipping effects on biomass

Similar to the inter-annual variation of ecosystem C fluxes, RB was significantly lower in 2013 (by a value of 50.2%) than in 2012, whereas there was no difference in AGB over the two growing seasons (Fig. 5).

Warming marginally increased AGB (p = 0.053) and clipping significantly reduced AGB (p<0.001), but there was no significant effect on RB (p = 0.26 and p = 0.16 for warming and clipping, respectively, Table 1). When analyzed separately by year using a three-way ANOVA, warming had no significant effect on AGB or RB in either year, whereas clipping significantly reduced AGB in

2013 (Table 2). The reduction in AGB by clipping was higher in unwarmed (14.4%) plots than in warmed plots (10.3%) in 2013.

Impacts of biotic and abiotic factors on ecosystem C fluxes

Over the two growing seasons, there was no clear relationship between ER, NEE, GEP and soil moisture, but there was a quadratic relationship between these variables and soil moisture when May and September data were excluded (Table 3). The optimal soil moisture for ER, NEE and GEP was about 15% (Table 3). When plotted separately, ER, NEE, and GEP decreased linearly with increasing soil moisture in 2012, and increased linearly with increasing soil moisture in 2013 (Fig. 6).

The temperature response curves for Rs, ER, NEE and GEP in 2012 were quite similar to those recorded in 2013. However, the slope of relationship between Rs and soil temperature was higher in 2012 than in 2013, but that between NEE and soil temperature was smaller in 2012 than in 2013 (Fig. 7).

The statistical interaction term of above-ground biomass index and soil moisture showed polynominal relationship only with Rs in 2012 (Fig. 8a), whereas it linearly correlated with all the ecosystem C fluxes in 2013 (Fig. 8). The interaction term of above-ground biomass and soil temperature explained more variation in ecosystem C fluxes than did the interaction term of above-ground

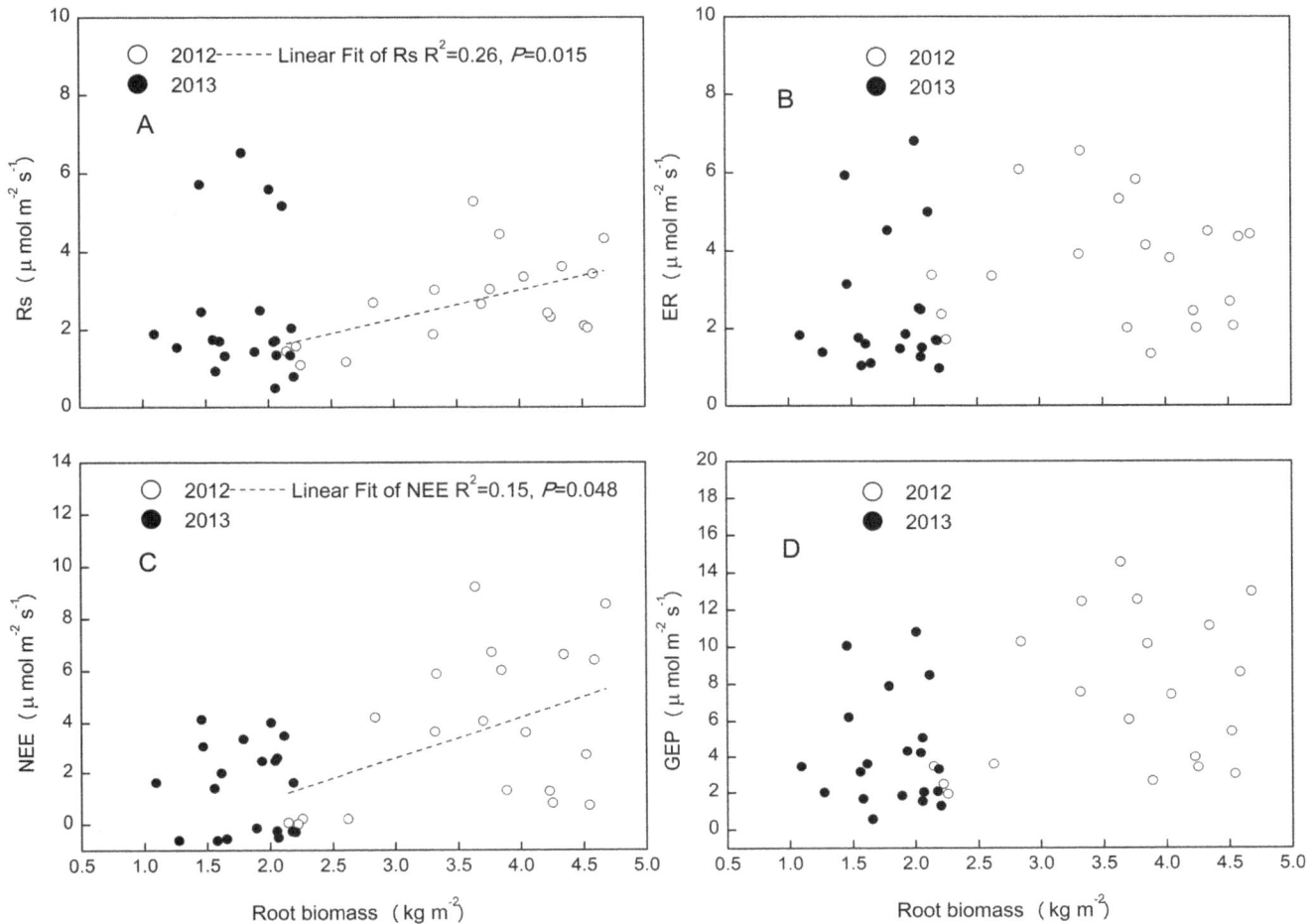

Figure 10. Relationship between root biomass and ecosystem C fluxes: soil respiration (Rs), ecosystem respiration (ER), net ecosystem exchange (NEE) and gross ecosystem productivity (GEP).

biomass and soil moisture (Figs. 8 and 9). The fitting slope between Rs and the interaction of above-ground biomass and soil temperature was higher in 2013 than in 2012 (Fig. 9a), but that between NEE and the interaction of above-ground biomass and soil temperature was smaller in 2013 than in 2012 (Fig. 9c). Root biomass only had effect on ecosystem C fluxes in wet year (Fig. 10).

Discussion

C fluxes and their inter-annual variation

In this alpine meadow ecosystem studied in the QTP, a higher uptake of C (GEP) than release (ER) resulted in a net C sink (Figs. 3–4). This result is similar to that reported in seasonally frozen areas in the QTP [23,24], and in some arctic ecosystems [7].

Higher precipitation and associated higher soil moisture in 2012 than in 2013, and the similar MAT between the two years suggest that drought reduced annual ER, NEE and GEP in 2013. Our results are in agreement with those from temperate grassland ecosystems [37,40,41]. The quadratic relationship between ecosystem C fluxes and soil moisture (Table 3) at the temporal scale support the above findings, as ecosystem C fluxes were positively related to soil moisture in 2013 (Fig. 6). There was greater fluctuation in GEP than ER: GEP was 34% lower in 2013

than in 2012, compared to ER, which was 22% lower in 2013 than in 2012. These results are consistent with those from boreal and temperate forests [42] and temperate grasslands [40], which indicate that GEP is more sensitive to inter-annual climatic variation than ER in alpine meadow ecosystems. Differences in the magnitudes of the inter-annual variation in GEP and ER could be explained by differences in the slopes between these variables when plotted against soil moisture (Fig. 6). The greater dependence of GEP than ER on soil moisture across both years suggests that GEP is more sensitive to changes in soil moisture than ER. Despite the significant inter-annual difference in ER, NEE and GEP, Rs did not differ significantly between the two years. This indicates that annual variations in Rs may be controlled by other factors, such as soil temperature, as Rs had the highest slope when plotted against this variable (Fig. 7).

Drought typically reduces aboveground biomass in grasslands [43,44]. However, in our study, no significant reduction in AGB was observed in 2013. One reason could be the various reactions of different species to drought [45,46], and this compensation may hold community AGB constant. The 50.2% decrease in RB in 2013 could be attributed to the reduction in soil moisture because RB in alpine ecosystems is positively correlated with annual precipitation [47]. The relative reduction rate of RB in our study (50.2% reduction in RB and 2.1 m^3 m^{-3} reduction in soil moisture) was higher than in a temperate grassland ecosystem

(23% reduction in RB and 5.2 m^3 m^{-3} reduction in soil moisture) [48]. The divergent responses of RB in the surface and deep soil layers [46,49] could cancel each other out and therefore lead to a lower relative decrease in RB across the whole soil profile.

However, only one wet and dry year was included in this study. The wet year was followed directly by a dry year, which might lowers the effect of drought because soil drought might lag the meteorological drought.

Main effects of warming

Experimental warming stimulated GEP more than ER (1.49 µ mol m^{-2}s^{-1} vs. 0.79 µ mol m^{-2}s^{-1}), leading to an increase in NEE in the warming treatment of this alpine meadow ecosystem in a permafrost area of the QTP. Warming effects on ecosystem C exchange are likely modulated by soil water regimes [7,37]. For example, Oberbauer (2007) reported that higher soil moisture in wet tundra limited increases in ER relative to increases in GEP under warming conditions, indicating the dependence of the warming effect on hydrological conditions. In the current alpine meadow ecosystem, differences in the responses of NEE to warming between 2012 and 2013 (Table 2) differed from results from a temperate steppe, in which NEE demonstrated no change under a warming treatment over two hydrologically contrasting years [40]. Soil moisture showed positive impacts on C fluxes in 2013 and negative impacts on fluxes in 2012 (Fig. 6). ABG and RB were positively correlated with ecosystem C fluxes (Figs. 8, 9). As there were no significant effects of warming on AGB or RB in either year (Tables 1, 2), the significant increase in NEE in warmed plots in 2012 could be attributed to the higher stimulation of GEP (2.09 µ mol m^{-2}s^{-1}) than ER (1.07 µ mol m^{-2}s^{-1}).

The insensitivity of NEE to warming in 2013 could be attributed to the effect of the soil moisture deficit on GEP and ER (Table 2, Fig. 6). The positive responses of GEP and ER to warming (Table 1) are consistent with those in a tundra ecosystem [7], but differ from those in a subalpine meadow ecosystem, where soil moisture stress induced by warming reduced ER [27,40]. ER is composed of Rs and respiration of AGB. Therefore, the significant increase in ER in 2012 could be attributed mainly to the stimulation in Rs (Table 2), as AGB was insensitive to warming (Table 2). This indicates that the response of ER to warming was determined by Rs even though AGB respiration is the major component of ER in alpine meadow ecosystems [50]. Rs is composed of root respiration and microbial decomposition of soil organic matter [51]. There was no significant change in RB at a depth of 0–10 cm in the warming treatment (Tables 1, 2), which suggests that the response of ecosystem C emission to warming is determined by soil organic matter decomposition. The non-significant response of ER to warming in 2013 likely resulted from lower soil moisture (less than 15%, the optimal soil moisture for ecosystem C fluxes, Table 3), and the warming-induced reduction in soil moisture (Fig. 2). This is because the negative effects of drought and warming induced soil water stress on Rs and ER, which could override the positive effect of warming on these variables, which has been shown for a Montane meadow [52] and a subalpine meadow ecosystem in the QTP [27]. The marginal increase in GEP in 2013 (Table 2) likely resulted from a change in the species composition, which was observed in an open top chamber warming experiment nearby our study site, where coverage of grass and sedges decreased but that of forbs increased with warming [46]. The increased forbs biomass could ameliorate the negative impact of warming-induced soil moisture stress and the effect of lower AGB on GEP [40].

Although experimental warming tends to have a positive effect on plant productivity across ecosystems, experiments in grasslands indicate that clear increases in plant productivity in response to warming are relatively rare [53]. We did not detect a significant change in RB. We attributed this to the fact that we sampled the RB at a depth of 0–10 cm, which was constrained by a reduction in soil water, whereas the RB at a depth of 10–50 cm was significantly stimulated by warming. In contrast to the response of forbs, sedges and grass [28] may have cancelled each other out, leading to the non-significant change in AGB in this alpine meadow ecosystem. AGB and RB were positively correlated with ecosystem C fluxes (Figs. 8, 9). There was no significant change in AGB or RB with warming, whereas ecosystem C fluxes were significantly stimulated by warming (Tables 1, 2). It is possible that this resulted from the large seasonal variation in AGB and RB compared to the relatively smaller warming-induced changes in these biomass pools.

Main effects of clipping

There was no significant effect of clipping on C fluxes, which contrasted with results from other studies where increases in GEP, ER, and NEE have been reported for a temperature steppe [54] and tallgrass prairie [13], and decreases in GEP, ER, and NEE have been reported for a Swiss grassland [55]. The negative impact of clipping on ecosystem C fluxes is attributed to the grass being cut in the middle of the growing season, which may reduce the green leaf area and thus C fluxes [55]. Positive effects of clipping on C fluxes may result primarily from improved light conditions with the removal of standing litter [54] and compensatory growth from clipping [56]. In the current study, we clipped the plants in late September once they had started to senesce, and this could be one reason for the non-significant effect of clipping on GEP, NEE and ER. In addition, soil temperature has been found to influence CO_2 exchange in alpine meadow ecosystems [24], and we did not detect a significant effect of clipping on soil temperature (Fig. 2). Besides temperature, biomass also affects C fluxes in an alpine meadow ecosystem [24], as was observed in our study (Fig. 8). The significant decrease in AGB in 2013 under the clipping treatment with a non-significant change in C fluxes, suggests that soil temperature is the major factor controlling the response of ecosystem C fluxes to clipping.

Conclusion

Ecosystem C fluxes responded positively to elevated temperature, with a higher relative increase in GEP than in ER, leading to a net C gain in this alpine meadow ecosystem. Clipping and its interaction with warming had no significant effect on ecosystem C fluxes because clipping did not significantly affect soil temperature. In addition, this study was conducted during two hydrologically contrasting years (wet in 2012 and dry in 2013), which provided a unique opportunity to understand how drought affects ecosystem C fluxes and their response to warming and clipping in an alpine meadow ecosystem. In the dry year, positive effects of warming on ecosystem C fluxes were cancelled by lower soil moisture. However, we caution that our study encompassed only a single wet and dry year, and thus our inferences of drought need to be supported by future research. Our findings will improve our understanding of the response of ecosystem C fluxes to the combined effects of climate change factors and human activities in an alpine meadow ecosystem in the permafrost region of the QTP.

Acknowledgments

Authors thank Prof. Yongzhi, Liu, Hanbo Yun, Guilong Wu, and Yuanwu Yang for their help in setting up the field experiment.

Author Contributions

Conceived and designed the experiments: FP XX. Performed the experiments: FP QY MX JG. Analyzed the data: FP. Contributed to the writing of the manuscript: FP TW XX.

References

1. IPCC (2007) Climate change 2007: The physical Science Basis Contributin of Working group I to the Fourth Assessment Report of the Intergovernmental Panel on Climate Change. In: S S., D Qin, M Manning, Z Chen, M Marquis, K. B Averyt, M Tignor and H. L Miller, editors. Cambridge, United Kingdom/ New York, NY USA: Cambridge University Press. 749–766.
2. Luo YQ, Wan SQ, Hui DF, Wallance LL (2001) Acclimatization of soil respiration to warming in a tall grass prairie. Nature 413: 622–625.
3. Melillo JM, Steudler PA, Abler JD (2002) Soil warming and carbon-cycle feedbacks to the climate system. Science 298: 2173–2175.
4. Friedlingstein P, Cox P, Betts R, Bopp L, von Bloh W, et al. (2006) Climate-Carbon Cycle Feedback Analysis: Results from the C4MIP Model Intercomparison. J. Climate 19: 3337–3353.
5. Rustad LE, Campbell JL, Marion GM, Norby RJ, Mitchell MJ, et al. (2001) A meta-analysis of the response of soil respiration, net nitrogen mineralization, and aboveground plant growth to experimental ecosystem warming. Oecologia 126: 543–562.
6. Peñuelas J, Gordon C, Llorens L, Nielsen T, Tietema A, et al. (2004) Nonintrusive Field Experiments Show Different Plant Responses to Warming and Drought Among Sites, Seasons, and Species in a North-South European Gradient. Ecosystems 7: 598–612.
7. Oberbauer SF, Tweedie CE, Welker JM, Fahnestock JT, Henry GHR, et al. (2007) Tundra CO2 fluxes in response to experimental warming across latitudinal and moisture gradients. Ecol. Monogr. 77: 221–238.
8. Kirschbaum MF (1995) The temperature dependence of soil organic mater decomposition, and the effect of global warming on soil organic C storage. Soil Biolo. Biochem. 27: 753–760.
9. Cramer W, Bondeau A, Woodward FI, Prentice IC, Betts RA, et al. (2001) Global response of terrestrial ecosystem structure and function to CO2 and climate change: results from six dynamic global vegetation models. Global Change Biolo. 7: 357–373.
10. Canadell JG, Le Quéré C, Raupach MR, Field CB, Buitenhuis ET, et al. (2007) Contributions to accelerating atmospheric CO2 growth from economic activity, carbon intensity, and efficiency of natural sinks. PNAS 104: 18866–18870.
11. Wu ZT, Dijkstra P, Koch GW, PeÑUelas J, Hungate BA (2011) Responses of terrestrial ecosystems to temperature and precipitation change: a meta-analysis of experimental manipulation. Global Change Biolo. 17: 927–942.
12. Lu M, Zhou XH, Yang Q, Li H, Luo YQ, et al. (2013) Responses of ecosystem carbon cycle to experimental warming: a meta-analysis. Ecology.
13. Niu SL, Sherry RA, Zhou XH, Luo YQ (2013) Ecosystem carbon fluxes in responses to warming and clipping in a tallgrass prairie. Ecosystems.
14. Bahn M, Knapp M, Garajova Z, Pfahringer N, Cernusca A (2006) Root respiration in temperate mountain grasslands differing in land use. Global Change Biolo. 12: 995–1006.
15. Ross DJ, Tate KR, Scott NA, Feltham CW (1999) Land-use change: effects on soil carbon, nitrogen and phosphorus pools and fluxes in three adjacent ecosystems. Soil Biolo. Biochem. 31: 803–813.
16. Klein J, Harte J, Zhao X (2004) Experimental warming causes large and rapid species loss, dampened by simulated grazing, on the Tibetan Plateau. Ecol. Lett. 7: 1170–1179.
17. Luo CY, Xu GP, Chao ZG, Wang SP, Lin XW, et al. (2010) Effect of warming and grazing on litter mass loss and temperature sensitivity of litter and dung mass loss on the Tibetan plateau. Global Change Biol. 16: 1606–1617.
18. Derner JD, Boutton TW, Briske DD (2006) Grazing and ecosystem carbon storage in the North American Great Plains. Plant Soil 280: 77–90.
19. Niu SL, Sherry RA, Zhou XH, Wan SQ, Luo YQ (2010) Nitrogen regulation of the climate-carbon feedback: evidence from a long-term global change experiment. Ecology 91: 3261–3273.
20. Tarnocai C, Canadell JG, Schuur EAG, Kuhry P, Mazhitova G, et al. (2009) Soil organic carbon pools in the northern circumpolar permafrost region. Global Biogeochem. Cy. 23: GB2023.
21. Schuur EAG, Bockheim J, Canadell JG (2008) Vulnerability of permafrost carbon to climate change:implication for the global carbon cycle. BioScience 58: 701–714.
22. Harden JW, Sundquist ET, Stallard RF, Mark RK (1992) Dynamics of soil carbon during deglaciation of the Laurentide ice sheet. Science 258: 1921–1924.
23. Kato T, Tang Y, Gu S, Cui X, Hirota M, et al. (2004) Carbon dioxide exchange between the atmosphere and an alpine meadow ecosystem on the Qinghai-Tibetan Plateau, China. Agr. Forest Meteorol. 124: 121–134.
24. Kato T, Tang Y, Gu S, Hirota M, Du M, et al. (2006) Temperature and biomass influences on interannual changes in CO2 exchange in an alpine meadow on the Qinghai-Tibetan Plateau. Global Change Biol. 12: 1285–1298.
25. Liu XD, Chen BD (2000) Climatic warming in the Tibetan Plateau during recent decades. Inter. J. Climatol. 20: 1729–1742.
26. Zheng D, Zhang QS, Wu SH (2000) Mountain Geoecology and sustainable development of the Tibetan Plateau. Dordrecht, Netherlands: Kluwer Academic Publishers.
27. Lin XW, Zhang ZH, Wang SP, Hu YG, Xu GP, et al. (2011) Response of ecosystem respiration to warming and grazing during the growing seasons in the alpine meadow on the Tibetan plateau. Agr. Forest Meteorol. 151: 792–802.
28. Li N, Wang GX, Yang Y, Gao YH, Liu GS (2011) Plant production, and carbon and nitrogen source pools, are strongly intensified by experimental warming in alpine ecosystems in the Qinghai-Tibet Plateau. Soil Biol. Biochem. 43: 942–953.
29. Lu Z, Wu Q, Yu S, Zhang L (2006) Heat and water difference of active layers beneath different surface conditions near Beiluhe in Qinghai-Xizang Plateau. J. Glaciol. Geogryol. 28: 642–647.
30. Wang G, Wang Y, Li Y, Cheng H (2007) Influences of alpine ecosystem responses to climatic change on soil properties on the Qinghai-Tibet Plateau, China. Catena 70: 506–514.
31. Pang Q, Cheng G, Li S, Zhang W (2009) Active layer thickness calculation over the Qinghai-Tibet Plateau. Cold Regions Science and Technology 57: 23–28.
32. Wu QB, Liu YZ (2004) Ground temperature monitoring and its recent change in Qinghai-Tibet Plateau. Cold Reg. Sci. Technol. 38: 85–92.
33. Yang ZL, Yang GH (2000) Potential productivity and livestock carrying capacity of high-frigid grassland in China. Resources Sci. 22: 72–77.
34. Pavelka M, Acosta M, Marek MV, Kutsch W, Janous D (2007) Dependence of the Q10 values on the depth of soil temperature measuring point. Plant Soil 292: 171–179.
35. Zhou XH, Wan SQ, Luo YQ (2007) Source components and interannual variability of soil CO2 efflux under experimental warming and clipping in a grassland ecosystem. Global Change Biol. 13: 761–775.
36. Steduto P, Çetinkökü Ö, Albrizio R, Kanber R (2002) Automated closed-system canopy-chamber for continuous field-crop monitoring of CO2 and H2O fluxes. Agr. Forest Meteorol.111: 171–186.
37. Niu S, Wu M, Han Y, Xia J, Li L, et al. (2008) Water-mediated responses of ecosystem carbon fluxes to climatic change in a temperate steppe. New Phytol. 177: 209–219.
38. Lawrence BF, Bruce GJ (2005) Interacting effects of temperature, soil moisture and plant biomass production on ecosystem respiration in a northern temperate grassland. Agr. Forest Meteorol. 130: 237–253.
39. Yang Y, Fang J, Ji C, Han W (2009) Above- and belowground biomass allocation in Tibetan grasslands. J. Veg. Sci. 20: 177–184.
40. Xia JY, Niu SL, Wan SQ (2009) Response of ecosystem carbon exchange to warming and nitrogen addition during two hydrologically contrasting growing seasons in a temperate steppe. Global Change Biol. 15: 1544–1556.
41. Williams M, Law BE, Anthoni PM, Unsworth MH (2001) Use of a simulation model and ecosystem flux data to examine carbon-water interactions in ponderosa pine. Tree Physiol. 21: 287–298.
42. Barr AG, Griffis TJ, Black TA, Lee X, Staebler RM, et al. (2002) Comparing the carbon budgets of boreal and temperate deciduous forest stands. Can. J. Forest Res. 32: 813–822.
43. Kahmen A, Perner J, Buchmann N (2005) Diversity-dependent productivity in semi-natural grasslands following climate perturbations. Funct. Ecol. 19: 594–601.
44. Gilgen AK, Buchman N (2009) Responses of tempeartue grasslands at different altitudes to simulated summer drought differed but scaled with annual precipitation. Biogeosciences 6: 2525–2539.
45. Sebastià M-T (2007) Plant guilds drive biomass response to global warming and water availability in subalpine grassland. J. Appl. Ecol. 44: 158–167.
46. Li N, Wang GX, Yang Y, Gao YH, Liu LA, et al. (2011) Short-term effects of temperature enhancement on community structure and biomass of alpine meadow in the Qinghai-Tibet Plateau. Acta Ecol. Sinica 31: 895–905.
47. Li XJ, Zhang XZ, Wu JS, Shen ZX, Zhang YJ, et al. (2011) Root biomass distribution in alpine ecosystem of the northern Tibet Plateau. Environ. Earth Sci. 64: 1911–1919.
48. De Boeck HJ, Lemmens CMHM, Gielen B, Bossuyt H, Malchair S, et al. (2007) Combined effects of climat warming and plant diversity loss on above- and below-ground productivity. Environ. Exp.l Bot. 60: 95–104.
49. Xu MH, Peng F, Xue X, You QG, Guo J (2014) All-year warming and autumnal clipping lead to the downward movement of the root biomass, carbon and total nitrogen in the soil of an alpine meadow. Environ. and Exp. Bot.: in press.
50. Zhang PC, Tang YH, Hirota M, Yamamoto A, Mariko S (2009) Use of regression method to partition sources of ecosystem respiration in an alpine meadow. Soil Biol. Biochem. 41: 663–670.
51. Hanson PJ, Edwards NT, Garten CT, Andrews JA (2000) Separating root and soil microbial contributions to soil respiratin: a review of methods and observations. Biogeochem. 48: 115–146.

52. Saleska S, Harte K, Torn M (1999) The effect of experimental ecosystem warming on CO_2 fluxes in a montane meadow. Global Change Biol. 5: 125–141.

53. Dukes JS, Chiariello NR, Cleland EE, Moore LA, Shaw MR, et al. (2005) Responses of Grassland Production to Single and Multiple Global Environmental Changes. PLoS Biol 3: e319.

54. Niu SL, Wu M, Han Y, Xia J, Zhang Z, et al. (2010) Nitrogen effects on net ecosystem carbon exchange in a temperate steppe. Global Change Biol. 16: 144–155.

55. Rogiers N, Eugster W, Furger M, Siegwolf R (2005) Effect of land management on ecosystem carbon fluxes at a subalpine grassland site in the Swiss Alps. Theor. Appl. Climatol. 80: 187–203.

56. Zhao W, Chen S-P, Lin G-H (2008) Compensatory growth responses to clipping defoliation in Leymus chinensis (Poaceae) under nutrient addition and water deficiency conditions. Plant Ecol. 196: 85–99.

Effect of Abandonment on Diversity and Abundance of Free-Living Nitrogen-Fixing Bacteria and Total Bacteria in the Cropland Soils of Hulun Buir, Inner Mongolia

Huhe[1¤], Shinchilelt Borjigin[1], Yunxiang Cheng[2], Nobukiko Nomura[1], Toshiaki Nakajima[1], Toru Nakamura[1], Hiroo Uchiyama[1]*

1 Graduate School of Life and Environmental Sciences, University of Tsukuba, Tsukuba, Ibaraki, Japan, 2 State Key Laboratory of Grassland Agro-Ecosystems, College of Pastoral Agriculture Science and Technology, Lanzhou University, Lanzhou, China

Abstract

In Inner Mongolia, steppe grasslands face desertification or degradation because of human over activity. One of the reasons for this condition is that croplands have been abandoned after inappropriate agricultural management. The soils in these croplands present heterogeneous environments in which conditions affecting microbial growth and diversity fluctuate widely in space and time. In this study, we assessed the molecular ecology of total and free-living nitrogen-fixing bacterial communities in soils from steppe grasslands and croplands that were abandoned for different periods (1, 5, and 25 years) and compared the degree of recovery. The abandoned croplands included in the study were natural restoration areas without human activity. Denaturing gradient gel electrophoresis and quantitative PCR (qPCR) were used to analyze the *nifH* and 16S rRNA genes to study free-living diazotrophs and the total bacterial community, respectively. The diversities of free-living nitrogen fixers and total bacteria were significantly different between each site ($P < 0.001$). Neither the total bacteria nor *nifH* gene community structure of a cropland abandoned for 25 years was significantly different from those of steppe grasslands. In contrast, results of qPCR analysis of free-living nitrogen fixers and total bacteria showed significantly high abundance levels in steppe grassland ($P < 0.01$ and $P < 0.03$, respectively). In this study, the microbial communities and their gene abundances were assessed in croplands that had been abandoned for different periods. An understanding of how environmental factors and changes in microbial communities affect abandoned croplands could aid in appropriate soil management to optimize the structures of soil microorganisms.

Editor: Bas E. Dutilh, Radboud University Medical Centre, NCMLS, Netherlands

Funding: This work was supported by a Grant-in-Aid for Scientific Research (B) (No. 21310050 to HU) and CREST program of Japan Science and Technology Corporation (JST; http://www.jst.go.jp/EN/index.html). The funders had no role in study design, data collection and analysis, decision to publish, or preparation of the manuscript.

Competing Interests: The authors have declared that no competing interests exist.

* Email: uchiyama.hiroo.fw@u.tsukuba.ac.jp

¤ Current address: Institute of Soil and Fertilizer and Save Water Agricultural, Gansu Academy of Agricultural Sciences, Anning District, Lanzhou, Gansu, China

Introduction

Steppe grasslands are distributed over vast areas in the arid and semiarid regions of the Eurasian continent [1]. The Inner Mongolia steppe is an important part of Eurasia and has been used by pastoral nomads for long periods of time. However, vast areas were converted to cropland, and farmers have increased the size of these croplands during the past 40 years because of a rapid increase in the human population. Subsequently, many of the croplands were abandoned because of soil degradation and desertification caused by inappropriate agricultural management [2,3]. Further, some croplands were abandoned to restore natural vegetation, as in the case where the Chinese government proposed to restore farmlands to grasslands or forests [4]. Some plant species, including grasses, shrubs, and trees, are introduced into the abandoned croplands to restore the vegetation in some locations [5–8]. In most areas, the plant community of abandoned cropland is likely to be restored naturally by controlling human activities [9]. Many researchers have evaluated the abandoned

cropland ecosystem changes by studying the vegetation changes [9–12]. There are interactions between plants and microorganisms. Plants exude diverse compounds, such as organic acids, enzymes, and polysaccharides, from the roots. Further, plants can recognize microbe-derived compounds and adjust their defense and growth responses according to the type of microorganisms encountered. Conversely, microorganisms can detect suitable plant hosts and initiate their colonization strategies in the rhizosphere by producing canonical plant growth-regulating substances such as auxins or cytokinins [13]. However, most studies have not evaluated changes in soil microbial community and its gene abundance in croplands that have been abandoned for different periods of time.

Cultivated soil or grassland soil contains an estimated 2×10^9 prokaryotic cells per gram [14]. Soil microbial communities are important factors of agriculturally managed systems because they are responsible for most nutrient transformations in soil and influence the aboveground plant diversity and productivity [15]. Next to water, nitrogen is the second-most limiting factor for plant growth [16]. Nitrogen cycling in natural ecosystems and during traditional agricultural production rely on nitrogen fixation of diazotrophic bacteria [17]. Diazotrophs are highly diverse and widely distributed across bacterial and archaeal taxa [18]. Approximately 80% of biological nitrogen fixation is performed by diazotrophs in symbiosis with legumes [19]. However, under specific conditions, the free-living bacteria in soil (e.g., cyanobacteria, *Pseudomonas*, *Azospirillum*, and *Azotobacter*), may fix significant amounts of nitrogen $(0-60 \text{ kg} \cdot \text{N} \cdot \text{ha}^{-1} \cdot \text{year}^{-1})$ [20,21]. This may be particularly important in abandoned field soils, where legume plants have not been cultivated and there are very few symbiotic plants (e.g., Leguminosae, *Azolla*, *Myrica*, and *Alnus*.).

In this study, the diazotrophic population was monitored by PCR-denaturing gradient gel electrophoresis (DGGE) exploiting the *nifH* gene. The *nifH* gene is the most conserved gene in the *nif* operon and encodes the Fe subunit of the nitrogenase enzyme [22]. Because of the conserved nature of the *nifH* gene, it has been possible to identify primer sets that can be used to analyze nitrogen fixers so that this community can be analyzed by a PCR-DGGE-based technique [20,23–25]. We have tested the diversity and abundances of free-living nitrogen fixers and the total bacterial population changes that occurred over time in soils belonging to artificially disrupted environments (abandoned cropland soils).

Materials and Methods

Ethics statement

No specific permissions are required for our conducting field survey in this area, since land in China belongs to the public and our field studies did not involve any endangered or protected plant species within.

Site description and sample collection

The study area is located in the Hulun Buir grassland $(115°31'–126°04'$ E, $47°05'–53°20'$ N) in northeastern Inner Mongolia, China (Figure 1). The Hulun Buir grassland area is about $2.6 \times 10^5 \text{ km}^2$, with a west to east distribution of arid steppe, semi-arid steppe, and meadow steppe. The study area is located in the semi-arid areas.

We established 4 sites, 3 abandoned croplands and a light-grazing steppe grassland (LGSG) (Figure 1) that had an intensity of about 1.4 sheep ha^{-1}. The 3 croplands were abandoned for 1, 5, and 25 years (Y1, Y5, and Y25, respectively), and the control area was a LGSG. In the Y1, Y5, and Y25 sites, *Zea mays*, *Helianthus annuus*, and *Elymus cylindricus* were rotated for approximately 40 years. The sites were subsequently abandoned because of land degradation; soil fertility including both organic C and total N had decreased by approximately 70%.

Plant surveying and soil sampling were conducted in August 2010. All of the sites were selected for their similar topography (flat). Each site contained 5 replicates in a randomized plot $(1 \times 1 \text{ m})$ design: Y1 (site 1), plots 1–5; Y5 (site 2), plots 6–10; Y25 (site 3), plots 11–15; and LGSG (site 4), plots 16–20. The coordinates and elevations of the sampled sites are as follows: Y1, 48° 38′ 43″ N, 116° 57′ 56″ E, 545 m; Y5, 48° 38′ 50″ N, 117° 00′ 48″ E, 550 m; Y25, 48° 38′ 45″ N, 117° 01′ 56″ E, 545 m and LGSG, 48° 32′ 00″ N, 116° 40′ 18″ E, 568 m. The mean temperature and precipitation from 2000 to 2009 for each site [26] were as follows: LGSG, Y1, Y5, and Y25, 1.6°C and 213 mm. In each plot, the species composition was recorded. Plant communities were classified on the basis of their differential species [27,28]; all species were identified and measured for cover, height, and density, and Shannon-Wiener diversity index was calculated. Soil moisture was measured with a TRIME-FM (Ettlingen, Germany). Above-ground plant biomass was also determined by clipping the plants at ground level, sorting by species, drying at 60°C for 48 h, and weighing the samples (Table 1 and Table S1).

In each plot, the soil samples were collected from 5 randomly selected points (0 to 10 cm deep) and mixed into 1 sample. After carefully removing the surface organic materials and fine roots, each mixed sample was divided into 2 parts. One part was air-dried for the analysis of soil physicochemical properties. The other was sifted through a 2-mm sieve, sealed in sample vials, kept on ice for transport to the laboratory, and stored at $-20°C$ for microbial assays.

The soil texture was determined by mechanical analysis using the pipette method [29], and the soil texture was used to estimate the saturated hydraulic conductivity [30] for each site (Table S2). The soil texture was classified according to the International

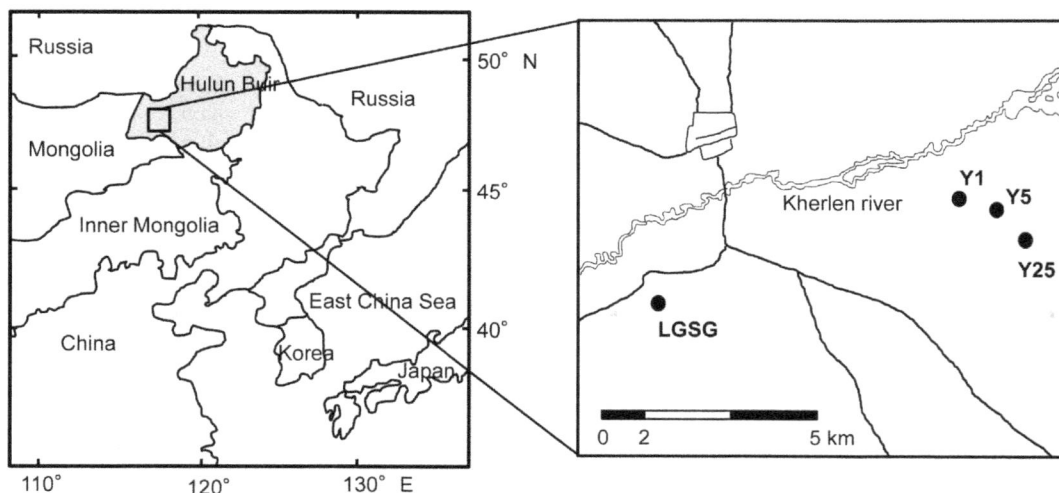

Figure 1. Map of the study area and the 4 study sites (·) in Hulun Buir.

Table 1. Changes to pH, available NO_3-N, available NH_4-N, available P, soluble Fe, organic C, total N, plant diversity (P-H′), plant biomass (P-B), soil moisture, and hydraulic conductivity (HC) across the field trial and Pearson's product-moment correlation analysis comparing data to *nifH* and 16S rRNA diversity and gene copies.

Abandoned Cropland or significance parameter	pH	NO_3-N (mg kg^{-1})	NH_4-N (mg kg^{-1})	P (mg kg^{-1})	Fe (g kg^{-1})	Organic C (g kg^{-1})	Total N (g kg^{-1})	H_2O (%)	P-H′	P-B	HC (×10^{-3} cm s^{-1})
Abandoned Cropland*											
Y1	7.72±0.13 a	3.35±0.42 a	1.59±0.64 a	12.38±4.09 a	0.43±0.06 a	6.55±0.96 a	0.7±0.09 a	8.0±0 a	1.63±0.34 a	48.5±9.77 a	2.93±0.07 a
Y5	8.57±0.37 b	3.99±0.67 b	0.76±0.04 a	12.92±2.68 a	0.56±0.04 a	9.16±1.51 a	0.94±0.12 b	9.06±0.13 ab	1.71±0.43 a	17.16±4.44 b	2.95±0.03 a
Y25	7.82±0.17 a	3.54±0.53 a	1.09±0.21 a	12.5±1.8 a	0.46±0.04 b	7.23±0.64 a	0.75±0.06 ab	10.7±1.09 b	1.82±0.21 a	79.65±13.5 c	2.79±0.02 b
LGSG	6.21±0.14 c	4.21±0.06 c	4.84±0.37 b	24.48±5.67 b	1.12±0.06 c	24.84±3.09 b	2.3±1.71 c	12.12±1.4 bc	0.94±0.02 b	141.9±5.54 d	2.75±0.02 b
Correlation (P**)											
with *nifH* DGGE H′	++	++	NS	+++	NS	NS	NS	NS	NS	NS	NS
with *nifH* copy number	NS	NS	++	+++	+++	+++	++	++	—	++	NS
with 16S rRNA DGGE H′	NS	NS	NS	NS	+++	++	++	+++	NS	++	—
with 16S rRNA copy number	NS	NS	+++	NS	NS	++	++	NS	NS	++	NS

*Y1, Y5 and Y25 mean field abandoned for 1 year, 5 years and 25 years, respectively. LGSG is light grazing steppe grassland. The values shown for management factors are means ± standard errors.
**P, Pearson's product-moment correlation coefficient; NS, not significant; +++/−−, significant positive or negative correlation at P<0.05; ++/−, significant positive or negative correlation at P<0.01. Significant differences are indicated by different letters.

Society of Soil Science (ISSS) classification system. The soils of all 4 sites were sandy loam.

Concentrations of NO_3-N and NH_4-N in the KCl extracts were determined with the zinc reduction-naphthylethylenediamine method for NO_3-N [31] and the indophenol blue colorimetric method for NH_4-N [32]. The soil phosphorus content was determined by the Truog method [33]. Ferrous iron was measured by the o-phenanthroline method [34,35]. The organic C and total N were determined by the dry combustion method using an NC analyzer (Sumigraph NC-900; Sumika Chemical Analysis Service, Tokyo, Japan). The soil samples were previously treated with acid to eliminate water and inorganic carbonates [36]. The soil pH was obtained by measuring the equilibrium pH of soil pastes containing 1 g of soil homogenized in 1 mL of H_2O (Table 1).

DNA extraction and PCR

DNA from each of the 20 samples (4 sites×5 replicates) was extracted in 3 subsamples from 0.5 g of soil with the FastDNA Spin Kit for soil (MP Biomedicals, Illkirch, France) according to the manufacturer's protocol. The quality and quantity of the DNA extracts were checked with a SmartSpec Plus spectrophotometer (Bio-Rad Laboratories, United States). The samples were pooled and stored at -20°C until use.

A fragment of the *nifH* gene (approximately 360 bp) was amplified with a nested PCR strategy. First-round reactions were performed with the primers nifH32F (TGAGACAGATAGCTA-TYTAYGGHAA) and nifH623R (GATGTTCGCGCGGCAC-GAADTRNATSA) as described previously [37]. The genomic DNA extract (15–40 ng) was added to PCR mixtures containing 5 µL of 10×ExTaq buffer (Takara, Madison, WI, United States), 4 µL of a mix of deoxynucleoside triphosphates (2.5 mM each), 37.5 µL of water, 0.5 µL of 100 µM nifH32F, 0.5 µL of 100 µM nifH623R, and 0.5 µL of ExTaq DNA polymerase (5 U/µL; Takara, Madison, WI, United States). The reaction mixtures were amplified by 1 denaturation step (5 min at 94°C), followed by 30 cycles of 94°C for 1 min, 50°C for 1 min, and 72°C for 1 min, and 1 final 7 min extension cycle at 72°C. For the second round of the nested amplification, 1 µL of this reaction mixture was used as the template in a 50 µL reaction mixture containing the same reagent mixture described above, but with 39 µL of water, 0.25 µL of 100 µM nifH1-GC (in order to clamp the products for DGGE, the primer nifH1 [CTGYGAYCCNAARGCNGA] was added to the GC-clamp [CGCCCGCCGCGCGCGGCGGGCGGGGCGG-GGGCACGGGGGG] on the 5′ side), and 0.25 µL of 100 µM nifH2 (ADNGCCATCATYTCNCC) [38]. The thermal cycling protocol for the nested reactions was the same as above except that the annealing temperature was raised to 57°C.

Fragments of approximately 200 bp, corresponding to the V3 region of the 16S rRNA gene [39], were amplified using a reaction mixture that contained the same reagent mixture described above, except with 0.25 µL of 100 µM 357F-GC (CCTACGGGAGG-CAGCAG-GC-clamp) and 0.25 µL of 100 µM 518R (GTAT-TACCGCGGCTGG); products were amplified using a touch-down thermocycling program [39]. All of the PCR amplicons were electrophoresed on an agarose gel to ascertain the sizes and purified using the UltraClean PCR Clean-Up Kit (MO BIO Laboratories, Carlsbad, CA, United States).

DGGE

DGGE was performed using the D-Code system (Bio-Rad Laboratories, Hercules, CA, United States) as described by Baxter and Cummings [40]. The polyacrylamide concentration in the gel was 8%, and the linear denaturing gradient was 30% to 60% (100% denaturant corresponds to 7 M urea and 40% deionized formamide). The gel was run at 36 V for 18 h at 60°C in 0.5×TAE buffer. The gel was then stained for 30 min with 1:10,000 (v/v) SYBR Gold, rinsed with 0.5×TAE, and scanned on a transilluminator. Bands were identified, and relative intensities were calculated based on the percentage of intensity of each band in a lane. This was done with an image-analyzing system (Image Master; Amersham Pharmacia Biotech, Uppsala, Sweden). Shannon-Wiener diversity index (H') was calculated by the formula $H' = -\Sigma\, p_i \ln(p_i)$, where p_i is the ratio of relative intensity of band i compared with the relative intensity of the lane. The used of different gradient gel (30% to 70%) to assessed the reproducibility of the DGGE results, similar results were obtained (data not shown).

Real-time PCR assay

Reactions were set up using SYBR green (Bio-Rad Laboratories, the Netherlands) according to Baxter and Cummings [41] with the LightCycler 1.5 system (Roche Applied Sciences, Indianapolis, IN, United States). Reaction mixtures were heated to 95°C for 15 min to denature the DNA before completing 40 cycles of denaturation (95°C for 45 s/15 s [*nifH*/16S rRNA]), annealing (55°C for 45 s/65°C for 15 s [*nifH*/16S rRNA]), and extension (72°C for 45 s/15 s [*nifH*/16S rRNA]). Soil DNA extracts were diluted 1:100 to prevent inhibition of PCR by soil contaminants (e.g., by co-extracted humic substances), and each run included triplicate reactions for each DNA sample, the standard curve, and the no template control. The average copy number was converted into copies of the gene per gram of soil. nifHF (AAAGGYGGWATCGGYAARTCCACCAC) and nifHR (TTGTTSGCSGCR TACATSGCCATCAT) primers [42] were used for *nifH* quantitative PCR (qPCR), and 357F and 518R primers were used for total bacteria qPCR. Dilution series of pGEM-T Easy vector (Promega, Madison, WI, United States) DNA with cloned bacterial *nifH* (*Azospirillum brasilense* ATCC 29729) and 16S rRNA gene (*Pseudomonas aeruginosa* PAO1) fragments were used to generate standard curves ranging from 10^1 to 10^7 gene copies·$µL^{-1}$ for DNA quantification. The specificity of the amplified products was checked by the observation of a single melting peak and the presence of a unique band of the expected size in a 2% agarose gel stained with ethidium bromide. The standard curve produced was linear ($r^2 > 0.98$), and the PCR efficiency (Eff = $10^{(-1/\text{slope})}$-1) was >0.90.

Statistical analysis

In all tests, significant effects/interactions were those with a P value that was <0.05. Statistical analysis was performed using SPSS statistical software package (version 19.0; SPSS, Inc., Chicago, IL, United States). Variables of each group must be normally distributed to perform Pearson's product-moment correlations and analysis of variance (ANOVA). The normal distribution of the residuals was evaluated using the Kolmogorov-Smirnov test. If the requirement was not met, data were log-transformed prior to analysis. The homogeneity of the variances was checked by Levene's test. For pairwise comparison of means, Tukey's test was applied. Differences between main effects were tested by ANOVA. Correlations between *nifH* and 16S rRNA diversity and gene copies and environmental factors were tested by Pearson's product-moment correlations.

The choice between a linear or unimodal species response model depends on the underlying gradient length, which is measured in standard deviation units along the first ordination axis and can be estimated by detrended correspondence analysis (DCA). It is recommended to use linear methods when the gradient length is <3, unimodal methods when it is >4, and any method for intermediate gradient lengths [43]. The DCA gradient length for *nifH* gene patterns was 1.77, and that for 16S rRNA patterns was 1.11. Therefore, linear species response models such as partial least-squares regression (PLSR) and redundancy analysis (RDA) were used for multivariate statistical analysis. PLSR is an extension of multiple regression analysis in which the effects of linear combinations of several predictors on a response variable (or multiple response variables) are analyzed. PLSR is especially useful when the number of predictor variables is similar to or higher than the number of observations and/or predictors are highly correlated [44,45]. For PLSR model generation, the software assigns a value known as the variable influence on projection (VIP) to each environmental variable. The VIP indicates the relative importance to the model. Significance was assessed using the VIP parameter (a VIP>1 indicated a significant contribution of the environmental factor to the statistical model) [46,47]. RDA can be considered an extension of principal component analysis (PCA) in which the main components are constrained to be linear combinations of the environmental variables. RDA not only represents the main patterns of species variation as much as they can be explained by the measured environmental variables but also displays correlations between each species and each environmental variable in the data [48]. DCA, RDA, and PCA were performed using the Canoco program for Windows 4.5 (Biometris, Wageningen, the Netherlands). PLSR was performed using the Simca-P 11.0 (Umetrics AB, Umea, Sweden).

Results

Diversity of *nifH*

DGGE gels are shown in Figure S1A. Analysis of the *nifH* DGGE Shannon-Wiener diversity index values for the whole data set indicated that the abandoned period significantly affected the *nifH* diversity ($P<0.001$) by a separate ANOVA. For these reasons, multiple comparisons of the *nifH* DGGE Shannon-Wiener diversity index values were conducted.

Soils of Y1 (2.8±0.02) and Y5 (2.83±0.06) showed significantly higher *nifH* diversity than soils of Y25 (2.72±0.08) and LGSG (2.72±0.1) (Y1×Y25, $P=0.04$; Y1×LGSG, $P=0.02$; Y5×Y25, $P=0.009$; Y5×LGSG, $P=0.025$) (Figure 2A). The Y25 and

LGSG soils did not show significant differences in *nifH* diversity. Pearson's product-moment correlation indicated that pH, NO_3-N, and P were positively correlated with the *nifH* Shannon-Wiener diversity index (Table 1).

qPCR of the *nifH* gene

We used qPCR to compare copy numbers of the functional gene (*nifH*) at the 4 sites (Figure 3A). The *nifH* gene copy number in LGSG soil (5.5×10^5 copies·g^{-1} of soil) was the higher compared to the soils of the other 3 abandoned croplands (average number of copies·g^{-1} of soil, 1.9×10^5 in Y1, 3.4×10^5 in Y5, and 2.3×10^5 in Y25). In the abandoned cropland soils, the *nifH* gene copy number did not significantly change over time ($P>0.05$) (Figure 3A).Using Pearson's product-moment correlation, we found that NH_4-N, P, Fe, organic C, total N, soil moisture, and plant biomass were positively correlated, and the Shannon-Wiener diversity index was negatively correlated with the *nifH* copy number (Table 1).

Total bacterial diversity

There are clearly differences in the nitrogen-fixing communities of soil from long and short abandoned periods. In order to ensure that these factors are affecting the nitrogen-fixing community specifically and not the bacterial community as a whole, the 16S rRNA gene diversity and abundance were also analyzed. DGGE gels showing the diversity of total bacteria are shown in Figure S1B. As with the free-living nitrogen-fixing community, ANOVA results for the Shannon-Wiener diversity index of the 16S rRNA gene indicated that the abandoned period ($P<0.001$) was a significant factor. Therefore, each sample was analyzed separately by multiple comparisons. Soils of Y25 and LGSG (both almost equal, 3.17±0.07) showed higher 16S rRNA diversity than soils of Y1 (2.8±0.04) and Y5 (3.06±0.01) (Figure 2B). Pearson's product-moment correlation indicated that organic C, total N, Fe, soil moisture, and plant biomass were positively correlated, and hydraulic conductivity was negatively correlated with the 16S rRNA gene Shannon-Wiener diversity index (Table 1).

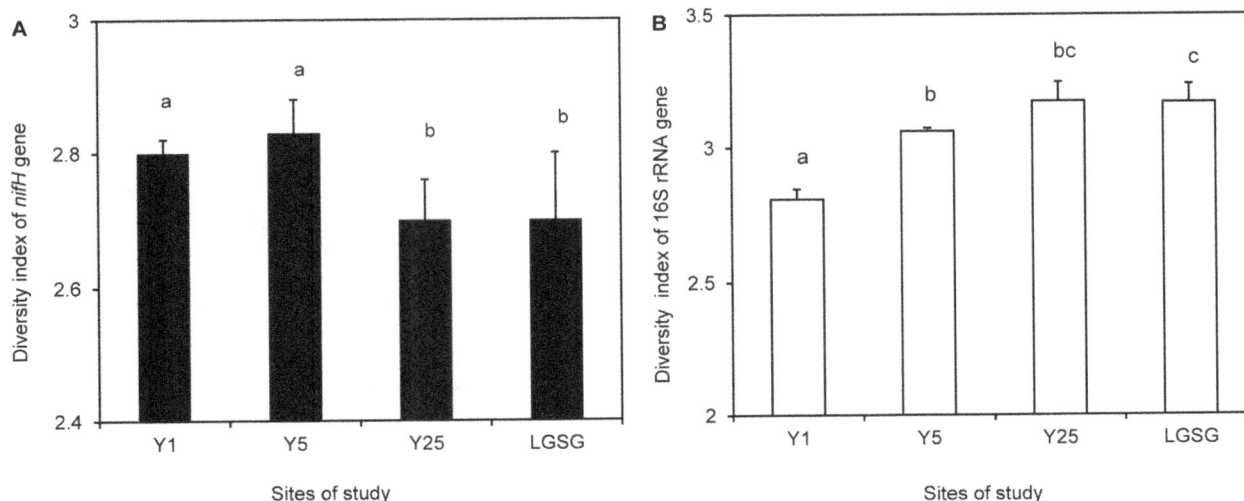

Figure 2. Shannon-Wiener diversity index values for *nifH* (A) and 16S rRNA genes (B). Data sets and results of analysis of variance (ANOVA) in the abandoned cropland (Y1, Y5, and Y25) and light-grazing steppe grassland (LGSG) soils (n=5; error bars represent standard deviations). Significant differences are indicated by different letters.

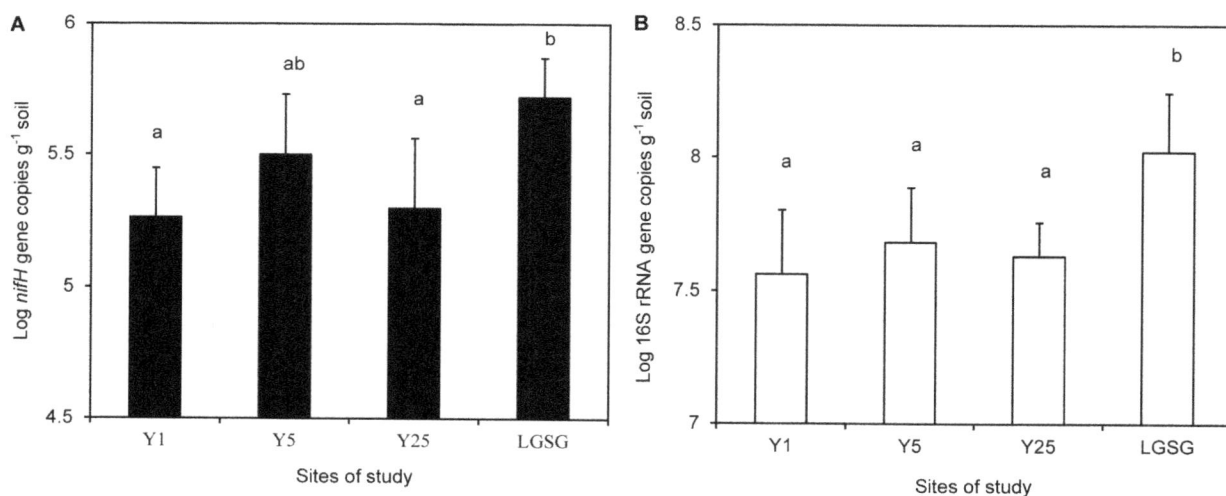

Figure 3. Copy numbers of the *nifH* **(A) and 16S rRNA (B) genes.** Results of analysis of variance (ANOVA) in the abandoned cropland (Y1, Y5, and Y25) and light-grazing steppe grassland (LGSG) soils (n = 5; error bars represent standard deviations). Significant differences are indicated by different letters.

qPCR of the 16S rRNA gene

By performing the qPCR analysis, we aimed to compare the differences between each site rather than attain an absolute quantification. Similar to *nifH* results presented above, the gene copy numbers of 16S rRNA in LGSG soil (1.2×10^8 copies·g^{-1} of soil) was higher than that of 3 other abandoned cropland soils (average numbers of copies·g^{-1} of soil, 8.2×10^7 in Y1, 5.2×10^7 in Y5, and 4.4×10^7 in Y25) (Figure 3A and B). The 16S rRNA gene copy numbers did not change significantly over time in the abandoned cropland soils (P>0.05) (Figure 3B). Pearson's product-moment correlation indicated that NH_4-N, organic C, total N, and plant biomass were positively correlated with the 16S rRNA gene copy number (Table 1).

Effects of environmental variables on the microbial community structure

Because of multicollinearity among environmental variables (Table S3), potential effects of environmental variables on free-living nitrogen-fixing bacteria and the total bacteria community composition were assessed by PLSR (Table 2). Plant biomass, hydraulic conductivity, pH, Fe, and soil moisture, and NH4-N predominantly affected on both free-living nitrogen-fixing bacteria and the total bacteria community composition in this study area. Moreover, C and N also exerted effects on the total bacteria community composition (Table 2).

These correlations of microbial diversity with environmental parameters were also supported by RDA (Figure S2). The ordination plots for the *nifH* and 16S rRNA genes are given in Figure S2A and S2B, respectively. In these figures, the first 2 canonical axes explained 43.6% and 59.1% of the variance of the species data (functional genes) and 51.0% and 63.4% of the variance of the species–environment relationship, respectively. The projection of environmental variables with respect to the free-living nitrogen-fixing community composition revealed that the first canonical axis is positively correlated with soil moisture, Fe, NH4-N, and plant biomass, and it is negatively correlated with pH and hydraulic conductivity (Figure S2A). With respect to the total bacteria community composition, the first canonical axis is positively correlated with Fe, NH4-N, plant biomass, C, and N,

and it is negatively correlated with pH and hydraulic conductivity (Figure S2B).

Discussion

This study allowed a detailed analysis of the effects of key components of abandoned cropland with the same land use history (permanent grassland was turned into arable land by cultivation about 40 years ago and had rotated crops of *Zea mayscorn*, *Helianthus annuus* and *E. cylindricus*) and light-grazing steppe grassland systems on soil bacterial and free-living nitrogen-fixing bacterial population structure and gene copy number. Cultivation imparted a strong effect on both total and free-living nitrogen-fixing bacterial population structures (measured by DGGE profiles) and abundances (measured by gene copy numbers).

The copy numbers of *nifH* per gram of LGSG soil was higher than that in Y1, Y5, and Y25 soil. In this study, markedly lower levels of nutrient (e.g., organic C, P, and Fe) in the soil of abandoned cropland (Table 1) may suppress the growth of free-living nitrogen-fixing bacteria. Organic C, P, and Fe and *nifH* copy number have a significant positive Pearson's product-moment correlation (Table 1). The source of carbohydrate is important to allow N_2 fixation activity, which requires large amounts of energy and reducing equivalents [49]. Phosphorus availability can increase nitrogen fixation because nitrogen fixation process requires large amounts of adenosine triphosphate (ATP) [50]. The [4Fe-4S] cluster of the *nifH* gene requires the Fe [22,51]; therefore, a low copy number of *nifH* might be related to the content of organic C, P, and Fe. In addition, Coelho et al. [52,53] found that 30% more free-living diazotrophs could be isolated from soil in the presence of low levels of nitrogen fertilizer than from soil in the presence of high levels of nitrogen fertilizer. In contrast, low copy number of *nifH* was obtained from the abandoned cropland of low nitrogen levels (NO_3-N and NH_4-N) (Table 1). The result may be caused by nutrients content of abandoned cropland soils.

Shannon-Wiener diversity index of *nifH* of LGSG was lower than that of Y1 and Y5 (Figures 2A and 3A). Previous studies indicated that pH is an important factor for the structure of the bacterial community [54,55]. Belnap [56] thought that most

Table 2. Results of partial least squares regression (PLSR) analysis with the explanatory capacity of the first component and the variable influence on projection (VIP) of each predictor within each component to estimate significant predictors.

Gene	*nifH*	**16S rRNA**
Explained variance in fingerprinting pattern (%)	15.2	22.5
Explained variance of component predictors (%)	7.80	15.40
Environment variables*	**Variable Influence on Projection (VIP)**	
P-B	1.29	1.23
HC	1.17	1.07
pH	1.15	1.07
Fe	1.13	1.17
H_2O	1.07	1.00
NH_4-N	1.04	1.02
ELE	0.97	0.95
N	0.94	1.01
C	0.93	1.01
P	0.77	0.73
P-*H'*	0.66	0.75
NO_3-N	0.61	0.70

*All environmental variables are shown and include pH, NO_3-N, NH_4-N, organic carbon (C), total nitrogen (N), available phosphorus (P), soluble iron (Fe), elevation (ELE), hydraulic conductivity (HC), soil moisture (H_2O), plant biomass (P-B), and plant diversity (P-*H'*).

nitrogen-fixing microorganisms have an optimum soil pH of 7 or above. In this study, the samples from abandoned croplands with pH>7 had a higher diversity of free-living nitrogen-fixing microorganisms than the samples from the LGSG with pH<7.

The *nifH* gene profile showed there was a higher variability between replicate samples than that of the 16S rRNA gene profile. Therefore, this result suggested that the free-living nitrogen-fixing bacteria community was susceptible in this study area. An explanation for this may be the local changes at the microsite/aggregate scale because of spatial and temporal variation in exudation along plant roots [57–59]. Similar approaches have demonstrated that there are differences in ammonia oxidizer populations in sediment and soil [60].

The RDA revealed that the distribution of the factors that influence both the free-living nitrogen-fixing bacteria and the broader bacterial community distribution are very similar. This suggests that soil factors that affect the free-living nitrogen fixers are likely to affect the community as a whole in steppe grassland soil. Larkin et al. [61] reported that plant effects are the most important drivers of soil microbial community characteristics within a given site and soil type. In our study, PLSR and RDA showed that plant biomass strongly influenced (Table 2) and was positively correlated with the first axis (Figure S2) in both the free-living nitrogen-fixing bacteria and the broader bacterial community distribution. This is caused by interactions between plants and microorganisms. Above-ground net primary productivity was expected to increase soil carbon input by enhancing the turnover of plant biomass and enhancing root exudation and may therefore influence carbon-limited microbial communities in the soil [62,63]. Concurrently, microorganisms also affect plants by producing canonical plant growth-regulating substances such as auxins or cytokinins [13]. However, the Shannon-Wiener diversity index was not significantly influenced in either the free-living nitrogen-fixing bacteria or the broader bacterial community distribution in this study area. Pearson's product-moment corre-

lations showed that plant diversity was not significantly correlated with *nifH* gene diversity, or 16S rRNA gene diversity and copy number, but did negatively correlate with the copy number of the *nifH* gene (Table 1). Carney and Matson [64] found that plant diversity had a significant effect on the microbial community composition through alterations in microbial abundance rather than community composition. However, several studies report that plant diversity has little direct effect on bacterial community composition [65,66]. Therefore, this effect may depend on the type of plant community examined.

A number of additional factors such as hydraulic conductivity, pH, Fe, soil moisture, and NH4-N significantly influenced both the free-living nitrogen-fixing bacteria and the broader bacterial community distribution (Table 2). In our study, the soil from different sites was the same type, but the water holding capacity function of each research soil was different. The soil moisture content is in the order of LGSG>Y25>Y5>Y1 (Table 1). This suggests that a balance of macropores and micropores has not been fully recovered in the abandoned crop soils. A balance of macropores and micropores in soil influences the permeability and water holding capacity of soil [67]. In addition, the biological decomposition of organic materials produces natural glues, which bind and strengthen soil aggregates [67], and helps soils hold water and nutrients, which may change the balance of macropores and micropores. Organic matter also is a long-term, slow-release storehouse of nitrogen, phosphorus, and sulfur [67]. Accordingly, organic matter significantly influenced the total bacterial community distribution (Table 2), which may have resulted because of the significant difference in soil organic matter content in the abandoned cropland and LGSG ($P<0.05$) (Table 1). Additionally, the soil texture and hydraulic conductivity is closely correlated, and soil texture influences the balance between macropores and micropores [67]. The hydraulic conductivities of sites Y1 and Y5 were significantly different from those of sites Y25 and LGSG (Table 1), and the RDA also indicated that soil moisture was negatively correlated with hydraulic conductivity (Figure S2). Therefore, the hydraulic conductivity of soil also showed that the water holding capacity function of soils from the research sites is different.

The significance of NH4-N indicates that the fixed nitrogen of the free-living nitrogen-fixing bacteria will be the main nitrogen source in the soil of steppe grasslands, where there are few plants such as legumes. Nitrogen cycling in natural ecosystems and traditional agricultural production relies on biological nitrogen fixation primarily by diazotrophic bacteria [16].

In our study, higher copy number/diversity of the $nifH$ gene/16S rRNA gene was observed in the LGSG soil, where the Fe content was significantly higher, than in the abandoned cropland soil. According to Pearson's product-moment correlation and PLSR, the Fe content affects the nitrogen-fixing community. This may result because Fe is a required material for the [4Fe-4S] cluster of the $nifH$ gene [22,51], and it is possible that the Fe content affects the 16S rRNA gene diversity in other microorganisms with the iron-containing enzyme [68-70] as it does the nitrogen-fixing bacteria.

Gaby and Buckley [71] comprehensively evaluated primers of $nifH$ gene and found that $nifH$ gene different primer sets amplified different groups in the $nifH$ phylogeny. In this study, we used primer sets for DGGE and qPCR analyses of the $nifH$ gene that targeted different sequence positions, thus the results obtained were not comparable. However, environmental factors differed greatly between the abandoned cropland and LGSG soils, thus having differential effects on the soil microbial communities. Furthermore, the trends in the changing copy number and diversity of the $nifH$ gene were similar to that of the 16S rRNA gene. Therefore, the results we obtained regarding the copy number and diversity of $nifH$ gene are able to explain the dynamic change in the trend of free-living nitrogen-fixing bacteria communities in abandoned cropland during different periods. Further research on molecular ecology could provide more detail analysis of microbial diversity, such as by sequencing of DGGE band and high-throughput sequencing technology etc., which would contribute to more accurate insights into the black box of the dynamic change of microbial communities.

We discovered that microbial communities of the abandoned cropland in this study area are strongly influenced by plant biomass, soil moisture, Fe, and NH4-N. Robust information about the mechanisms that regulate the diversity, structure, and composition of natural communities is urgently needed to help conserve ecosystem function and mitigate biodiversity loss from current and future environmental changes. Conversely, the results of this study suggest that advances in desertification may be prevented by adjusting environmental factors of the abandoned cropland, such as the soil moisture content, Fe, and NH4-N, which will enhance the function of a microbial community and possibly increasing plant biomass production. In addition, $nifH$ gene copy number had significant positive Pearson's product-moment correlation with soil moisture, organic C, P, and Fe etc., therefore, by adjusting these environmental factors may also increase the abundance of free-living nitrogen-fixing bacteria in abandoned cropland.

Supporting Information

Figure S1 Denaturing gradient gel electrophoresis (DGGE) profiles of the $nifH$ (A) and 16S rRNA genes (B). For all images, the numbers refer to the plot numbers in the sample areas.

Figure S2 Redundancy analysis (RDA) of $nifH$ (A) and 16S rRNA (B) genes data. Ordination plots of $nifH$ (A) and 16S rRNA (B) genes associated with abandoned croplands for different times: Y1 (•), Y5 (▲), and Y25 (■), and *light-grazing* steppe grassland (LGSG, ○). The plots were generated by redundancy analysis (RDA) of the denaturing gradient gel electrophoresis (DGGE) profiles. All environmental variables are shown, including pH, NO_3-N, NH_4-N, organic carbon (C), total nitrogen (N), available phosphorus (P), soluble iron (Fe), elevation (ELE), hydraulic conductivity (HC), soil moisture (H_2O), plant biomass (P-B), and plant species richness (P-H'). Values on the axes indicate the percentages of total variation explained by each axis.

Table S1 Total plant cover, plant Shannon's diversity index, plant biomass, plant types, and their coverage in each research plot.

Table S2 Soil texture (clay, silt, and sand) content and saturated hydraulic conductivity of soil samples at each site.

Table S3 Collinearity among environmental parameters as determined by Spearman's rank correlation coefficient rho. Significantly correlated parameters show Spearman's rank correlation coefficient rho>0.6 and <−0.6 in bold. All environmental variables are shown, such as pH, NO_3-N, NH_4-N, organic carbon (C), total nitrogen (N), available phosphorus (P), soluble iron (Fe), elevation (ELE), hydraulic conductivity (HC), soil moisture (H_2O), plant biomass (P-B), and plant diversity (P-H').

Acknowledgments

We thank Dr. Kenji Tamura (University of Tsukuba) for his valuable suggestions on the research content and soil sampling, and Takashi Kanda (University of Tsukuba) for excellent technical support in measuring organic C and total N concentrations.

Author Contributions

Conceived and designed the experiments: H SB HU. Performed the experiments: H SB. Analyzed the data: H YXC. Contributed reagents/materials/analysis tools: HU NN T. Nakajima T. Nakamura. Wrote the paper: H.

References

1. Archibold OW (1995) Ecology of World Vegetation. London: Chapman & Hall. pp. 1–522.
2. He C, Zhang Q, Li Y, Li X, Shi P (2005) Zoning grassland protection areas using remote sensing and cellular automata modeling – A case study in Xilingol steppe grassland in northern China. J Arid Environ 63: 814–826.
3. Tong C, Wu J, Yong S, Yang J, Yong W (2004) A landscapescale assessment of steppe degradation in the Xilin River basin, Inner Mongolia, China. J Arid Environ 59: 133–149.
4. Li JH, Fang XW, Jia JJ, Wang G (2007) Effect of legume species introduction to early abandoned field on vegetation development. Plant Ecol 191: 1–9.
5. Van Der Putten WH, Mortimer SR, Hedlund K, Van Dijk C, Brown VK, et al. (2000) Plant species diversity as a driver of early succession in abandoned fields: a multi-site approach. Oecologia 124: 91–99.
6. Xue ZD, Hou QC, Han RL, Wang SQ (2002) Trails and research on ecological restoration by *Sophora viciifolia* in Gullied Rolling Loess Region. J Northwest Forestry Univ 17: 26–29.
7. Zhang JD, Qiu Y, Chai BF, Zheng FY (2000) Succession analysis of plant communities in Yancun low middle hills of Luliang Mountains. J Plant Resour. Environ 9: 34–39.

8. Zou HY, Chen JM, Zhou L, Hongo A (1998) Natural recoverage succession and regulation of the Prairie vegetation on the Loess Plateau. Res. Soil. Water Conserv 5: 126–138.

9. Cheng YX, Nakamura T (2007) Phytosociological study of steppe vegetation in east Kazakhstan. Grassland Sci 53: 172–180.

10. EI-Sheikh MA (2005) Plant succession on abandoned fields after 25 years of shifting cultivation in Assuit, Egypt. J Arid Environ 61: 461–481.

11. Prévosto B, Kuiters L, Bernhardt-Römermann M, Dölle M, Schmidt W, et al. (2011) Impacts of land abandonment on vegetation: successional pathways in European habitats. Folia Geobot 46: 303–325.

12. Štolcová J (2002) Secondary succession on an early abandoned field: vegetation composition and production of biomass. Plant Protection Sci 38: 149–154.

13. Ortiz-Castro R, Contreras-Cornejo HA, Macias-Rodriguez L, Lopez-Bucio J (2009) The role of microbial signals in plant growth and development. Plant Signal. Behav 4: 701–712.

14. Daniel R (2005) The metagenomics of soil. Nat Rev Microbiol 3: 470–478.

15. Van Der Heijden M, Bardgett R, van Straalen N (2008) The unseen majority: soil microbes as drivers of plant diversity and productivity in terrestrial ecosystems. Ecol Lett 11: 296–310.

16. Vitousek PM, Aber J, Howarth RW, Likens GE, Matson PA, et al. (1997) Human alteration of the global nitrogen cycle: causes and consequences. Ecol Appl 7: 737–750.

17. Orr CH, James A, Leifert C, Cooper JM, Cummings SP (2011) Diversity and activity of free-living nitrogen-fixing bacteria and total bacteria in organic and conventionally managed soils. Appl Environ Microbiol 77: 911–919.

18. Dixon R, Kahn D (2004) Genetic regulation of biological nitrogen fixation. Nat Rev Microbiol 2: 621–631.

19. Peoples MB, Herridge DF, Ladha JK (1995) Biological nitrogen fixation: an efficient source of nitrogen for sustainable agricultural production? Plant Soil 174: 3–28.

20. Burgmann H, Widmer F, Von Sigler W, Zeyer J (2004) New molecular screening tools for analysis of free-living diazotrophs in soil. Appl Environ Microbiol 70: 240–247.

21. Kahindi JHP, Woomer P, George T, de Souza Moreira FM, Karanja NK, et al. (1997) Agricultural intensification, soil biodiversity and ecosystem function in the tropics: the role of nitrogen-fixing bacteria. Appl Soil Ecol 6: 55–76.

22. Roeselers G, Stal LJ, van Loosdrecht MCM, Muyzer G (2007) Development of a PCR for the detection and identification of cyanobacterial nifD genes. J Microbiol Methods 65: 550–556.

23. Poly F, Monrozier LJ, Bally R (2001) Improvement in the RFLP procedure for studying the diversity of nifH genes in communities of nitrogen fixers in soil. Res Microbiol 152: 95–103.

24. Rosado AS, Duarte GF, Seldin L, Van Elsas JD (1998) Genetic diversity of nifH gene sequences in Paenibacillus azotofixans strains and soil samples analyzed by denaturing gradient gel electrophoresis of PCR-amplified gene fragments. Appl Environ Microbiol 64: 2770–2779.

25. Widmer F, Shaffer BT, Porteous LA, Seidler RJ (1999) Analysis of nifH gene pool complexity in soil and litter at a Douglas fir forest site in the Oregon Cascade Mountain range. Appl Environ Microbiol 65: 374–380.

26. Matsuura K, Willmott CJ (2009) Terrestrial precipitation: 1900-2010 gridded monthly time series. Available: http://climate.geog.udel.edu/~climate/. Accessed 2014 Aug 27.

27. Braun-Blanquet J (1964) Pflanzensoziologie, 3rd revised edn. New York: Springer-Verlag. pp. 1–865.

28. Mueller-Dombois D, Ellenberg H (1974) Aims and Methods of Vegetation Ecology. New York: John Wiley and Sons. pp. 1–547.

29. Day PR (1965) Particle fraction and particle-size analysis. In: Black CA, editor. Methods of soil analysis. Madison: American Society of Agronomy. pp. 545–566.

30. Siosemarde M, Byzedi M (2011) Studding of number of dataset on precision of estimated saturated hydraulic conductivity. World Academy Sci, Eng Technol 74: 521–524.

31. Leonardo M (2009) Development and validation of a method for determination of residual nitrite/nitrate in foodstuffs and water after zinc reduction. Food Anal Methods 2: 212–220.

32. Motsara MR, Roy RN (2008) Guide to laboratory establishment for plant nutrient analysis. FAO Fertilizer and Plant Nutrition Bulletin No. 19, FAO, Rome. Pp.17–76.

33. Truog E, Meyer AH (1929) Improvements in the deniges colorimetric method for phosphorus and arsenic. Indus, and Engin Chem, Analyt Ed 1: 136–139.

34. Saywell LG, Cunningham BB (1937) Determination of iron: colorimetric o-phenanthroline method. Ind Eng Chem Anal 9: 67–69.

35. Sugio T, Taha T, Kanao T, Takeuchi F (2007) Increase in Fe²⁺-producing activity during growth of Acidithiobacillus ferrooxidans ATCC 23270 on sulfur. Biosci Biotechnol Biochem 71: 2663–2669.

36. Schumacher BA (2002) Methods for the determination of total organic carbon (TOC) in soils and sediments. U.S. Environmental Protection Agency, Washington, Ecological Risk Assesment Support Center.

37. Steward GF, Zehr JP, Jellison RP, Montoya JP, Hollibaugh JT (2004) Vertical distribution of nitrogen-fixing phylotypes in a meromictic, hypersaline lake. Microb Ecol 47: 30–40.

38. Zehr JP, McReynolds LA (1989) Use of degenerate oligonucleotides for amplification of the nifH gene from the marine cyanobacterium Trichodesmium thiebautii. Appl Environ Microbiol 55: 2522–2526.

39. Muyzer G, De Waal EC, Uitterlinden AG (1993) Profiling of complex microbial populations by denaturing gradient gel electrophoresis analysis of polymerase chain reaction-amplified genes coding for 16S rRNA. Appl Environ Microbiol 59: 695–700.

40. Baxter J, Cummings SP (2006) The impact of bioaugmentation on metal cyanide degradation and soil bacteria community structure. Biodegradation 17: 207–217.

41. Baxter J, Cummings SP (2008) The degradation of the herbicide bromoxynil and its impact on bacterial diversity in a top soil. J Appl Microbiol 104: 1605–1616.

42. Rösch C, Mergel A, Bothe H (2002) Biodiversity of denitrifying and dinitrogen-fixing bacteria in an acid forest soil. Appl Environ Microbiol 68: 3818–3829.

43. ter Braak CJF, Smilauer P (2002) CANOCO reference manual and CanoDraw for Windows user's guide: software for canonical community ordination (version 4.5). Ithaca, USA (url: www.canoco.com): Microcomputer Power. 500 p.

44. Carrascal LM, Galvan I, Gordo O (2009) Partial least squares regression as an alternative to current regression methods used in ecology. Oikos 118: 681–690.

45. Naether A, Foesel BU, Naegele V, Wüst PK, Weinert J, et al. (2012) Environmental factors affect acidobacterial communities below the subgroup level in grassland and forest soils. Appl Environ Microbiol 78: 7398–7406.

46. Umetrics AB (2005) SIMCA-P v. 11 Analysis Advisor. Umeå, Sweden: Umetrics AB.

47. Tremaroli V, Workentine ML, Weljie AM, Vogel HJ, Ceri H, et al. (2009) Metabolomic investigation of the bacterial response to a metal challenge. Appl Environ Microbiol 75: 719–728.

48. Ramette A (2007) Multivariate analyses in microbial ecology. FEMS Microbiol Ecol 62: 142–160.

49. Chan YK, Barraquio WL, Knowles R (1994) N₂-fixing pseudomonas and related soil bacteria. FEMS Microb Rev 13: 95–118.

50. Reed SC, Seastedt TR, Mann CM, Suding KN, Townsend AR, et al. (2007) Phosphorus fertilization stimulates nitrogen fixation and increases inorganic nitrogen concentrations in a restored prairie. Appl Soil Ecol 36: 238–242.

51. Gavini N, Burgess BK (1992) FeMo cofactor synthesis by a nifH mutant with altered MgATP reactivity. J Biol Chem 267: 21179–21186.

52. Coelho MRR, de Vos M, Carneiro NP, Marriel IE, Paiva E, et al. (2008) Diversity of nifH gene pools in the rhizosphere of two cultivars of sorghum (Sorghum bicolor) treated with contrasting levels of nitrogen fertilizer. FEMS Microbiol Lett 279: 15–22.

53. Coelho MRR, Marriel IE, Jenkins SN, Lanyon CV, Seldin L, et al. (2009) Molecular detection and quantification of nifH gene sequences in the rhizosphere of sorghum (Sorghum bicolor) sown with two levels of nitrogen fertilizer. Appl Soil Ecol 42: 48–53.

54. Noll M, Wellinger M, (2008) Changes of the soil ecosystem along a receding glacier: Testing the correlation between environmental factors and bacterial community structure. Soil Biol Biochem 40: 2611–2619.

55. Wakelin SA, Macdonald LM, Rogers SL, Gregg AL, Bolger TP, et al. (2008) Habitat selective factors influencing the structural composition and functional capacity of microbial communities in agricultural soils. Soil Biol Biochem 40: 803–813.

56. Belnap J (2001) Factors influencing nitrogen fixation and nitrogen release in biological soil crusts. In: Belnap J, Lange OL, editors. Biological soil crusts: structure, function, and management. Berlin: Springer-Verlag. pp. 241–261.

57. Clayton SJ, Clegg CD, Murray PJ, Gregory PJ (2005) Determination of the impact of continuous defoliation of Lolium perenne and Trifolium repens on bacterial and fungal community structure in rhizosphere soil. Biol Fertil Soils 41: 109–115.

58. Marilley L, Vogt G, Blanc M, Aragno M (1998) Bacterial diversity in the bulk soil and rhizosphere fractions of Lolium perenne and Trifolium repens as revealed by PCR restriction analysis of 16S rDNA. Plant Soil 198: 219–224.

59. Marschner P, Yang CH, Liebere R, Crowley DE (2001) Soil and plant specific effects on bacterial community composition in the rhizosphere. Soil Biol Biochem 33: 1437–1445.

60. Stephen JR, McCaig AE, Smith Z, Prosser JI, Embley TM (1996) Molecular diversity of soil and marine 16S rRNA gene sequences related to beta-subgroup ammonia-oxidizing bacteria. Appl Environ Microbiol 62: 4147–4154.

61. Larkin RP, Honeycutt CW (2006) Effects of different 3-year cropping systems on soil microbial communities and Rhizoctonia diseases of potato. Phytopathology 96: 68–79.

62. Niklaus PA, Alphei J, Ebersberger D, Kampichler C, Kandeler E, et al. (2003) Six years of in situ CO₂ enrichment evokes changes in soil structure and soil biota of nutrient-poor grassland. Glob Chang Biol 9: 585–600.

63. Zak DR, Holmes W, White DC, Peacock A, Tilman D (2003) Plant diversity, soil microbial communities, and ecosystem function: are there any link? Ecology 84: 2042–2050.

64. Carney KM, Matson PA (2005) Plant communities, soil microorganisms, and soil carbon cycling: does altering the world belowground matter to ecosystem functioning? Ecosystems 8: 928–940.

65. Kennedy N, Brodie E, Connolly J, Clipson N (2004) Impact of lime, nitrogen and plant species on bacterial community structure in grassland microcosms. Environ Microbiol 6: 1070–1080.

66. Nunan N, Daniell TJ, Singh BK, Papert A, McNicol JW, et al. (2005) Links between plant and rhizoplane bacterial communities in grassland soils, characterized using molecular techniques. Appl Environ Microbiol 71: 6784–6792.

67. Cogger C (2000) Soil Management for Small Farms. EB 1895. Pullman, WA: Washington State University Cooperative Extension. 24 p.

68. Sze IS, Dagley S (1984) Properties of salicylate hydroxylase and hydroxyquinol 1,2-dioxygenase purified from *Trichosporon cutaneum*. J Bacteriol 159: 353–359.

69. Conway T, Ingram LO (1989) Similarity of *Escherichia coli* propanediol oxidoreductase (fucO product) and an unusual alcohol dehydrogenase from *Zymomonas mobilis* and *Saccharomyces cerevisiae*. J Bacteriol 171: 3754–3759.

70. Drennan CL, Heo J, Sintchak MD, Schreiter E, Ludden PW (2001) Life on carbon monoxide: X-ray structure of Rhodospirillum rubrum Ni-Fe-S carbon monoxide dehydrogenase. Proc Natl Acad Sci USA 98: 11973–11978.

71. Gaby JC, Buckley DH (2012) A comprehensive evaluation of PCR primers to amplify the nifH gene of nitrogenase. PLoS ONE 7: e42149.

12

Effects of Traditional Flood Irrigation on Invertebrates in Lowland Meadows

Jens Schirmel*, Martin Alt, Isabell Rudolph, Martin H. Entling

Institute of Environmental Science, University of Koblenz-Landau, Landau, Germany

Abstract

Lowland meadow irrigation used to be widespread in Central Europe, but has largely been abandoned during the 20th century. As a result of agri-environment schemes and nature conservation efforts, meadow irrigation is now being re-established in some European regions. In the absence of natural flood events, irrigation is expected to favour fauna typical of lowland wet meadows. We analysed the effects of traditional flood irrigation on diversity, densities and species composition of three invertebrate indicator taxa in lowland meadows in Germany. Unexpectedly, alpha diversity (species richness and Simpson diversity) and beta diversity (multivariate homogeneity of group dispersions) of orthopterans, carabids, and spiders were not significantly different between irrigated and non-irrigated meadows. However, spider densities were significantly higher in irrigated meadows. Furthermore, irrigation and elevated humidity affected species composition and shifted assemblages towards moisture-dependent species. The number of species of conservation concern, however, did not differ between irrigated and non-irrigated meadows. More variable and intensive (higher duration and/or frequency) flooding regimes might provide stronger conservation benefits, additional species and enhance habitat heterogeneity on a landscape scale.

Editor: Judi Hewitt, University of Waikato (National Institute of Water and Atmospheric Research), New Zealand

Funding: The study was funded by the research initiative 'AufLand' of the federal state of Rhineland-Palatinate' (Germany) and the research and development fund of the University of Koblenz-Landau. The funders had no role in study design, data collection and analysis, decision to publish, or preparation of the manuscript.

Competing Interests: The authors have declared that no competing interests exist.

* Email: schirmel@uni-landau.de

Introduction

Semi-natural grasslands are key habitats for biodiversity conservation and an integral part of the Central European cultural landscape [1–3]. They are among the most species-rich habitats and serve as refuges for several rare and endangered species [2–4]. Regular disturbance due to traditional management permits the coexistence of numerous species in semi-natural grasslands [2]. During the last decades however, semi-natural grassland have dramatically declined in Central Europe and further declines to less than 50% of the current area are predicted [5,6]. Major causes are agricultural intensification and the abandonment of traditional management. The latter is mainly due to the reduced cost-effectiveness of traditional land use practices [5,7]. Agricultural intensification practices for semi-natural meadows include higher fertilizer and herbicide applications, earlier and more cuts per year, and the use of modern mowing techniques [8]. This results in eutrophic, structurally poor, and homogeneous meadows with negative impacts on diversity, species composition, and ecosystem processes [9].

Until the early 20th century, meadow irrigation was widespread in Central Europe to increase hay yield [10]. For example, around 1900 in some regions of Germany, irrigated meadows made up about 60% of the total grassland [11]. The main effects of irrigation were nutrient input, topsoil humidification, and extension of the vegetation period. Nowadays irrigation practices are mostly abandoned and traditionally irrigated meadows with their associated species are restricted to few remnant areas [12]. Thanks to agri-environment schemes (e.g. in form of compensation payments), nature conservation efforts, and due to mitigation and compensation measures, the traditional irrigation practices could be maintained or re-established in some European regions [13]. However, the value of agri-environment schemes is under debate and further analyses of management strategies are necessary [14,15]. Therefore, it is of growing interest to determine, how traditional irrigation practices affect biological diversity. In this context, Riedener et al. [16] recently showed that changes in irrigation techniques have influenced some aspects of plant and gastropod diversity in Swiss mountain hay meadows. However, knowledge of the influence of traditional meadow irrigation on invertebrate diversity and composition is still poorly understood, and this is especially true for flood irrigation in lowland regions.

The objective of this study was to analyse whether traditional flood irrigation in lowland meadows has an effect on invertebrate diversity and species composition. Irrigation is assumed to create small-scale differences in moisture and sediment conditions which may increase microhabitat and vegetation heterogeneity [16]. In accordance to the habitat-heterogeneity-hypothesis irrigation might therefore have positive effects on local species richness [17–20]. Moreover, flood irrigation in our study area is conducted in a similar way among irrigated sites, but differs in timing and intensity. This may lead to non-uniform moisture conditions among irrigated meadows with heterogeneous species compositions and higher beta diversity. To investigate these predictions we conducted a field survey in the 'Queichtal' in Rhineland-Palatine, Germany. We compared traditionally flood-irrigated meadows with meadows, where there has been no irrigation for at least thirty years. We focused on orthopterans, carabids and spiders, which are found at different trophic levels within grassland food-webs and occur in different vegetation layers. Orthopterans (Orthoptera) are mostly grass-dwelling herbivores, where they are often both the main invertebrate consumers and the main food source [21]. Most carabids (Coleoptera: Carabidae) are ground-dwelling predators, but some are scavengers and herbivores [22]. Spiders (Araneae) inhabit both the ground and field layer, often in high abundances, and are predatory [23]. All three arthropod groups have been used as indicators of ecosystem conditions and habitat quality (orthopterans: [24,25]; carabids: [26,27]; spiders: [28,29]).

We addressed the following hypotheses: (i) Flood irrigation increases the local diversity of orthopterans, carabids, and spiders compared to non-irrigated lowland meadows. (ii) Flood irrigation leads to higher beta diversity relative to non-irrigated meadows. (iii) Flood irrigation shifts species assemblages towards more moisture-dependent species and those of higher conservation concern than species in non-irrigated meadows. Based on our findings we discuss if traditional flood irrigation can be useful for conserving biodiversity of semi-natural grassland species.

Materials and Methods

Ethics statement

Invertebrates were collected with the permission 42/553-254 from the Struktur- und Genehmigungsdirektion Süd (federal state authority of Rhineland-Palatine, Germany). Additionally, we obtained permissions from all private farmers and landowners to conduct the field work on their meadows.

Study sites

The study was conducted in 2012 in the 'Queichtal' in the Upper Rhine valley in Rhineland-Palatine, Germany (Fig. 1).

With a length of ~51 km, the river Queich is an important drainage system of the adjacent low mountain range 'Pfälzerwald' into the Rhine. Soils of the alluvial sediments are sandy to loamy [30]. Annual mean temperature in this region is 10.5°C (station Neustadt) and annual mean precipitation is 667 mm (station Landau; German Weather Service). The studied section of the Queichtal covers about 700 ha, is part of the NATURA 2000 network, and is thus protected by the EU habitats Directive [31].

Due to the predominance of moist soils with low cation availability, land use in the Queichtal is dominated by forest and grassland with different management and irrigation regimes. The formerly widespread traditional flood irrigation of lowland meadows was almost totally abandoned after the Second World War and is nowadays only applied in a few remnant areas. For flood irrigation the water of the river Queich (or the tributaries 'Spiegelbach' and 'Fuchsbach') is dammed by weirs (Fig. 1) and delivered to the meadows by open ditches where it slowly flows over the ground ('lowland irrigation type'; [10]). Meadows are irrigated on average four times per year between April and August and each irrigation event lasts for 1–3 days.

A total of 32 meadows were selected stratified to meadow irrigation practice (yes or no) and fertilization (yes or no) (Table S1). Half of the meadows were traditionally irrigated and days of irrigation ranged from 4 to 12 days per year. On the other 16 meadows irrigation ceased more than 30 years ago. Half of the irrigated and not irrigated meadows were fertilized (with a maximum of ~60 kg N•ha-1•yr-1). Meadows were normally mown twice per year and extensive winter grazing by sheep occured on all meadows.

On each meadow we selected a 50×50 m plot with a minimum distance of 100 m from the nearest plot and 10 m to the next ditch to minimise edge effects. Irrigated and non-irrigated meadows did not differ significantly in mean distance to the nearest forest (t-test: $t_{30} = 0.563$, $P = 0.578$) and to the nearest permanent water (t-test: $t_{30} = 0.529$, $P = 0.601$). Permanent water was defined as any standing and flowing water body which permanently contained water.

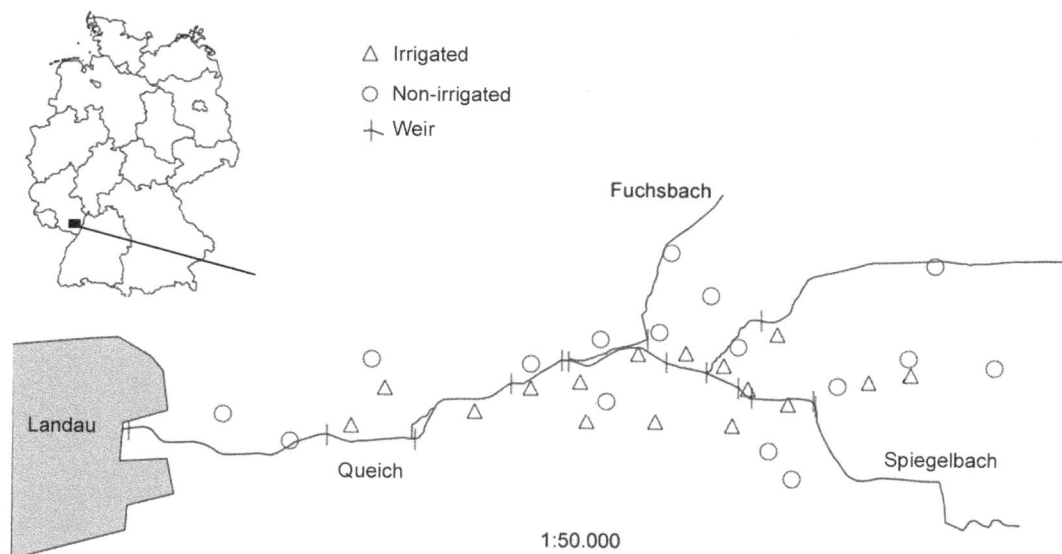

Figure 1. Position of the study area 'Queichtal' in Germany (small figure) and of the 32 study sites.

Management and environmental parameters

Land use data on irrigation practice (yes or no) and fertilization (yes or no) were collected through on-site observations and interviews with landowners and local farmers. Plant species were recorded in three randomly selected 3×3 m subplots per plot in May/June before the first cut (unpublished data). For subsequent analyses, data of the three subplots were averaged and we calculated the unweighted mean Ellenberger indicator values for moisture (in the following 'humidity' to avoid confusion with animal species moisture indicator values) and nitrogen for a description of local habitat conditions. As species can be influenced by patch isolation [32], we calculated the distance (m) to permanent water and to forest for each sampling location in Google Earth [33].

Invertebrate sampling

Orthopterans were sampled once per plot during their main activity period in August with a box-quadrat. The box-quadrat is a very effective method for sampling orthopteran densities [34]. The box-quadrat we used had an area of 2 m^2 (1.41×1.41 m) with white gauze-covered sides of 0.8 m height and was randomly dropped at 20 different locations per plot (total sampled area = 40 m^2 per plot). Collected individuals were determined to species level directly in the field using Bellmann [35] and then released.

Carabids and spiders were sampled using pitfall traps (6.5 cm in diameter, 7 cm deep) filled to one third with a 50% propylene-glycol solution. Per plot, four pitfall traps (N = 160 traps) were randomly installed with a minimum distance of 5 m to each other. Traps were exposed for two sampling periods from 03 to 24 April and again from 12 to 28 June. Carabids and spiders were determined to species level using the identification keys of Müller-Motzfeld [36] (carabids) and Roberts [37] (spiders). The four traps per plot were treated as a unit and data from both sampling periods were combined to obtain one dataset for further analyses. Due to loss and damage of some pitfall traps, we finally included 28 plots (N = 14 per meadow type each with N = 7 fertilized) in the data analyses of carabids and spiders.

Data analysis

Species were classified as species of conservation concern when they were listed in regional red lists (all species belonging to the categories '1', '2', '3', '4', and 'V'; orthopterans: [38]; carabids: [39]; spiders: [40]). For species moisture dependence we used published moisture indicator values. For orthopterans, transformed moisture values were obtained from Maas et al. [41] (Table S2). The values range from '1' (strongly xerophilic) to '5' (strongly hygrophilic). For carabids, moisture values range from '0' (most xerophilic) to '9' (most hygrophilic) according to Irmler and Gürlich [42]. For spiders, we used the moisture values of Entling et al. [43]. For a better comparison to the other taxa we transformed values by 1−x, i.e. species with the lowest value '0' are most xerophilic and species with the highest value '1' are most hygrophilic. For each species we calculated the spearman rank correlation coefficient between species density and irrigation to express their 'species irrigation affinity' for our study area. Relationships between species irrigation affinities and species moisture indicator values (based literature data) were analysed using linear models.

The effect of irrigation on species richness and densities of orthopterans, carabids, and spiders were analysed using Poisson GLM's for count data. Similarly, the irrigation effect on the combined number of species of conservation concern of all taxa (N = 28 sites) was analysed. In cases of overdispersion, we

corrected the standard errors using a quasi-GLM model [44]. Differences in Simpson diversity (1−D) between irrigated and non-irrigated meadows were tested with non-parametric Wilcoxon rank sum test, because assumptions for a t-test were violated.

Community differentiation (beta diversity) among irrigated and non-irrigated meadows was analysed using the homogeneity of multivariate dispersions based on the Sørensen similarity of species presence-absence data (using the command 'betadisper' in the R package 'vegan') [45]. For each taxon, an ANOVA was used to test for differences between the multivariate dispersions of both meadow types.

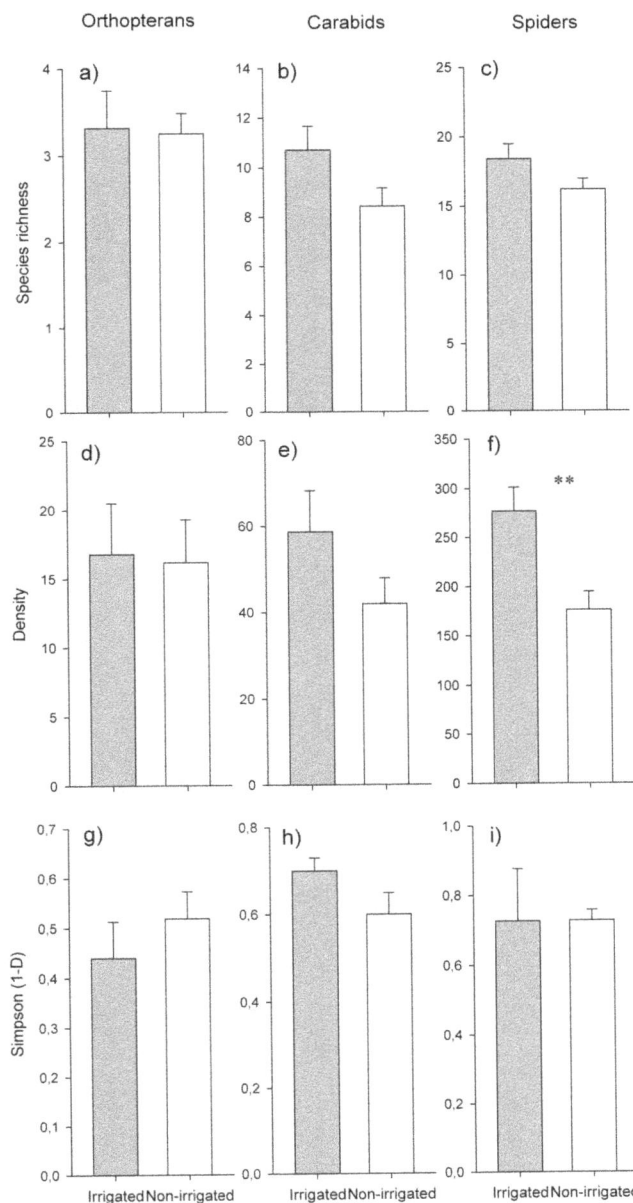

Figure 2. Comparison of species richness (a–c), densities (d–f), and Simpson diversity (g–i) of orthopterans, carabids, and spiders between irrigated and non-irrigated meadows (mean and SE). Differences of species richness and densities were tested with Poisson GLM's and of Simpson diversity (1−D) with non-parametric wilcoxon rang sum tests.

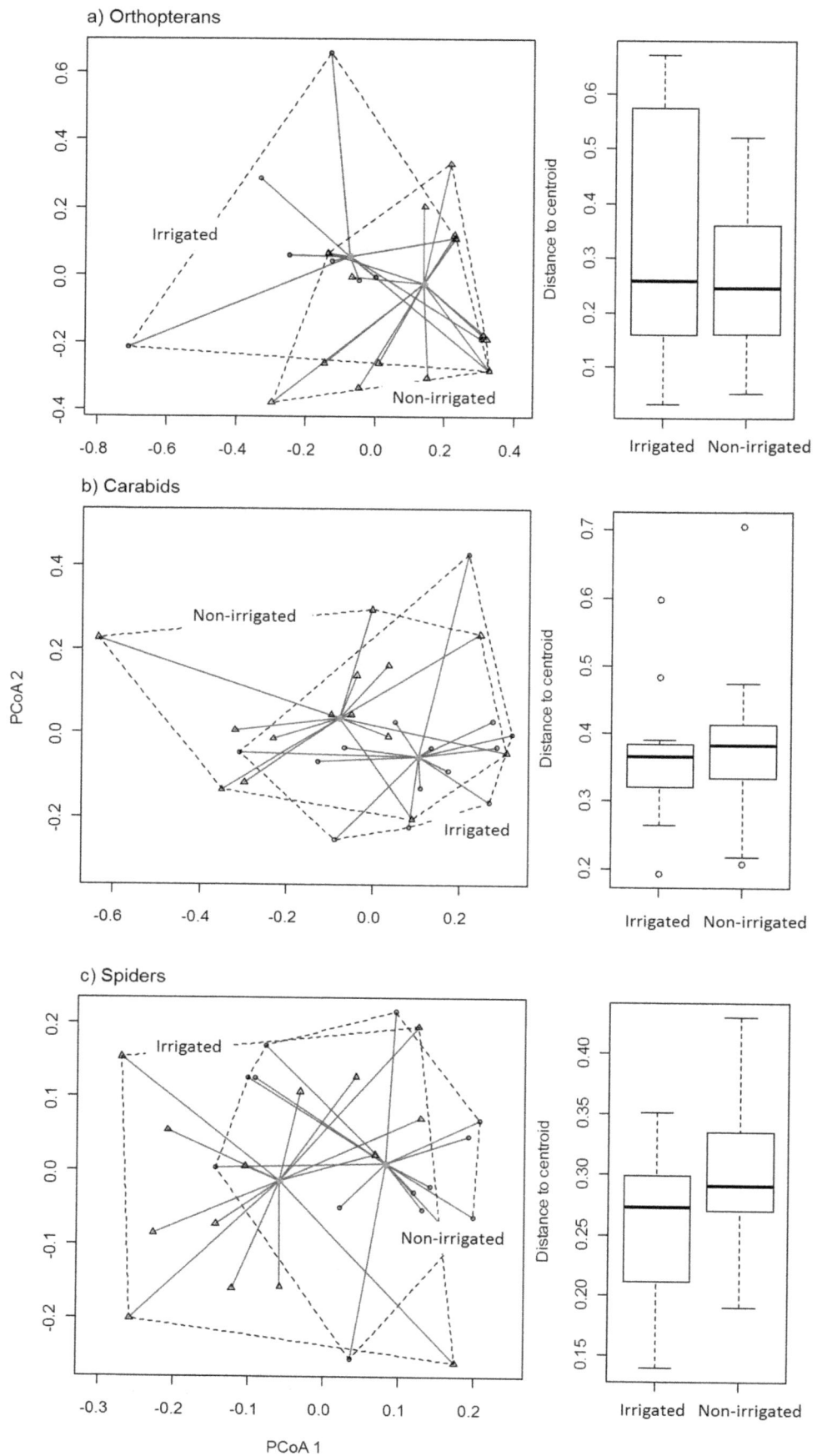

a) Orthopterans

b) Carabids

c) Spiders

Figure 3. Beta diversity (multivariate homogeneity of dispersions) of a) orthopterans, b) carabid and c) spider assemblages of irrigated and non-irrigated meadows. Distances (Sørensen similarity) are reduced to principal coordinates and distances to group centroids (irrigated or non-irrigated) are shown. Differences of mean distances between meadow types were tested by ANOVA.

Effects of the management and environmental variables on species composition of orthopterans, carabids, and spiders were analysed with a permutational multivariate ANOVA (command 'adonis' in R package 'vegan'; [46]). Predictor variables were the two factors irrigation (yes or no) and fertilization (yes or no), the two local habitat parameters humidity and nitrogen (mean Ellenberger indicator values), and the two landscape parameters distance to permanent water and distance to forest. As a distance measure the Bray-Curtis distance was used. Significance of environmental variables was tested with permutation tests (999 permutations) with pseudo-F ratios. Variation of species compositions were visualised using nonmetric multidimensional scaling (NMDS) with the command 'metaMDS' in R package 'vegan'. Again, the Bray-Curtis distance was used as a distance measure. All statistical analyses were done in R 2.12.2 [47].

Results

General results

A total of 7 orthopteran species (528 individuals), 47 carabid species (1,410 individuals), and 56 spider species (6,347 individuals) were found (Tables S3–S6). All 7 orthopteran species were detected in both meadow types (Fig. 2a). A total of 40 carabid species were found in irrigated meadows compared to 32 species in non-irrigated meadows (Fig. 2b). In total 49 spider species could be detected in irrigated and 46 species in non-irrigated meadows (Fig. 2c).

Local diversity

There was no significant effect of irrigation on species richness of orthopterans ($z = 0.098$, $P = 0.922$, Fig. 2a), carabids ($t = 1.950$, $P = 0.051$; Fig. 2b), and spiders ($z = 1.407$, $P = 0.160$, Fig. 2c). While densities of orthopterans ($t = 0.130$, $P = 0.898$, Fig. 2d) and carabids ($t = 1.484$, $P = 0.150$, Fig. 2e) did not significantly differ between both meadow types densities of spiders were significantly higher in irrigated meadows ($t = 3.266$, $P = 0.003$, Fig. 2f). Similar to species richness, Simpson diversity did not differ for orthopterans ($W_{30} = 147.5$, $P = 0.4731$, Fig. 2g), carabids ($W_{26} = 62$, $P = 0.104$, Fig. 2h), and spiders ($W_{26} = 87$, $P = 0.629$, Fig. 2i).

Community differentiation (beta diversity)

Beta diversity (multivariate dispersion) of all investigated taxa was not influenced by irrigation. Mean distances to centroids did not differ significantly between irrigated and non-irrigated meadows for orthopterans ($F = 1.237$, $P = 0.275$, Fig. 3a), carabids ($F = 0.287$, $P = 0.596$, Fig. 3b), and spiders ($F = 4.023$, $P = 0.055$, Fig. 3c).

Species composition

Irrigation (yes or no) and humidity were the only variables having a significant effect on species composition, while fertilization, nitrogen availability, distance to permanent water and distance to forest had no effect (Table 1, Fig. 4). Orthopteran species composition was significantly affected by irrigation ($F = 2.51$, $R^2 = 0.073$, $P = 0.019$) and humidity ($F = 2.93$, $R^2 = 0.085$, $P = 0.011$). Carabid species composition was affected only by humidity ($F = 2.49$, $R^2 = 0.088$, $P = 0.024$) while spider species composition was influenced by irrigation ($F = 2.31$, $R^2 = 0.080$, $P = 0.041$).

As hypothesised, irrigation favoured the occurrence of moisture-dependent species. For carabids ($r = 0.48$, $P = 0.002$, Fig. 5b) and spiders ($r = 0.44$, $P < 0.001$, Fig. 5c) there was a significant positive relationship between species irrigation affinity (expressed as the spearman rank correlation coefficient) and species moisture indicator value. For orthopterans no significant relationship was found, however this may be a result of the low number of $N = 7$ species (Fig. 5a). The combined number of species of conservation concern of all three taxa did not differ between irrigated (3.6 ± 0.6) and non-irrigated (2.4 ± 0.3) meadows ($z = 1.853$, $P = 0.064$).

Discussion

Unexpectedly, traditional flood irrigation had no significant effect on diversity and species of conservation concern of orthopterans, carabids, and spiders in lowland meadows. However, flood irrigation and the associated environmental parameter humidity influenced the species composition of all taxa and shifted species assemblages towards more moisture-dependent species.

Table 1. Effects of management and environmental variables on species composition of orthopterans, carabids, and spiders in irrigated and non-irrigated meadows in the Queichtal, Germany.

	Orthopterans		Carabids		Spiders	
	R^2	P	R^2	P	R^2	P
Factors						
Irrigation (yes or no)	0.078	**0.019**	0.050	0.171	0.080	**0.041**
Fertilization (yes or no)	0.028	0.451	0.012	0.976	0.023	0.710
Habitat parameters						
Humidity	0.085	**0.011**	0.088	**0.024**	0.060	0.102
Nitrogen	0.023	0.568	0.029	0.582	0.033	0.451
Landscape parameters						
Distance to water	0.032	0.370	0.048	0.175	0.012	0.924
Distance to forest	0.031	0.405	0.027	0.613	0.061	0.121

Significance was tested by permutational multivariate ANOVA (command 'adonis' in R package vegan). Significant results (P<0.05) are shown in bold.

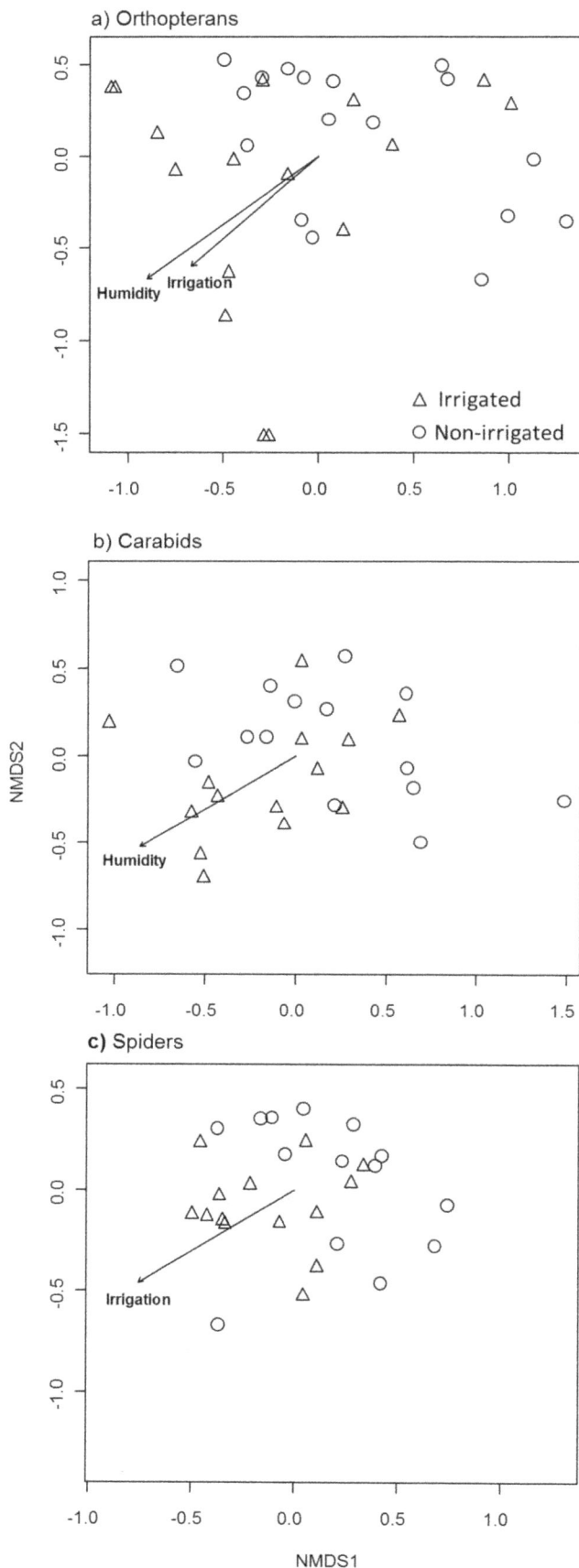

a) Orthopterans

b) Carabids

c) Spiders

Figure 4. NMDS ordinations of a) orthopterans, b) carabid and c) spider species composition of irrigated and non-irrigated meadows. Only significant environmental variables are shown (permutational multivariate ANOVA, for statistics see Table 1).

Local diversity

We assumed that traditional flood irrigation leads to small-scale differences in moisture and sediment conditions and consequently higher microhabitat and vegetation heterogeneity resulting in increased local diversity [16,48]. However, in contrast to our hypothesis, traditionally flood irrigated meadows did not contain a higher local diversity (in terms of species richness and Simpson diversity) of orthopterans, carabids, and spiders compared to non-irrigated meadows. Possibly, the traditional irrigation system in our study area - on average four flooding events with a maximum of twelve irrigation days between April and August - is not sufficient to induce (measurable) heterogeneity effects. Riedener et al. [16] assumed that effects of the irrigation technique on diversity are only effective in combination with other management factors such as mowing and grazing regimes. Additionally, landscape variables such as patch isolation can influence species [32]. We accounted for possible confounding effects of management (mowing frequency, fertilization) and landscape parameters (distance to permanent water, distance to forest), none of which differed significantly between irrigated and non-irrigated meadows. However, as in any observational study, we cannot rule out that additional unmeasured management or environmental parameters have influenced our results.

In the literature, several studies in riparian habitats were able to detect positive effects of (natural) flooding on diversity [48–51]. However, in contrast to our study system with no (non-irrigated) and low intensive (irrigated) flooding, flood intensities in these studies are mostly studied along gradients containing higher intensities. Gerisch et al. [50] found positive effects for carabids at the river Elbe (Germany) and explained this by higher resource diversity in frequently flooded habitats. At the river Meuse (Belgium/Netherlands), Lambeets [50] could show that flooding initially had a positive effect on carabid diversity, which peaked at intermediate flooding degrees. This is in line with findings of Pollack et al. [48] in riparian meadows where plant species richness was highest at intermediately flooded river banks because of increased microhabitat heterogeneity.

Similar to diversity, densities of orthopterans and carabids did not differ between meadow types. However, spider densities were higher in irrigated meadows. One explanation might be enhanced food availability, because short time flooding can enhance soil organisms [52], which present important food source especially for linyphiid spiders [53].

Community differentiation (beta diversity)

Irrigation did not influence community differentiation and, in contrast to our hypothesis, beta diversity of orthopterans, carabids, and spiders was not higher in irrigated compared to non-irrigated meadows. Although flood irrigation between irrigated meadows differed in time and intensity, these differences were obviously too weak to result in more diverse species assemblages. Moreover, the traditional flooding method in the region - where the dammed river water slowly streams into the meadows and back into the river through a system of open ditches - leads to a relatively homogenous water flow. This prevents stagnant moisture [12] and moisture conditions on irrigated sites might be more uniform than expected. Human-altered repetitive flood events are known to result in uniform species compositions due to a homogenization of habitat structure [54,55].

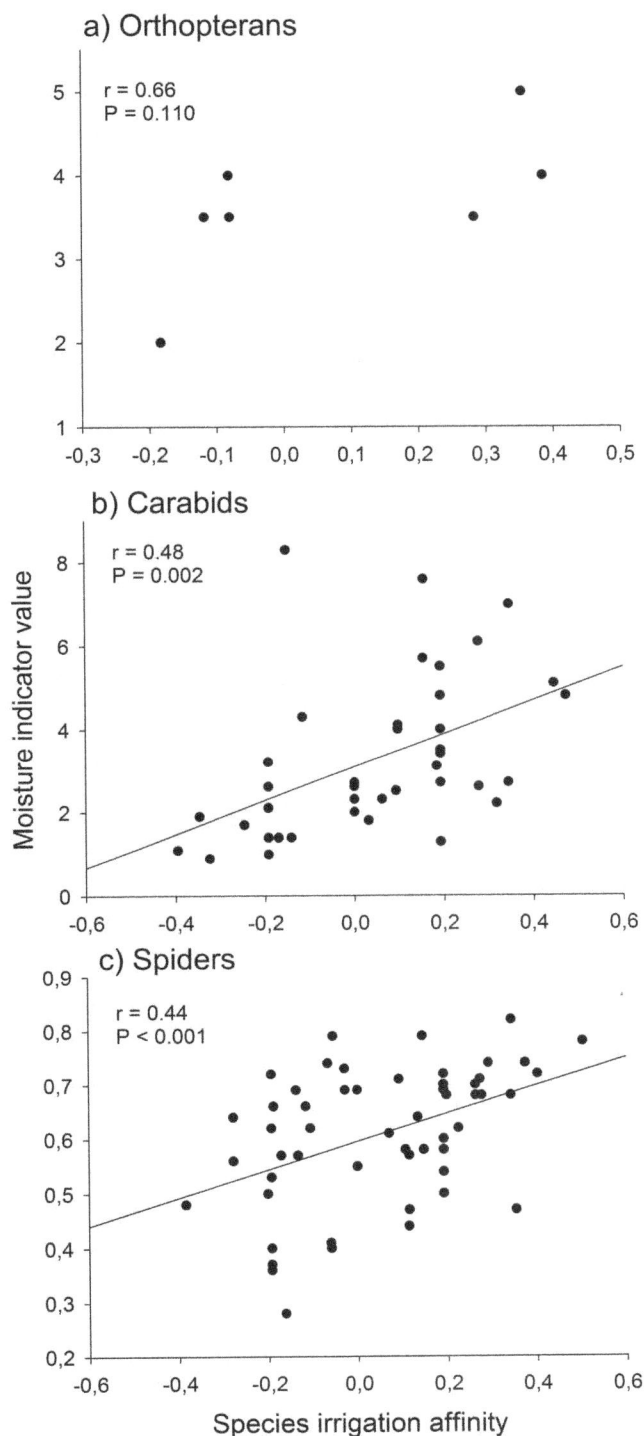

Figure 5. Relationship between species irrigation affinity (spearman rank correlation coefficient of species densities to irrigation) and moisture indicator value of a) orthopterans, b) carabids, and c) spiders.

Species composition

As hypothesized, meadow irrigation and the associated altered humidity conditions influenced species composition of orthopterans, carabids, and spiders. Irrigation may therefore increase beta diversity at the landscape scale and contribute to diverse grassland communities. Assemblages of irrigated meadows contained more moisture-dependent species compared to non-irrigated ones. This was reflected for carabids and spiders by the positive relationships between species irrigation affinity and their moisture indicator value. For orthopterans this effect was not significant (most likely because of the low number of species), but the two species with the highest moisture indicator value – *Mecostethus parapleurus* and *Stetophyma grossum* – were significantly more abundant in irrigated meadows (Table S3). For all three taxa, humidity is known to be one of the most influencing environmental parameter structuring species compositions [27,43,56]. Impacts of (natural) flood disturbance on species and trait composition of orthopterans were previously shown by Dziock et al. [57]. Bonn et al. [48] found that flood regime strongly influenced carabid species assemblages, which was also found by Lambeets et al. [50] for carabid and spider communities. Although fertilization can strongly influence arthropods [58,59] we could not detect an effect of fertilization. In general, fertilization application rates were low in the study area (0 to ~60 kg N•ha-1•yr-1), a range in which also plants showed no significant decrease in species richness (unpublished data). Similar to fertilization, the landscape parameters, distance to forest and to permanent water, did not differ between irrigated and non-irrigated meadows and had no effect on species compositions, respectively.

In contrast to our hypothesis, species compositions of flood irrigated meadows did not contain more species of conservation concern than non-irrigated meadows. This is in contrast to Bonn et al. [49] and Lambeets et al. [51], where anthropogenic alterations in flooding regimes not only have a strong influence on arthropod communities but also on the distribution of rare (and often endangered) riparian species. Again, this difference may be due to the low irrigation intensity in our study system.

Conclusion

Flood irrigation had no significant effect on local and beta diversity of orthopterans, carabids, and spiders in lowland meadows. In contrast, flood irrigation clearly changed species assemblages towards moisture-dependent species and probably increased beta diversity at the landscape scale. However, these species were mostly common species and assemblages of irrigated meadows did not contain more species of conservation concern compared to non-irrigated ones. More variable and intensive (higher duration and/or frequency) flooding regimes are likely to provide much stronger conservation benefits. Moreover, beneficial effects of flood irrigation might be more pronounce along the irrigation infrastructures (open ditches, drains, weirs) than in the open meadow, which will be tested in further studies.

Acknowledgments

We thank Lisa Gundall, Melanie Hilbert, Larissa Hermann, Franziska Möller, Max Pfirrmann, Stefan Pietrusky Tessa Schardt, Lisa Weigel and Kim Wilsdorf for help with the field work and species determination.

Author Contributions

Conceived and designed the experiments: JS MHE. Performed the experiments: JS MA IR MHE. Analyzed the data: JS. Contributed reagents/materials/analysis tools: JS MA MHE. Contributed to the writing of the manuscript: JS MHE.

References

1. Muller S (2002) Appropriate agricultural management practices required to ensure conservation and biodiversity of environmentally sensitive grassland sites designated under Natura 2000. Agric Ecosys. Environ 89: 261–266.

2. Poschlod P, WallisDeVries MF (2002) The historical and socioeconomic perspective of calcareous grasslands - lessons from the distant and recent past. Biol Conserv 104: 361–376.

3. Isselstein J, Jeangros B, Pavlu V (2005) Agronomic aspects of biodiversity targeted management of temperate grasslands in Europe – A review. Agron Res 3: 139–151.

4. Duelli P, Obrist MK (2003) Regional biodiversity in an agricultural landscape: the contribution of seminatural habitat islands. Basic Appl Ecol 4: 129–138.

5. Strijker D (2005) Marginal lands in Europe – causes of decline. Basic Appl Ecol 6: 99–106.

6. Rounsevell MDA, Ewert F, Reginster I, Leemans R, Carter TR (2005) Future scenarios of European agricultural land use - II. Projecting changes in cropland and grassland. Agric Ecosyst Environ 107: 117–135.

7. Baur B, Cremene C, Groza G, Rakosy L, Schileyko AA, et al. (2006) Effects of abandonment of subalpine hay meadows on plant and invertebrate diversity in Transylvania, Romania. Biol Conserv 132: 261–273.

8. Ellenberg H, Leuschner C (2012) Die Vegetation Mitteleuropas mit den Alpen. Stuttgart: UTB. 1334 p.

9. Rosenthal G, Hölzel N (2009) Renaturierung von Feuchtgrünland, Auengrün-land und mesophilem Grünland. In: Zerbe S, Wiegleb G, editors. Renaturierung von Ökosystemen in Mitteleuropa. Heidelberg: Spektrum Akademischer Verlag. pp. 283–316.

10. Leibundgut C (2004) Historical meadow irrigation in Europe - a basis for agricultural development. In: Rodda JC, Ubertini L, editors. The Basis of Civilization - Water Science? IHAS Publication 286: 77–87.

11. Schellberg S (2005) Meadow irrigation in the federal state Baden-Württemberg. Portrayal of a nearly forgotten land use system. Integrated Land and Water Resources Management in History, Schriften der DWhG, Sonderband 2: 123–131.

12. Hassler DM, Glaser K-H (1995) Wässerwiesen - Geschichte, Technik und Ökologie der bewässerten Wiesen, Bäche und Gräben in Kraichgau, Hardt und Bruhrain. Ubstadt-Weiher: Verlag Regionalkultur. 432 p.

13. Leibundgut C (2011) Die Wässermatten des Oberaargaus. Ein regionales Kulturerbe als Modell für Europa? Jahrb des Oberaargaus 54: 121–144.

14. Kleijn D, Baquero RA, Clough Y, Díaz M, De Esteban J, et al. (2006) Mixed biodiversity benefits of agri-environment schemes in five European countries. Ecol Lett 9: 243–254.

15. Kleijn D, Sutherland WZ (2003) How effective are European agri-environment schemes in conserving and promoting biodiversity? J Appl Ecol 40: 947–969.

16. Riedener E, Rusterholz H-P, Baur B (2013) Effects of different irrigation systems on the biodiversity of species-rich hay meadows. Agric Ecosyst Environ 164: 62–69.

17. MacArthur RH, MacArthur JW (1961) On bird species diversity. Ecology 42: 594–598.

18. MacArthur RH, Wilson EO (1967) The theory of island biogeography. Princeton: Princeton University Press. 224 p.

19. Sousa WP (1984) The role of disturbance in natural communities. Ann Rev Ecol Syst 15: 353–391.

20. Benton TG, Vickery ZA, Wilson JD (2003) Farmland biodiversity: is habitat heterogeneity the key? Trends Ecol Evol 18: 182–188.

21. Curry JP (1994) Grassland Invertebrates – Ecology, influence on soil fertility and effects on plant growth. London: Chapman and Hall. 437 p.

22. Thiele HU (1977) Carabid beetles in their environments. A study on habitat selection by adaptations in physiology and behaviour. Heidelberg, Berlin, New York: Springer. 369 p.

23. Wise D (1993) Spiders in ecological webs. Cambridge: Cambridge University Press. 328 p.

24. Báldi A, Kisbenedek T (1997) Orthopteransn assemblages as indicators of grassland naturalness in Hungary. Agric Ecosyst Environ 66: 121–129.

25. Schirmel J, Mantilla-Contreras J, Blindow I, Fartmann T (2011) Impacts of succession and grass encroachment on heathland Orthopterans. J Insect Conserv 15: 633–642.

26. Rainio J, Niemelä J (2003) Ground beetles (Coleoptera: Carabidae) as bioindicators. Biodivers Conserv 12: 487–506.

27. Schirmel J, Buchholz S (2011) Response of carabid beetles (Coleoptera: Carabidae) and spiders (Araneae) to coastal heathland succession. Biodivers Conserv 20: 1469–1482.

28. Scott AG, Oxford GS, Selden PA (2006) Epigeic spiders as ecological indicators of conservation value for peat bogs. Biological Conserv 127: 420–428.

29. Horvath R, Magura T, Szinetar C, Tothmeresz B (2009) Spiders are not less diverse in small and isolated grasslands, but less diverse in overgrazed grasslands: A field study (East Hungary, Nyirseg). Agric Ecosyst Environ 130: 16–22.

30. Briem E, Geiger M (2008) Das Fließgewässer Queich. In: Geiger M, editor. Haardt, Weinstraße und Queichtal: Ein Geo-Führer. Neustadt/Weinstraße: Pollichia. pp. 114–116.

31. Ssymank A, Hauke U, Rückriem C, Schröder E (1998) Das europäische Schutzgebietssystem NATURA 2000– BfN Handbuch zur Umsetzung der Fauna-Flora-Habitat-Richtlinie und der Vogelschutz-Richtlinie. Schriftenr Landschaftspfl Natursch 53: 1–560.

32. Bailey D, Schmidt-Entling MH, Eberhart P, Herrmann JD, Hofer G, et al. (2010) Effects of habitat amount and isolation on biodiversity in fragmented traditional orchards. J Appl Ecol 47: 1003–1013.

33. Google (2011) Google Earth 6.1.0.5001. Google Inc. Available: http://www.google.com/earth/index.html. Accessed 2013 Oct 19.

34. Gardiner T, Hill J, Chesmore D (2005) Review of the methods frequently used to estimate the abundance of Orthopterans in grassland ecosystems. J Insect Conserv 9: 151–173.

35. Bellmann H (2006) Der Kosmos Heuschreckenführer. Die Arten Mitteleuropas sicher bestimmen. Stuttgart: Kosmos. 350 p.

36. Müller-Motzfeld G (2006) Die Käfer Mitteleuropas. Band 2 Adephaga 1: Carabidae (Laufkäfer). Heidelberg/Berlin: Spektrum Akademischer Verlag. 536 p.

37. Roberts MJ (1996) Collins field guide. Spiders of Britain and Northern Europe. Berkshire: D & N Publishing. 383 p.

38. Pfeifer MA, Niehuis M, Renker C (2011) Die Fang- und Heuschrecken in Rheinland-Pfalz. Flora und Fauna in Rheinland-Pfalz 41: 1–678.

39. Schüle P, Persohn M (2000) Rote Liste der in Rheinland-Pfalz gefährdeten Laufkäfer (Coleoptera: Carabidae). Ministerium für Umwelt und Forsten. Mainz: Rhein Main Druck. 29 p.

40. Nährig D, Harms KH (2003) Rote Listen und Checklisten der Spinnen in Baden-Württemberg. Mannheim: Landesanstalt für Umweltschutz Baden-Württemberg, JVA Mannheim-Druckerei. 203 p.

41. Maas S, Detzel P, Staudt A (2002) Gefährdungsanalyse der Heuschrecken Deutschlands. Verbreitungsatlas, Gefährdungseinstufung und Schutzkonzepte. Münster: Landwirtschaftsverlag. 401 p.

42. Irmler U, Gürlich S (2004) Die ökologische Einordnung der Laufkäfer (Coleoptera: Carabidae) in Schleswig-Holstein. Faunistisch-Ökologische Mitt Suppl 32: 1–115.

43. Entling W, Schmidt MH, Bacher S, Brandl R, Nentwig W (2007) Niche properties of Central European spiders: shading, moisture and the evolution of the habitat niche. Global Ecol Biogeogr 16: 440–448.

44. Zuur AF, Ieno IN, Walker NJ, Saveliev AA, Smith GM (2009) Mixed effects models and extensions in ecology with R. Berlin, Heidelberg: Springer. 574 p.

45. Anderson MJ, Ellingsen KE, McArdle BH (2006) Multivariate dispersion as a measure of beta diversity. Ecol Lett 9: 683–693.

46. Oksanen J, Blanchet FG, Kindt R, Legendre P, O'Hara RB, et al. (2011) vegan: Community Ecology Package. R package version 1.17-11. http://CRAN.R-project.org/package=vegan.

47. R Development Core Team (2010) R: A language and environment for statistical computing. Vienna: R Foundation for Statistical Computing. www.R-project.org.

48. Pollock MM, Naiman RJ, Hanley TA (1998) Plant species richness in riparian wetlands - a test of biodiversity theory. Ecology 79: 94–105.

49. Bonn A, Hagen K, Wohlgemuth-von Reiche D (2002) The significance of flood regimes for carabid beetle and spider communities in riparian habitats – a comparison of three major rivers in Germany. River Res Appl 18: 43–64.

50. Gerisch M, Agostinelli V, Henle K, Dziock F (2012) More species, but all do the same: contrasting effects of flood disturbance on ground beetle functional and species diversity. Oikos 121: 508–515.

51. Lambeets K, Vandengehuchte ML, Maelfait JP, Bonte D (2008) Understanding the impact of flooding on trait-displacements and shifts in assemblage structure of predatory arthropods on river banks. J Animal Ecol 77: 1162–1174.

52. Plum N (2005) Terrestrial invertebrates in flooded grassland: a literature review. Wetlands 25: 721–737.

53. Nyffeler M (1999) Prey selection of spiders in the field. J Arachnol 27: 317–324.

54. Bonn A, Kleinwächter M (1999) Microhabitat distribution of spider and ground beetle assemblages (Araneae, Carabidae) on frequently inundated river banks of the River Elbe. Z Ökologie Naturschutz 8: 109–123.

55. Vanbergen AJ, Woodcock BA, Watt AD, Niemelä J (2005) Effect of land-use heterogeneity on carabid communities at the landscape scale. Ecography 28: 3–16.

56. Ingrisch S, Köhler G (1998) Die Heuschrecken Mitteleuropas. Magdeburg: Westarp-Wissenschaften. 460 p.

57. Dziock F, Gerisch M, Sieger M, Hering I, Scholz M, et al. (2011) Reproducing or dispersing? Using trait based habitat templet models to analyse Orthopterans response to flooding and land use. Agric Ecosyst Environ 145: 85–94.

58. Siemann E (1998) Experimental tests of effects of plant productivity and diversity on grassland arthropod diversity. Ecology 79: 2057–2070.

59. van Wingerden WKRE, van Kreveld AR, Bongers W (1992) Analysis of species composition and abundance of grasshoppers (Orth., Acrididae) in natural and fertilized grasslands. J Appl Entomol 113: 138–152.

13

Comparison of Phenology Models for Predicting the Onset of Growing Season over the Northern Hemisphere

Yang Fu[1], Haicheng Zhang[1], Wenjie Dong[1], Wenping Yuan[1,2]*

1 State Key Laboratory of Earth Surface Processes and Resource Ecology, Beijing Normal University, Beijing, China, **2** State Key Laboratory of Cryospheric Sciences, Cold and Arid Regions Environmental and Engineering Research Institute, Chinese Academy of Sciences, Lanzhou, Gansu, China

Abstract

Vegetation phenology models are important for examining the impact of climate change on the length of the growing season and carbon cycles in terrestrial ecosystems. However, large uncertainties in present phenology models make accurate assessment of the beginning of the growing season (BGS) a challenge. In this study, based on the satellite-based phenology product (i.e. the V005 MODIS Land Cover Dynamics (MCD12Q2) product), we calibrated four phenology models, compared their relative strength to predict vegetation phenology; and assessed the spatial pattern and interannual variability of BGS in the Northern Hemisphere. The results indicated that parameter calibration significantly influences the models' accuracy. All models showed good performance in cool regions but poor performance in warm regions. On average, they explained about 67% (the Growing Degree Day model), 79% (the Biome-BGC phenology model), 73% (the Number of Growing Days model) and 68% (the Number of Chilling Days-Growing Degree Day model) of the BGS variations over the Northern Hemisphere. There were substantial differences in BGS simulations among the four phenology models. Overall, the Biome-BGC phenology model performed best in predicting the BGS, and showed low biases in most boreal and cool regions. Compared with the other three models, the two-phase phenology model (NCD-GDD) showed the lowest correlation and largest biases with the MODIS phenology product, although it could catch the interannual variations well for some vegetation types. Our study highlights the need for further improvements by integrating the effects of water availability, especially for plants growing in low latitudes, and the physiological adaptation of plants into phenology models.

Editor: Ben Bond-Lamberty, DOE Pacific Northwest National Laboratory, United States of America

Funding: This study was supported by the National Science Foundation for Excellent Young Scholars of China (41322005), National Natural Science Foundation of China (41201078), Program for New Century Excellent Talents in University (NCET-12-0060) and the Fundamental Research Funds for the Central Universities. The funders had no role in study design, data collection and analysis, decision to publish, or preparation of the manuscript.

Competing Interests: The authors have declared that no competing interests exist.

* Email: yuanwpcn@126.com

Introduction

Phenology refers to the timing of recurring biological cycles, and is considered a sensitive indicator of climate change [1–3]. In particular, as research interest in global change increases, determining the beginning of the growing season (BGS) of land vegetation has become an important research subject [4]. Previous studies revealed that plant activity is more sensitive to climatic changes in spring than other seasons; and changes in the BGS would strongly impact the seasonal energy balance and net carbon dioxide (CO_2) flux of terrestrial ecosystems [5,6].

Large uncertainties, however, in present phenology models make accurate assessment of BGS a challenge. Two classes of process-based models have been developed for simulating the spring phenological phases. Models belonging to the first class, the 'one-phase' models, are the simplest and have been used in agronomy since the 18th century [7]. This kind of model implicitly assumes that bud dormancy is fully released after a fixed sum of accumulated temperatures has been reached. The second class of models, the 'two-phase' models, considers the breaking of two dormancy phases [8]. The first phase is a period when buds remain dormant due to plant endogenous factors, and the second

phase is a period when buds remain dormant due to unfavorable environmental factors [9]. Many studies have described the breaking of the first phase and overcoming the second phase in terms of chill accumulation to break the first phase followed by a period of forcing temperature to overcome the second phase [10,11]. The two-phase models are of more recent development, and are conceptually based on experimental studies which highlighted that chilling was the major factor responsible for dormancy release [12–15].

Many phenology observations have focused on cultivated rather than natural plants [16,17]. Geographically, most of the observations were conducted in North America and Europe [18–20]. Due to the limited availability of phenological observation data on a large scale, most phenology models are calibrated at local scales [21] and thus are unlikely to accurately predict BGS across different vegetation types. These phenology models might underestimate or overestimate the BGS when applied to a regional or global scale [22]. For example, a comparison of phenology models in 14 terrestrial biosphere models indicated that almost all models failed to track the phenology, and most predicted an earlier BGS, overestimating the gross ecosystem photosynthesis by 20% [23].

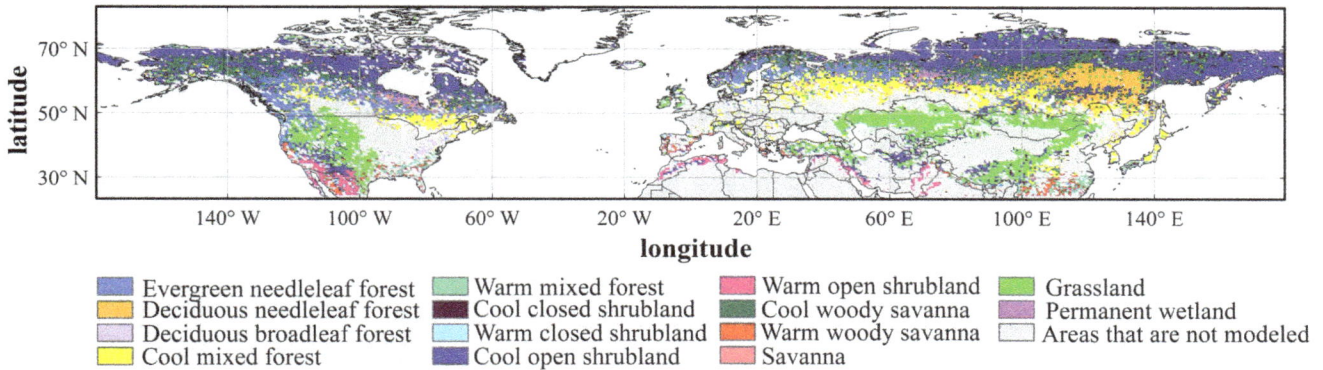

Figure I. Vegetation distribution map of the Northern Hemisphere retrieved from the V005 MODIS Land Cover Type Product (MCD12Q1). Grey areas are either excluded vegetation types like croplands, or areas with no seasonal cycle detectable by satellite.

Remote sensing data from satellites provide broad coverage of useful information on vegetation phenology for diverse ecosystems at various scales, and help to calibrate the phenology models [24–28]. For example, Yang et al. [22] parameterized three budburst models in New England using 11 years of remotely sensed phenology and climate data. Nowadays, remote sensing-based phenology has been significantly improved with the Moderate Resolution Imaging Spectroradiometer (MODIS) on board the Terra and Aqua satellites [29]. Since 2009, the latest version of the MODIS Land Cover Dynamics Product (MCD12Q2) has been available [30], which provides valuable phenology data for the present study.

Based on the global satellite-based phenological observations, the primary objectives of this study are to (1) calibrate four phenology models; (2) compare the relative strengths of four phenology models; and (3) assess the spatial pattern and interannual variability of BGS in the Northern Hemisphere.

Data and Methods

1. Satellite and meteorological data

The V005 MODIS Land Cover Dynamics (MCD12Q2) product (informally called the MODIS Global Vegetation Phenology product) was used to estimate the vegetation phenology of the study area. It identifies the vegetation growth, maturity, and senescence that mark seasonal cycles at global scales with a 500 ×500 m spatial resolution and is available from 2001 to 2010 [30].

This product is produced each year from the 8-day vegetation index EVI (Enhanced Vegetation Index) calculated from the NBAR reflectance (Nadir Bidirectional Reflectance Distribution Function-Adjusted Reflectance). More complete details regarding algorithm implementation are provided in Zhang et al. [29] and Ganguly et al. [30].

The V005 MODIS Land Cover Type Product (MCD12Q1) was used to identify land cover properties. It provides data characterizing five global land cover classification systems at annual time steps and 500 ×500 m spatial resolution for 2001-present. In this study, we chose the International Geosphere Biosphere Programme (IGBP) classification scheme, which includes 11 natural vegetation classes, three developed and mosaicked land classes, and three non-vegetated land classes. We excluded the evergreen broadleaf forest from our analysis as it has little or no leaf seasonal cycle. We also excluded croplands and crop/natural vegetation mosaics because human management practices strongly impact their phenology (e.g. irrigation, fertilization). In the classification of IGBP, a single vegetation type may exist in both subtropical and boreal regions (e.g. woody savannah in Figure 1). As plants in different regions require markedly different quantities of heat, it is necessary to subdivide vegetation types according to the climatic conditions in order to get the optimal model parameters. Therefore, we subdivided four vegetation types which are distributed across a wide range of latitudes, based on the climate criteria of Botta et al. [31]. Three meteorological variables were used to identify the vegetation types,

Table 1. Climate criteria used to subdivide the four vegetation types which are distributed across a wide range of latitudes.

Vegetation type	Subdivision	Climate criteria
Mixed forest	Cool mixed forest	$T_C < 0°C$
	Warm mixed forest	$T_C \geq 0°C$
Closed shrub	Cool closed shrub	$T_C < 0°C$
	Warm closed shrub	$T_C \geq 0°C$
Open shrub	Cool open shrub	$\Delta T > 20°C$ or $T_C < 5°C$
	Warm open shrub	$\Delta T \leq 20°C$ and $T_C \geq 5°C$
Woody savanna	Cool woody savanna	$T_C < 0°C$
	Warm woody savanna	$T_C \geq 0°C$

The climate criteria is gained from Botta et al. [31]. T_C and ΔT are respectively the minimum daily temperature of the year (T_C) and the difference between annual maximum (T_W) and minimum daily temperatures ($\Delta T = T_W - T_C$).

Table 2. Parameter values: the degree-day base temperature (T_{th_GDD}) is estimated for model GDD (Eq. (1)); the critical value of growing degree days (GDD_C) is estimated for model GDD (Eq. (2)); the degree-day base temperature (T_{th_BBGC}) is estimated for model BBGC; empirical coefficients (a and b) are estimated for model BBGC (Eq. (3)); the empirical coefficient (c) is estimated for model BBGC (Eq. (4)); the underdetermined soil temperature threshold (d) determining warm grasslands or cool grasslands is estimated for model BBGC (Eq. (4)); the proportion (k) of the average annual precipitation is estimated for model BBGC (Eq. (5)); the base temperature (T_{th_NGD}) and the critical number of growing days (NGD_C) are estimated for model NGD; the degree-day base temperature ($T_{th\_NCD-GDD}$) is estimated for model NCD-GDD (Eq. (6)); the chill day base temperature (T_{th_NCD}), and empirical coefficients (g, h, and w) are estimated for model NCD-GDD (Eq. (7)).

Biome	GDD		BBGC						NGD		NCD-GDD				
	T_{th_GDD} (°C)	GDD_C (°C*days)	T_{th_BBGC} (°C)	a	b	c	d	k	T_{th_NGD} (°C)	NGD_C (days)	T_{th_NCD} (°C)	$T_{th\_NCD-GDD}$ (°C)	g	h	w
Evergreen needleleaf forest	6	50	0	4.755	0.117	-	-	-	9	5	0	3	-300	400	-0.1
Deciduous needleleaf forest	2	52	0	4.755	0.117	-	-	-	5	6	0	-1	-300	400	-0.1
Deciduous broadleaf forest	-5	591	-5	5.505	0.085	-	-	-	6	21	0	-5	-100	700	-0.1
Cool mixed forest	5	52	0	4.63	0.101	-	-	-	8	6	-5	3	-300	400	-0.1
Warm mixed forest	7	236	-5	6.005	0.057	-	-	-	5	59	-5	5	-300	600	-0.1
Cool closed shrub	5	50	2	4.13	0.109	-	-	-	8	5	-5	3	-300	400	-0.1
Warm closed shrub	7	308	-5	6.63	0.041	-	-	-	5	77	-5	5	-300	700	-0.1
Cool open shrub	0	140	-3	5.505	0.057	-	-	-	5	11	-5	1	-300	400	-0.1
Warm open shrub	7	350	-5	6.005	0.069	-	-	-	10	41	-2	7	-300	600	-0.1
Cool woody savanna	3	58	0	-	-	15	119	0.05	7	6	-5	1	-300	400	-0.1
Warm woody savanna	7	296	-5	-	-	5	119	0.06	5	58	-5	5	-300	600	-0.1
Savanna	1	126	-2	-	-	11	209	0.16	5	12	-5	5	-300	400	-0.1
Grassland	-5	448	-5	-	-	15	369	0.05	6	12	-5	5	-300	400	-0.1
Permanent wetland	-5	378	-1	5.005	0.101	-	-	-	-4	47	-5	1	-300	400	-0.1

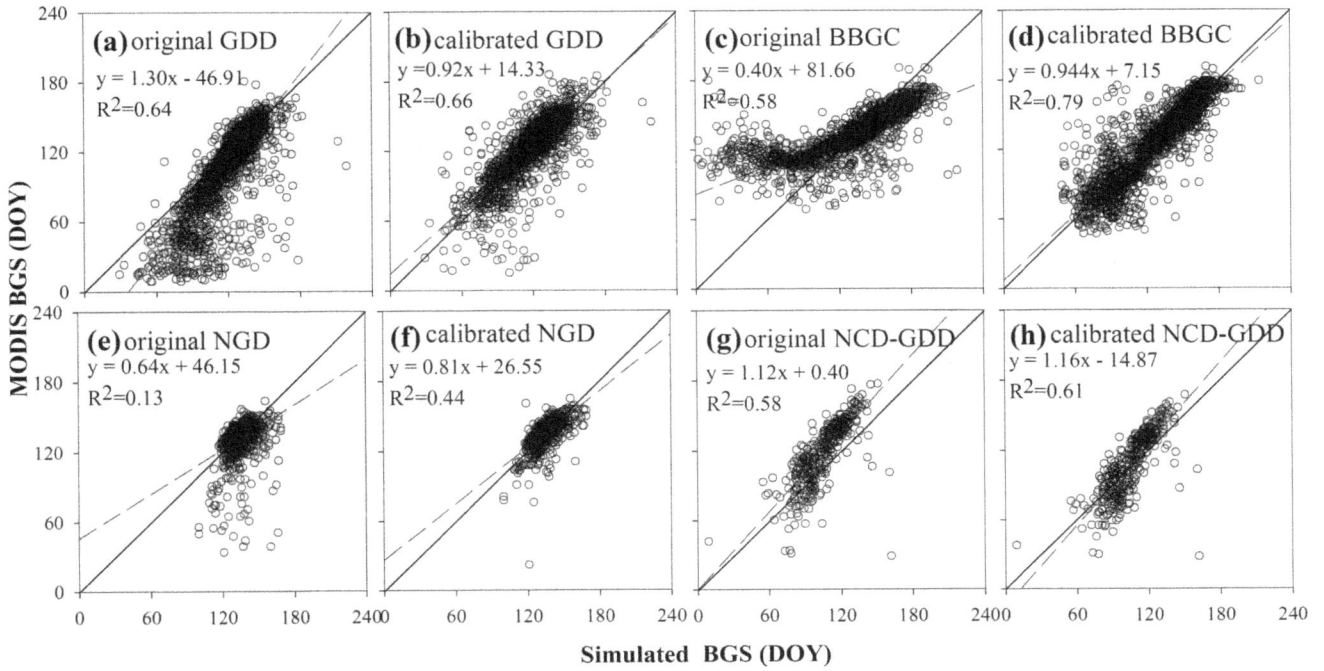

Figure 2. The correlations between MODIS BGS and simulated BGS. (a) and (b) show BGS simulations derived from GDD models with the original parameter values in IBIS model and calibrated parameters respectively; (c) and (d) show BGS simulations of deciduous forest and grassland derived from BBGC models with the original parameter values in Biome-BGC model and calibrated parameters respectively; (e) and (f) show BGS simulations of deciduous needle leaf forest derived from NGD models with the original parameter values in ORCHIDEE model and calibrated parameters respectively; (g) and (h) show BGS simulations of deciduous broadleaf forest derived from NCD-GDD models with the original parameter values in ORCHIDEE model and calibrated parameters respectively. The solid line is the 1:1 line and the short dashed lines are regression lines.

including the annual mean of daily temperature (T_{mean}), the minimum daily temperature of the year (T_c) and the difference between annual maximum (T_w) and minimum daily temperatures ($\Delta T = T_w - T_c$) (Table 1).

Daily meteorological data, including temperature and precipitation, were derived from the MERRA (Modern Era Retrospective-Analysis for Research and Applications) archive for 2001–2010. MERRA is a NASA reanalysis for the satellite era using a major new version of the Goddard Earth Observing System Data Assimilation System Version 5 (GEOS-5) [32]. MERRA uses data from all available surface weather observations globally every 3 hours. The GEOS-5 is used to interpolate and grid these point data on a short time sequence, and produces an estimate of climatic conditions for the world at 10 m above the land surface (i.e., approximating canopy height conditions). The resolution is 0.5° latitude by 0.67° longitude. The MERRA reanalysis dataset has been validated carefully at the global scale using surface meteorological data sets to evaluate the uncertainty of various meteorological factors (i.e. temperature, radiation, humidity, precipitation and energy balance). Detailed information on the MERRA dataset is available at the website (http://gmao.gsfc.nasa.gov/research/merra).

2. Phenology Models

In this study, we compared three one-phase phenological models for the beginning of growing season (BGS) including the Growing Degree Day model (GDD), the Biome-BGC phenology model (BBGC) and the Number of Growing Days model (NGD), and a two-phase phenological model (the Number of Chilling Days-Growing Degree Day model (NCD-GDD)) over the

Northern Hemisphere (Figure 1). We did not include the Southern Hemisphere and tropical regions because of the poor performance of the V005 MODIS Land Cover Dynamics (MCD12Q2) product over these regions [25].

The GDD model is a classical one-phase phenological model, and has been used to predict the timing of BGS in spring by a function of accumulated temperature [33,34]. After a starting date t_0 (usually January 1st), mean air temperature above a degree-day base temperature (T_{th_GDD}) is accumulated until a critical value (GDD_c) is exceeded; at that time (t_1) the prescribed growing season starts. The model can be described as follows:

$$GDD(t) = \sum_{t=t_0}^{t_1} Max(T - T_{th\_GDD}, 0) \qquad (1)$$

$$GDD(t_1) \geq GDD_C \qquad (2)$$

The BBGC model is integrated into the Biome-BGC (BioGeochemical Cycles) terrestrial ecosystem process model, described in White et al. [21]. The BBGC model divides vegetation phenology into two types: woody plants (i.e. trees and brush) and grasses [35]. For deciduous woody plants, the growing season begins when the running sum of the daily average soil temperatures (when the average soil temperature is above a degree-day base soil temperature (T_{th_BBGC})) is above a critical value defined by:

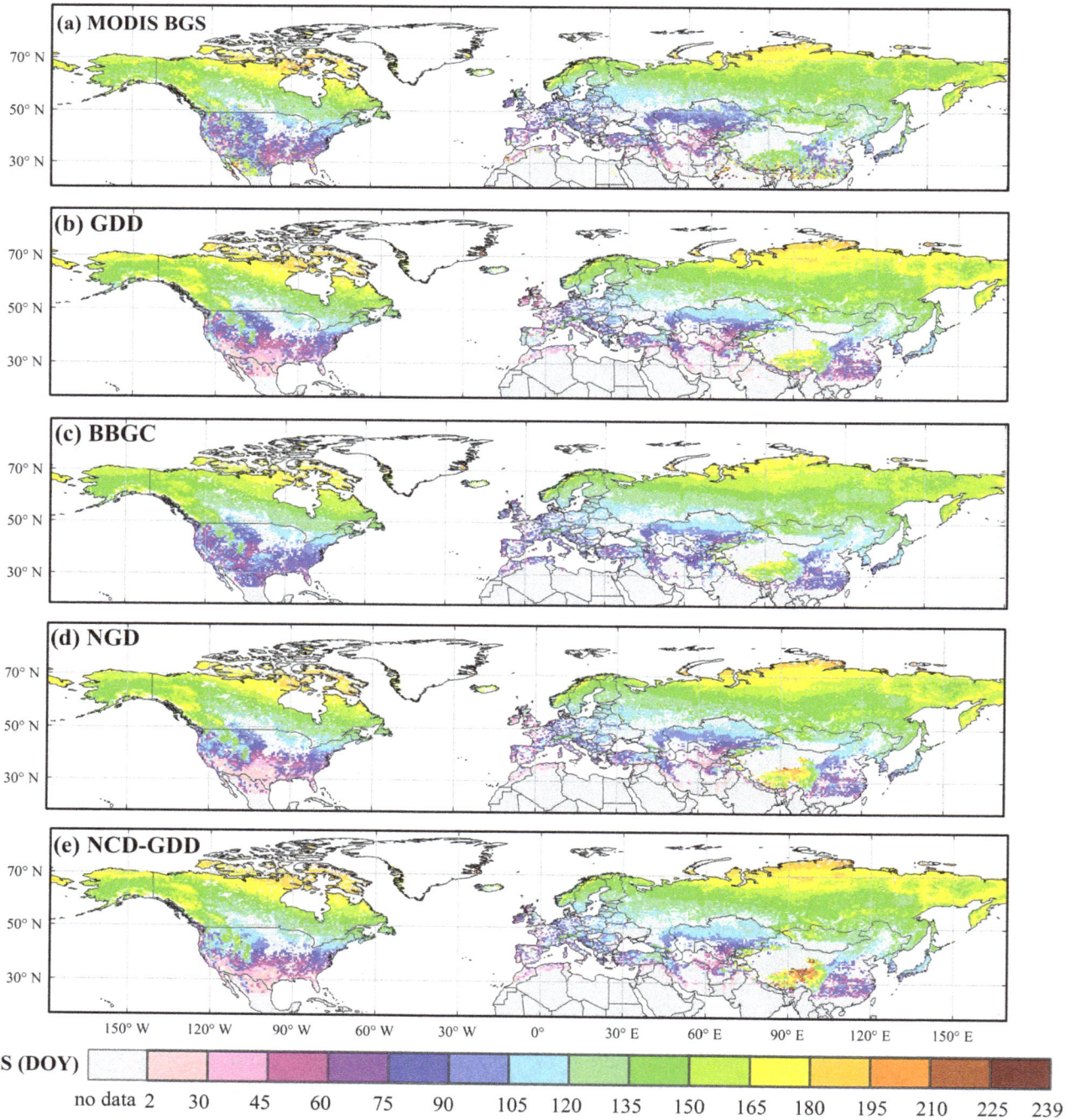

Figure 3. Spatial pattern of mean dates for the beginning of growing season (BGS) in the Northern Hemisphere during 2001–2010.
(a) The start dates derived from the MODIS product; (b)–(e) indicate the simulated start dates of the four phenology models.

$$\mathrm{TcritSum}_{\mathrm{woody}} = e^{a + b \times \mathrm{Tavg}} \qquad (3)$$

where $\mathrm{T}_{\mathrm{avg}}$ is the mean daily average temperature; a and b are empirical coefficients. Moreover, the model specifies that the day length must be longer than 39300 seconds for leaf out to occur.

For grasses, the BGS is controlled by both temperature and water availability. When both of the accumulated soil temperatures and the accumulated precipitation values are larger than or equal to the critical values, the growing season begins. The critical accumulated soil temperature value ($\mathrm{TcritSum}_{\mathrm{grass}}$) and the critical accumulated precipitation value ($\mathrm{PrcpCritSum}_{\mathrm{grass}}$) for grasses are defined as:

$$\mathrm{TcritSum}_{\mathrm{grass}} = c \times \left[\frac{e^{32.9 \times (\mathrm{Tavg\text{-}d})} - 1}{e^{32.9 \times (\mathrm{Tavg\text{-}d})} + 1} \right] + 900 \qquad (4)$$

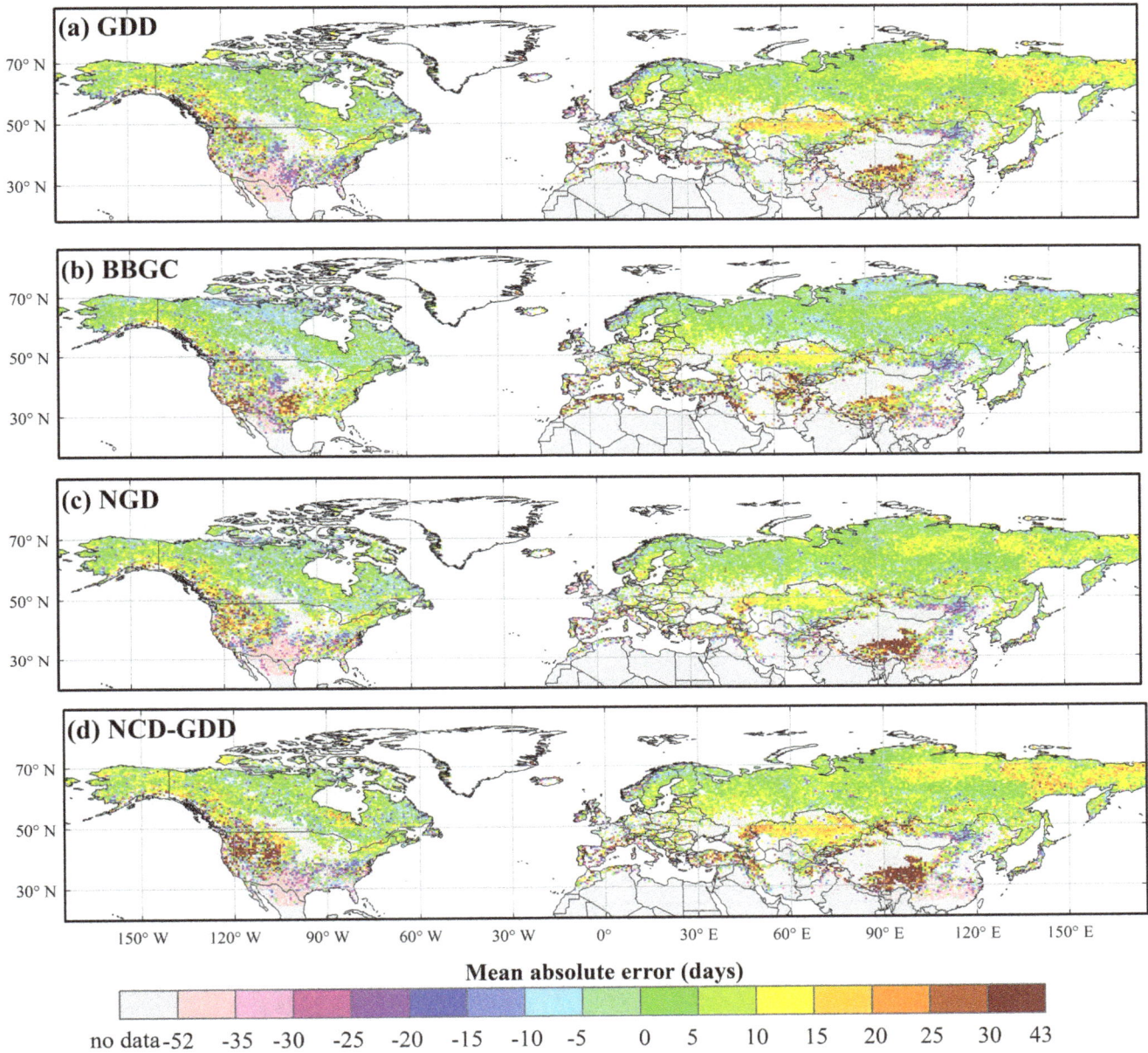

Figure 4. Spatial pattern of the mean absolute error (R$_A$) of the BGS simulations from four phenology models in the Northern Hemisphere. The mean absolute error (R$_A$) values are derived from the comparison of the MODIS vegetation product results with those of the parameterized models.

$$PrcpCritSum_{grass} = AvgAnnPrcp \times k \qquad (5)$$

where AvgAnnPrcp is the annual mean precipitation; c is an empirical coefficient; d is an underdetermined soil temperature threshold which determines warm grasslands or cool grasslands; k is a proportion of the average annual precipitation. The actual leaf onset day is 15 days prior to this calculated date to estimate the start of the growing season. Soil temperature is assumed to be the 11 day running average of daily average temperature [36]. Detailed information on the BBGC model is available at the website (http://www.ntsg.umt.edu/project/biome-bgc).

The NGD model, proposed by Botta et al. [31], determines the BGS when the NGD, defined as the number of days with temperature above a base temperature (T$_{th\_NGD}$), exceeds a critical number of growing days (NGD$_c$).

The NCD-GDD model is a two-phase model. Numerous experiments have confirmed that some plant species need to experience low temperatures to break physiological dormancy [37]. The NCD-GDD model defines the chilling days as the days with daily mean air temperature below a chill day base temperature threshold (T$_{th\_NCD}$). More chilling days can reduce the demand of plants for accumulated temperature [38]. The NCD-GDD model initiates bud burst if a certain relationship between the number of chilling days (NCD) since the leaves are

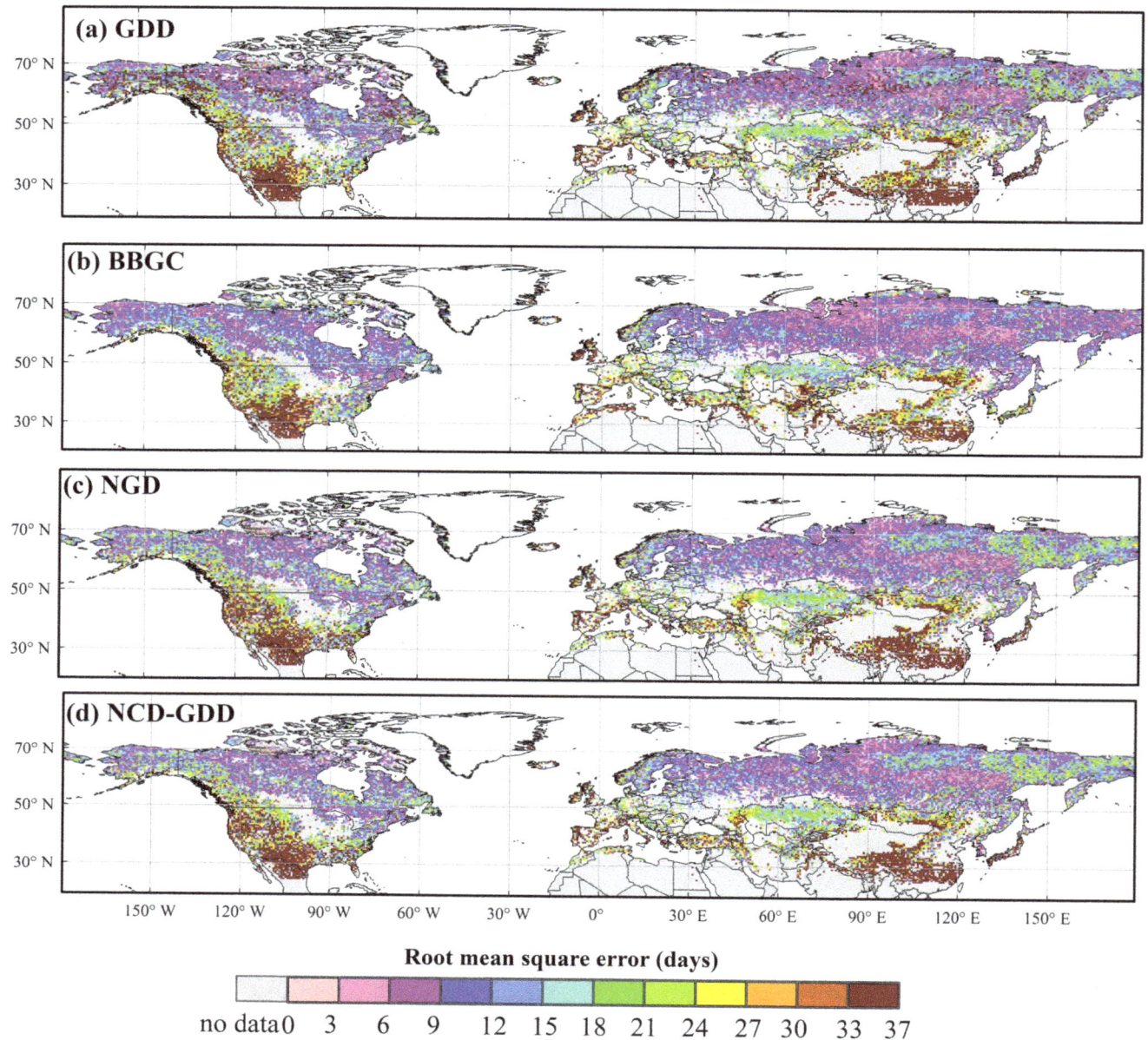

Figure 5. Spatial pattern of the root mean square error (RMSE) of the BGS simulations from four phenology models in the Northern Hemisphere. The RMSE values are derived from the comparison of the MODIS vegetation product results with those of the parameterized models.

lost, and the growing degree-days (GDD) since midwinter, is fulfilled, using the following empirical negative exponential law:

$$GDD(t) = \sum_{t=t_0}^{t_1} Max(T - T_{th\_NCD-GDD}, 0) \qquad (6)$$

$$GDD(t_1) \geq g + h \times e^{(w \times NCD_{Nov}(t))} \qquad (7)$$

where $T_{th\_NCD\text{-}GDD}$ is the degree-day base temperature; g, h and w are empirical coefficients. We used the method of Murray et al. [12], starting summation from fixed dates: November 1st for

the number of chilling days (NCD_{Nov}) to cover the major part of the dormant period, and January 1st for GDD [31].

3. Parameter Inversion

In each vegetation type, we randomly selected one half of the pixels to calibrate model parameters, and validated the models at the other half pixels. The nonlinear regression procedure (Proc NLIN) in the Statistical Analysis System (SAS, SAS Institute Inc., Cary, NC, USA) was applied to optimize the parameter values of the four phenology models. We used the Newton method to train the data and got the optimal model parameters when the error sum of squares was minimized. The other options were set as the default. The details of the calibrated parameter values of the four phenology models are found in Table 2.

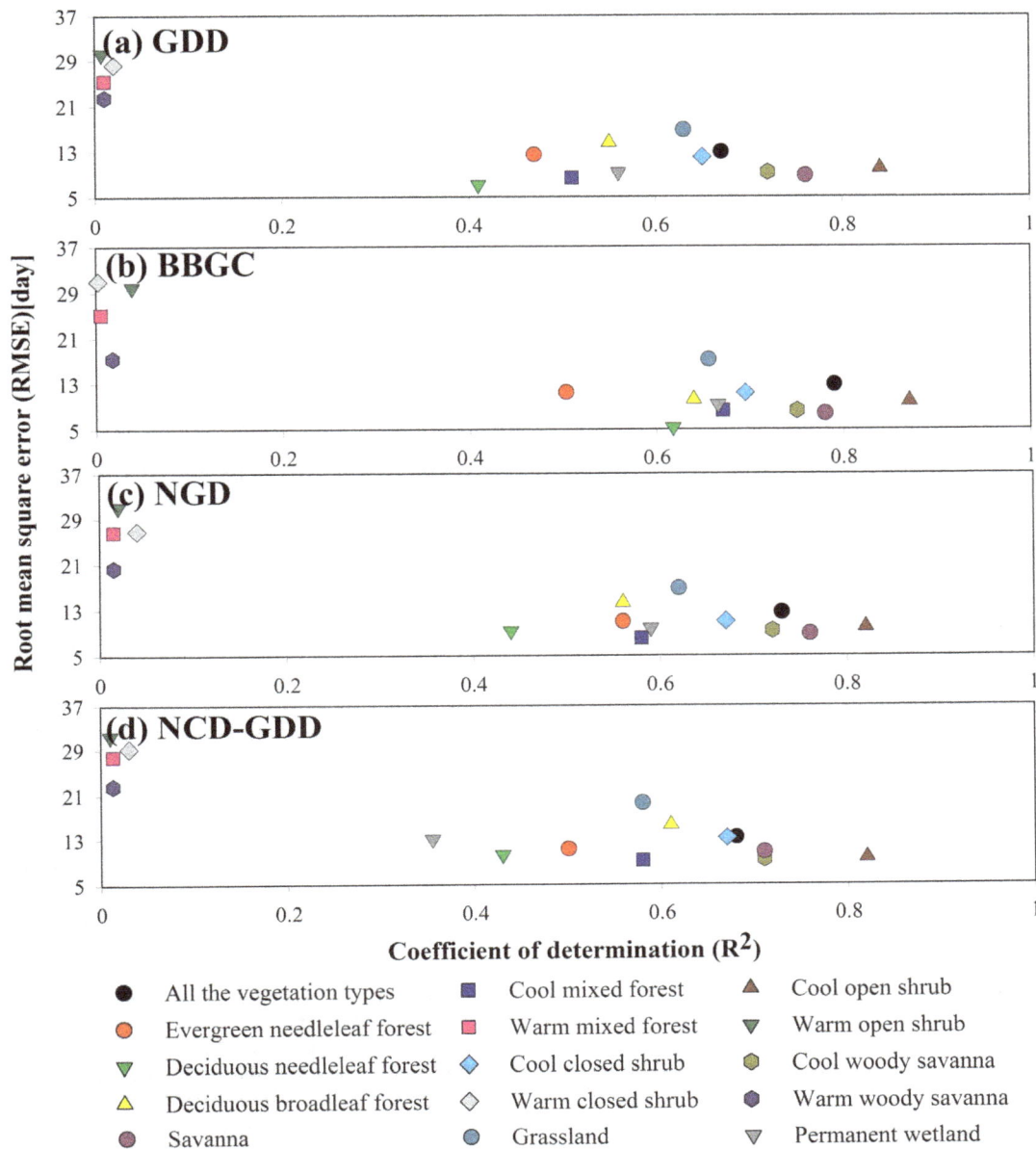

Figure 6. Coefficient of determination (R^2) and root mean square error (RMSE) for four phenology models at various vegetation types over the Northern Hemisphere. The R^2 and RMSE values are derived from the comparison of the MODIS vegetation product results with those of the parameterized models.

4. Model comparison

We made two comparisons in this study. First, we compared the original and calibrated models based on the start dates of phenology derived from the MODIS product for various of biomes. Second, we compared the start dates of phenology from calibrated models and the MODIS product over the northern hemisphere biomes. The performance of the parameterized and original phenology models is assessed by comparison with the results of the MODIS Land Cover Dynamics dataset. For the phenological data, all dates were transformed to days of the year (DOY) for convenience of data analysis.

Results

Model parameterization significantly improved performance of the four models. We calibrated and examined the four phenology models used in the global dynamic vegetation models based on satellite phenology observations over the Northern Hemisphere. The performance of the calibrated phenology models was better than that of the original models. For example, in the Integrated Biosphere Simulator (IBIS) model, the GDD phenology model is used to estimate the BGS for winter-deciduous forest, grassland and shrub [39]. For winter-deciduous forest, the original parameters values of T_{th_GDD} and GDD_C were set as $0°C$ and 100 degree-days, respectively. For grassland and shrub, the

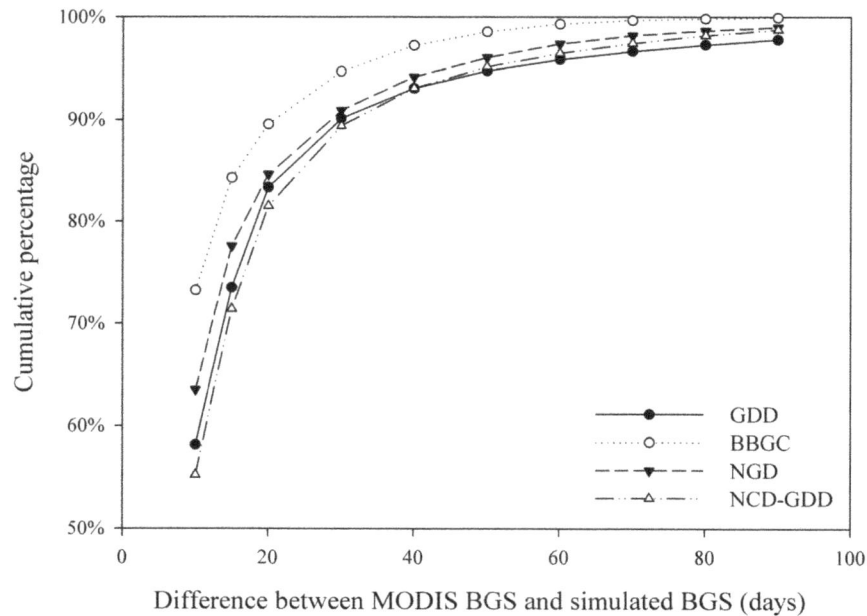

Figure 7. Cumulative percentage of absolute differences between MODIS BGS and simulated BGS from four models.

original parameters values of T_{th_GDD} and GDD_C were set as 5°C and 150 degree-days, respectively. The results of the calibrated simulation more accurately predicted the BGS, giving higher R^2 (Figure 2a and b). Similar results were also found in the respective phenology modules of the Biome-BGC and the Organising Carbon and Hydrology in Dynamic Ecosystems (ORCHIDEE) models (Figure 2). In the ORCHIDEE model, the NGD model and the NCD-GDD model are used in the deciduous needle leaf forest and deciduous broadleaf forest, respectively. Therefore, we compared the BGS simulations at these two vegetation types with the original parameter values and calibrated parameters in the ORCHIDEE model, respectively.

All of the four calibrated phenology models simulated similar spatial patterns of the BGS, which agreed very well with those of the MODIS BGS (Figure 3). A late BGS was found in the boreal and cool regions, intermediate BGS at temperate regions and early BGS in warm regions. In terms of the spatial patterns of the mean absolute error (R_A) and the root mean square error (RMSE), the four models showed good performance in most vegetation types (Figure 4 and Figure 5). The results showed low RMSE and R_A of the BBGC simulations in most boreal and cool regions. The average RMSE value in the whole study area was 16 ± 15 days (mean$\pm$1SD). The GDD, NGD and NCD-GDD models showed higher RMSE, with average values of 20 ± 19 days (mean$\pm$1SD), 19 ± 18 days (mean$\pm$1SD) and 22 ± 20 days (mean$\pm$1SD), respectively (Figure 5). In contrast, all of the four models showed poor performance at the four warm vegetation types (i.e. warm mixed forest, warm closed shrub, warm open shrub and warm woody savanna). The coefficient of determination (R^2) in warm mixed forest, warm closed shrub, warm open shrub and warm woody savanna regions were close to zero, and the average RMSE was in the range of 17–31 days (Figure 5 and Figure 6). In addition, all of the four phenology models predicted a later BGS in the grassland areas of the Qinghai–Tibet Plateau.

The results showed large differences in simulated BGS among the four phenology models (Figure 6). On average, they explained about 67% (GDD), 79% (BBGC), 73% (NGD) and 68% (NCD-

GDD) of the BGS variations over the Northern Hemisphere (Figure 6). According to the average R^2 and RMSE, the BBGC model showed the best performance with the highest R^2 (0.50–0.87) for the 9 vegetation types and lowest RMSE (5–11 days) (Figure 6b). The GDD and NGD models showed relatively similar performance in almost all vegetation types. In contrast, the NCD-GDD model showed a slightly lower R^2 (0.35–0.82) compared with the other models in most vegetation types, with RMSE ranging from 6 to16 days (Figure 6d). The cumulated frequencies of absolute difference between simulations and the MODIS BGS further demonstrated different simulation accuracy (Figure 7). On the whole, the best estimate was the BBGC model, which reproduced the timing of BGS for 73.2% of the pixels in the study areas within 10 days of the MODIS BGS, and for 84.3% within 15 days (Figure 7). Similarly, the NGD model reproduced the timing of BGS for 63.5% of the pixels within 10 days of the MODIS BGS, and for 77.6% within 15 days. The GDD and NCD-GDD models performed slightly worse and reproduced the timing of BGS for 58.2 and 55.3% within 10 days of the MODIS BGS, and for 73.5 and 71.4% within 15 days, respectively.

The magnitude and long-term change trends of the date of BGS differed significantly among the four phenology models (Figure 8). The Pearson's correlation coefficient (r) was used to quantify the performance of the four models in different vegetation types (Figure 8). The BBGC model had the highest r for almost all vegetation types, with the average value of 0.75. The GDD and NGD models showed relatively similar performance of r between simulations and the MODIS BGS, with the average values of 0.69 and 0.67, respectively. The NCD-GDD model had the lowest r, with an average value of 0.64.

Discussion

Vegetation phenology plays an important role in the functioning of the earth system as it steers the exchanges of carbon, water and energy between vegetation and the atmosphere [40,41]. The changes of phenology periods may significantly impact the ecosystem and climate system [42,43]. For example, an advanced

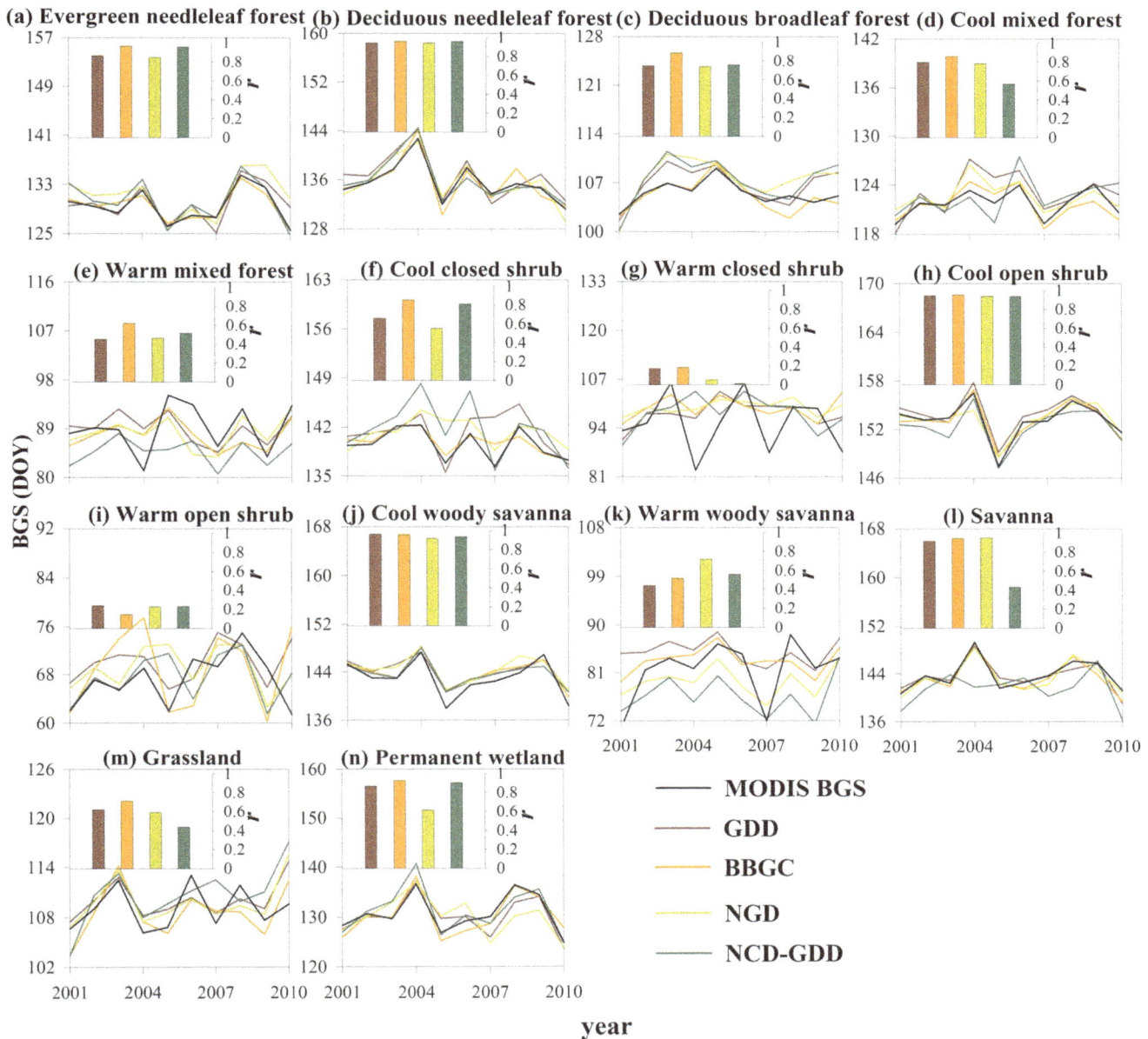

Figure 8. Interannual variability of the start dates of growing season from MODIS product and four phenology models from 2001 to 2010. The inset panels show the correlation coefficient (*r*) between BGS simulations of the four phenology models with MODIS BGS.

spring may enhance carbon sequestration and affect species interactions, and then alter the structure and function of ecosystems [44,45]. Therefore, the phenology module is one of the most important components of dynamic vegetation models and earth system models [46,47].

This study examined four phenology models, which have been widely integrated into various global dynamic vegetation models [39,48–50]. The major parameters of the original phenology models have not been carefully estimated or only calibrated over local scales [21,22,51]. For example, White et al. [21] used satellite data to calibrate a phenology model which was integrated into the Biome-BGC model, but this was only conducted at the North American not the global scale. Therefore, large biases in predicting the BGS exist among phenology modules, resulting in poor performance of these dynamic vegetation models [23]. This

study calibrated and examined the four phenology models used in the global dynamic vegetation models based on satellite phenology observations over the Northern Hemisphere. When the parameters were calibrated, the performance of the calibrated phenology models was better than that of the original models.

The four temperature-driven phenology models showed poor performance for vegetation in low latitude areas (i.e. warm mixed forest, warm closed shrub, warm open shrub and warm woody savanna). Experimental evidence indicates that plant growth may be largely controlled by precipitation and drought stress for these plant species [52]. For instance, Bernal et al. [53] studied the phenology of a Mediterranean shrub, *Erica multiflora*, and found that its growth was mainly driven by precipitation. Moreover, other studies also indicated that the plant phenology in low latitude areas was responsive to rainfall and water availability (e.g.

Peñuelas et al. [54]). However, many phenological models for the low-latitude plant species are found to be solely driven by temperature [21,39,50,55]. Thus, it is important to integrate water availability in plant phenology models when simulating the BGS of low latitudes.

Overall, the BBGC model showed better model performance than the other one-phase models (GDD and NGD). To account for this, we attributed two reasons. First, the BBGC model uses the mean annual temperature to determine the threshold of growing degree-days [21]. Under the local environmental conditions, vegetation phenology is the optimization of the plant activity and reproduction resulting from natural selection [56]. Plant species have adapted their temperature requirements to their local environment [10,57,58]. The BBGC model is essential in order to integrate the physiological adaptation of plants to the local temperature into the models and improve model performance at the global scale. Second, the BBGC model added the precipitation component to the start of the growing season calculation for grass biomes.

Moreover, the two-phase phenology model (NCD-GDD) did not perform better than the one-phase models in most vegetation types. Although it could simulate the interannual variations well for some vegetation types, it showed larger biases in the whole North Hemisphere. This result is consistent with other studies [11,59]. For example, Yuan et al. [11] analyzed the phenological characteristics of two dominant grass species for one-phase and two-phase models and also found better performance of the one-phase model. Leinonen and Kramer [33] also found that chilling was not important for good performance of models and proposed two explanations: first, with a boreal climate, winter temperatures are so low that the chilling requirement will always be fulfilled; second, the chilling requirement is observed to be lower for northern tree species and provenances compared to southern ones, i.e. relatively short exposure to low temperature is sufficient to break bud dormancy.

Our study was based on the MODIS Land Cover Dynamics (MCD12Q2) product, and the uncertainties from this product have a certain impact on the simulated results of phenology models. For example, a previous study compared the BGS derived from MODIS product with field measurements of forest canopy phenology at Harvard Forest for 2001–2006 and found differences of 1–17 days in each of the six years [30]. In addition, the MODIS BGS showed large uncertainties in the tropics [25]. Overall, ongoing efforts focusing on improving the precision of the phenology product are needed to improve phenology models.

Summary

In the present study, we calibrated four temperature-driven phenology models and compared their performances in the Northern Hemisphere. Although all of the four models indicated similar spatial patterns of the BGS, there were substantial differences among the models. The four models explained 67–79% of the variability in BGS. The BBGC model showed better performance than the other models. Conversely, the NCD-GDD model showed larger biases compared with the other three models in the whole North Hemisphere, although it could simulate the interannual variations well for some vegetation types. Moreover, all models showed good performance for most types in cool regions but poor performance in warm regions. Our study showed that it is necessary to integrate the effects of water availability into phenology models, especially for plants growing in low latitudes. Moreover, the thresholds used in phenology models to determine the BGS should be location dependent rather than a constant, as plants growing in different places show different physiological adaptabilities to environments (such as cold tolerance and drought tolerance).

Acknowledgements

This study was supported by the National Science Foundation for Excellent Young Scholars of China (41322005), National Natural Science Foundation of China (41201078), Program for New Century Excellent Talents in University (NCET-12-0060) and the Fundamental Research Funds for the Central Universities.

Author Contributions

Performed the experiments: YF. WD. Conceived and designed the experiments: YF WD WY. Analyzed the data: YF WY HZ. Contributed reagents/materials/analysis tools: WD WY. Wrote the paper: YF WY.

References

1. Schwartz MD (1998) Green-wave phenology. Nature 394: 839–840.
2. Menzel A, Fabian P (1999) Growing season extended in Europe. Nature 397: 659.
3. Zhu W, Tian H, Xu X, Pan Y, Chen G, et al. (2012) Extension of the growing season due to delayed autumn over mid and high latitudes in North America during 1982–2006. Global Ecology and Biogeography 21: 260–271.
4. Chen X, Tan Z, Schwartz MD, Xu C (2000) Determining the growing season of land vegetation on the basis of plant phenology and satellite data in Northern China. International Journal of Biometeorology 44: 97–101.
5. White MA, Running SW, Thornton PE (1999) The impact of growing-season length variability on carbon assimilation and evapotranspiration over 88 years in the eastern US deciduous forest. International Journal of Biometeorology 42: 139–145.
6. Cesaraccio C, Spano D, Snyder RL, Duce P (2004) Chilling and forcing model to predict bud-burst of crop and forest species. Agricultural and Forest Meteorology 126: 1–13.
7. Vitasse Y, François C, Delpierre N, Dufrêne E, Kremer A, et al. (2011) Assessing the effects of climate change on the phenology of European temperate trees. Agricultural and Forest Meteorology 151: 969–980.
8. Lang GA, Early JD, Martin GC, Darnell RL (1987) Endo-, para-, and ecodormancy: physiological terminology and classification for dormancy research. Horticultural Science 22(3): 371–377.
9. Cesaraccio C, Spano D, Snyder RL, Duce P (2004) Chilling and forcing model to predict bud-burst of crop and forest species. Agricultural and Forest Meteorology 126: 1–13.
10. Kramer K (1994) Selecting a Model to Predict the Onset of Growth of Fagus sylvatica. Journal of Applied Ecology 31: 172–181.

11. Yuan W, Zhou G, Wang Y, Han X, Wang Y (2007) Simulating phenological characteristics of two dominant grass species in a semi-arid steppe ecosystem. Ecological Research 22: 784–791.
12. Murray MB, Cannell MGR, Smith RI (1989) Date of Budburst of Fifteen Tree Species in Britain Following Climatic Warming. Journal of Applied Ecology 26: 693–700.
13. Falusi M, Calamassi R (1990) Bud dormancy in beech (Fagus sylvatica L.). Effect of chilling and photoperiod on dormancy release of beech seedlings. Tree Physiology 6: 429–438.
14. Heide OM (1993) Dormancy release in beech buds (Fagus sylvatica) requires both chilling and long days. Physiologia Plantarum 89: 187–191.
15. Faust M, Erez A, Rowland LJ, Wang SY, Norman HA (1997) Bud Dormancy in Perennial Fruit Trees: Physiological Basis for Dormancy Induction, Maintenance, and Release. HortScience 32: 623–629.
16. Chmielewski F-M, Müller A, Bruns E (2004) Climate changes and trends in phenology of fruit trees and field crops in Germany, 1961–2000. Agricultural and Forest Meteorology 121: 69–78.
17. Vezina PE, Grandtner MM (1965) Phenological observations of spring geophytes in Quebec. Ecology: 869–872.
18. Cleland EE, Chuine I, Menzel A, Mooney HA, Schwartz MD (2007) Shifting plant phenology in response to global change. Trends in ecology & evolution 22: 357–365.
19. Menzel A (2002) Phenology: its importance to the global change community. Climatic change 54: 379–385.
20. Morin X, Lechowicz MJ, Augspurger C, O'Keefe J, Viner D, et al. (2009) Leaf phenology in 22 North American tree species during the 21st century. Global Change Biology 15: 961–975.

21. White MA, Thornton PE, Running SW (1997) A continental phenology model for monitoring vegetation responses to interannual climatic variability. Global Biogeochemical Cycles 11: 217–234.

22. Yang X, Mustard JF, Tang J, Xu H (2012) Regional-scale phenology modeling based on meteorological records and remote sensing observations. Journal of Geophysical Research: Biogeosciences 117: G03029.

23. Richardson AD, Anderson RS, Arain MA, Barr AG, Bohrer G, et al. (2012) Terrestrial biosphere models need better representation of vegetation phenology: results from the North American Carbon Program Site Synthesis. Global Change Biology 18: 566–584.

24. Brown ME, de Beurs KM, Marshall M (2012) Global phenological response to climate change in crop areas using satellite remote sensing of vegetation, humidity and temperature over 26years. Remote Sensing of Environment 126: 174–183.

25. Hmimina G, Dufrêne E, Pontailler JY, Delpierre N, Aubinet M, et al. (2013) Evaluation of the potential of MODIS satellite data to predict vegetation phenology in different biomes: An investigation using ground-based NDVI measurements. Remote Sensing of Environment 132: 145–158.

26. Karlsen SR, Tolvanen A, Kubin E, Poikolainen J, Høgda KA, et al. (2008) MODIS-NDVI-based mapping of the length of the growing season in northern Fennoscandia. International Journal of Applied Earth Observation and Geoinformation 10: 253–266.

27. Kross A, Fernandes R, Seaquist J, Beaubien E (2011) The effect of the temporal resolution of NDVI data on season onset dates and trends across Canadian broadleaf forests. Remote Sensing of Environment 115: 1564–1575.

28. White MA, De Beurs KM, Didan K, Inouye DW, Richardson AD, et al. (2009) Intercomparison, interpretation, and assessment of spring phenology in North America estimated from remote sensing for 1982–2006. Global Change Biology 15: 2335–2359.

29. Zhang X, Friedl MA, Schaaf CB, Strahler AH, Hodges JCF, et al. (2003) Monitoring vegetation phenology using MODIS. Remote Sensing of Environment 84: 471–475.

30. Ganguly S, Friedl MA, Tan B, Zhang X, Verma M (2010) Land surface phenology from MODIS: Characterization of the Collection 5 global land cover dynamics product. Remote Sensing of Environment 114: 1805–1816.

31. Botta A, Viovy N, Ciais P, Friedlingstein P, Monfray P (2000) A global prognostic scheme of leaf onset using satellite data. Global Change Biology 6: 709–725.

32. Rienecker MM, Suarez MJ, Gelaro R, Todling R, Bacmeister J, et al. (2011) MERRA: NASA's Modern-Era Retrospective Analysis for Research and Applications. Journal of Climate 24: 3624–3648.

33. Leinonen I, Kramer K (2002) Applications of Phenological Models to Predict the Future Carbon Sequestration Potential of Boreal Forests. Climatic Change 55: 99–113.

34. Kramer K, Leinonen I, Loustau D (2000) The importance of phenology for the evaluation of impact of climate change on growth of boreal, temperate and Mediterranean forests ecosystems: an overview. International Journal of Biometeorology 44: 67–75.

35. Biome-BGC (2010) Biome BGC version 4.2: Thoeretical framework of BIOME-BGC. Available: http://www.ntsg.umt.edu/project/biome-bgc. Accessed 2014 Aug 4.

36. Zheng D, Hunt Jr ER, Running SW (1993) A daily soil temperature model based on air temperature and precipitation for continental applications. Climate Research 2: 183–191.

37. Orlandi F, Garcia-Mozo H, Ezquerra LV, Romano B, Dominguez E, et al. (2004) Phenological olive chilling requirements in Umbria (Italy) and Andalusia (Spain). Plant Biosyst 138: 111–116.

38. Cannell MGR, Smith RI (1983) Thermal Time, Chill Days and Prediction of Budburst in Picea sitchensis. Journal of Applied Ecology 20: 951–963.

39. Foley JA, Prentice IC, Ramankutty N, Levis S, Pollard D, et al. (1996) An integrated biosphere model of land surface processes, terrestrial carbon balance, and vegetation dynamics. Global Biogeochemical Cycles 10: 603–628.

40. Zhang H, Yuan W, Dong W, Liu S (2014) Seasonal patterns of litterfall in forest ecosystem worldwide. Ecological Complexity.

41. Yuan W, Liang S, Liu S, Weng E, Luo Y, et al. (2012) Improving model parameter estimation using coupling relationships between vegetation production and ecosystem respiration. Ecological Modelling 240: 29–40.

42. Schwartz MD, Reiter BE (2000) Changes in North American spring. International Journal of Climatology 20: 929–932.

43. Richardson AD, Keenan TF, Migliavacca M, Ryu Y, Sonnentag O, et al. (2013) Climate change, phenology, and phenological control of vegetation feedbacks to the climate system. Agricultural and Forest Meteorology 169: 156–173.

44. Walther G-R, Post E, Convey P, Menzel A, Parmesan C, et al. (2002) Ecological responses to recent climate change. Nature 416: 389–395.

45. Yuan W, Liu S, Liang S, Tan Z, Liu H, et al. (2012) Estimations of Evapotranspiration and Water Balance with Uncertainty over the Yukon River Basin. Water Resources Management 26: 2147–2157.

46. Zhao M, Peng C, Xiang W, Deng X, Tian D, et al. (2013) Plant phenological modeling and its application in global climate change research: overview and future challenges. Environmental Reviews 21: 1–14.

47. Cai W, Yuan W, Liang S, Zhang X, Dong W, et al. (2014) Improved estimations of gross primary production using satellite-derived photosynthetically active radiation. Journal of Geophysical Research: Biogeosciences 119: 2013JG002456.

48. Kucharik CJ, Foley JA, Delire C, Fisher VA, Coe MT, et al. (2000) Testing the performance of a dynamic global ecosystem model: Water balance, carbon balance, and vegetation structure. Global Biogeochemical Cycles 14: 795–825.

49. Thornton PE, Law BE, Gholz HL, Clark KL, Falge E, et al. (2002) Modeling and measuring the effects of disturbance history and climate on carbon and water budgets in evergreen needleleaf forests. Agricultural and Forest Meteorology 113: 185–222.

50. Sitch S, Smith B, Prentice IC, Arneth A, Bondeau A, et al. (2003) Evaluation of ecosystem dynamics, plant geography and terrestrial carbon cycling in the LPJ dynamic global vegetation model. Global Change Biology 9: 161–185.

51. Chuine I (2000) A unified model for budburst of trees. J Theor Biol 207: 337–347.

52. Llorens L, Peñuelas J, Estiarte M, Bruna P (2004) Contrasting Growth Changes in Two Dominant Species of a Mediterranean Shrubland Submitted to Experimental Drought and Warming. Annals of Botany 94: 843–853.

53. Bernal M, Estiarte M, Peñuelas J (2011) Drought advances spring growth phenology of the Mediterranean shrub Erica multiflora. Plant Biology 13: 252–257.

54. Peñuelas J, Filella I, Zhang X, Llorens L, Ogaya R, et al. (2004) Complex spatiotemporal phenological shifts as a response to rainfall changes. New Phytologist 161: 837–846.

55. Verseghy DL, McFarlane NA, Lazare M (1993) Class—A Canadian land surface scheme for GCMS, II. Vegetation model and coupled runs. International Journal of Climatology 13: 347–370.

56. Chuine I (2010) Why does phenology drive species distribution? Philosophical Transactions of the Royal Society B: Biological Sciences 365: 3149–3160.

57. Chuine I, Cour P (1999) Climatic determinants of budburst seasonality in four temperate-zone tree species. New Phytologist 143: 339–349.

58. Chuine I, Beaubien EG (2001) Phenology is a major determinant of tree species range. Ecology Letters 4: 500–510.

59. Fu YH, Campioli M, Deckmyn G, Janssens IA (2012) The Impact of Winter and Spring Temperatures on Temperate Tree Budburst Dates: Results from an Experimental Climate Manipulation. PLoS ONE 7: e47324.

14

How Ecosystem Services Knowledge and Values Influence Farmers' Decision-Making

Pénélope Lamarque[1]*, Patrick Meyfroidt[2,3], Baptiste Nettier[4], Sandra Lavorel[1]

1 Laboratoire d'Ecologie Alpine, Unité Mixte de recherche 5553, Centre National de la Recherche Scientifique, Université Joseph Fourier, Grenoble, France, 2 Fonds de la recherche scientifique (F.R.S.-FNRS), Louvain-La-Neuve, Belgium, 3 Earth and Life Institute, Georges Lemaître Centre for Earth and Climate Research, University of Louvain, Louvain-la-Neuve, Belgium, 4 Irstea centre de Grenoble, unité de recherche Développement des territoires montagnards, Grenoble, France

Abstract

The ecosystem services (ES) concept has emerged and spread widely recently, to enhance the importance of preserving ecosystems through global change in order to maintain their benefits for human well-being. Numerous studies consider various dimensions of the interactions between ecosystems and land use via ES, but integrated research addressing the complete feedback loop between biodiversity, ES and land use has remained mostly theoretical. Few studies consider feedbacks from ecosystems to land use systems through ES, exploring how ES are taken into account in land management decisions. To fill this gap, we carried out a role-playing game to explore how ES cognition mediates feedbacks from environmental change on farmers' behaviors in a mountain grassland system. On a close to real landscape game board, farmers were faced with changes in ES under climatic and socio-economic scenarios and prompted to plan for the future and to take land management decisions as they deemed necessary. The outcomes of role-playing game were complemented with additional agronomic and ecological data from interviews and fieldwork. The effects of changes in ES on decision were mainly direct, i.e. not affecting knowledge and values, when they constituted situations with which farmers were accustomed. For example, a reduction of forage quantity following droughts led farmers to shift from mowing to grazing. Sometimes, ES cognitions were affected by ES changes or by external factors, leading to an indirect feedback. This happened when fertilization was stopped after farmers learned that it was inefficient in a drought context. Farmers' behaviors did not always reflect their attitudes towards ES because other factors including topographic constraints, social value of farming or farmer individual and household characteristics also influenced land-management decisions. Those results demonstrated the interest to take into account the complete feedback loop between ES and land management decisions to favor more sustainable ES management.

Editor: Kurt O. Reinhart, USDA-ARS, United States of America

Funding: This work was funded by the Agence National de la Recherche through the FP6 BiodivERsA Eranet VITAL project and by the French Ministry responsible for environment through the GICC-2 SECALP project. The funders had no role in study design, data collection and analysis, decision to publish, or preparation of the manuscript.

Competing Interests: The authors have declared that no competing interests exist.

* Email: penelope.lamarque@gmail.com

Introduction

Assessing the consequences of ecosystem change on ecosystem services (ES), defined as the outputs of ecosystems [1] from which people derive benefits, is of primary importance. In agro-ecosystems, flows of ES are directly affected by farmers' behaviors and land management decisions [2]. ES stress the need to integrate ecological and social science to study coupled human and natural systems [3], and therefore require to explicitly address the complex feedback loops formed by reciprocal interactions between people and nature [4]. These feedbacks depend on how changes in management affect ES and how, in turn, these changes in ES are perceived by land managers [5]. Nevertheless, while numerous studies consider various dimensions of the interactions between ecosystems and land use via ES, integrated research addressing the complete feedback loop between biodiversity, ES and land use has remained mostly theoretical. Most published frameworks (e.g., [6,7]) investigate the interactions between ecosystems, ES and human well-being by considering values generated for people, and close the loop by exploring changes and future trends in ES according to scenarios, with possible institutional responses. The full cascade of ES from ecosystem processes to benefits [1] is sometimes considered (e.g., [6,8]) but the feedbacks effects from ES to human actions and the consequences on ecosystem processes are rarely taken into account [5]. The main research themes in which ES are related to decision-making concern: (i) studies on payments for ES, i.e. financial incentives to sustain management of resources which maintain or enhance ES delivery [9,10]; (ii) economic valuation is used to raise decision-makers' awareness of the importance of ES through the costs associated with their loss [11,12]; and (iii) ES mapping as a decision tool for landscape planning [13]. Other studies explored how ES could fit into formal institutional arrangements [14]. However, how people perceive ecosystems and their ability to provide values affects choices about how to manage the environment [6]. Psychology, decision sciences and behavioral economics show that individuals are not necessarily utility maximizers or financially rational [15], and individual preferences are evolving [16]. Economic valuation methods do not adequately address these complexities linked to attitudes and motivational systems, and their effects on behaviors

(Kumar and Kumar, 2008). Recent reviews [5,17] underline the interest of using mental models to explore mechanisms by which individual decisions are made and thereby enhance sustainable management of land and natural resources. A wide range of theories and models based on psycho-social constructs such as attitudes, beliefs and values can help to understand how environmental change can influence decision-making [5,18]. Two of the most popular theories are the Theory of Planned Behavior (TPB) model [19] and the Value-Belief-Norm Theory (VBN) [20]. Both are based on the premise that individuals' behavior towards the environment is influenced by what they feel and think with respect to the environment. The TPB is based on self-interest and rational choice deliberation, while the VBN focuses primarily on the role of values and moral norms. The main limitation of these theories is that they do not explain the formation of the cognitions (beliefs, values, preferences, attitudes) that are used in a complete feedback loop of decision-making process, and which are crucial to understand adaptation to non-linear and rapid environmental change [5,18]. Studies that have explored stakeholders' perceptions [21–24] or preferences and values [25,26] in terms of ES have shown the diversity of stakeholders' knowledge and/or values attributed to different ES. Other studies have investigated farmers' decision-making process [27,28], sometimes taking into account interactions between environmental perceptions and actions [29–31], but few of these use the ES framework [32].

To fill this gap, this paper studies how ES are taken into account in land-use decisions in the context of mountain grasslands management. The study area in the Central French Alps is typical of extensive grassland management systems found in drier European mountains, and is mainly composed of permanent grasslands used for livestock farming. We focused on behavior of farmers since they are the key decisional actors in this system as they are the ones determining land management for most of the area. Three main types of land management change affecting ES were previously identified: (1) manuring *versus* not, (2) mowing *versus* grazing, (3) early *versus* late mowing [33–35]. We tested the hypothesis that these three land management behaviors are driven by farmers' motivation to benefit from ES. Previous studies on farmers' behavior have stressed the need to consider multiple potential explanatory factors (e.g. biophysical, economic, political, sociological) and the relationships among them in order to address the complexity of the social-ecological system [28]. This led us to analysing the influence of multiple ES as well as a broader context of climate and socio-economic change. We built on the theoretical frameworks of Meyfroidt [5] and Vignola et al. [36] to explore the feedbacks between ES and behaviors through farmers' cognitions. First, we describe the cognitive model underpinning our analysis. We then present the methodology used to document how ES are taken into account in farmers' decisions and describe results for each component of the cognitive model. Finally, the discussion explores the feedback loop between ES and land-use through farmers' cognitive processes.

Conceptual Framework

Land-management decisions are determined by cognitive factors regarding ES, and other contextual factors (Figure 1). Contextual factors then influence whether decisions are indeed carried on through land management behaviors.

Thus, $(B = f(D, C))$ and $(D = f(K, V, C))$, where:

- Behavior (B) refers to a series of actions (here the land-use/agricultural practices) selected among possible alternatives [28].

Behaviors follow decisions (D) except when contextual factors (C) force the agent to deviate from the preferred alternative;

- Decisions (D) refer to the preferred action selected among alternatives, taking into account the knowledge (K) and values (V) about ES, as well as the influence of contextual factors (C).

- Knowledge (K) focuses here specifically on farmers' knowledge about contributions of ecosystem functioning to ES, and on effects of their practices on these ecosystem functioning;

- Values (V) correspond to general assessments about things (here, ES) that are seen as desirable [37];

- Contextual factors (C) are factors external to farmers' cognition that can influence decisions by affecting the valence attributed to different options, or make behaviors easier or more difficult to carry out, e.g. climatic conditions, social or political context [28]. Contextual factors are also referred to in other frameworks as drivers, driving forces [38] or pressures [39]. Our focus here is on ES, which are thus presented in a separate box from contextual factors. Yet, in other studies with other objectives, ES might be considered as contextual factors themselves. We considered as ES, "the aspects of ecosystems utilized (actively or passively) to produce human well-being". [40]. Two types of ES are distinguished here; those that can be turned directly into benefits (called 'final ecosystem services') and those that support other services (called 'intermediate ecosystem service'). Before being used, consumed or enjoyed by human beneficiaries, ES should only be considered as potential ES" [1].

In the following we examine evidence for the different components of the framework and assemble them in order to understand the mechanisms of mountain farmers' decisions.

Social-Ecological System and Methods

a. Study area

The study site (45°03′ N, 6°24′ E, 13 km²) is part of the Ecrins National Park in the Central French Alps, and located on the south-facing slopes of Villar d'Arène (Figure S1a). The climate is subalpine with a strong continental influence. Mean annual rainfall is 956 mm and mean monthly temperatures range between −4.6°C in January and 11°C in July (at 2050 m a.s.l.). However, the last decade has seen several drought episodes that may be considered as warnings of future climate change. Most of the upper slopes of Villar d'Arène (above 2200 m, further called "Alpine meadows") have been extensively grazed continuously for centuries. Since the 20th century, the lower slopes (1650–2000 m), that were formerly terraced, ploughed and used for cropping (henceforth "terraces"), are cut for hay during summer or grazed during spring and autumn, and some are manured [41]. Intermediate unterraced grasslands (1800–2500 m) (henceforth "unterraced grasslands") have been managed for hay production since the 1700s, but since the 1970's mowing has gradually ceased over 75% of the area, which is now lightly grazed in early summer (Figure S1a). Management practices are extensive, with low stocking rates and manure inputs (every two or three years), and a single annual hay cut. Trajectories of land-use changes have shaped the landscape into a mosaic of land management types resulting in distinct patterns of fertility, floristic and functional composition, and associated ecosystem functioning [42,43].

A key element of farmers' strategy is fodder self-sufficiency. Farmers typically cannot afford to purchase the feed (i.e. hay) necessary to maintain livestock during the long winter period (6–7 months). Thus, farmers are strongly motivated to avoid purchasing feed and instead harvest and stockpile their own hay. This strategy has been challenged by recent droughts and a vole outbreak in 2009–10 which decrease fodder yield and quality. The eight

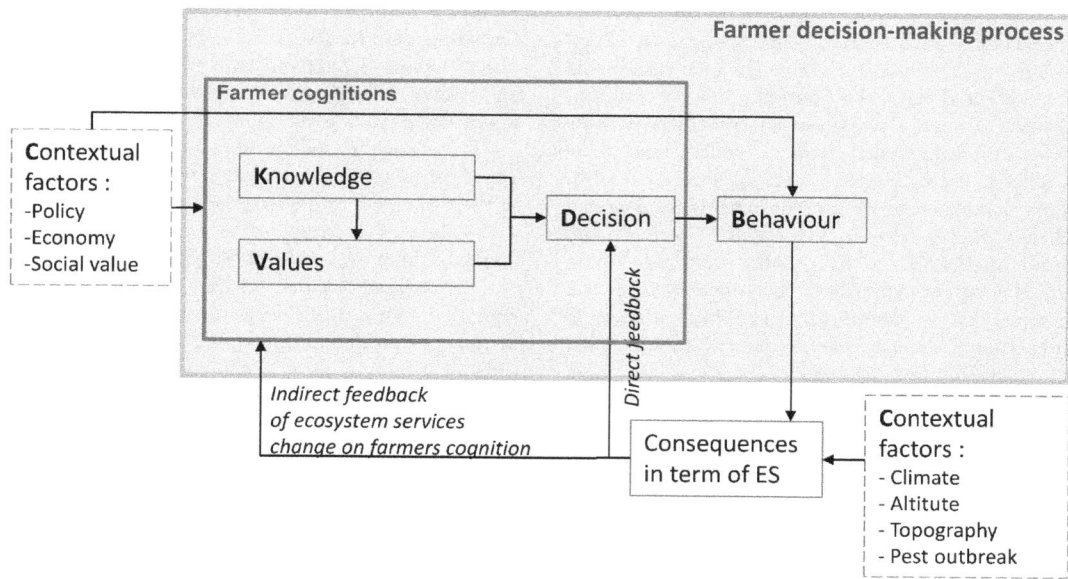

Figure 1. Socio-cognitive conceptual model of ecosystem services feedbacks on farmer behavior. Feedback from changes in ES supply to farmers' cognitions and behaviors can be either direct, affecting only the perceived parameters of decision, or indirect, affecting the different cognitive components underlying the behavior [5].

farmers managing the study area can be classified in three categories according to their production: (1) 3 sheep farmers producing lambs (mean = 21 livestock units (LU), 19 ha); (2) 3 cattle farmers breeding calves and heifers for dairy farms situated in neighbouring areas (mean = 67 LU, 55 ha), (3) 2 farmers raising both sheep and cattle (mean = 54 LU, 48 ha). During summer, most of the alpine meadows are grazed by a shepherd who manages local farmers' sheep along with his/her own flock (around 1400 sheep in total). The remaining alpine meadows are divided into paddocks for cattle grazing.

These farms are part of a "Less Favored Area" due to the combination of a short growing season (April-September) because of high altitude, and steep slopes (from 0 to more than 50°). Compensations for low productivity from European subsidies and agri-environmental measures to conserve mowing practices and related biodiversity constitute a significant share of farmers' income. Grasslands are collectively managed through an association called "AFP" (Association foncière pastorale) created in 1975 in which agricultural parcels of all landowners are pooled and allocated among farmers, in order to lower constraints (ex. production costs, accessibility to parcel) and increase the average size of parcels and secure land access on long-term.

In addition to agriculture, tourism is a dominant economic activity in this region recognized for its aesthetic, cultural and conservation value and recreation opportunities [44].

b. Ethics Statement

This analysis is partially based on survey results. The interviewees were voluntary, and their answers are confidential and analyzed anonymously. Farmers surveyed are familiar with the Central French Alps Long-Term Socio-Ecological Research (LTSER) site, with whom they have been participating in research since 2003. When asking them to participate we explained that the survey contents would remain confidential and anonymous, and would not be used beyond our study. We also committed to communicate to them the results of the study, which was done in March 2013. Farmers consented verbally to these conditions,

given that written consents are not enforced in the Central French Alps LTSER, nor there is an institutional review board for this study area. No specific permissions were required for research studies in this study site (45°03'N, 6°24'E). Our activities did not involve endangered or protected species.

c. Data collection

Qualitative and quantitative data to understand the different components of the farmers' decision-making process (Figure 1 and Table 1) were collected using a role-playing game (called hereafter the "feedback game") which took place with seven (out of the 8) farmers of the site in January 2012. The "feedback game" aimed at understanding how ES and other factors are taken into account in land management decisions in different contexts regarding levels of ecosystem service provision, socio-economic and drought conditions. The role-playing game methodology was used (i) to put farmers in an experimental situation of decisions with the help of different supports; (ii) to distinguish between what people say ('espoused theory') and what they do ('theory in use') [17] and (iii) to present to farmers how their adaptive management responding to climate and socio-political change affected ES delivery.

The "feedback game" is a role-playing game made of a two-dimensional board game composed of cells representing a simplified landscape where farmers playing their own role were asked to place pieces representing their land management (quantity of cattle, fodder harvested and manure) according to rules translating the effects of scenarios on ES for each type of grasslands (for more information on the material see [35]). The board game retained the actual grassland types managed by each farmer, with the same proportions of grassland types, which makes it possible to compare game results with the actual land use map.

The "feedback game" was built on the outcomes of a first role-playing game (called hereafter the "scenario game") with the same farmers (April 2011) which identified and mapped farmers' land-management adaptations to integrated climate and socio-economic change scenarios [35]. Farmers' management adaptations to alternative contexts in the "scenario game" were used to design

Table 1. Data collection and analysis of the different components of the conceptual model of the farmer decision-making process (Figure 1).

Objectives	Decision-making process components	Data collection	Data types	Data Analysis	Results
Role-playing game inputs	Contextual factors	Three socio-economic and climatic scenarios and initial boards corresponding to the three sessions of the "feedback game"	Qualitative and quantitative	"Scenarios game	
	Consequence on ES	"Feedback game" rules (number of pieces allowed in each land use type)	Qualitative and quantitative	"Scenarios game"	Table 1
Modelling farmers cognitions	Knowledge	"Feedback game" discussions	Qualitative	Retranscription	Figure 2; Table 6
	Values	"Feedback game" questionnaire	Quantitative	Likert-scale	Table 4; Table 6
	Decision	"Feedback game" discussions	Qualitative	Retranscription	Table 8; Table 6
	Behavior	"Feedback game" board game	Qualitative and quantitative	Digitalisation	Table 7
	Links between components	"Feedback game"	Qualitative and quantitative	Process tracing approach	Table 6
Validation of game results	Knowledge	Farmers individual + group interviews	Qualitative	Qualitative analysis of description of current practices	
	Values	Farmers individual interviews + Field function mapping	Quantitative	Anova and Chi-squared	Table 5
	Decision	Farmers individual interviews + scenarios game	Quantitative Quantitative	Regression	Table 5
	Behavior	Farmer participatory photo mapping	Quantitative	Regression	Table 5
	Contextual factors	Ancillary data	Quantitative	Regression	Table 5

the initial land management board of the "feedback game" and the same equipment (board and pieces) was used.

The "feedback game" was played in three sessions lasting each between 1 h and 1 h30 during the same day. Each session was composed of one round corresponding to a year where farmers were projected in a 2030-like situation according to one scenario (Table 2), in order for them to consider the effects of the adaptations they had made in the "scenario game". These scenarios and the initial board game were used to identify the main contextual factors considered in this study (climate, socio-economic and political context from the scenarios; altitude, topography, distance to farm from the board game). The scenarios also provided a quantification of the consequences on ES delivery of land management in the "scenario game" (Table 3).

The three scenarios used to vary the levels of ES and other contextual factors, and analyzed their effects on farmers' behaviors are: (1) the "drastic and local" scenario were repeated droughts occurring during four consecutive years with a return period of four years combined with a socio-economic context assuming demand for local products and area-related agricultural subsidies; (2) the "intermittent and international" scenario alternating favorable climatic years and droughts combined with a globalized socio-economic context; (3) the "drastic anticipation" scenario with repeated droughts as in "drastic and local" scenario, but with the current socio-economic context (Table 2). At the beginning of each session, information on ES change by land-management type (percentage increase or decrease between the current situation and 2030) was given for a set of ESs previously shown to be important for these farmers and regional experts [23]: forage quality, forage quantity, date of flowering onset, litter quantity, plant diversity,

aesthetics, water quality, nitrate leaching and carbon storage (Table 3). ES changes were calculated using spatially-explicit models predicting the response of ecosystem functioning to drought and management based on plant and microbial traits, and abiotic parameters [35].

Before starting the game, each farmer ranked the value of each ES or service on a five levels Likert-scale (Table 4). This was followed by a group discussion on the attribution of values and a discussion of each service. Then, three game sessions were conducted corresponding each to one scenario. To document individual decisions in addition to collective discussions during the game, at the end of each session farmers were asked to write the reasons for adopting a given practice for each cell of the board game. The pieces placed by farmers on the board game illustrated the behaviors that they adopted. The game finished with a general debriefing where decisions and behaviors were discussed. Knowledge about ecosystem services was extracted from the discussion during the presentation of ecosystem services change before the game, through the discussions during the game and the debriefing.

Due to the complexity of the socio-ecological system, plausibility of the "feedback game" results were cross-checked and completed with information from multiple sources on actual land management behavior (Table 5). Firstly, farmers' actual land management behaviors were documented from semi-directed individual interviews conducted with the eight farmers in summer 2009 about farm structure and features of the herd, forage resource and management practices [45]. These interviews included a participatory photomapping where interviewees outlined their parcels over aerial photographs, and described the management (i.e. mowing, grazing, manuring, dates and stocking rate) as well as

Table 2. Drivers and related assumptions describing the four scenarios combining climatic and socio-economic alternatives (adapted from [61]).

Drivers	Climate alternatives	
	"Drastic"	**"Intermittent"**
Season of drought and occurrence	Spring drought during four consecutive years	Spring or summer drought every two years
Effects on vegetation	Change in species composition. Development of species adapted to drought (eg. *Festuca paniculata, Carex sempervirens*)	No change
Effects on biomass production	Decrease by more than 50%	Decrease by 15% during drought years
Effects on water quantity (springs)	Decreased flow of all springs, even quenching of the less productive ones	Decreased flow of the springs
	Socio-economic alternatives	
	"Local"	**"International"**
Consumption demand	Local and high quality products	Cheapest prices
Aim of agricultural subsidies	To maintain both an agriculture with quality production and a high level of ecosystem services and biodiversity conservation. High subsidies but more restrictive in term of expected outcomes than in the "International" alternative.	To maintain open landscapes and production of environmental services such as carbon sequestration. Lower subsidies than on the local alternative, but less restrictive. A minimal income is guaranteed to farmers
Subsidies	20% of CAP pillar 1 support: no minimum guaranteed; Agri-environmental measures (AEM): Bonus for biodiversity with commitment to results (e.g. maintain plant diversity): 210€/ha (maximum 10000€/farm) c).; Strengthening of eco-conditionality requirements for funding (e.g. manure control)	20% of CAP pillar 1 support: subsidies generally decoupled but minimum guaranteed (1 yearly minimum wage); Agri-environmental measures (AEM): Bonus for maintaining grasslands; Carbon credits: 76€/ha (maximum 76000€/farm)

each parcel's "field function", i.e. the main parcel value assigned by farmers [46]. Field functions can be interpreted as levels of ES provision and were coded according to the expectations on output: (i) both quantity and quality of fodder being expected, (ii) only quantity expected, (iii) only quality expected. We used these data to compare perceptions and actual behaviors of farmers to game results. Secondly, data about knowledge, values and decision were obtained from other surveys conducted from 2009 to 2011: (i) semi-directed individual interviews about knowledge and adaptations to past droughts [34], (ii) the "scenario game" on adaptations to future climatic and socio-economic change under four scenarios [35], and (iii) a group interview conducted in January 2010 with 3 farmers and other inhabitants to elicit their perceptions of biodiversity and ES related to management of mountain grasslands [23].

Finally, ancillary spatial data was used to study the effects of contextual factors (i.e. altitude, topography, distances), on land management behaviors: a land-use map of the site constructed using a combination of cadastral (1810 to 2009) and aerial photographic data (since 1952) [41], a 10 m×10 m Digital Elevation Model, and settlements, farms and road digitized from the 1:25000 topographic map (IGN).

d. Data analysis

Interviews and game discussions were recorded, typed and coded by themes using Nvivo 9 (QSR International) to extract the different components of farmers' cognitions (Table 6; Figure 1: values, knowledge, decisions) for each ES (Table 4) and draw the mental map (Figure 2). The "feedback game" questionnaires about values were analyzed using the Likert-scale data (Table 4). Land management type and their location on the boards of the "feedback game" sessions were recorded (photography and GIS) and analysed for each farmer to collect data on behavior (Table 7).

Maps resulting from participatory photomapping were digitized and georeferenced (with ARCGIS, ESRI), to overlay with the other maps. In order to test whether relations between behaviors and other elements discussed during the game process were consistent with those in the real life, we performed statistical analyses (ANOVA and regressions) of the relationships between actual land management behaviors (mowing, grazing or manuring), as dependent variables, and ES (expected values in term of quality and quantity identified by farmers (field functions)) or potential contextual factors (listed in Table 5) as explanatory variables (Table 5) using R (R Development Core Team 2008). These statistical analyses provide one additional element to test the main hypothesis, i.e. the effect of ES knowledge and values on farmers' decisions.

The entire feedback loop from change in ES supply to farmers' behaviors was then analyzed by combining all this data and using the « process-tracing » approach [47] to explore individually each component of the conceptual model before considering links between them (Figure 1) (as in [31]). This method attempts to identify the causal chain and mechanisms between independent variables (cognitive factors such as knowledge, values and decisions, and contextual factors such as climate change or the socio-economic context) and the outcome of the dependent variable (farmers' behaviors). Tracing all the steps in the process chain linking knowledge and values to behavior and the consequences in term of ES – Figure1), and all the necessary implications of the main hypothesis (farmers' land management behaviors are driven by their willingness to benefit from ES) provides evidence to test this main hypothesis. Meanwhile, the alternative paths through which the same outcome could have occurred, without being influenced by perception of ES (e. g. through effects of external factors on behavior) were identified and tested, also by being decomposed as a series of steps [47].

Table 3. Change of potential ecosystem services (decrease (↘) and increase (↗) greater than 10%) between practices in each category of grassland, for the drastic and local scenario (column "D") and the intermittent and international scenario (column "I") (data from [61]).

	Carbon storage		Nitrate leaching		Forage quantity		Litter quantity		Forage quality		Plant diversity		Flowering onset	
	D	I	D	I	D	I	D	I	D	I	D	I	D	I
Manuring vs. not manuring														
Mown terraces					↖			↗					↖	
Grazed terraces					↖		↖							↗
Mown unterraced grasslands		↗						↖	↖				↖	
Grazed unterraced grasslands											↗	↗	↗	
Grazing vs. mowing														
Manured terraces						↗		↖		↗	↗	↗	↗	
Not manured terraces			↖		↗	↖	↖	↖	↗	↗	↗	↗		
Not manured unterraced grasslands				↖		↖		↖		↗		↗		↖

Results

This section presents successively the different components of the decision-making framework: (a) cognitive variables (knowledge and values) about ES and practices, (b) behaviors and their explanations provided by farmers (decisions) and influences of environmental cognitions and ES on farmers' land management behaviors. Finally, we explore (c) factors other than ES which influenced farmers' behavior.

a. Farmers' environmental cognitions

Knowledge. This section describes farmers' understanding and perceptions of: (1) each ES and its relationships to others, (2) relationships between ES and agricultural practices (Figure 2), and (3) effects of contextual factors on ES.

The ES described by researchers (see Figure 2 for the list of ES) were all known by farmers except nitrate leaching and carbon storage which required more explanations. Farmers were knowledgeable about ES, even without calling them "ecosystem services" [23,34]. Several types of influences among services were recognized: (i) services mutually influencing each other (e.g. plant diversity, flowering onset and litter quantity), and (ii) some ES are only influenced by other services but do not themselves have influence on others (aesthetic, forage quantity, forage quality) (Table 6, quotes 1.1 to 1.4). Finally, some relationships, e.g. between nitrate leaching or carbon storage and other services were not mentioned, even after our explanations.

Regarding the mutual influences between ES and practices, farmers did not perceive manuring to affect nitrate leaching in their agricultural system (Table 6, quotes 2.1 and 2.2). Nevertheless some farmers were mindful of nitrate leaching because of legislation. Farmers considered that manuring unterraced grasslands may increase forage quantity and quality, and also plant diversity, but did not wish to apply manure more often or in greater quantity than under current conditions (Table 6, quotes 2.4 and 2.5). Spreading manure in autumn was considered more efficient compared to spring, and soiling fodder was avoided (Table 6, quotes 2.1 and 2.2). Mowing was considered to increase plant diversity but also directly aesthetics (Table 6, quotes 2.7 and 2.8). Moreover, farmers asserted that the decision to mow was influenced by productivity in a given year: some of them do not mow when forage quantity is not considered worthwhile. The timing of mowing influenced the forage quality (expected to decrease with late mowing date) and quantity (expected to increase with late mowing date) of forage harvests, leading to a trade-off between both services. But a late mowing date was also perceived to increase plant diversity, thus indirectly forage quality on long-term. Mowing date was in part motivated by the date of flowering onset of dominant grasses. Farmers agreed that lower parcels (terraces) are mown at the beginning of July, and higher parcels not before the 10th August in years with early vegetation onset, and some years not even before the 20th August (Table 6, quotes 2.9 to 2.12). Finally, ES influencing the decision to shift to grazing were not mentioned, but they usually grazed if there is no enough forage quantity to mown. Grazing was mentioned to have negative effects on aesthetics and plant diversity, and to decrease litter quantity, though less than mowing.

The effects of additional factors, such as climate, altitude or a recent vole outbreak on ES and practices were also discussed. Rainfall influenced forage quality and quantity (Table 6, quotes 3.1 to 3.5). Forage quantity was also known to be influenced by temperature and altitude. Altitude was perceived as influencing the date of flowering onset more than plant diversity. The presence of snow was considered to affect litter decomposition. The effects

Table 4. Ecosystem services with their values attributed by farmer (number indicates the number of farmers giving this value to a service), sorted by decreasing order of average value.

	Very low	Low	Medium	High	Very high
Forage quality				2	5
Plant diversity conservation				5	2
Forage quantity			2	3	2
Water quality (ES related to nitrate leaching EF)		1	3		3
Aesthetics	2		2	1	2
Litter quantity	2	1	2	1	1
Flowering onset	1	2	3	1	
Nitrate leaching	2	1	3	1	
Carbon storage	2	2	2	1	

of these external factors on practices will be presented in the "alternative hypotheses" section (see also the alternative hypothesis section of Table 6 and the contextual factors section of Table 8).

Values. By averaging the scores of importance given in ranking tables filled individually, ES were ranked by decreasing value and desirability (Table 4). Higher values were attributed to final ES from which farmers benefit directly [1], including forage quality and quantity, while intermediate services to the production of final ES, except plant diversity, received lower values. Farmers consistently attributed high values to some services like plant diversity, forage quality and forage quantity, while there were more heterogeneity in values attributed to other services.

The reasons for these rankings were then expressed by farmers during a collective discussion following the Likert-scale rating exercise before the "feedback game". Forage quality was considered as highly desirable for herd welfare or for parts of the herd with higher needs such as lambs or dairy cows, and for some farmers was complementary to forage quantity (Table 6, quotes 4.1 to 4.5). At the same time, forage quality and quantity were also factors contributing to farm economy. In addition to the information on value, Figure S1b shows the location of parcels where quality and/or quantity were expected according to the field functions mapped by farmers (photomapping interviews). Plant diversity was also highly valued by farmers for its contribution to forage quality, to aesthetics, or both, consistent with the indirect links suggested between these functions (Figure 2). Litter quantity received very varying values, considering on one hand a positive short-term effect on vegetation re-growth due to protection against frost and a fertilizing effect of litter when mown every couple of years, and on the other hand a negative long-term effect as litter chokes out vegetation and then decreases forage quantity and quality. Some farmers considered only the negative or positive effects of litter quantity. Low scores of carbon storage were probably due to a lack of knowledge rather than lack of interest. Nitrate leaching received a low value, probably because farmers did not feel concerned by nitrate leaching, or because it was seen as having a negative influence on water quality which was generally highly valued.

b. Farmers' behaviors and explanations

For each practice, this section describes, (1) behaviors adopted by farmers during each "feedback game" session based on board game analyses (Table 7); (2) actual farmers' behaviors reported on maps of actual practices during the 2009 participatory photomap-

ping (Table 7 and Figure S1); (3) the explanations given by farmers during the "feedback game" of the influences of ES on their behaviors (Table 6 and Table 8); and finally (4) validation of farmers' behaviors during the "feedback game" by analysing the relationships between actual land management behaviors and field functions interpreted as ES values (ANOVA Table 5).

Manuring vs. non manuring. In the "drastic and local" scenario most farmers stopped fertilizing terraces and unterraced grasslands. By contrast, during the "intermittent and international" scenario, they all increased the number of terraces manured, except one who stopped manuring them. Finally, in the "drastic anticipation" scenario, two farmers stopped manuring terraces, one of them started to manure mown grasslands, and two others manured grazed unterraced grasslands.

In the actual practices in 2009, farmers did not manure all their land, but only some mown terraces (Figure S1c). Grazed unterraced grasslands were not manured. Sheep farmers did not manure at all and farmers having both sheep and cattle used only cattle manure (sheep manure is given to a compost making firm).

Farmers did not manure when this did not maintain or increase forage quantity (Table 6, quote 5.2), forage quality or plant diversity (as in the "drastic-local" scenario), while they manured when it did (as in the "intermittent-international" scenario). Manuring of parcels to increase forage quality coincided with the desire to increase forage quantity both in reality and in the "feedback game" sessions. One farmer manured terraces and another one unterraced grasslands to increase carbon storage and hence receive carbon credits as proposed in the "intermittent-international" scenario. Nitrate leaching was never mentioned as a reason to adapt manure practices.

The comparison between the maps of actual practices (Figure S1c) and of expected quantity and quality of fodder (Figure S1b) showed that manure was applied mostly on parcels where changes in forage quantity were expected more than changes in forage quality (Table 5, chi-squared test).

Mowing vs. grazing. In the "Drastic and local" scenario, terraces were mainly grazed at the expense of mowing. Half of the farmers stopped mowing unterraced grasslands, in order to feed herds during grazing seasons. Then they had to buy fodder for winter. During the "Intermittent and international" scenario, terraces were mown, and mowing was resumed on some grazed terraces. Unterraced grasslands were mainly grazed. Only two farmers manured some unterraced grasslands. In "drastic anticipation", one farmer continued mowing them and another farmer

Table 5. Summary of the statistical analyses at parcel level (excluding alpine meadows).

	Actual behavior (Dependent variables)			ES Values		Contextual factors					
	Manuring[1a]	Mowing[1a]	Mowing date[1a]	Expected forage quality[1b]	Expected forage quantity[1b]	Slope[3]	Elevation[3]	Distance to road[4]	Distance to farm[4]	Intercept	Test result
Description	Presence/absence of application of manure in the parcel	Mowing vs. grazing practice in the parcel	Mowing date (day) (for the year 2009)	Parcels where forage quality is expected by farmers. Quality only, or together with quantity	Parcels where forage quantity is expected by farmers. Quantity only, or together with quality	Log 10 of mean slope of the parcel (degree).	Log 10 of mean elevation of the parcel (m)	Log 10 of Euclidian distance from the middle of the parcel to the road or track suitable for vehicles (m)	Log 10 of Euclidian distance from the middle of the parcel to the farm (m)		
Chi-square test	X				X						$X^2 = 20{,}07$ (1), $p<0{,}001$ ***
ANOVA A			X (early mowing)	X							$F = 12{,}89$ (2), $p<0{,}001$ ***
ANOVA B			X (late mowing)		X						$F = 12{,}17$ (1), $p <0{,}001$ ***
Linear regression			X				$233{,}38$*** ($p<0{,}001$)	$6{,}49$ ($p=0{,}06$)	$12{,}52$*** ($p<0{,}001$)	$-690{,}33$*** ($p<0{,}001$)	Adjusted $R^2 = 0{,}35$
Logistic regression A	X					$-0{,}148$** ($p=0{,}005$)	$-15{,}70$* ($p=0{,}03$)	$0{,}93.$ ($p=0{,}08$)	$0{,}17$ ($p=0{,}71$)	$50{,}59$* ($p=0{,}03$)	D^2 (Pseudo R^2) $=0{,}10$
Logistic regression B		X				$0{,}43$*** ($p<0{,}001$)	$12{,}51$ ($p=0{,}20$)	$0{,}03$ ($p=0{,}96$)	$-1{,}96$** ($p=0{,}001$)	$-43{,}65$ ($p=0{,}16$)	D^2 (Pseudo R^2) $=0{,}42$

Variables used in each analysis are depicted by "X". The three behavioral variables (manuring, mowing and mowing date) are dependent variables, the others are explanatory variables. ANOVA and Chi-square tests discriminate pairs of variables depicted by "X". Regression results presented for each variables are parameter estimates and p-value. Significance levels:
* (0,05);
** (0,01);
*** (0,001), N = 217 parcels.
Data sources:
[1a]Land managements and [1b] field functions from participatory photo mapping;
[2]Digital elevation model;
[3]Land use map;
[4]ArcGIS Euclidian distance based on Land use and topographic maps.

Table 6. Representative quotes extracted mainly from farmers' discussions and the debriefing of the "feedback game" (7 farmers, January 2012).

	Knowledge
	Quotes about relations between ecosystem services:
1.1	"A beautiful grassland with a lot of flowers, it's more beautiful than a grassland with only Festuca paniculata"
1.2	"diversity corresponds to quality"
1.3	"after one year, we can see the effect during the spring on plant re-growth on grassland which are grazed only a little bit. It's protected from frost"
1.4	"in a grassland that I have not grazed a lot, in autumn and even next spring nothing regrows"
	Quotes about relations between ecosystem services and practices:
2.1	"We manure only with natural fertilizer (manure). It is not certified organic, but we do not use mineral fertilizer, so we are far from this kind of problem"
2.2	"Rain or snow seep manure into the soil. There is no leaching"
2.3	"I take that into account because I have a plan for spreading manure agreed with the authority"
2.4	"Today, everything shows me that fertilizing increases forage quantity"
2.5	"Do not manure beyond some limits, because after you change the flora"
2.6	"In the autumn manure rots better than during spring when it's dry. After we have it on fodder"
2.7	"Farmers are all aware that the floristic diversity will change if we stop mowing"
2.8	"It (mowing) maintains an open landscape"
2.9	"We wait as long as possible until plants are at flowering stage. We maximize quantity. But that's not the best (for quality)"
2.10	"Mowing too early, before July the 20th, doesn't leave plants time to set seed and then decreases the number of species"
2.11	"The sheep do not put their head into (Festuca paniculata), but cows manage to pull out a few"
2.12	"The quality of fodder is linked to farmer's work. The way the grassland is managed every day"
	Quotes about the effects of additional factors on ecosystem services
3.1	"In 1988, rain occurred throughout June, there was so much fodder that we could not give it all, we wasted a lot because it was hard, coarse and the sheep didn't want to eat it"
3.2	"Summer rains lead to a bit of second growth"
3.3	"With this spring drought we did not have a lot of forage"
3.4	"If vegetation starts to grow too late, at 2000 meters of altitude if a cold snap occurs, the vegetation does not restart"
3.5	"Snow is also needed to rot plant litter"
	Values
	Quotes about values of ecosystem services
4.1	"There is a difference between fodder, and a palatable fodder consumed by cows"
4.2	"That's nice to have fodder in quantity but if it's crap fodder … you have only crap fodder"
4.3	"It's the balance between quality and quantity that is interesting"
4.4	"If we do not have quality fodder, we will have to buy quality fodder to compensate" (all farmers)
4.5	"In the cost of one hectare of mowing grassland, there is also the result in terms of forage quantity and quality to take into account"
	Behaviors and explanations
	Quotes about land management practices
5.1	"Our herd is our business, therefore we keep our herd and we adapt the rest (the area mowed) on the herd"
5.2	"We will not manure if this does not bring quantity"
5.3	"I am perturbed. This grassland in altitude (unterraced grasslands) should stay mown. It's better to have a spread in fodder than have fodder at a single altitude"
	Quotes about changes in environmental cognitions (knowledge and values) arising as indirect feedbacks from changes in ecosystem services
6.1	"During years of crisis, we look first at quantity and quality, before considering colours of flowers and all these things. If you asked us the same question some years ago, we would probably not have answered the same thing"
6.2	"Some years ago, I was more or less independent for fodder. I was looking mainly for quality to have a specific fodder for lambs and calves"
6.3	"Due to the vole outbreak, we had bad fodder because soil was collected along with fodder. This led us to think differently"
	Quotes about contextual factors affecting the decisions and alternative hypotheses
7.1	"We manure the best and flattest parcels". "Manured and grazed … how is it possible? Only (one farmer) has flat land… anyway"
7.2	"We will not manure where there is drinking water extraction"
7.3	"I do not take into account distance to stream, because we have a lot of streams and with 30 meters we are far into the parcel"
7.4	"In the lower part, I have to continue to fertilize, because I have to empty the manure pit …"
7.5	"In this parcel we cannot load the hay. We need to bring it down to the road"
7.6	"Here, mowing currently grazed parcels is not possible. There is no lands were a return to mowing is possible"

Table 6. Cont.

Behaviors and explanations	
Quotes about contextual factors affecting the decisions and alternative hypotheses	
7.7	*"The problem with grazing is that we need water supply. A cow consumes 40 liters per day and there is not always an access to carry water"*
78	*"We will try to continue to mow as far as we can by respect towards elderly people ... but on mechanisable parcels"*
7.9	*"To respect their work, the terraces they built"*
7.10	*"We have to respect land. Not entering when it's wet, and not grazing when mowing is possible. When a terrace is grazed it's due to an accessibility issue"*
7.11	*"Grazing instead of mowing is another system; the agreement of the landowner would be needed".*
7.12	*"What we will do during summer if we stop mowing? We will have a lot of time. We are not shepherds"*
7.13	*"If land becomes available, I will stop to mow over the entire landscape and do it near my farm, to waste as little time as possible. Even if I will need to increase by two the hours per day or to take an additional worker during a few days"*

Quotes illustrate the different components of the conceptual model presented in Figure 1.

even mowed and manured previously grazed unterraced grasslands.

In the current practices, farmers organized their land management around spring grazing and mowing, because summer alpine meadows are large enough to ensure flexibility in forage resources (Figure S1c). During autumn, the herd grazed extensively on the re-growth of mown or spring grazed grassland. Areas of grazed versus mown unterraced grasslands were adjusted to herd size (Table 6, quote 5.1), while the remaining area was used to mow, leading three farmers to harvest part of their fodder in other municipalities.

Some farmers attributed their decision in the RPG to mow terraces to its positive effect on forage quality and on the reduction of litter (in the "drastic-local" scenario), or to its benefits for plant diversity and forage quality (in the "intermittent-international" scenario). Date of flowering was also cited once as a factor influencing decision to mow terraces. Unterraced grasslands were also mown by some farmers to increase or maintain plant diversity and decrease litter quantity, in the "drastic-local" scenario. By contrast, these grasslands were grazed to increase carbon storage in the "intermittent-international" scenario.

Maps of expected ES (field functions) showed that in mown parcels farmers often expected to obtain high fodder quality as well as large fodder quantity (Figure S1b). Mown parcels where quality was expected were concentrated on the lower part of the slope, mixed with parcels where only quantity was required.

Date of mowing. Dates of mowing were not discussed during "feedback game" sessions. During interviews and participatory photo mapping, farmers explained that dates of mowing are spread between 1st July and mid-September (FigureS1d).

One farmer indicated that, by choosing to graze unterraced grasslands to increase carbon storage (Table 6) and gain credits as proposed in the "intermittent and international" scenario (Table 2), dates of mowing on his parcels were disturbed as he had lost the possibility of later mowing in unterraced grasslands (Table 6, quote 5.3). This farmer faced a trade-off between maintaining a spread in mowing date or receiving carbon credits and continued to graze according to the "intermittent and international" scenario.

Map comparison revealed associations between early mowing (current practice) and expected quality or late mowing and expected quantity (Table 5, ANOVA) (Figure S1b and S1d). However there were no significant relationships between date of mowing and actual date of flowering onset, or plant diversity (Simpson index) and date of mowing.

c. Contextual factors affecting the decisions and alternative hypotheses

This section describes contextual factors (internal or external to the farm) influencing farmers' decisions, using explanations by farmers (during games or interviews) and statistical analyses between spatial factors and practices (Table 5). These factors can explain divergences between farmers' behaviors and attitudes, or in case of consistent behaviors, constitute alternatives to our main hypothesis that farmers' land management behaviors are driven by their willingness to benefit from ES.

Manuring vs. non manuring. As mowing and manuring are mechanised, constraints on the mechanisation of parcels such as slope and accessibility came out as a recurrent theme in farmers' discussions (Table 6, quote 6.1). Distance to farm was mentioned as a factor influencing manuring due to price of fuel and travel time to the parcel (except by one farmer who rents a truck bringing manure close to the most remote parcels). Other characteristics of parcels were sometimes considered such as proximity to dwellings, streams and water springs (Table 6, quote 6.2 and 6.3). These aspects were mainly considered because of legislation and policy support including fertilisation management plans, which impose quantity, date and distance thresholds. Sheep farmers usually did not use their manure and gave it to a specialized company, which does not take liquid manure or slurry. Therefore, the capacity of slurry storage pits forced farmers to manure when it was full ((Table 5, quote 6.3). Individual parcels were manured only once every two or three years. Finally, for some parcels, the short time between autumn grazing and snow (around 1st November on average) limited manuring. Mineral fertilisation was considered as too costly.

The effects of some contextual factors mentioned by farmers were confirmed by statistical analyses on land use maps. The logistic regression of factors influencing manuring showed that manuring was mainly applied on gentle slopes, but distance to farm, distance to road and altitude were not significant (Table 5, logistic regression A). In addition, we estimated the maximum area which could be manured according to the amount of manure produced depending on farm characteristics. This theoretical calculation considered farm herd size, an average annual production of manure of 4.5 T per livestock unit, a theoretical average of 15 T/ha of manure per spreading and a frequency of

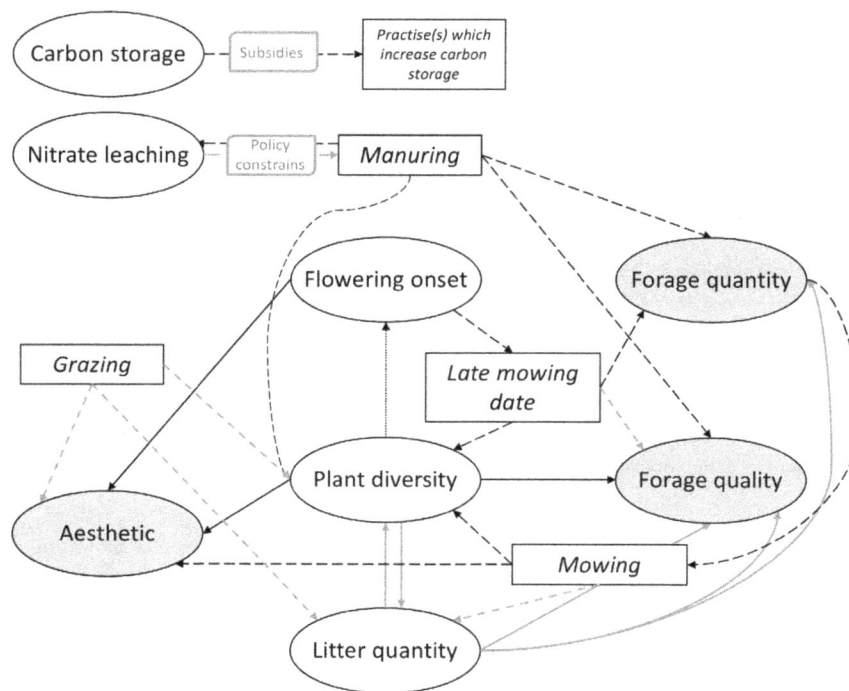

Figure 2. Farmers' ecosystem services values and knowledge. Conceptual representation based on farmers' discourses on values and knowledge about the relationship between ES and land-management practices. Rectangle boxes indicate practices and ellipses indicate ES. Dashed arrows indicate links between practices and ES and plain arrows indicate links between ES. Grey arrows indicate a negative effect and black arrows a positive effect. Except for the effect of litter quantity on forage quantity, farmers agree on all the relationships. Note that ES in grey are seen as final ES by farmers while the others are considered as intermediate ES [40].

manuring of every two or three years for each parcel. The results suggest that all farmers could potentially fertilize almost all their mown grasslands at the selected frequency, or that they could increase frequency of manuring on usually fertilized parcels.

Mowing vs. grazing. Mechanisation of parcels was mentioned as an important determinant of conversion from mowing to grazing (Table 6, quote 6.5). At the time of data collection farmers considered that they were mowing all the mechanisable parcels (Table 6, quote 6.6). They regarded mowing equipment adapted to mountain terrain as too expensive. Factors related to farms' economy, such as cost of mowing considering investment into equipment (purchase cost, depreciation and maintenance), appeared often in farmers' discourses. This was even more prominent when the farmer had acquired new equipment to meet the norms of the European Union and/or when farm debt level was important. When asked to rate the importance of equipment investments compared to ES in their behavior, farmers responded "very high" for cost of mowing and "high" for ES. Some equipment such as manure spreaders was shared between farmers to decrease costs, but each farmer usually owned their personal cutting equipment whose use could not be spread through time. Agri-environmental measures provided subsidies to mow unterraced grasslands and a possibility of extra subsidy to mow less mechanisable parcels with a pedestrian mower. While this kind of subsidies substantially contributed to the farms' economy, farmers argued several times during the "feedback game" that, although policy supports were carefully taken into consideration (balancing the financial amount and constraints), their amount were insufficient to compensate the actual cost of maintaining mowing practices.

Farmers discussed different elements favouring conversion from mowing to grazing. They explained that mown parcels surrounded by grazed parcels belonging to other farmers could be more prone to conversion, to avoid risks like trampling by cattle or fence removal. The altitude of the site did not allow multiple uses of parcels throughout the season (e.g. grazing before mowing) because vegetation re-grew only at the end of summer, and in small quantities. Parcels close to the farm were needed to turn out the herd to grass during the first weeks of spring. Grazing also required the presence of water or the possibility to install a trough (Table 6, quote 6.7). Future opportunities to mow elsewhere or to acquire parcels from retiring farmers might arise, allowing remaining farms to increase their land and then fodder production, or to restrict mowing to the more mechanisable parcels and to graze the others. By consolidating land among farmers, the "Association foncière pastorale" has allowed them to increase the average size of their parcels and decrease their production costs, but also to solve conflicts between families about some parcels, allowing to manage them effectively. This could in turn favour mowing because some farmers perceived a social pressure to properly manage land, and thus to prevent shrub encroachment, especially in terraces, which have a high cultural value and suffer from trampling by stock. This is directly linked with farmers' perceptions of the social value of farming and of social pressure from landowners, other farmers and/or inhabitants (Table 6, quote 6.8 to 6.11). Mowing also appeared as an important aspect of the farming profession: in the discussions, the possibility of completely stopping mowing was always a source of laughter (Table 6, quote 6.12).

The logistic regression on mapped data confirmed farmers' explanations that mowing, in contrast to grazing, was preferen-

Table 7. Farmers' behaviors in reality and in each scenario ("feedback game" session) for each type of grasslands: Mown terraces; Grazed terraces, Mown unterraced grassland, Grazed unterraced grassland.

	Manuring				Mowing (vs. Grazing)			
	Mown terraces	Grazed terraces	Mown unterraced grassland	Grazed unterraced grasslands	Mown terraces	Grazed terraces	Mown unterraced grassland	Grazed unterraced grasslands
Current practices (2009 data)	Y/N	Y/N	Y/N	Y/N	Y	N	Y	Y/N
Scenario "drastic and local"	-	Y/N	N	N	-	N	N	N
Scenario "intermittent and international"	Y	Y/N	Y/N	Y/N	Y	Y/N	Y/N	N
Scenario "drastic anticipation"	Y/N	N	Y	Y/N	Y	N	Y	Y/N

"_" means: no information. "Y" means they adopted the behavior and "N" they didn't. Y/N indicates that both behaviors were adopted by farmers of the area.

tially located on parcels with gentler slopes and further from farms. Altitude and distance to roads were not significant in the model (Table 5, logistic regression B).

Early vs. late mowing. For farmers having contracted agri-environmental measures, mowing cannot occur before the 1st July on unterraced grasslands. This subsidy was perceived as a constraint by some farmers, depending on the inter-annual variability in flowering date. Having parcels spread across the landscape increased the time necessary to mow them all (although re-parcelling between farmers decreased that constraint), but the altitudinal span of parcels allowed them to stagger the mowing over the summer season. This was seen by some farmers as an opportunity, while others perceived it more as a constraint and argued that ideally it would be preferable to have the entire mown area around their farms (Table 6, quote 6.13). All farmers worked alone on their farm, and additional labour was hired exclusively from family when needed, e.g. for mowing. On parcels enclosed within other farmers' parcels, farmers of the enclosed parcel had to wait for surrounding parcels to be mown first, before being able to harvest their own parcel. Around the villages and campsites, some parcels were mown earlier than desired to avoid trampling by tourists. Finally, mowing dates depended on rain as dry weather is necessary to harvest fodder.

According to spatial data, early mowing occurred mainly on the lower part of the area and late mowing on the upper part, except for some parcels (Figure S1d). A linear regression model confirmed that not only parcel elevation but also distance to farm had an influence on the date of mowing (Table 5, linear regression).

Discussion

First, results are discussed looking at how knowledge and values about ES influence behavior, and how contextual factors can change farmers' cognitions or decisions. Second, methodological relevance is discussed. Third, implications of findings for future studies on ES are examined.

a. Role of ES in farmers' decision-making process

Visions of ES differed between farmers and scientists [48]. Farmers explained that for them, *"ES are neither numbers nor upward or downward trends"*, but are part of a more complex system of decision-making. Here we discuss the correspondence between farmers' willingness to adopt different behaviors according to their values and knowledge towards ES, and farmers' behaviors in the "feedback game" or actual life. Returning to the conceptual chain (Figure1), three configurations emerge, explaining whether ES were taken into account in farmers decisions or not. First, some ES were not part of farmers' knowledge or far away from their interest and therefore had low values. This was the case for carbon storage and nitrate leaching, which were thus in principle not considered by farmers when making decisions. Yet, institutional mechanisms may lead some farmers to consider these services [49], as demonstrated in the second "feedback game" session where contractual carbon credits could be allocated to farmers, which indeed modified some decisions. Second, some ES were known by farmers but had a low value, or farmers considered not having enough knowledge to include this ES in their decisions. For example, farmers perceived an influence of mowing date on the date of flowering onset, itself affecting forage quantity and quality, but they did not consider themselves capable of obtaining a desirable ecosystem service delivery by adjusting mowing dates. Additional reasons (section 3.3, e.g. distance to farm, surroundings of the parcels, weather) also constrained mowing dates. Third, in some cases farmers had both knowledge

Table 8. Factors influencing farmers' decisions to adopt a practice during the "feedback game", according to farmers accounts and discussions.

	Manuring	Mowing	Date of mowing
Ecosystem services			
Forage quality	X	X	
Plant diversity conservation	X	X	
Forage quantity	X		
Litter quantity		X	
Flowering onset			X
Nitrate leaching			
Carbon storage	X	X	
Contextual factors			
Steep slope	−	−	
Altitude			+
Proximity to farm	+	−	−
Low accessibility		−	+
Proximity of dwellings or streams	−	−	
Parcels surroundings	−	−	−/+
Availability in manure	+		
Equipment costs	−	−	
Social conflicts		+	
Social value of farming		+	+
Subsidies amount		+	+
Policy or legislation constraints	−	−	+
Availability of land		+	
Snow or rain	−		+

The first part of the table presents ecosystem services, with a X when a given service is said to influence a given practice (manuring, mowing, late mowing). The second part presents other contextual factors, with their positive (+) or negative (−) influence on the decision to adopt a behavior corresponding to alternative hypotheses.

and high values for given ES. "Feedback game" sessions showed that more parcels were manured when it enabled increased forage quantity and secondarily forage quality or plant diversity. According to their knowledge late mowing could be favored to improve forage quantity, as well as plant diversity and thus indirectly forage quality on long-term. But, because late mowing also decreased forage quality directly, trade-offs had to be considered. In this case, farmers could opt for the most highly valued ES and adapt their behavior accordingly. Whatever the behavior adopted, in this case farmers took multiple ES into account in their decision. These results suggest that knowledge and/or values were necessary but not always decisive in farmers' decisions.

Results also suggest that both direct and indirect feedbacks (Figure 1) explain how ES were taken into account in farmers decisions. Most changes in ES during the "feedback game" had direct feedback effects on farmers' decisions because these represented changes that farmers face frequently (e.g. change in fodder quality or quantity due to weather conditions). But an indirect feedback was also observed with the case of carbon storage. Farmers were not aware of carbon storage before the "feedback game", and some changed their values and knowledge about it, so that for some this factor entered in their decisions. This was also the case for their knowledge about the effects of manuring on forage quantity during drastic drought. While in the previous "scenario game" analysing farmers' adaptation to climate change, farmers increased manuring to face droughts [35], in the first

session of the "feedback game" some farmers considered the results presented to them on ES impacts (Table 3) and, realizing the inefficiency of this practice under drought, stopped manuring. Occurrence and amount of change in ES could also influence the feedback type. Short-term or small changes in ES affected farmers' behaviors mainly through direct feedbacks leading to tactical adaptation (e.g. conversion of mowing to grazing on a given year). By contrast, greater or frequent changes in ES supply (e.g. repeated drought decreasing forage quantity, or a vole outbreak during several years) could lead to change in values (Table 6, quotes 7.1 to 7.3).

Nevertheless, it would be naive to consider that ES fully drive farmers' behaviors. Indeed, behaviors did not always correspond to their attitudes regarding ES and this could depend on the parcels considered (section 4c). Other factors have been shown to influence land use practices in European mountain systems: parcel characteristics (e.g. topography, location, size, land-locked position, proximity of water supply), market prices (e.g. input prices and output prices of production), policy support (e.g. types and amount of subsidies), climate (e.g. drought, frost, rain) and pest outbreaks (e.g. voles, grasshoppers) [27,50,51]. Social factors also matter, including structure of the farm business (e.g. farm type, farm size, farm economy), farmer characteristics (e.g. age, gender, education, and personality), household characteristics (e.g. level of pluri-activity, work pattern of the spouse), and structure of the social environment (e.g. local culture, social capital, information flows) [52]. On the one hand, our results confirmed that, although the influence of ES is not negligible, in some cases these other factors outweigh ES in farmers' decisions. For example, the difference in manuring between repeated and intermittent drought conditions ("feedback game" sessions on "drastic and local" and intermittent and international" scenarios) confirmed that forage quantity influenced farmers' decisions, because they fertilised only when it was efficient. Nevertheless, some parcels were not manured (in "feedback game" sessions on "intermittent and international" and "drastic and local" scenarios) because some of them were not mechanisable, too far from the farm resulting in high transport cost, or at unauthorised distance from streams or settlements imposed by regulations. By contrast, despite the inefficiency of manuring in repeated drought conditions, some farmers still manured some parcels because they had to use their manure. During the "feedback game" farmers chose to mow some parcels to increase forage quantity and/or quality, plant diversity or decrease litter quantity but sometimes grazing was preferred due to the cost of mowing, accessibility and possibility to mechanise. Financial incentives also had importance in farmers' decision as shown by the farmer grazing a parcel to receive carbon storage subsidies although mowing would have been preferred for forage quality. Farmers took into account date of mowing to increase forage quantity or quality, but some parcels around villages were mown earlier than expected to avoid trampling. Similarly, unterraced grasslands were mown too late because of the time needed to mow the other parcels nearer to the farm. On the other hand, even when behaviors were found to be consistent with attitudes towards ES, alternative reasons could also have driven farmers to adopt these behaviors. ES would thus be only one factor among others contributing to farmers' decisions. For example, some farmers mowed parcels to increase plant diversity, but this could also have been favored by financial support from agri-environmental measures and social value attributed to mowing as part of the farming profession. Agri-environmental measures could also favor late mowing, as they impose a date threshold.

b. Methodological relevance

The results presented are valid for our study located in a high mountain farmer community of the French Alps, where agriculture is very extensive. Although to our knowledge no previous study has analyzed the complete feedback loop, some studies suggested that farmers in other contexts have detailed and accurate knowledge about the relationship between their practices and the functioning of their agro-ecosystem [27,51]. Therefore, our results might hold for other mountain agricultural social-ecological systems where farmers opportunities (policy support, economy) and constraints (e.g. topography, weather conditions, higher cost of productions) to adopt different behaviors are to a large extent similar [50,51].

The results of this study are based on data gathered from eight farmers which represent the total farming population of the area. This small number did not allow statistical analyses on decision-making following established paradigms such as the Theory of Planned Behavior (TPB) or the Value-Belief-Norm Theory (VBN) [53] but enabled deeper analysis of the reasons behind the relationships between the different components of farmers' decision-making process and of their effects on behavior. The direct involvement of farmers during the role-playing game and their adaptations in a realistic situation of change allowed us to observe and to discuss how they decided to adopt a given behavior [35]. The coherence of results was increased by cross-checking data from different sources of information. This led us to use a combination of qualitative and quantitative methods which have both strengths and weaknesses.

Spatial analyses (photomapping of actual behaviors) provide robust quantitative data on decisions, but do not give information to interpret them in terms of ES values and knowledge. Interviews provide useful qualitative data to interpret the reason behind decisions, but give little quantitative information to assess whether rationales given by the farmers are effectively implemented. Role-playing game provide a crucial link between quantitative spatial data and qualitative data from interviews, by providing explicit, quantitative data on decisions, though less robust than spatial analyses of actual behaviors, and less self-explanatory than interviews.

Participatory simulation, of which role-playing game constitute one tool, is a very useful approach in decision-making processes in complex systems [54,55]. However, this approach implies different forms of learning such as the learning related to the stakes involved, to the other players or to the technical aspects [55]. The addition of a protocol to assess farmers' learning as an outcome of the iterative process of our approach could be beneficial.

c. Implications for ES research

Our results show the importance of considering stakeholders' perception and use of ecosystems rather than focusing only on potential ES supply [56]. Stakeholders' perceptions and potential ES delivery sometime differ [1] and the latter may not coincide with stakeholders' needs. Nevertheless, most of the potential ES presented to farmers were indeed considered by them as services, except for water quality/nitrate leaching and carbon storage of which they were not aware. Studying systemic representation of potential ES by farmers (Figure 2) also allowed us to show that potential ES can benefit to farmers (i) either individually (forage quality, forage quantity, litter quantity, date of flowering onset, plant diversity and aesthetics) or as tradeoffs (forage quantity and quality), and (ii) directly (i.e. final ES: forage quantity, forage quality, aesthetics) and/or indirectly (i.e. intermediate ES: flowering onset, plant diversity, litter quantity) [40]. For example, plant diversity was not considered by farmers for its intrinsic value but rather for its value to contribute to forage quality and aesthetics. Quality and quantity of "green" forage (vegetation) were seen as services deriving directly from the ecosystem, while quality and quantity of "dry" fodder (harvested) were considered as benefits, acknowledging that manufactured and/or human capital (e.g. farmer know-how) are required to generate a valued good from ES (Table 6, quote 2.12) [1,6]. Moreover, our results showed the importance to not only consider perception or valuation of ES by stakeholders but also their effective uses. ES potentially supplied, ES perceptions and ES actually used can differ according to individuals and to the spatial and temporal contexts. For example farmers might not manage a parcel with high plant diversity towards quality fodder because of topography limiting access, or because of the context forcing them to maximise quantity at the expense of quality. For the purpose of ES conservation, it is also important to consider farmers' awareness, willingness and/or ability to adopt a practice maintaining or enhancing ES delivery in the social-ecological system as a whole. Results of this study confirmed and illustrated the utility of decomposing ES along the conceptual cascade from ecosystem processes to values [1,57,58].

Although all components of agro-ecosystems cannot be translated in terms of ES, research focusing on ES can complement agronomic studies for several reasons. Firstly, the ES framework translates ecological complexity into common language easily understandable by researchers from different disciplines as well as farmers, stakeholders or policy makers [59]. This study showed that the concept was rapidly understood, even by farmers who had never heard the term before [23]. Secondly, it emphasizes human-environment interactions which have been generally overlooked in previous research, increasing the awareness of the dependence of society on biodiversity and ecosystems [60]. Thirdly, it allows identifying and arbitrating trade-offs and priorities at the farm, municipality or even larger scales involving beneficiaries having different interests. In this study, the national park aims at maintaining mowing in unterraced grasslands to conserve this rare agro-ecosystem and the related biodiversity. In contrast, farmers are interested in maximising other ES and are induced to stop mowing by other contextual factors including profitability or topography. The ES framework could help design public policies to reconcile the interests of different stakeholders.

Conclusion

To our knowledge this is the first ES study exploring the feedback between multiple ES and stakeholders' behavior through the decisions making process. By demonstrating the causal chain and mechanisms leading farmers to adopt a behavior, our study shows that farmer's knowledge about ES and willingness to benefit from these services are taken into account in their decisions, but do not constitute the main factor of decision. ES constitute necessary but not sufficient conditions in explaining behaviors, as other key factors were determinant (i.e. altitude, topography, parcels location, policies or social value). Such an approach should be

tested at other sites with a greater set of ES and/or other beneficiaries and land managers, as well as in other natural resource management systems.

Supporting Information

Figure S1   (a) Study site map with grassland management types and location of farms and roads (modified from [62]). Maps made by farmers during the 2009 interviews: (b) farmers' expectations about forage quality and quantity (colours) for mown (plain) or grazed parcels (shaded); (c) current practices and (d) current date of mowing.

Acknowledgments

We thank all participants of interviews, workshops and role-playing games for their contributions and the time they devoted to our study. We also thank Fabien Quétier, Aloïs Artaux, Eric Deboeuf and Coline Byczek for their help during the game sessions, workshop and/or interviews, Cécile Barnaud for her advices for the role-playing games development and the Joseph Fourier Alpine Station for logistical support. This research was conducted on the long term research site "Zone Atelier Alpes", a member of the ILTER-Europe network.

Author Contributions

Conceived and designed the experiments: PL PM SL. Performed the experiments: PL SL BN. Analyzed the data: PL BN PM. Contributed reagents/materials/analysis tools: PL. Wrote the paper: PL PM SL.

References

1. Lamarque P, Quetier F, Lavorel S (2011) The diversity of the ecosystem services concept and its implications for their assessment and management. Comptes Rendus Biologies 334: 441–449.

2. Foley JA, DeFries R, Asner GP, Barford C, Bonan G, et al. (2005) Global Consequences of Land Use. Science 309: 570–574.

3. Diaz S, Quetier F, Caceres DM, Trainor SF, Perez-Harguindeguy N, et al. (2011) Linking functional diversity and social actor strategies in a framework for interdisciplinary analysis of nature's benefits to society. Proceedings of the National Academy of Sciences, USA 108: 895–902.

4. Liu J, Dietz T, Carpenter SR, Folke C, Alberti M, et al. (2007) Coupled Human and Natural Systems. AMBIO: A Journal of the Human Environment 36: 639–649.

5. Meyfroidt P (2013) Environmental cognitions, land change, and social-ecological feedbacks: an overview. Journal of Land Use Science 8: 341–367.

6. Mace GM, Bateman I, (ed.) (2011) Conceptual Framework and Methodology. In: The UK National Ecosystem Assessment Technical Report. UK National Ecosystem Assessment, UNEP-WCMC, Cambridge.

7. Hein L, van Koppen K, de Groot RS, van Ierland EC (2006) Spatial scales, stakeholders and the valuation of ecosystem services. Ecological Economics 57: 209–228.

8. Reyers B, Biggs R, Cumming GS, Elmqvist T, Hejnowicz AP, et al. (2013) Getting the measure of ecosystem services: A social-ecological approach. Frontiers in Ecology and the Environment 11: 268–273.

9. Banerjee S, Secchi S, Fargione J, Polasky S, Kraft S (2013) How to sell ecosystem services: a guide for designing new markets. Frontiers in Ecology and the Environment 11: 297–304.

10. Robertson M, BenDor TK, Lave R, Riggsbee A, Ruhl JB, et al. (2014) Stacking ecosystem services. Frontiers in Ecology and the Environment 12: 186–193.

11. Costanza R, d'Arge R, de Groot R, Farber S, Grasso M, et al. (1997) The value of the world's ecosystem services and natural capital. Nature 387: 253–260.

12. TEEB (2009) The Economics of Ecosystems and Biodiversity for Policy Makers Summary: Responding to the Value of Nature. Available at: www.teebweb.org, accessed 10 june 2010.

13. von Haaren C, Albert C (2011) Integrating ecosystem services and environmental planning: limitations and synergies. International Journal of Biodiversity Science, Ecosystem Services & Management 7: 150–167.

14. Salzman J, Thompson Jr BH, Daily GC (2001) Protecting ecosystem services: Science, economics, and law. Stan Envtl 20: 309.

15. St John FAV, Edwards-Jones G, Jones JPG (2011) Conservation and human behaviour: lessons from social psychology. Wildlife Research 37: 658–667.

16. Kumar M, Kumar P (2008) Valuation of the ecosystem services: A psycho-cultural perspective. Ecological Economics 64: 808–819.

17. Jones NA, Ross H, Lynam T, Perez P, Leitch A (2011) Mental Models: An Interdisciplinary Synthesis of Theory and Methods. Ecology and Society 16 (1) 46.

18. Wise RM, Fazey I, Stafford Smith M, Park SE, Eakin HC, et al. (in press) Reconceptualising adaptation to climate change as part of pathways of change and response. Global Environmental Change.

19. Ajzen I (1991) The theory of planned behavior. Organizational behavior and human decision processes 50: 179–211.

20. Stern PC, Dietz T, Abel T, Guagnano GA, Kalof L (1999) A value-belief-norm theory of support for social movements: The case of environmentalism. Human ecology review 6: 81–98.

21. O'Farrell PJ, Donaldson JS, Hoffman MT (2007) The influence of ecosystem goods and services on livestock management practices on the Bokkeveld plateau, South Africa. Agriculture, Ecosystems & Environment 122: 312–324.

22. Lewan L, Soderqvist T (2002) Knowledge and recognition of ecosystem services among the general public in a drainage basin in Scania, Southern Sweden. Ecological Economics 42: 459–467.

23. Lamarque P, Tappeiner U, Turner C, Steinbacher M, Bardgett RD, et al. (2011) Stakeholder perceptions of grassland ecosystem services in relation to knowledge on soil fertility and biodiversity. Regional Environmental Change 11: 791–804.

24. Fontaine CM, Dendoncker N, De Vreese R, Jacquemin I, Marek A, et al. (2013) Towards participatory integrated valuation and modelling of ecosystem services under land-use change. Journal of Land Use Science: 1–26.

25. Duguma LA, Hager H (2011) Farmers' assessment of the social and ecological values of land uses in central Highland Ethiopia. Environmental management 47: 969–982.

26. Martín-López B, Iniesta-Arandia I, García-Llorente M, Palomo I, Casado-Arzuaga I, et al. (2012) Uncovering ecosystem service bundles through social preferences. PLoS ONE 7: e38970. doi:38910.31371/journal.pone.0038970.

27. Nettier B, Dobremez L, Seres C, Pauthenet Y, Orsini M, et al. (2011) Biodiversity conservation by livestock farmers: advantages and shortcomings of the agri-environmental scheme 'Prairies fleuries'. Fourrages: 283–292.

28. Feola G, Binder CR (2010) Towards an improved understanding of farmers' behaviour: The integrative agent-centred (IAC) framework. Ecological Economics 69: 2323–2333.

29. Tengö M, Belfrage K (2004) Local management practices for dealing with change and uncertainty: a cross-scale comparison of cases in Sweden and Tanzania. Ecology and Society 9.

30. Lauer M, Aswani S (2010) Indigenous knowledge and long-term ecological change: detection, interpretation, and responses to changing ecological conditions in pacific Island Communities. Environmental management 45: 985–997.

31. Meyfroidt P (2013) Environmental cognitions, land change and social-ecological feedbacks: local case studies of the forest transition in Vietnam. Human ecology 41: 367–392.

32. Poppenborg P, Koellner T (2013) Do attitudes toward ecosystem services determine agricultural land use practices? An analysis of farmers' decision-making in a South Korean watershed. Land Use Policy 31: 422–429.

33. Quétier (2006) Vulnérabilité des écosystèmes semi-naturels européens aux changements d'utilisations des terres. Montpellier: Ecole supérieure Agronomique de Montpellier. 269 p.

34. Nettier B, Dobremez L, Coussy JL, Romagny T (2010) Attitudes of livestock farmers and sensitivity of livestock farming systems to drought conditions in the French Alps. Revue De Geographie Alpine-Journal of Alpine Research 98: 383–400.

35. Lamarque P, Artaux A, Barnaud C, Dobremez L, Nettier B, et al. (2013) Taking into account farmers' decision making to map fine-scale land management adaptation to climate and socio-economic scenarios. Landscape and Urban Planning 119: 147–157.

36. Vignola R, Koellner T, Scholz RW, McDaniels TL (2010) Decision-making by farmers regarding ecosystem services: Factors affecting soil conservation efforts in Costa Rica. Land Use Policy 27: 1132–1142.

37. Dietz T, Fitzgerald A, Shwom R (2005) Environmental values. Annu Rev Environ Resour 30: 335–372.

38. Geist HJ, Lambin EF (2002) Proximate causes and underlying driving forces of tropical deforestation. Bioscience 52: 143–150.

39. Svarstad H, Petersen LK, Rothman D, Siepel H, Watzold F (2008) Discursive biases of the environmental research framework DPSIR. Land Use Policy 25: 116–125.

40. Fisher B, Turner RK, Morling P (2009) Defining and classifying ecosystem services for decision making. Ecological Economics 68: 643–653.

41. Girel J, Quétier F, Bignon A, Aubert S (2010) Histoire de l'agriculture en Oisans. Hautes Romanche et pays faranchin. Villar d'Arène, Hautes-Alpes (History of agriculture in Oisans. Villar d'Arène, Hautes-Alpes). Grenoble, France: Station Alpine Joseph Fourier. 79 p.

42. Quétier F, Thebault A, Lavorel S (2007) Plant traits in a state and transition framework as markers of ecosystem response to land-use change. Ecological Monographs 77: 33–52.

43. Diaz S, Lavorel S, de Bello F, Quétier F, Grigulis K, et al. (2007) Incorporating plant functional diversity effects in ecosystem service assessments. Proceedings of the National Academy of Sciences, USA 104: 20684–20689.

44. Quétier F, Rivoal F, Marty P, de Chazal J, Thuiller W, et al. (2010) Social representations of an alpine grassland landscape and socio-political discourses on rural development. Regional Environmental Change 10: 119–130.

45. Deboeuf E (2009) Adaptabilité des systèmes d'élevage de haute-montagne à des aléas. Le cas de Villar d'Arène: Enita de Clermont-Ferrand, France. 91 p.

46. Fleury P, Dubeuf B, Jeannin B (1996) Forage management in dairy farms: A methodological approach. Agricultural Systems 52: 199–212.

47. George AL, Bennett A (2005) Case studies and theory development in the social sciences. Cambridge, Massachusetts: The MIT Press.

48. Lugnot M, Martin G (2013) Biodiversity provides ecosystem services: scientific results versus stakeholders' knowledge. Regional Environmental Change 13: 1145–1155.

49. Vatn A (2010) An institutional analysis of payments for environmental services. Ecological Economics 69: 1245–1252.

50. Mottet A, Ladet S, Coque N, Gibon A (2006) Agricultural land-use change and its drivers in mountain landscapes: A case study in the Pyrenees. Agriculture Ecosystems & Environment 114: 296–310.

51. von Glasenapp M, Thornton T (2011) Traditional Ecological Knowledge of Swiss Alpine Farmers and their Resilience to Socioecological Change. Human Ecology 39: 769–781.

52. Edwards-Jones G (2006) Modelling farmer decision-making: concepts, progress and challenges. Animal Science 82: 783–790.

53. Kaiser FG, Hubner G, Bogner FX (2005) Contrasting the theory of planned behavior with the value-belief-norm model in explaining conservation behavior. Journal of Applied Social Psychology 35: 2150–2170.

54. Voinov A, Bousquet F (2010) Modelling with stakeholders. Environmental Modelling & Software 25: 1268–1281.

55. Etienne Mc (2010) La modélisation d'accompagnement: Une démarche participative en appui au développement durable. Versailles, France: Quae Editions. 368 p.

56. Termorshuizen JW, Opdam P (2009) Landscape services as a bridge between landscape ecology and sustainable development Landscape Ecology Volume 24: 1037–1052.

57. Haines-Young R, Potschin M, Raffaelli DG, Frid CLJ (2010) The links between biodiversity, ecosystem services and human well-being Ecology E, editor: Cambridge University Press.

58. TEEB (2010) The Economics of Ecosystems and Biodiversity: Ecological and Economic Foundations. Edited by Pushpam Kumar. Earthscan, London and Washington.

59. Barnaud C, Antona M, Marzin J (2011) Vers une mise en débat des incertitudes associées à la notion de service écosystémique. Vertigo 11.

60. Vihervaara P, Rönkä M, Walls M (2010) Trends in Ecosystem Service Research: Early Steps and Current Drivers. Ambio 39: p. 314–324.

61. Lamarque P, Lavorel S, Mouchet M, Quétier F (2014) Plant trait-based models identify direct and indirect effects of climate change on bundles of grassland ecosystem services. Proc Natl Acad Sci USA, 10.1073/pnas.1216051111

62. Lavorel S, Grigulis K, Lamarque P, Colace M-P, Garden D, et al. (2011) Using plant functional traits to understand the landscape distribution of multiple ecosystem services. Journal of Ecology 99: 135–147.

15

The Soil Biota Composition along a Progressive Succession of Secondary Vegetation in a Karst Area

Jie Zhao[1,2,9], Shengping Li[1,2,39], Xunyang He[1,2], Lu Liu[1,2,3], Kelin Wang[1,2]*

1 Key Laboratory of Agro-ecological Processes in Subtropical Region, Institute of Subtropical Agriculture, Chinese Academy of Sciences, Changsha, Hunan, China, 2 Huanjiang Observation and Research Station for Karst Ecosystems, Chinese Academy of Sciences, Huanjiang, Guangxi, China, 3 Graduate University of Chinese Academy of Sciences, Beijing, China

Abstract

Karst ecosystems are fragile and are in many regions degraded by anthropogenic activities. Current management of degraded karst areas focuses on aboveground vegetation succession or recovery and aims at establishing a forest ecosystem. Whether progressive succession of vegetation in karst areas is accompanied by establishment of soil biota is poorly understood. In the present study, soil microbial and nematode communities, as well as soil physico-chemical properties were studied along a progressive succession of secondary vegetation (from grassland to shrubland to forest) in a karst area in southwest China. Microbial biomass, nematode density, ratio of fungal to bacterial biomass, nematode structure index, and nematode enrichment index decreased with the secondary succession in the plant community. Overall, the results indicated a pattern of declines in soil biota abundance and food web complexity that was associated with a decrease in soil pH and a decrease in soil organic carbon content with the progressive secondary succession of the plant community. Our findings suggest that soil biota amendment is necessary during karst ecosystem restoration and establishment and management of grasslands may be feasible in karst areas.

Editor: Dafeng Hui, Tennessee State University, United States of America

Funding: This study was supported by the Chinese Academy of Sciences Action Plan for the Development of Western China (KZCX2-XB3-10), the National Key Technology Research and Development Program of the Ministry of Science and Technology of China (2010BAE00739-02) given to Kelin Wang. The study was also supported by a Western Light Program from CAS and a NSFC program (31300448) given to Jie Zhao. The funders had no role in study design, data collection and analysis, decision to publish, or preparation of the manuscript.

Competing Interests: The authors have declared that no competing interests exist.

* Email: kelin@isa.ac.cn

9 These authors contributed equally to this work.

Introduction

Karst landscapes account for about 12% of the Earth's land surface [1]. Ecosystems in karst areas are fragile and suffering extreme degradation in many regions of the world, such as southwest China, Southeast Asia (e.g., Vietnam and Thailand), and Central America (e.g., Mexico, Puerto Rico, Dominican Republic, and Cuba) [2–7]. Karst areas in southern and central Europe (e.g., Austria, Bosnia, Herzegovina, Croatia, Germany, Italy, Russia, and Spain [not all land cover of these countries are karst but there are karst areas in them]) also are facing significant land degradations [3]. One important example of this degradation is "karst rocky desertification", which is characterized in part by shallow soils and exposed bedrock [1]. Karst rocky desertification results largely from anthropogenic activities such as irrational cultivation and extensive logging. Since the 1980s, karst rocky desertification has been recognized as a critical factor limiting the social and economic development of the karst region in southwest China [8–9].

The Chinese central and local governments have promoted the ecological restoration of karst areas in southwest China, and as a consequence many farmlands and plantations have been abandoned and allowed to recover naturally. Several studies have documented the succession of plant communities in the karst region in southwest China. For example, Zeng et al. (2007) identified the following six plant communities occurring during succession following the natural revegetation of areas suffering from rocky desertification: sparse grass, grass, shrub, liana-shrub, deciduous broad-leaved forest, and mixed evergreen and deciduous broad-leaved forest [8]. An earlier study had also identified six vegetation types: grass, grass-shrub, shrub, shrub-tree, tree forest, and climax mixed evergreen and deciduous broad-leaved forest [10]. Although the classification of the plant communities is somewhat different in these two studies, the succession from grass to shrub to forest is considered typical in this area [11–12].

The restoration of degraded karst areas in southwest China is currently evaluated based on the degree of forest vegetation establishment but the restoration of ecosystem services will also depend on the recovery of other organisms including soil organisms. Whether a secondary vegetation succession can promote the increment of biomass and abundance of soil biota in degraded karst areas is largely unknown, and this lack of knowledge limits our ability to understand and predict the functioning of such ecosystems.

In studying the succession of plant communities and soil biota, researchers frequently use space-for-time substitution, i.e., chronosequences [13]. In addition, researchers recognize two general types of succession, which are primary and secondary. Secondary succession is widespread and occurs at a high rate relative to primary succession [14]. In the past two decades, a number of studies have explored the relationship between secondary vegetation succession and soil biota. For example, vegetation succession can influence the composition of soil biota through resource and habitat changes [15–17], and the soil biota can affect the vegetation succession by selective feeding and by altering nutrient availability [18–19]. Few studies, however, have explored the changes in soil biota (especially nematodes) during secondary succession from grassland to other ecosystem types, and these studies have yielded inconsistent results concerning the relationship between vegetation succession and soil biota succession. For example, soil microbial biomass increased during the secondary succession of abandoned fields ranging in age from 1 to 60 years in east-central Minnesota [17] but declined across a secondary vegetation succession from short grassland to tall grassland to shrub-land in northwest France [20]. In addition, soil microbial biomass did not differ between early and late stages in a secondary succession from shrubland to forest in New Zealand [16].

In this study, we determined whether the progressive succession of the plant community is accompanied by the establishments of soil microbial and nematode communities in karst shallow soils; by progressive succession, we mean that the community becomes more complex, with increased species richness, community biomass, and food web trophic links. We explored the compositions of soil microbial and nematode communities along a secondary vegetation succession from grassland to shrubland to forest in the karst region in southwest China, where ecosystems have been severely disrupted and where the soils are very shallow. We test the hypothesis that progressive succession of the plant community should improve the soil habitat (because of increases in soil organic matter and nutrients) and thereby support the establishments of the soil microbial and nematode communities.

Materials and Methods

No specific permits were required for the described field studies. No specific permissions were required for these locations/activities. The location is not privately owned or protected in any way and the field studies did not involve endangered or protected species.

Site description

This study was carried out at the Huanjiang Observation and Research Station for Karst Ecosystems (107°51′–108°43′E, 24°44′–25°33′N), Chinese Academy of Sciences (CAS), Guangxi Province, China. The climate is subtropical monsoon with a distinct wet (from April to September) and dry season (from October to March). The mean annual temperature and precipitation are 18.5°C and 1,389 mm, respectively. The soil developed from a dolostone base and is calcareous.

Three plant communities (or ecosystems) were identified on the hillsides around the Huanjiang Observation and Research Station: grassland, shrubland, and forest. The main criteria used in selecting and designating these vegetation types were as follows: the plant community composition should be different, the disturbance history of the site should be known (based on information obtained from local farmers), and the environmental conditions (soil origin, altitude, slope orientation, and slope gradient) should be as similar as possible. All three vegetation types were situated in the middle of south-facing slopes with gradients ranging from 30 to 50°. The altitude ranged from 280 to 464 m. The grasslands were developed from managed forests

Table 1. Soil physico-chemical properties at two depths in a grassland, shrubland, and forest in the degraded karst region of southwest China.

Property	Depth (cm)	Vegetation type		
		Grassland	Shrubland	Forest
SWC (%)	0–5	47.3 (1.4)	54.9 (4.4)	44.7 (8.3)
	5–10	43.2 (2.9)	45.8 (2.6)	40.1 (6.3)
pH_WATER	0–5	8.17 (0.02) a^a	7.53 (0.32) b	6.89 (0.42) c
	5–10	8.20 (0.03)	7.60 (0.31)	6.85 (0.33)
TN (g/kg)	0–5	4.79 (0.22)	4.29 (0.20)	5.25 (0.79)
	5–10	4.33 (0.07)	3.56 (0.14)	4.29 (0.73)
SOC (g/kg)	0–5	96.9 (5.0) a	67.8 (9.6) b	56.5 (8.8) b
	5–10	87.9 (6.1)	50.6 (13.6)	39.4 (6.9)
C:N	0–5	20.4 (1.9) a	16.1 (3.1) a	10.8 (0.5) b
	5–10	20.4 (1.7)	14.5 (4.3)	9.2 (0.5)

^{a}Means for a property (averaged over both soil depths) followed by different letters are significantly different ($p<0.05$) according to the LSD test.
SWC (%), pH, TN (g kg^{-1} dry soil), SOC (g kg^{-1} dry soil), and C:N stand for soil water content, soil pH, soil total nitrogen, soil organic carbon, and ratio of soil organic carbon to total nitrogen, respectively.

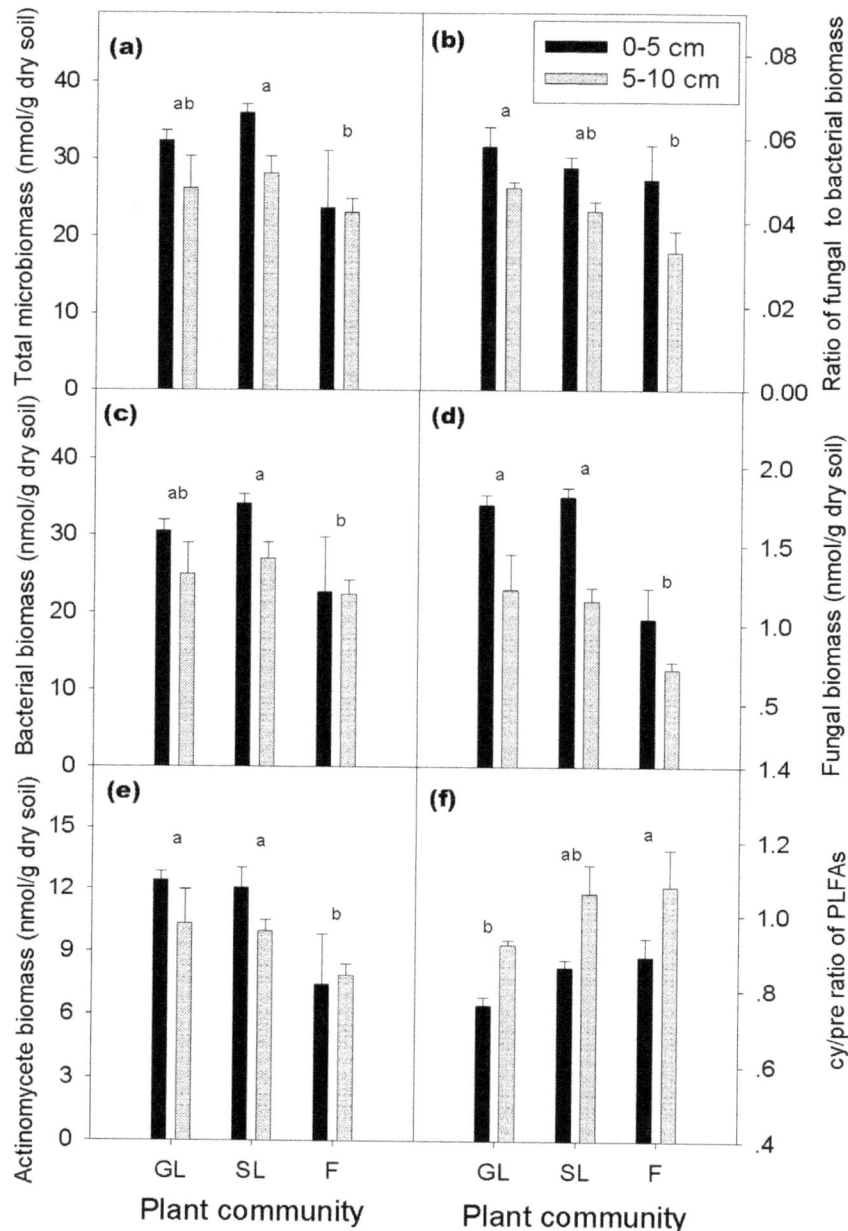

Figure 1. Characteristics of microbial communities (as determined by PLFA analysis) in the grassland (GL), shrubland (SL), and forest (F) at 0–5 and 5–10 cm soil depths. (a) Microbial biomass; (b) Ratio of fungal biomass to bacterial biomass; (c) Bacterial biomass; (d) Fungal biomass; (e) Actinomycetes biomass; and (f) cy/pre ratio. Bars indicate standard errors of means. Within each vegetation type, values with the same letters are not significantly different ($p > 0.05$) according to the LSD test.

which had been abandoned for 9 to 10 years, and wildfire had occurred three times since all of the trees had been harvested and replaced by grasses. The dominant plant in the grassland was *Miscanthus floridulu*. The shrublands were developed from managed forests which had been abandoned for 25 to 30 years, at which time all of the trees were removed (i.e., all of the wood was harvested). The common shrub species were *Loropetalum chinensis*, *Litsea coreana* lvl. var. *Sinensis*, and *Pittosporum tonkinense Gagnep*. The forest sites had previously been managed forests and had been abandoned 50 to 55 earlier. The dominant overstory species in the forest was *Cyclobalanopsis glauca*; the

common understory species included *Litsea coreana* levl. var. *Sinensis* and *Loropetalum chinensis*.

In 2011, the three identified plant communities at nine sites were selected, among which each plant community including three sites (replicates). A 10×10 m plot was established at each of the grassland and shrubland site, while a 20×30 m plot was established at each of the forest site. Because of anthropogenic disturbances in this area over the past decades, no primary forest ecosystem on calcareous soil that developed from a dolomite base existed near the research station. Therefore, no primary forest ecosystems were included in the present study.

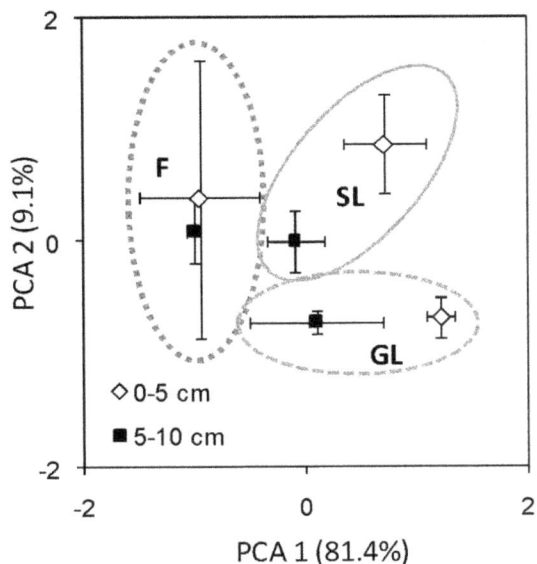

Figure 2. Principal component analysis (PCA) of microbial PLFA biomarkers of the grassland (GL), shrubland (SL), and forest (F) at 0–5 and 5–10 cm soil depth.

Soil sampling and analysis

Soil was sampled in September 2012. Soil cores (2.5 cm in diameter, 5 cm in length) were taken at 0–5 cm and 5–10 cm depths from 15 random locations within each of the nine plots. The 15 cores of the same depth from each plot were combined to form one composite sample, giving three replicate samples for each combination of vegetation type and depth. The surface litter was carefully removed before soil cores were collected. Samples were brought to the laboratory in insulated boxes. Then, the soil was sieved to pass through a 2 mm sieve to remove the roots. The soil water content, microbial and nematode communities were determined on a fresh subsample. Another subsample was thereafter air-dried and used for determination of soil pH, soil organic C and total nitrogen.

Soil pH was determined in 1:2.5 (w/v) soil: water suspensions, and soil water content (SWC %, g of water per 100 g dry soil) was measured by oven-drying the soil for 48 h at 105°C. Soil organic C (SOC, g kg^{-1} dry soil) was determined by dichromate oxidation, and total N (TN, g kg^{-1} dry soil) was measured with an ultraviolet spectrophotometer after Kjeldahl digestion [21].

Phospholipid fatty acids (PLFA) were extracted from 8 g fresh soil and were analyzed according to Bossio and Scow (1998) [22]. Concentrations of each PLFA were calculated relative to 19:0 internal standard concentrations. Bacterial biomass was considered to be represented by 10 PLFAs (i15:0, a15:0, 15:0, i16:0, 16:1ω7, i17:0, a17:0, 17:0, cy17:0, and cy19:0), fungal biomass was considered to be represented by the PLFA 18:2ω6,9, and actinomycete biomass was considered to be represented by three PLFAs (10 Me 16:0, 10 Me 17:0, and 10 Me 18:0) [23–25]. Microbial biomass was considered to be represented by the 10 bacterial PLFAs and the one fungal PLFA. Other PLFAs such as 16:1ω5, 18:1ω7, and 18:1ω9 were also used to analyze the soil microbial community.

Nematodes were extracted from 50 g of fresh soil using the Baermann funnel method [26]. After fixation in a 4% formalin solution, nematodes were counted with a differential interference contrast (DIC) microscope (ECLIPSE 80i, Nikon), and the first 100 individuals encountered were identified to genus. All nematodes were identified to genus when the sample contained fewer than 100 individuals. Nematodes were assigned to main trophic groups (bacterivores, fungivores, herbivores, omnivores, and predators) [27] and colonizer-persister guilds [28].

The microbial data were used to calculate a ratio of fungal biomass to bacterial biomass (F:B) and a cy/pre ratio of PLFAs [(cy17:0+cy19:0)/(16:1ω7c+18:1ω7c)]. The nematode data were used to calculate an enrichment index (EI), a structure index (SI), a maturity index (MI), and a plant-parasitic index (PPI) for each sample [29–30].

Data analysis

Two-way ANOVAs were used to examine the main and interaction effects of plant community and soil depth on dependent variables describing soil physico-chemical properties and soil microbial and nematode communities. Soil physico-chemical properties and microbial community composition were also analyzed by transforming the data to their principal components (PCA) and analyzing these using ANOVAs [31–32]. Data were transformed (natural log, square root, or rank) to meet assumptions of normality and homogeneity of variance. Statistical significance was determined at $p<0.05$. ANOVAs and PCAs were performed using SPSS software (SPSS Inc., Chicago, IL). LSD was used to test differences among treatment means. Tamhane's T2 was used to test differences among treatments when variances of transformed data were unequal. Redundancy analysis (RDA) was used to determine the relationship between soil physico-chemical properties and microbial and nematode community indices, microbial biomass, or nematode abundance. Soil layer was converted to a nominal variable (0, 1) and was treated as a covariable [33–35]. Monte Carlo permutation tests were used to compute statistical significance. RDA was performed with CANOCO 4.5 software [33].

Results

Soil physico-chemical properties

Soil pH was highest in the grassland, lowest in the forest, and intermediate in the shrubland (Table 1). Soil organic carbon (SOC) was greater in the grassland than in the shrubland and forest. Soil C/N was greater in the grassland and shrubland than in the forest. Soil water content (SWC) and total nitrogen (TN) content, however, did not differ among the plant communities (Table 1).

Soil microbial community

Microbial biomass and bacterial biomass were higher in the shrubland than in the forest (Fig. 1a and c). Fungal biomass and actinomycete biomass were higher in the grassland and shrubland than in the forest (Fig. 1d and e). The ratio of fungal to bacterial biomass (F:B) decreased gradually from grassland to shrubland to forest and was significantly higher in the grassland than in the forest (Fig. 1b). The cy/pre ratio of PLFAs increased from grassland to shrubland to forest (Fig. 1f) and was significantly higher in the forest than in the grassland ($p<0.05$) and tended to be higher in the shrubland than in the grassland ($p = 0.058$). In addition, ANOVA results of the transformed principal components of the soil microbial community showed that the microbial community structure of the forest differed from that of the grassland and shrubland ($p<0.01$) (Fig. 2). Moreover, the soil microbial community structure tended to differ between the grassland and shrubland ($p = 0.069$) (Fig. 2).

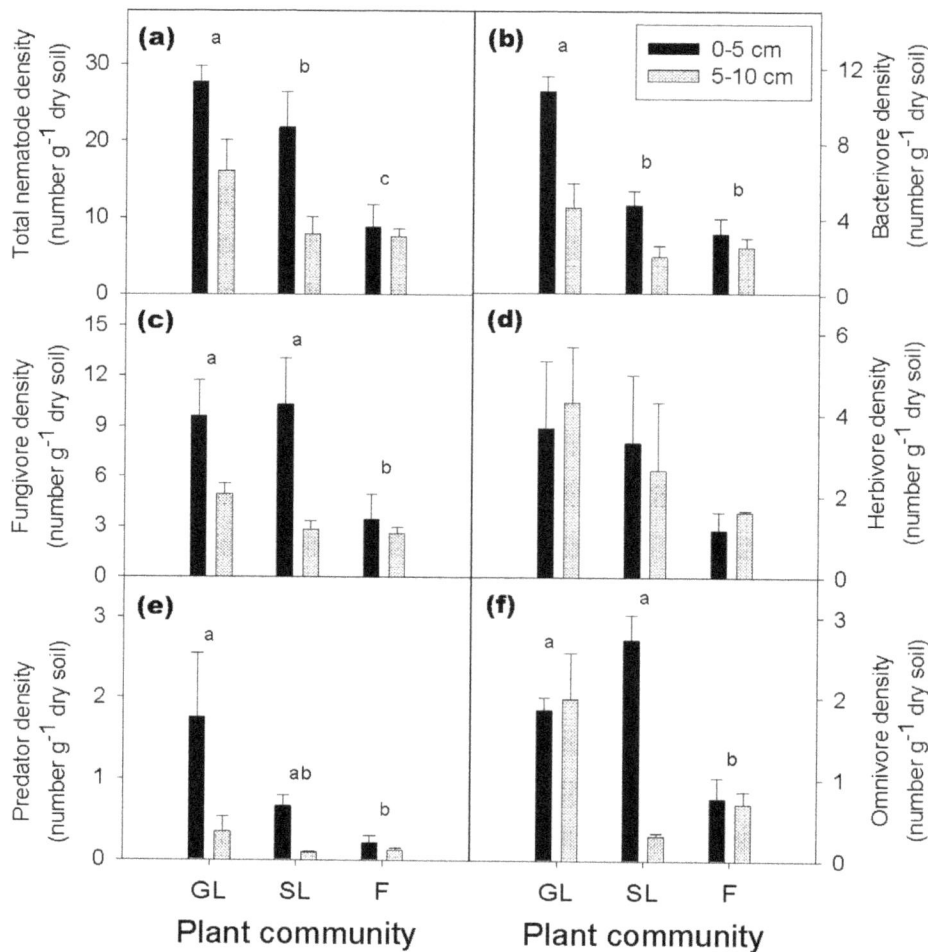

Figure 3. Densities of total nematodes and different nematode trophic groups in the grassland (GL), shrubland (SL), and forest (F) at 0–5 and 5–10 cm soil depth. (a) Total nematodes; (b) Bacterivores; (c) Fungivores; (d) Herbivores; (e) Predators; and (f) Omnivores. Bars indicate standard errors of means. Within each panel, values (averaged across both depths for each plant community) with the same or no letters are not significantly different ($p > 0.05$) according to the LSD test.

Soil nematode community

Fungivores and bacterivores were the most abundant trophic groups (Fig. 3b and c). Total nematode density declined gradually from grassland to shrubland to forest at both 0–5 and 5–10 cm soil depths (Fig. 3a). Bacterivore density was greater in the grassland than in the shrubland and forest (Fig. 3b). Fungivore density and omnivore density were greater in the grassland and shrubland than in the forest (Fig. 3c and f); omnivore density in the shrubland, however, was very low at 5–10 cm depth (Fig. 3f). Predator density declined gradually from grassland to shrubland and forest and was significantly greater in the grassland than in the forest ($p = 0.023$) (Fig. 3e). Herbivore density declined gradually from grassland to shrubland to forest but did not significantly differ among the three ecosystems (Fig. 3d).

The weighted soil nematode fauna analysis showed that the nematode communities of all three ecosystems at both 0–5 and 5–10 cm soil depths projected onto quadrat III of the nematode fauna plot (Fig. 4a). The composition of the soil nematode community differed among the three ecosystems, and the value of the nematode structure index (SI) was greater in the grassland than in the forest ($p = 0.003$) (Fig. 4b). The SI value of the shrubland was intermediate to that of the grassland and forest, and

tended to differ from that of the grassland ($p = 0.092$) and of the forest ($p = 0.086$). The nematode enrichment index (EI) was not significantly influenced by ecosystem type.

Relationships between soil physico-chemical properties and soil biota

The first two canonical axes explained about 71.6%, 51.9%, and 62.1% of the total variance of microbial and nematode community indices (Fig. 5a), microbial community biomass (Fig. 5b), and nematode community density (Fig. 5c), respectively. The microbial and nematode community indices were related to soil water content ($p = 0.012$, F = 7.99), SOC ($p = 0.004$, F = 7.46), and soil pH ($p = 0.040$, F = 3.44) (Fig. 5a). For each index, SI and MI were mainly and positively correlated with SOC; EI and F:B were mainly and positively correlated with pH; cy/pre was mainly and negatively correlated with pH; and PPI was mainly and positively correlated with TN (Fig. 5a). The microbial community biomass tended to be correlated with soil pH ($p = 0.052$, F = 4.02) (Fig. 5b). Total microbial biomass, fungal biomass, bacterial biomass, and actinomycetes biomass were mainly and negatively correlated with TN (Fig. 5b). Nematode community density was significantly related to SOC ($p = 0.010$, F = 8.19) (Fig. 5c). Total

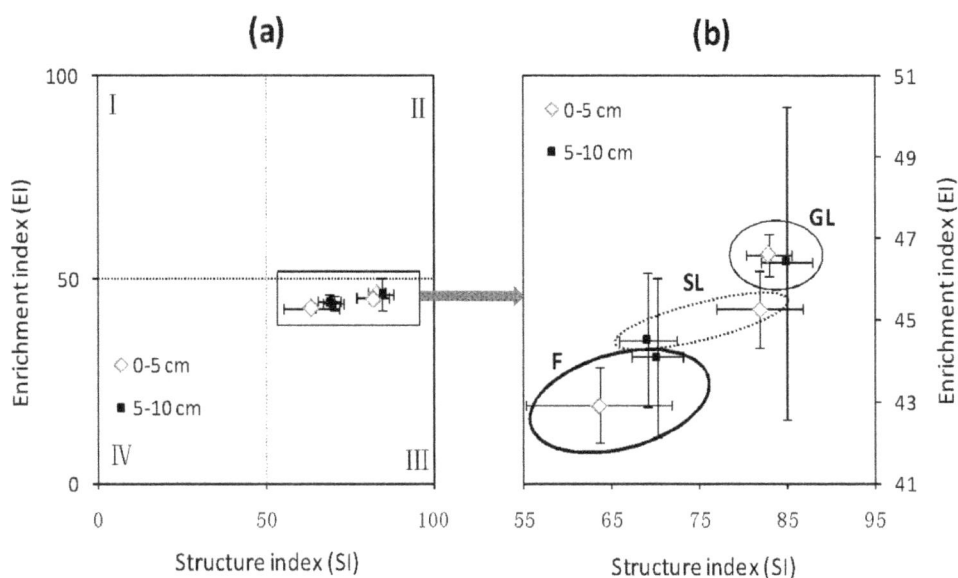

Figure 4. Weighted soil nematode faunal analysis at 0–5 depth and 5–10 cm depth in three vegetation types (GL, grassland; SL, shrubland; and F, forest). (a) Bi-plot with four quadrats; (b) Magnified quadrate-area of bi-plot A. Bars indicate the standard errors of means.

nematode density, omnivore density, and predator density were mainly and positively correlated with pH and SOC; fungivore density was mainly and positively correlated with soil water content and C/N; and herbivore density and bacterivore density were positively correlated with TN, pH, and SOC (Fig. 5c).

Discussion

Succession of soil microbial communities

In this study of a degraded karst region, the progressive secondary succession in vegetation was generally accompanied by a decline of the soil microbial community in that microbial biomass and other microbial variables decreased as the plant community succeeded from grassland to shrubland to forest. Consistent with our results, microbial biomass declined along a secondary succession from grassland to early stages of forest in calcareous soils in northwestern France [20]. In contrast, a previous study reported that succession from grassland to forest increased the soil microbial biomass and biodiversity in karst calcareous soils in southwest China [36]. Another study reported that bacterial phylogenetic diversity was higher but that bacterial metabolic diversity was lower in a grassland than in a forest in karst calcareous soils in southwest China [11]. Conversely, soil microbial biomass decreased with intensity of land degradation from forest to shrubland and grassland in karst calcareous soils in southwest China [37]. In these previous studies in karst areas in southwest China, the responses of soil microbial communities (mainly total microbial biomass) to vegetation succession could be ascribed to the shifts in soil organic carbon, i.e., microbial biomass was positively correlated with soil organic carbon. Although this was also the case in this study, soil organic carbon declined with plant succession. It has been well established that soil organic carbon is usually greater in grasslands than in forests [20,38], and land use changes from grassland to forest have decreased soil carbon stocks worldwide [39]. It is also possible that the relationship between microbial communities and soil organic carbon differs among ecosystems. For instance, the relationship between the F/B ratio and soil organic carbon was positive in

agricultural soils but negative in prairie soils at Fermilab, Batavia, in Illinois [40].

In the present study, the increase in the cy/pre ratio with secondary succession might indicate that the soil habitat was becoming more harsh with plant succession. Many previous studies have documented that increasing cy/pre ratios is associated with stress and starvation (e.g., anaerobic conditions, nutrient stress, and water stress) [22,41–42]. Moreover, soil microbial communities in the current study were primarily influenced by soil pH, and the cy/pre ratio was negatively correlated with soil pH. Therefore, soil acidification and the deterioration of the soil habitat occurring with secondary vegetation succession might be the primary environmental factors affecting soil microbial communities. Previous studies had demonstrated that soil acidification significantly reduced the soil microbial biomasses and/or activities and soil nematode abundances [43–45]. In addition, many studies have reported that succession from grassland to forest decreased the soil pH [15,46–48]. Soil pH in the present study was not only negatively associated with the cy/pre ratio but was also positively associated with the F:B ratio. In agreement with our results, the F:B ratio increased slightly with increasing pH as reported by [49]. However, the relationship between the F:B ratio and soil pH is inconsistent [24,50]. Although soil total nitrogen was not influenced by vegetation type, total microbial biomass, bacterial biomass, fungal biomass, and actinomycetes biomass tended to be negatively correlated with soil TN. The reasons for this unexpected result are unclear.

Succession of soil nematode communities

Like soil microbial biomass, soil nematode abundance showed declined patterns with respect to the progressive successional patterns of the plant communities in this study. Few studies have explored the soil nematode communities associated with plant community succession from grassland to forest or from forest to grassland. One study reported that soil nematode densities and biodiversity (i.e., plant feeders and non-plant feeders) increased with secondary succession from savanna to forest in the tropical semi-arid zone of West Africa (Senegal) [51]. Although many

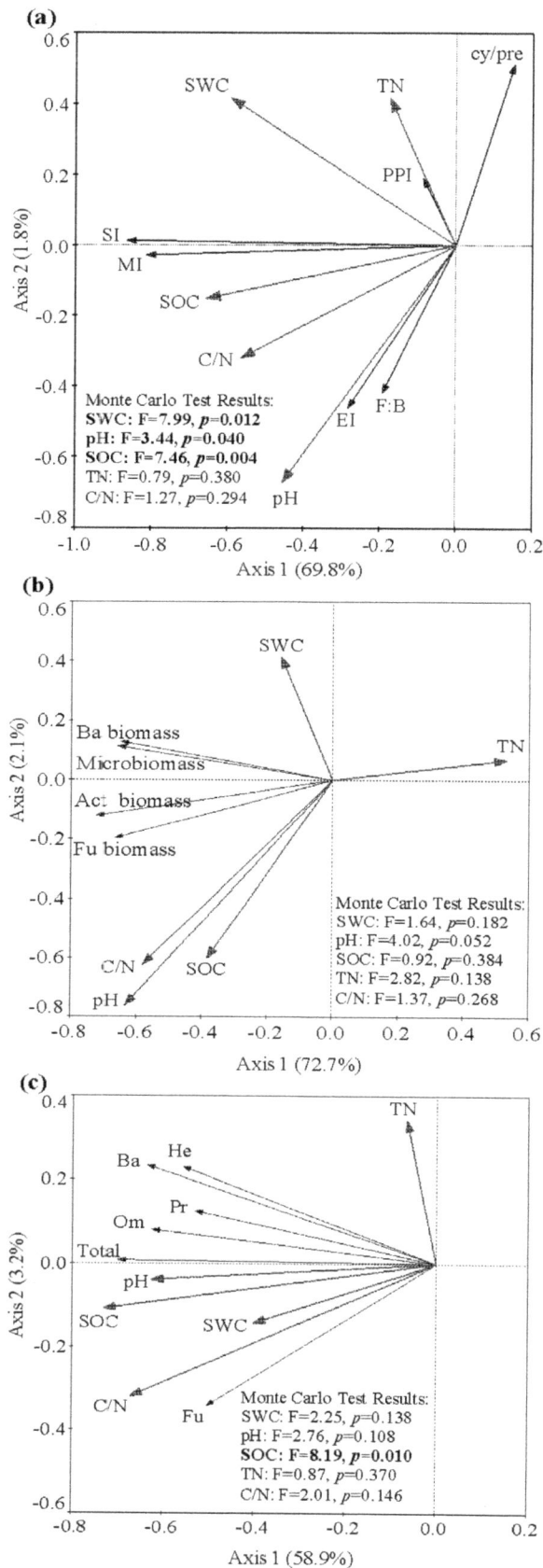

(a)

(b)

(c)

Figure 5. Redundancy analysis (RDA) of microbial and nematode community indices (a), microbial community biomass (b), and nematode community density (c). Ordination diagrams present species scores and environmental factor scores (vectors). Environmental factors include soil water content (SWC), soil pH, soil total nitrogen (TN), soil organic carbon (SOC), and ratio of soil organic carbon to total nitrogen (C:N). Microbial and nematode community indices include ratio of fungal biomass to bacterial biomass (F:B), cy/pre ratio of PLFAs (cy/pre), nematode maturity index (MI), plant parasite index (PPI), structure index (SI), and enrichment index (EI). Variables of microbial community biomass include total microbial biomass (Microbiomass), fungal biomass (Fu biomass), bacterial biomass (Ba biomass), and actinomycetes biomass (Act biomass). Variables of nematode community density include densities of total nematodes (Total), bacterivores (Ba), fungivores (Fu), herbivores (He), omnivores (Om), and predators (Pr).

previous studies have compared the compositions of soil nematode communities of grasslands, shrublands, and forests, it is unclear how inferences based on ecosystem type can be applied to ecosystem succession. For example, Wasilewska (1979) reported that densities of herbivorous, omnivorous, carnivorous, and total nematodes and number of nematode genera were higher in grassland than in forest ecosystems but that densities of bacterivorous and fungivorous nematodes were similar in grassland and forest ecosystems [52]. In another study, land-use change from grassland to shrub heathland suppressed densities of nematode trophic groups (i.e., herbivores, bacterivores, fungivores, omnivores, and predators) in northwest Europe [47].

The weighted soil nematode fauna analysis showed that the nematode communities of all three ecosystems projected onto quadrat III in the nematode fauna plot, which demonstrated that the structures of the soil food webs of all three ecosystems were relatively similar [29]. However, the values of the nematode structure index (SI) decreased from grassland to shrubland to forest ecosystems, which demonstrated that the complexity of the soil food webs decreased in the order grassland > shrubland > forest. In addition, soil organic carbon was the main factor associated with total nematode density and the density of each trophic group. In agreement with our results, many studies have reported that soil organic carbon affects nematode communities [53–55]. Soil nematodes are also affected by spatial heterogeneity, plant community composition, and the soil environment [56]. Although the nematode plant-parasitic index (PPI) was positively correlated with soil total nitrogen, and although the enrichment index (EI) was positively correlated with pH, soil total nitrogen and EI did not significantly change with the secondary vegetation succession. Therefore, soil nematode communities might be more affected by the main soil resource (i.e., soil organic carbon) than by pH or other components of the soil habitat.

Potential implications for land conservation and vegetation recovery in karst areas

The aim of traditional ecosystem restoration in degraded karst areas has usually been to restore forest ecosystems [4–5,8,10]. Our study clearly shows that ecosystem restoration in degraded karst areas might not necessarily be favored by secondary succession to forests. Based on the results of the current study, soil carbon storage and soil biodiversity are greater in grasslands and shrublands than in forests in degraded karst areas. The low soil organic carbon content and biota biomass associated with the forest might be explained in part by karst soils, which are generally very shallow and which can provide only low levels of nutrients and water [57]. The limitations of nutrient and water supply can be exacerbated when the live plant biomass increases during succession [58]. In addition, because, the live biomass is greater for forests than for grasslands or shrublands, natural and/or anthropogenic disruptions (e.g., disease and pest outbreak, drought, harvest, and contamination) may cause greater material loss from forests than from grasslands or shrublands. Consequently, natural and/or anthropogenic disruptions may be more damaging in forests than in grassland or shrubland ecosystems. Therefore, soil carbon and biota ought to be amended in degraded karst areas during vegetation succession or vegetation recovery.

Karst areas of southwest China, Southeast Asian, and Central America have large human populations [1–2]. The challenge is to provide the basic needs of these populations while protecting the fragile karst ecosystems. Agricultural development is the cause of soil erosion and rocky desertification in karst areas [1,12]. As noted in the previous paragraph, disruptions of karst forest ecosystems might be especially destructive. Thus, the cultivation of crops and forestry does not seem to be suitable for maintaining karst ecosystems. In contrast, herbivorous animal husbandry in karst areas of southwest China has provided economic benefits to local farmers while reducing soil erosion [9]. In addition, the perennial C4 grass *Miscanthus* (i.e., the dominant genus in the grassland of the current study) has produced substantial biomass crops worldwide [59]. Another perennial C4 grass, hybrid napiergrass (*Pennisetum hydridum*), has also been widely cultivated and is mainly used for raising livestock in karst areas in southwest China and in other subtropical and tropical areas [60].

Most restoration studies have focused on soil nutrient recovery rather than on soil biota recovery, and many previous studies have used a "space-for-time substitution approach" to explore the soil nutrient status during succession from grassland to forest in karst areas. Most of these studies have reported that soil nutrients accumulate with secondary succession [11,36–37] but that was not

the case in the current study. This difference might be explained by the differences in the grasslands. In contrast to grasslands in the earlier studies, the studied grassland was much older and had experienced repeated wildfires.

Conclusions

Contrary to our hypothesis, the progressive succession of vegetation occurring in karst sites in southwest China was accompanied by declines in soil biota abundances and food web complexity. The declines of the soil microbial and nematode communities might be explained by changes in soil habitat (i.e., reduced pH) and resource availability (i.e., reduced soil organic carbon). Because sustainable ecosystem management should integrate aboveground and belowground ecosystem properties and functions [61–62], our results indicate that soil biota amendment is necessary in karst areas during ecosystem restoration and management and establishment of grasslands in degraded karst areas are feasible. This finding provides a scientific basis for the development of herbivorous animal husbandry in karst areas that are under population pressure.

Acknowledgments

We thank Bruce Jaffee for his help in preparing the manuscript. We also thank the editor and two anonymous reviewers for their comments.

Author Contributions

Conceived and designed the experiments: JZ XH KW. Performed the experiments: JZ SL LL. Analyzed the data: JZ SL. Contributed reagents/materials/analysis tools: XH LL. Wrote the paper: JZ SL KW.

References

1. Yuan D (2002) Geology and Geohydrology of Karst and its Relevance to Society; Paris. UNESCO. pp. 15–18.
2. Tuyet D (2001) Characteristics of karst ecosystems of Vietnam and their vulnerability to human impact. Acta Geologica Sinica 75: 325–329.
3. Parise M, De Waele J, Gutierrez F (2009) Current perspectives on the environmental impacts and hazards in karst. Environmental Geology 58: 235–237.
4. Rivera LW, Aide TM (1998) Forest recovery in the karst region of Puerto Rico. Forest Ecology and Management 108: 63–75.
5. Rivera LW, Zimmerman JK, Aide TM (2000) Forest recovery in abandoned agricultural lands in a karst region of the Dominican Republic. Plant Ecology 148: 115–125.
6. Querejeta JI, Estrada-Medina H, Allen MF, Jiménez-Osornio JJ (2007) Water source partitioning among trees growing on shallow karst soils in a seasonally dry tropical climate. Oecologia 152: 26–36.
7. Latinne A, Waengsothorn S, Herbreteau V, Michaux J (2011) Thai limestone karsts: an impending biodiversity crisis. The 1st Environment Asia International Conference. Bangkok, Thailand. pp. 176–187.
8. Zeng F, Peng W, Song T, Wang K, Wu H, et al. (2007) Changes in vegetation after 22 years' natural restoration in the Karst disturbed area in northwestern Guangxi, China. Acta Ecologica Sinica 27: 5110–5119.
9. Wang K, Su Y, Zeng F, Chen H, Xiao R (2008) Ecological process and vegetation restoration in karst region of southwest China. Research of Agricultural Modernization 29: 641–645 (in Chinese with English abstract).
10. Yu L, Zhu S, Wei L, Chen Z, Xiong Z (1998) Study on the natural restoration process of degraded Karst communities–Successional sere. Journal of Mountain Agriculture and Biology 2: 71–77. In Chinese with English abstract.
11. Xiong Y, Xia H, Li Za, Cai Xa, Fu S (2008) Impacts of litter and understory removal on soil properties in a subtropical *Acacia mangium* plantation in China. Plant and Soil 304: 179–188.
12. Qi X, Wang K, Zhang C (2013) Effectiveness of ecological restoration projects in a karst region of southwest China assessed using vegetation succession mapping. Ecological Engineering 54: 245–253.
13. Walker LR, Wardle DA, Bardgett RD, Clarkson BD (2010) The use of chronosequences in studies of ecological succession and soil development. Journal of Ecology 98: 725–736.
14. Walker LR (1999) Ecosystems of disturbed ground. Amsterdam: Elsevier.
15. Schipper LA, Degens BP, Sparling GP, Duncan LC (2001) Changes in microbial heterotrophic diversity along five plant successional sequences. Soil Biology and Biochemistry 33: 2093–2103.
16. Wardle D (1993) Changes in the microbial biomass and metabolic quotient during leaf litter succession in some New Zealand forest and scrubland ecosystems. Functional Ecology 7: 346–355.
17. Zak D, Grigal D, Gleeson S, Tilman D (1990) Carbon and nitrogen cycling during old-field succession: Constraints on plant and microbial biomass. Biogeochemistry 11: 111–129.
18. van der Heijden MGA, Bardgett RD, van Straalen NM (2008) The unseen majority: soil microbes as drivers of plant diversity and productivity in terrestrial ecosystems. Ecology Letters 11: 296–310.
19. De Deyn GB, Raaijmakers CE, Zoomer HR, Berg MP, de Ruiter PC, et al. (2003) Soil invertebrate fauna enhances grassland succession and diversity. Nature 422: 711–713.
20. Chabrerie O, Laval K, Puget P, Desaire S, Alard D (2003) Relationship between plant and soil microbial communities along a successional gradient in a chalk grassland in north-western France. Applied Soil Ecology 24: 43–56.
21. Liu G (1996) Analysis of soil physical and chemical properties and description of soil profiles. Beijing: China Standard.
22. Bossio DA, Scow KM (1998) Impacts of carbon and flooding on soil microbial communities: Phospholipid fatty acid profiles and substrate utilization patterns. Microbial Ecology 35: 265–278.
23. Joergensen RG, Wichern F (2008) Quantitative assessment of the fungal contribution to microbial tissue in soil. Soil Biology and Biochemistry 40: 2977–2991.
24. Frostegård A, Bååth E (1996) The use of phospholipid fatty acid analysis to estimate bacterial and fungal biomass in soil. Biology and Fertility of Soils 22: 59–65.
25. Frostegård Å, Tunlid A, Bååth E (2011) Use and misuse of PLFA measurements in soils. Soil Biology and Biochemistry 43: 1621–1625.
26. Barker KR, Carter CC, Sasser JN (1985) An advanced treatise on Meloidogyne. Volume II: Methodology. Raleigh: United States Agency for International Development. vi + 223 pp. p.
27. Yeates GW, Bongers T, Degoede RGM, Freckman DW, Georgieva SS (1993) Feeding habits in soil nematode families and genera-an outline for soil ecologists. Journal of Nematology 25: 315–331.
28. Bongers T, Bongers M (1998) Functional diversity of nematodes. Applied Soil Ecology 10: 239–251.

29. Ferris H, Bongers T, de Goede RGM (2001) A framework for soil food web diagnostics: extension of the nematode faunal analysis concept. Applied Soil Ecology 18: 13–29.

30. Neher DA (2001) Role of nematodes in soil health and their use as indicators. Journal of Nematology 33: 161–168.

31. Zhao J, Wan S, Li Z, Shao Y, Xu G, et al. (2012) *Dicranopteris*-dominated understory as major driver of intensive forest ecosystem in humid subtropical and tropical region. Soil Biology and Biochemistry 49: 78–87.

32. Zhao J, Wan S, Fu S, Wang X, Wang M, et al. (2013) Effects of understory removal and nitrogen fertilization on soil microbial communities in *Eucalyptus* plantations. Forest Ecology and Management 310: 80–86.

33. Lepš J, Šmilauer P (2003) Multivariate analysis of ecological data using CANOCO. New York: Cambridge University Press. xii + 269 pp. p.

34. Zhao J, Neher D (2013) Soil nematode genera that predict specific types of disturbance. Applied Soil Ecology 64: 135–141.

35. Zhao J, Shao Y, Wang X, Neher DA, Xu G, et al. (2013) Sentinel soil invertebrate taxa as bioindicators for forest management practices. Ecological Indicators 24: 236–239.

36. Zhu H, He X, Wang K, Su Y, Wu J (2012) Interactions of vegetation succession, soil bio-chemical properties and microbial communities in a Karst ecosystem. European Journal of Soil Biology 51: 1–7.

37. Zhang P, Li L, Pan G, Ren J (2006) Soil quality changes in land degradation as indicated by soil chemical, biochemical and microbiological properties in a karst area of southwest Guizhou, China. Environmental Geology 51: 609–619.

38. Evrendilek F, Celik I, Kilic S (2004) Changes in soil organic carbon and other physical soil properties along adjacent Mediterranean forest, grassland, and cropland ecosystems in Turkey. Journal of Arid Environments 59: 743–752.

39. Guo LB, Gifford RM (2002) Soil carbon stocks and land use change: a meta analysis. Global Change Biology 8: 345–360.

40. Allison VJ, Miller RM, Jastrow JD, Matamala R, Zak DR (2005) Changes in soil microbial community structure in a tallgrass prairie chronosequence. Soil Science Society of America Journal 69: 1412–1421.

41. Schmitt A, Glaser B (2011) Organic matter dynamics in a temperate forest soil following enhanced drying. Soil Biology and Biochemistry 43: 478–489.

42. Kieft TL, Ringelberg DB, White DC (1994) Changes in ester-linked phospholipid fatty acid profiles of subsurface bacteria during starvation and desiccation in a porous medium. Applied and Environmental Microbiology 60: 3292–3299.

43. Räty M, Huhta V (2003) Earthworms and pH affect communities of nematodes and enchytraeids in forest soil. Biology and Fertility of Soils 38: 52–58.

44. Chen D, Lan Z, Bai X, Grace JB, Bai Y (2013) Evidence that acidification-induced declines in plant diversity and productivity are mediated by changes in below-ground communities and soil properties in a semi-arid steppe. Journal of Ecology.

45. Rousk J, Brookes PC, Baath E (2010) The microbial PLFA composition as affected by pH in an arable soil. Soil Biology & Biochemistry 42: 516–520.

46. Dmowska E, Ilieva-Makulec K (2006) Secondary succession of nematodes in power plant ash dumps reclaimed by covering with turf. European Journal of Soil Biology 42: S164–S170.

47. Holtkamp R, Kardol P, van der Wal A, Dekker SC, van der Putten WH, et al. (2008) Soil food web structure during ecosystem development after land abandonment. Applied Soil Ecology 39: 23–34.

48. Robertson GP, Vitousek PM (1981) Nitrification potentials in primary and secondary succession. Ecology: 376–386.

49. Bååth E, Anderson T-H (2003) Comparison of soil fungal/bacterial ratios in a pH gradient using physiological and PLFA-based techniques. Soil Biology and Biochemistry 35: 955–963.

50. Rousk J, Brookes PC, Bååth E (2009) Contrasting soil pH effects on fungal and bacterial growth suggest functional redundancy in carbon mineralization. Applied and Environmental Microbiology 75: 1589–1596.

51. Pate E, Ndiaye-Faye N, Thioulouse J, Villenave C, Bongers T, et al. (2000) Successional trends in the characteristics of soil nematode communities in cropped and fallow lands in Senegal (Sonkorong). Applied Soil Ecology 14: 5–15.

52. Wasilewska L (1979) The structure and function of soil nematode communities in natural ecosystems and agrocenoses. Polish Ecological Studies 5: 97–145.

53. Yeates G, Bongers T (1999) Nematode diversity in agroecosystems. Agriculture, Ecosystems & Environment 74: 113–135.

54. Liang W, Lou Y, Li Q, Zhong S, Zhang X, et al. (2009) Nematode faunal response to long-term application of nitrogen fertilizer and organic manure in Northeast China. Soil Biology and Biochemistry 41: 883–890.

55. Freckman D (1988) Bacterivorous nematodes and organic-matter decomposition. Agriculture, Ecosystems & Environment 24: 195–217.

56. Viketoft M (2013) Determinants of small-scale spatial patterns: Importance of space, plants and abiotics for soil nematodes. Soil Biology and Biochemistry 62: 92–98.

57. Zhang X, Wang K (2009) Ponderation on the shortage of mineral nutrients in the soil-vegetation ecosystem in carbonate rock-distributed mountain regions in Southwest China. Earth and Environment 37: 337–341 (in Chinese with English abstract).

58. Guo K, Liu C, Dong M (2011) Ecological adaptation of plants and control of rocky-desertification on karst region of Southwest China. Chinese Journal of Plant Ecology 35: 991–999 (in Chinese with English abstract).

59. Matlaga DP, Davis AS (2013) Minimizing invasive potential of Miscanthus × giganteus grown for bioenergy: identifying demographic thresholds for population growth and spread. Journal of Applied Ecology 50: 479–487.

60. Zhao J, Zhang W, Wang K, Song T, Du H (2014) Responses of the soil nematode community to management of hybrid napiergrass: The trade-off between positive and negative effects. Applied Soil Ecology 74: 134–144.

61. Bardgett RD, Bowman WD, Kaufmann R, Schmidt SK (2005) A temporal approach to linking aboveground and belowground ecology. Trends in Ecology & Evolution 20: 634–641.

62. Smith RS, Shiel RS, Bardgett RD, Millward D, Corkhill P, et al. (2003) Soil microbial community, fertility, vegetation and diversity as targets in the restoration management of a meadow grassland. Journal of Applied Ecology 40: 51–64.

16

Impacts of Land Cover Data Selection and Trait Parameterisation on Dynamic Modelling of Species' Range Expansion

Risto K. Heikkinen[1]*, **Greta Bocedi**[2], **Mikko Kuussaari**[1], **Janne Heliölä**[1], **Niko Leikola**[1], **Juha Pöyry**[1], **Justin M. J. Travis**[2]

1 Finnish Environment Institute, Natural Environment Centre, Helsinki, Finland, **2** Institute of Biological Sciences, University of Aberdeen, Aberdeen, United Kingdom

Abstract

Dynamic models for range expansion provide a promising tool for assessing species' capacity to respond to climate change by shifting their ranges to new areas. However, these models include a number of uncertainties which may affect how successfully they can be applied to climate change oriented conservation planning. We used RangeShifter, a novel dynamic and individual-based modelling platform, to study two potential sources of such uncertainties: the selection of land cover data and the parameterization of key life-history traits. As an example, we modelled the range expansion dynamics of two butterfly species, one habitat specialist (*Maniola jurtina*) and one generalist (*Issoria lathonia*). Our results show that projections of total population size, number of occupied grid cells and the mean maximal latitudinal range shift were all clearly dependent on the choice made between using CORINE land cover data vs. using more detailed grassland data from three alternative national databases. Range expansion was also sensitive to the parameterization of the four considered life-history traits (magnitude and probability of long-distance dispersal events, population growth rate and carrying capacity), with carrying capacity and magnitude of long-distance dispersal showing the strongest effect. Our results highlight the sensitivity of dynamic species population models to the selection of existing land cover data and to uncertainty in the model parameters and indicate that these need to be carefully evaluated before the models are applied to conservation planning.

Editor: Francesco de Bello, Institute of Botany, Czech Academy of Sciences, Czech Republic

Funding: This work was supported by the SCALES project (Securing the Conservation of biodiversity across Administrative Levels and spatial, temporal, and Ecological Scales) funded by the European Commission as a Large-scale Integrating Project within FP 7 under grant 226 852 and the Finnish Research Programme on Climate Change (FICCA) project A-LA-CARTE (Assessing limits of adaptation to climate change and opportunities for resilience to be enhanced), funded by the Academy of Finland (decision no. 140846). The funders had no role in study design, data collection and analysis, decision to publish, or preparation of the manuscript.

Competing Interests: The authors have declared that no competing interests exist.

* Email: risto.heikkinen@ymparisto.fi

Introduction

One of the challenges in conservation and management planning is developing robust assessments of the impacts of climate change on species' ranges. To date, such assessments have relied on static 'bioclimatic envelope' ('BEMs'), or 'environmental niche' models ('ENMs') [1,2], which relate the species' distributions to current climate and then project future ranges by fitting the derived models to different climate scenarios. However, the capacity of BEMs to provide useful guidelines for climate change oriented conservation planning is limited. First, their outputs are rather coarse-scaled and provide little understanding of potential differences in species' responses in different parts of the study region [3,4]. Second, BEMs generally do not account for the fact that a species' range expansion depends on the characteristics of the landscape over which individuals disperse [5–7]. Importantly, connectivity of the habitat network has a critical role in species' range dynamics [8,9].

Dynamic models for species' range shifts are a promising tool for conservation biology providing improved possibilities for assessing species' abilities to track the changing climate and persist in a habitat network [6,10,11]. There are a few example of such models, with applications to habitat networks developed at local [12,13], regional [3,14,15] and national scale [16]. However, although these models hold much promise, they have potential caveats which need to be explored to avoid their uncritical implementation [12,17,18].

In this study we address two potentially important sources of uncertainty: the selection of habitat maps and the parameterisation of species' life-history traits. Modelling studies conducted over large areas face the challenge of obtaining sufficiently robust data on the distribution of suitable habitats for the species [19]. This is because more accurate spatial data on species' habitats are often available only for some intensively surveyed localised areas. Land cover databases gathered over large areas, such as national CORINE databases in Europe, rely often on remote sensing and other data sources and are thus likely to show substantially more within-land-cover-type variation in habitat quality than the habitat maps based on intensive field surveys. Recent studies suggest that this lack of fine resolution in habitat classification is a feature of European-wide databases including CORINE 2000 and 2006 and that this may cause biases in modelling [20]. A key problem is that coarse resolution classification can result in larger areas being classified as suitable habitat for a species than there is in reality, especially in the case of habitat specialists. This topic has been surprisingly poorly investigated in the context of models for projecting range expansion, although a few exceptions [12,16] suggest that varying the amount of habitat in the landscape can have a significant impact on the outputs of dynamic models.

A second main challenge and source of uncertainty is to develop accurate estimates for species' dispersal abilities and demographic parameters [21–23]. The confidence in the species' parameters employed in simulations for range expansion is often very limited. Thus, to be useful for conservation, dynamic simulation models should provide estimates of the extent to which model outputs are sensitive to these uncertainties. Sensitivity analyses provide means for addressing this problem and for giving more robust confidence intervals to the projections. A number of studies employing simulated landscapes have shown that projections of species' expansion rates may be rather sensitive to the parameter values for certain key life-history traits [6,24,25]. However, corresponding studies carried out for real species on real landscapes have addressed model sensitivity to various degrees. Some of them have scrutinised the impact of varying several species' parameter values [12,15,16], while others have assessed the model sensitivity to only one or very few life-history parameters [e.g. 13], referred to a priori tests [e.g. 14], or otherwise provided limited information on the sensitivity of model projections to species' parameter selection [e.g. 26].

Here we investigate the impacts of land cover data selection and parameterisation of species' traits on the projected species' range expansion dynamics by using two butterfly species (*Maniola jurtina* and *Issoria lathonia*) inhabiting different types of grasslands in Finland. Butterflies are useful model species for studying range expansion and ecological sufficiency of habitat networks because they have the potential to respond rapidly to climate change [27–29]. Our focal study environment, unimproved grasslands, represents important habitat for nature conservation throughout Europe [30,31]. These habitats are threatened due to agricultural intensification and abandonment of marginal areas [32–34], which is likely to hamper the range expansion of grassland specialist species [28,35].

Our main objective in this study is to compare the degree of uncertainty associated with the land cover data selection with that stemming from the species' life history parameterisation. The range dynamics and population persistence capacity of our two example butterfly species is explored using RangeShifter, a novel dynamic modelling platform for species' range dynamics [36].

Both species are reliant in Finland on the network of grassland biotopes. We constructed representations of this grassland network using two different extensive sources of land cover and land use data sets. The first data set is the European-wide CORINE land cover database, while the second data set is a combination of three sources, the National Survey of Valuable Traditional Rural Biotopes, grassland sites managed based on the Agri-environment scheme (AES) [37,38], and data on distribution of all types of grasslands in Finland, gathered in the SLICES land cover database [39]. For species traits, we focused on the separate impacts of four key dispersal and demographic parameters which are likely to affect the model outcomes: population growth rate, carrying capacity, mean dispersal distance and probability of long-distance dispersal events [6,12,16,40].

Materials and Methods

Study species

Our two model species were Meadow Brown *Maniola jurtina* (Lepidoptera, Nymphalidae) (Linnaeus, 1758) and Queen of Spain Fritillary *Issoria lathonia* (Lepidoptera, Nymphalidae) (Linnaeus, 1758). *Maniola jurtina* is a grass-feeding species which behaves as a grassland habitat specialist in Finland, where it occurs at its northern range boundary [cf. 41]. In the butterfly transect monitoring surveys in Finland, the species has been found to favour managed (dry) unimproved grasslands over other types of grasslands [42]. We acknowledge that the species is a common grassland generalist in other parts of Europe, especially areas south of Finland [43–45]. In contrast, *Issoria lathonia* is a violet-feeding generalist fritillary and behaves in Finland similarly as in other parts of Europe [46]. It is capable of inhabiting many different grassland types, including lower quality grasslands such as set-asides and grassy strips along field margins [43,47,48]. We focus on these two butterfly species because they provide useful examples of ecologically contrasting species inhabiting the grassland habitat network. Moreover, the current range of both species is limited to southern Finland, from where they can be expected to move northwards following the warming climate, making them realistic model species for simulating range expansion dynamics.

The known occurrence records for the two study species were extracted from the National Butterfly Recording Scheme in Finland (NAFI). The NAFI is based on observations made by professional and volunteer amateur lepidopterists using a uniform 10×10 km grid system for the whole country [49,50]. We divided these records into two time periods, 1991–2000 and 2001–2011, and used the data from the first period, 1991–2000, to select the areas for initialising the simulations (i.e. the 10×10 km with records of species occurrences; see Figure 1). It should be noted that the butterfly occurrence records for the whole study area (Figure 1) were available only at this resolution although solitary records have been made using finer resolution mapping. Therefore, in our simulations, we were constrained to initialise the butterfly populations at a resolution of 10×10 km. Moreover, as all the simulations were run at a of 200×200 m (see below), all the 200×200 m cells included in a 10×10 km with known occurrence records were seeded.

Land cover data

The first main source of land cover data employed was the CORINE 2000 Land Cover database. We opted for these data due to their complete spatial coverage of Finland, because of the shared methodology at the pan-European level (EU countries) and because they are widely used in studies on impacts of land use on

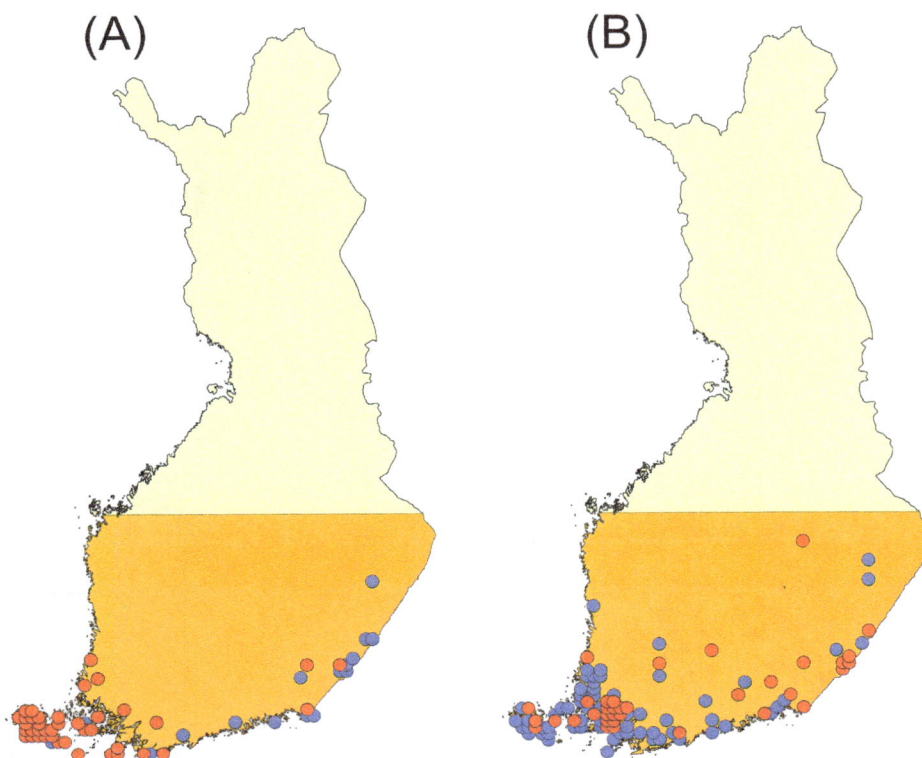

Figure 1. 10×10 km grid cells with known occurrences for (A) *Maniola jurtina* **and (B)** *Issoria lathonia*. The occurrence records in the 10×10 km grid cells in Finland were divided into the two time periods, 1991–2000 (red dots) and 2001–2011 (blue dots). Area where the range expansion simulations were performed is shown with orange.

species distributions [e.g. 51,52]. The classification of land cover into CORINE classes in Finland is based on automated interpretation of Landsat ETM satellite images and subsequent data integration with existing digital maps on land use and soil information [53]. The resolution of the CORINE data is 25×25 m in Finland and the classification of land cover includes four hierarchical levels. However, as we ran our simulations at a resolution of 200×200 m, we scaled-up the CORINE data for the relevant land cover types by simply summing-up their cover based on the sixty-four 25×25 m resolution cells embedded in each of the 200×200 m grid cells. This was done throughout the study area, which comprised southern Finland up to approximately the latitude of 63°N (Figure 1).

For the grassland specialist species, *Maniola jurtina*, we calculated the cover at 200×200 m resolution of the CORINE categories 'Pastures' (2.3.1 in CORINE classification) and 'Natural grassland' (3.2.1 in CORINE). For *Issoria lathonia*, the grassland generalist, the CORINE categories 2.4.3 ('Land principally occupied by agriculture, with significant areas of natural vegetation') and 2.1.1.2 ('Abandoned arable land') were additionally included, together with field margins measured based on the CORINE class 2.1.1 (arable land), when assessing the total amount of suitable habitat. Field margins' cover was estimated by multiplying their length in a 200×200 m cell by an effective width of 1 meter, based on empirical observations from monitoring schemes of grassland butterflies. For both the study species, a given 200×200 m grid cell was considered to be potentially suitable "habitat" for the species if it contained some amount of the above listed CORINE classes (thus there was no threshold for the amount of particular type of grassland required for a 200×200 m

cell to be considered habitat). The percentage habitat cover determined the cell total carrying capacity for each species. For example, the carrying capacity K for *Issoria lathonia* was estimated to be approximately 60 individuals/ha (see *Species parameterisation and dynamic range expansion modelling*); thus for a 200×200 m cell with 100% cover of suitable habitat, the maximal potential total carrying capacity was 240 individuals (for more details see [36]).

Simulations conducted with the CORINE-based grassland network were compared with those ran using more detailed data for grasslands, obtained by combining three different national grassland databases. For *Maniola jurtina*, we calculated the summed cover of all open grasslands in each of the 200×200 m grid cells mapped in (1) the National Survey of Valuable Traditional Rural Biotopes [54], together with the cover of open grassland sites included in (2) the national Agri-environment scheme (AES) in Finland [see 39]. Because both National Survey and AES-based managed grasslands initially included also wooded sites, we used (3) the SLICES land cover database to dissect the open grasslands from the wooded ones [39]. The SLICES database, which is compiled by the National Land Survey of Finland, shows the distribution of all types of common treeless grasslands in Finland. For *Issoria lathonia*, we added the non-overlapping SLICES grasslands to the open AES-managed grasslands and National Survey grasslands, in order to construct an estimate of the total habitat available for a grassland generalist species. All the habitat analyses and calculations were done by using ArcView Spatial Analyst (Version 3.2, ESRI Inc., Redland, CA, USA).

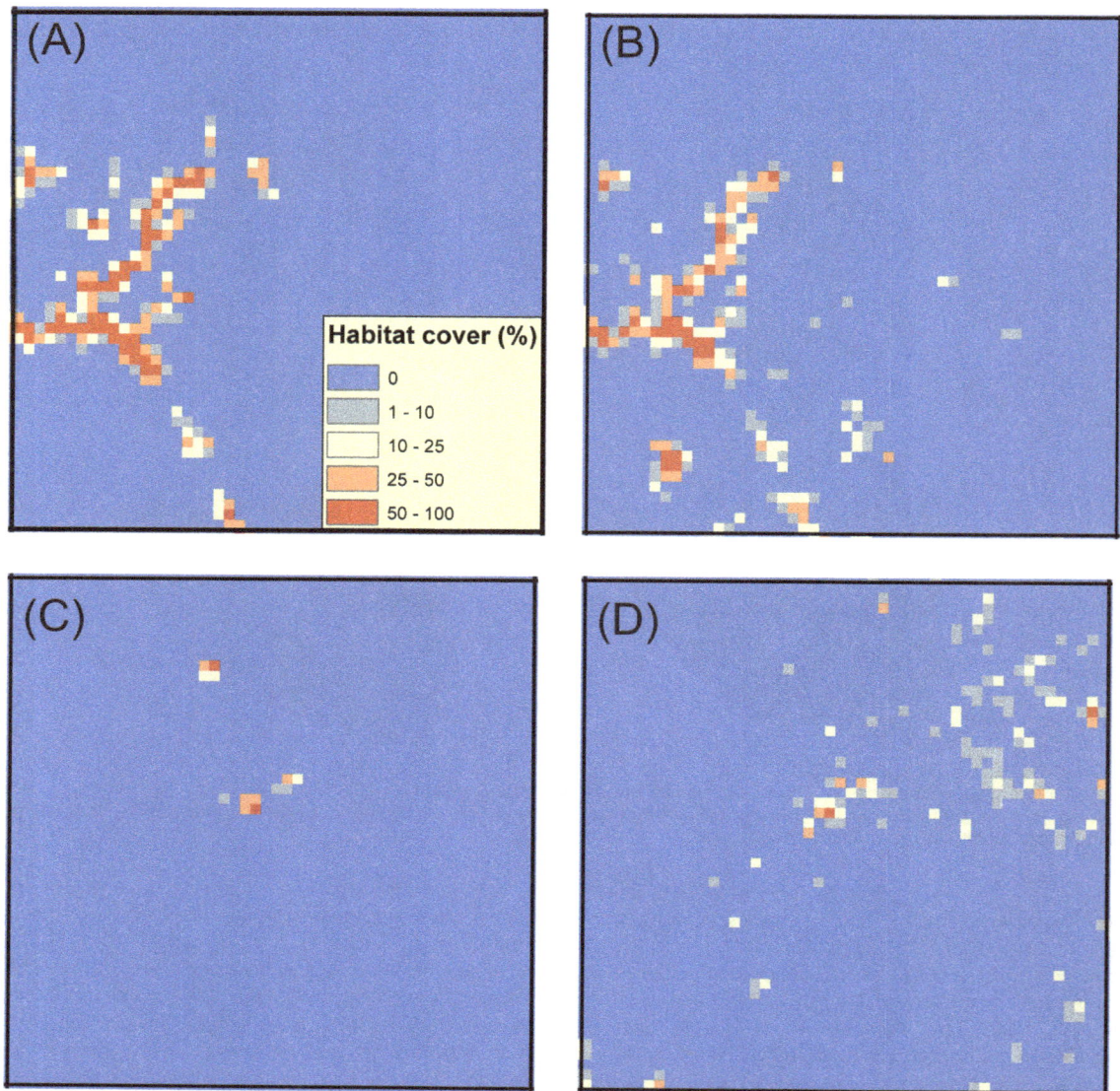

Figure 2. Variation in the estimated amount of suitable grasslands for *Maniola jurtina*, a grassland specialist butterfly. The amount of suitable habitat is shown for two exemplary 10×10 km grid cells and it was calculated based of the two different sources of grassland data. A–B: the first example 10×10 km cell; C–D: the second example 10×10 km cell. A and C: summed cover of open grasslands included in the National Survey and grasslands managed via Agri-environment Scheme (AES) in the 200×200 m cells; B and D: summed cover of CORINE classes 'Pastures' and 'Natural grassland' in the 200×200 m cells.

Additionally, for illustrative purposes we calculated the difference in the amount of suitable habitat estimated with the two approaches (i.e. the CORINE database vs. the AES-National Survey-SLICES databases). For this, we calculated for each of the 10×10 km grid cells included in our study area, the amount of habitat classified as suitable using each of the two methods, and report the spatial distribution of differences between the two methods.

Species parameterisation and dynamic range expansion modelling

Range expansion modelling was conducted using RangeShifter v1.0, a platform for individual-based dynamic modelling of single species' ecological and evolutionary dynamics [36]. At the heart of RangeShifter is the explicit modelling of population dynamic and

dispersal, the latter divided into its three fundamental phases of emigration, transfer and settlement.

From the options available within RangeShifter we chose to use a female-only and non-overlapping generations population model [55], which requires the population intrinsic growth rate (R_{max}) and carrying capacity (K). We assumed one reproductive season per year [cf. 56]. After reproduction all adults die and each offspring have a density-dependent probability of dispersing given by the following equation:

$$d = \frac{D_0}{1 + e^{-\left(N_{i,t}/K_i - \beta\right)\alpha}}$$

where D_0 is the maximum emigration probability, β is the inflection point and α is the slope of the curve at the inflection

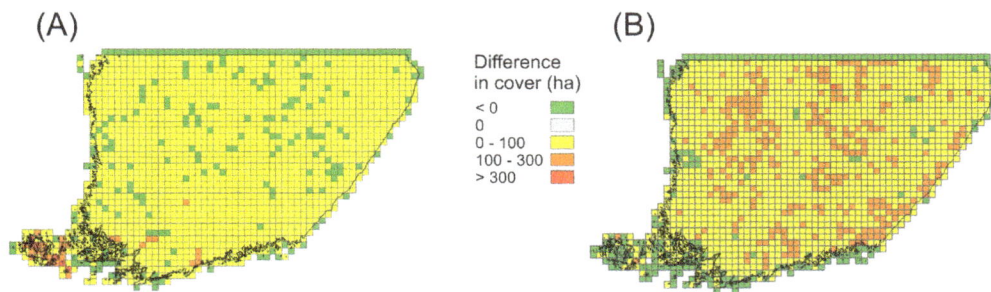

Figure 3. Difference in the amount of estimated suitable grasslands between the two land cover datasets. The distribution of differences was calculated by subtracting the amount of open grasslands in the National Survey-AES(-SLICES) databases from the amount of CORINE land cover types deemed as suitable for the two study species. The differences are shown in hectares across the 10×10 km grid cells of the simulation area. (A) *Maniola jurtina*: National Survey-AES grasslands were subtracted from the summed cover of the CORINE types 'Pastures' and 'Natural grassland'; (B) *Issoria lathonia*:, National Survey-AES-SLICES grasslands were subtracted from the CORINE types 'Pastures', 'Natural grassland,'Land principally occupied by agriculture, with significant areas of natural vegetation', 'Abandoned arable land' and field margins measured based on the CORINE class 'arable land'.

point. $N_{i,t}$ represents the number of individuals in cell i at time t, while K_i is the carrying capacity of the cell. We fixed the above parameters to $D_0 = 0.4$, $\beta = 1.0$ and $\alpha = 5.0$ (for the derived emigration probability curve see Figure S1). For those individuals who disperse, the distance is sampled from a double negative exponential distribution [57,58]. This is composed from two negative exponential distributions with different means and different probabilities of occurrence: the first, more common and with shorter mean ('dispersal I'); the second, less common and with longer mean in order to simulate relatively rare and long distance dispersal events ('dispersal II'). The dispersing individual is displaced at the sampled distance in a random direction. If the arrival cell is unsuitable, the individual is either displaced in one of the eight nearest neighbouring cells, if any of those is suitable, or assumed to die. We assume no additional dispersal mortality.

We conducted an extensive literature search to specify values for the model parameters for our study species. To supplement information from the literature, we used data extracted from long-term butterfly monitoring surveys carried out in Finland, in particular the transect count data from the Finnish Butterfly Monitoring Scheme [59], as well as expert-knowledge-based assessments on the study species' biology. Where required, dispersal and demographic parameter values were further adjusted based on studies on ecologically similar species and, more generally, on how life-history traits have been observed or estimated to vary among grassland butterfly species [cf. 14,60]. A more in-depth description of the study species' parameterisation process is included in Text S1. Following we provide the key information.

For the four focal life-history parameters in our model, i.e. carrying capacity K, maximum population growth rate R_{max}, mean dispersal distances and probability of long-distance dispersal events, an intermediate 'default' value and lower and higher alternative values were determined for both of the model species. We estimated carrying capacity from the data from the Finnish Butterfly Monitoring Scheme [59] and selected literature [43]. $K = 250$ individuals/ha was employed as carrying capacity value, while $K = 200$ and $K = 300$ as the lower and higher alternative for *Maniola jurtina*, and $K = 60$ individuals/ha as default value and $K = 30$ and $K = 90$ as the two alternatives for *Issoria lathonia*. The amount of grassland habitat deemed suitable for the study species was employed to assess the maximal potential size of the population in each of the 200×200 m cells which were either initially seeded or modelled to be colonised during the simulation,

and this assessment was conducted based on the three different values of K for both the species (lower alternative for K resulted in lower estimates of 200×200 cell population size, and higher alternative to higher estimates, respectively).

The population growth rates were determined based on measurements on ecologically similar (and dissimilar) butterfly species from the literature, and expert judgements based on field observations. These suggested that both *Maniola jurtina* and *Issoria lathonia* are likely to show intermediate population growth rates. Thus, we used $R_{max} = 2.0$ as the default value and $R_{max} = 1.5$ and $R_{max} = 2.5$ as the two alternatives for both species [15].

We used only one value for the mean short-distance dispersal distance: 150 m for *Maniola jurtina* [21,61–64], and 300 m for the more mobile *Issoria lathonia* [47,62,65,66]. For the mean long-distance dispersal distance, we used 3 km as the intermediate default value and 1.5 and 5 km as alternatives for *Maniola jurtina*, and 3, 5 and 10 km for *Issoria lathonia*. Based on the observations of Öckinger and Smith (2007) [62] on *Maniola jurtina* movements, we set the probability of individuals dispersing with the first, short distance dispersal kernel to either 0.80, 0.90 (default) or 0.95.

We assumed no environmental stochasticity because our focus was examining the potential impacts of the four key species life-history traits on range expansion simulation results. However, it should be noted that RangeShifter inherently incorporates two other key sources of stochasticity, demographic stochasticity and stochasticity in dispersal [67,68].

Species distribution data for the years 1991–2000 were used as a starting point for the simulations. All the 200×200 m grid cells with some suitable grassland habitat and located in the occupied 10×10 km species' distribution cells were initialised with a number of individuals equal to the cell total carrying capacity, determined by the habitat percentage cover in the cell. This initialisation approach very likely produced an exaggerated abundance for the species as a starting point. However, pilot runs showed that there was only a 2–5 year burn-in phase in the simulations during which the initialised cells with too little habitat or too isolated in space lost their individuals, after which the total simulated population size either remained constant or started to increase. All the simulations were run over a 50-year time window.

Varying the parameters as described above resulted in 9 different simulations for both of the study species which were conducted on the two alternative landscape maps. For each simulation, 100 replicate runs were conducted. Here we focus on

CORINE

AES - National survey

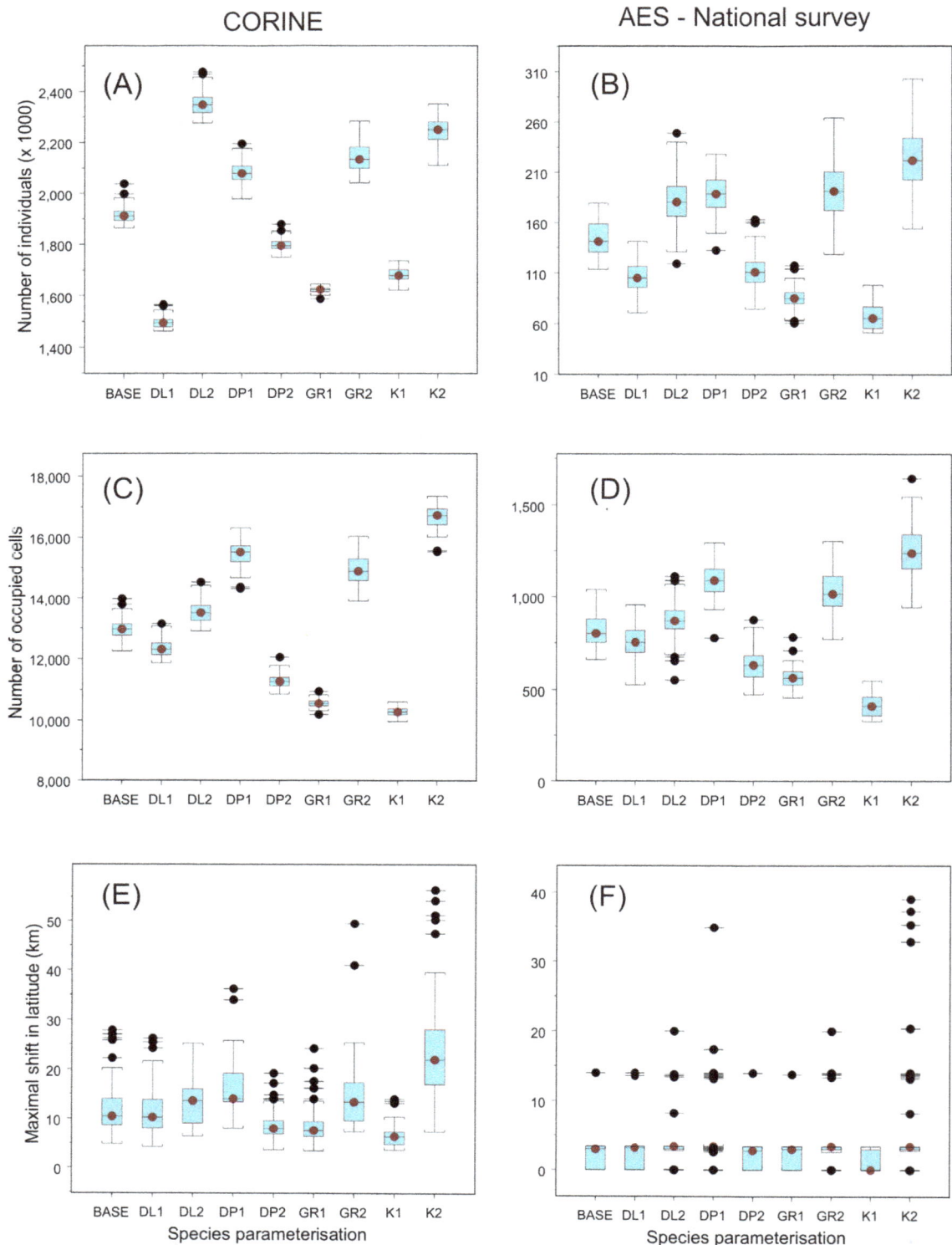

Figure 4. The total population abundance and range dynamic statistics for *Maniola jurtina*. The statistics include projected total number of *Maniola jurtina* individuals (A–B), number of 200×200 m cells occupied (C–D) and maximal range shift of the butterfly (E–F) at the end of a 50 year dynamic simulation period. Simulations were conducted using RangeShifter, a dynamic and individual-based modelling platform, and either summed cover of COFINE classes 'Pastures' and 'Natural grassland' (A, C, E) vs. open grasslands included in the National Survey of Traditional Rural Biotopes and grasslands managed via Agri-environment Scheme (B, D, F). All nine different simulation settings included 100 replicate runs. Species parameterisation: BASE = the default model parameterisation (K = 250; DL = 3000 m; DP = 0.90; GR = 2.0); alternative values for mean distance of long-distance dispersal events (DL1 = 1500 m, DL2 = 5000 m), probability of short-distance events (DP1 = 0.80, DP2 = 0.95), growth rate (GR1 = 1.5, GR2 = 2.5) and carrying capacity (K1 = 200, K2 = 300).

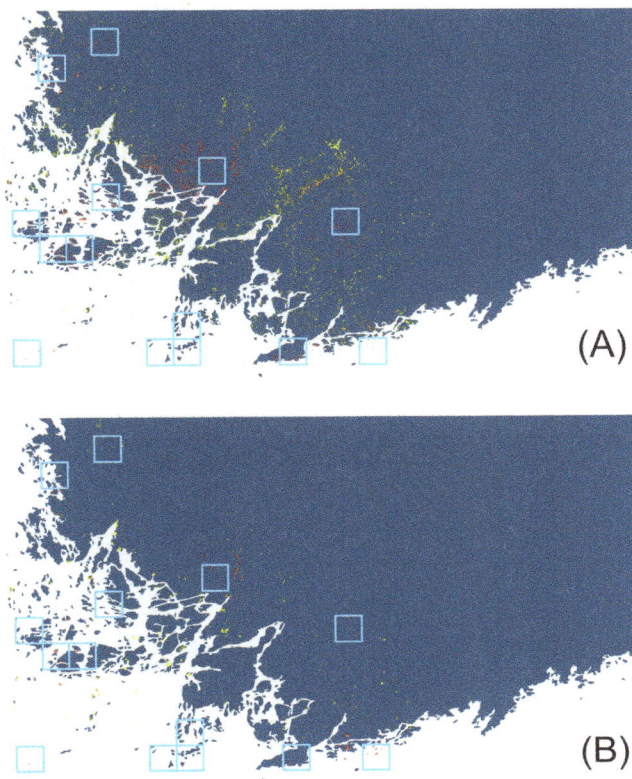

Figure 5. Example output maps for the simulated range expansion of *Maniola jurtina* in SW Finland. The maps for range expansion were produced by RangeShifter. Probability of a 200×200 m grid cell to be occupied after a 50-year simulation run is depicted with a colour ramp from red (high) to orange (intermediate) and yellow (low), with areas in dark blue having a probability of zero. Probability of a cell having a population was assessed based on 100 replicate simulations. Light blue squares indicate 10×10 km grid cells where the simulations were seeded. Simulations were done using default values for species traits and (A) CORINE data and (B) AES – National Survey data.

the simulation results dealing with the species' projected ranges: (1) total numbers of individuals, (2) total numbers of occupied cells and (3) species' range extents and projected shifts in range, measured as the maximum latitude at the year 0 vs. the maximum latitude year 50 (east–west range shifts were not examined).

Results

Mapping habitat suitability using the alternative land cover data

Comparison of the total amount of suitable grassland habitat in the study area, i.e. the CORINE database vs. the AES-National Survey-SLICES databases, revealed substantial differences in both the studied butterfly species. These differences were greater for *Maniola jurtina*, the grassland specialist. In total, 7,705 ha are classified as suitable for *Maniola jurtina* when using AES-National Survey while it increases to 33,951 ha when using the CORINE database. For *Issoria lathonia* there is again greater total amount of suitable habitat when using CORINE database compared to AES-National Survey-SLICES databases (160,075 ha versus 60,035 ha). In addition, there is substantial spatial variation in the extent of the difference obtained using the two alternative datasets. An illustration for *Maniola jurtina* for two 10×10 km

example grid cells shows one area where the CORINE-based habitat availability pattern is broadly similar to those based on data from AES-managed grasslands and grasslands included in the National Survey (Figure 2A vs. B), and another area where the CORINE data suggests that much more suitable habitat occurs in the landscape than AES - National Survey data (Figure 2C vs. D). Figure 3A illustrates the overall distribution of differences across all 10×10 km cells of the study region in Southern Finland. Greatest differences occur in SW archipelago where much more of the landscape is designated as suitable when using CORINE. A similar general pattern in differences (though more subtle) is found for *Issoria lathonia* in the two 10×10 km example grid cells (Figure S2), but the areas where the differences in the 10×10 km grid cells are greatest occur now in different regions (Figure 3B).

Maniola jurtina – the grassland specialist

Varying the four life-history parameters had notable impacts on the projected number of individuals, number of occupied grid cells and the maximal latitudinal range shift of *Maniola jurtina*. In the analysis where the default parameter values and the CORINE land cover data were used, the projected mean (± standard deviation) number of *Maniola jurtina* individuals was 1,913,603±30,162 individuals after 50 years. The strongest change in this baseline result occurred when changing the mean length for long-distance dispersal (Figure 4A). In contrast, the corresponding results from the simulations based on the more detailed land cover data, i.e. the AES-based managed grasslands and the National Survey grasslands, indicated highest importance for carrying capacity and growth rate (Figure 4B). However, the most striking result was the notable difference in the projected number of individuals in the simulations based on the two land cover data sources: in the simulations with the default demographic parameters 1,913,603±30,162 (CORINE) vs. 144,264±17,156 (AES-National Survey) individuals.

The impact of varying the four life-history traits on the projected total number of occupied 200×200 m grid cells was qualitatively similar in CORINE-based vs. AES-National Survey data based simulations. Here, the largest life-history trait based impact was related to alternative carrying capacities (Figure 4C and 4D), and the quantitative difference in the number of occupied cells between simulations based on the two land cover data sources was clear, reflecting the conspicuous difference in amount of suitable habitat between the two landscape maps.

The mean (± s.d.) projected latitudinal range shift was 12.7±7.2 km in the simulations based on the CORINE data, and 3.1±4.1 km in the National Survey – AES data based simulations, respectively (Figure 4E and 4F). The largest range shifts were observed for the higher carrying capacity, but also increasing the growth rate and the probability and mean distance of long-distance dispersal events caused an increase in projected latitudinal range shifts. In very few cases, the maximal range shifts obtained exceeded 50 km in the CORINE data based results. Figure 5 shows the spatial differences in the projected occupancy probability of the 200×200 m cells in the SW coastal area of Finland. In this comparison there are clear spatial differences between the model outputs from CORINE data vs. National Survey – AES data, whereas the corresponding differences stemming from varying the four species traits were more subtle (results not shown).

Issoria lathonia – the grassland generalist

The corresponding simulations for *Issoria lathonia* showed qualitatively similar patterns (Figure 6). There was a substantial quantitative difference between the results obtained for the

CORINE

AES - National survey - SLICES

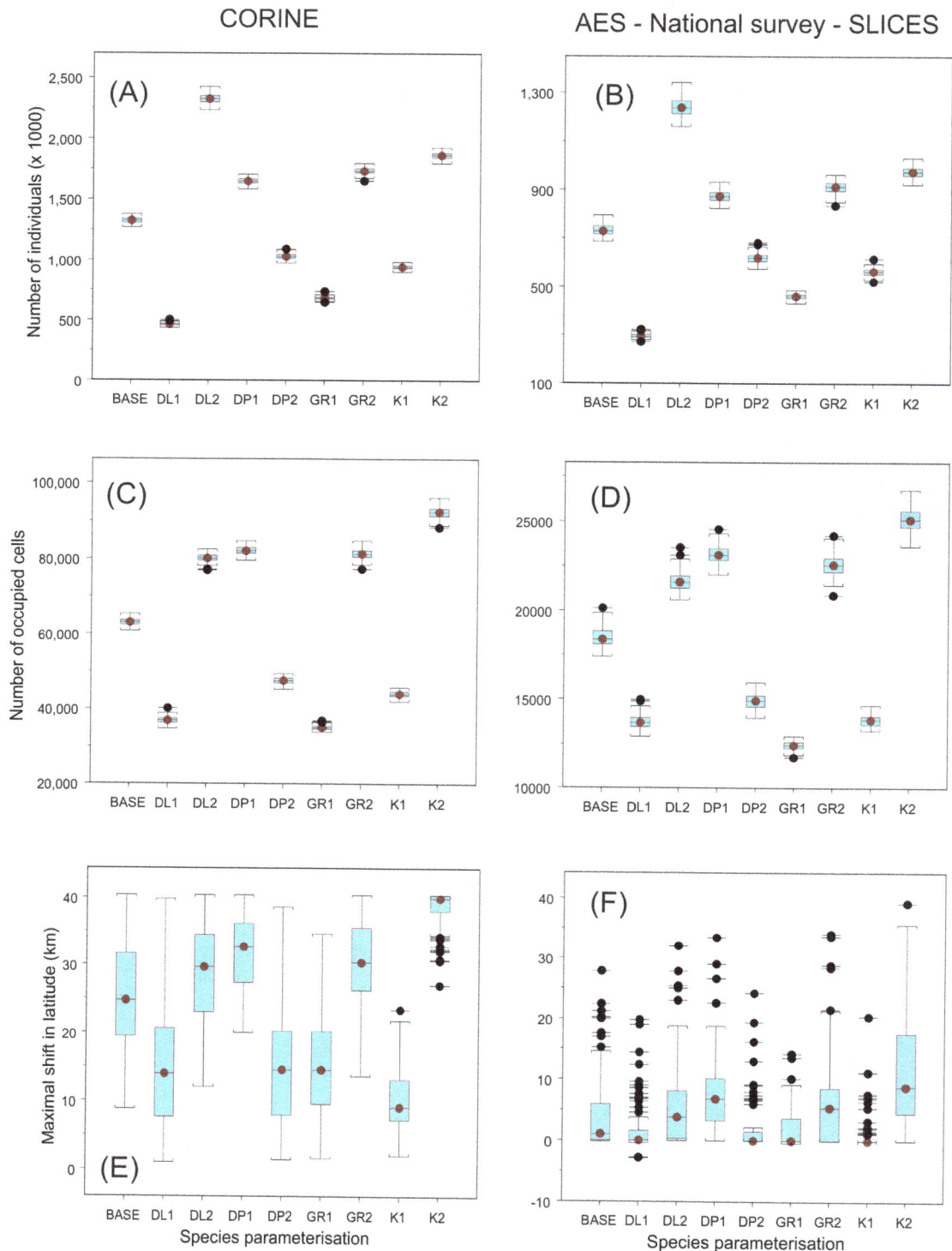

Figure 6. The projected total abundance and range dynamic statistics for *Issoria lathonia*. The projected total number of *Issoria lathonia* individuals (A–B), number of 200×200 m cells occupied (C–D) and maximal range shift of the butterfly (E–F) at the end of a 50 year dynamic simulation period. Simulations were conducted using either summed cover of CORINE classes 'Pastures' and 'Natural grassland', 'Land principally occupied by agriculture, with significant areas of natural vegetation', 'Abandoned arable land' and field margins (A, C, E) vs. open grasslands included in the National Survey, SLICES database and those managed via AES (B, D, F). For species parameterisation see Figure 3.

different landscape datasets. The projected number of individuals and occupied cells were clearly higher in the CORINE results (mean ± s.d.): 1,144,264±23,470 vs. 733,457±24,782 individuals,

and 62,988±993 vs. 18,444±574 occupied 200×200 m cells, in CORINE data vs. AES – National Survey – SLICES data based results, respectively (Figure 6, simulations with default values).

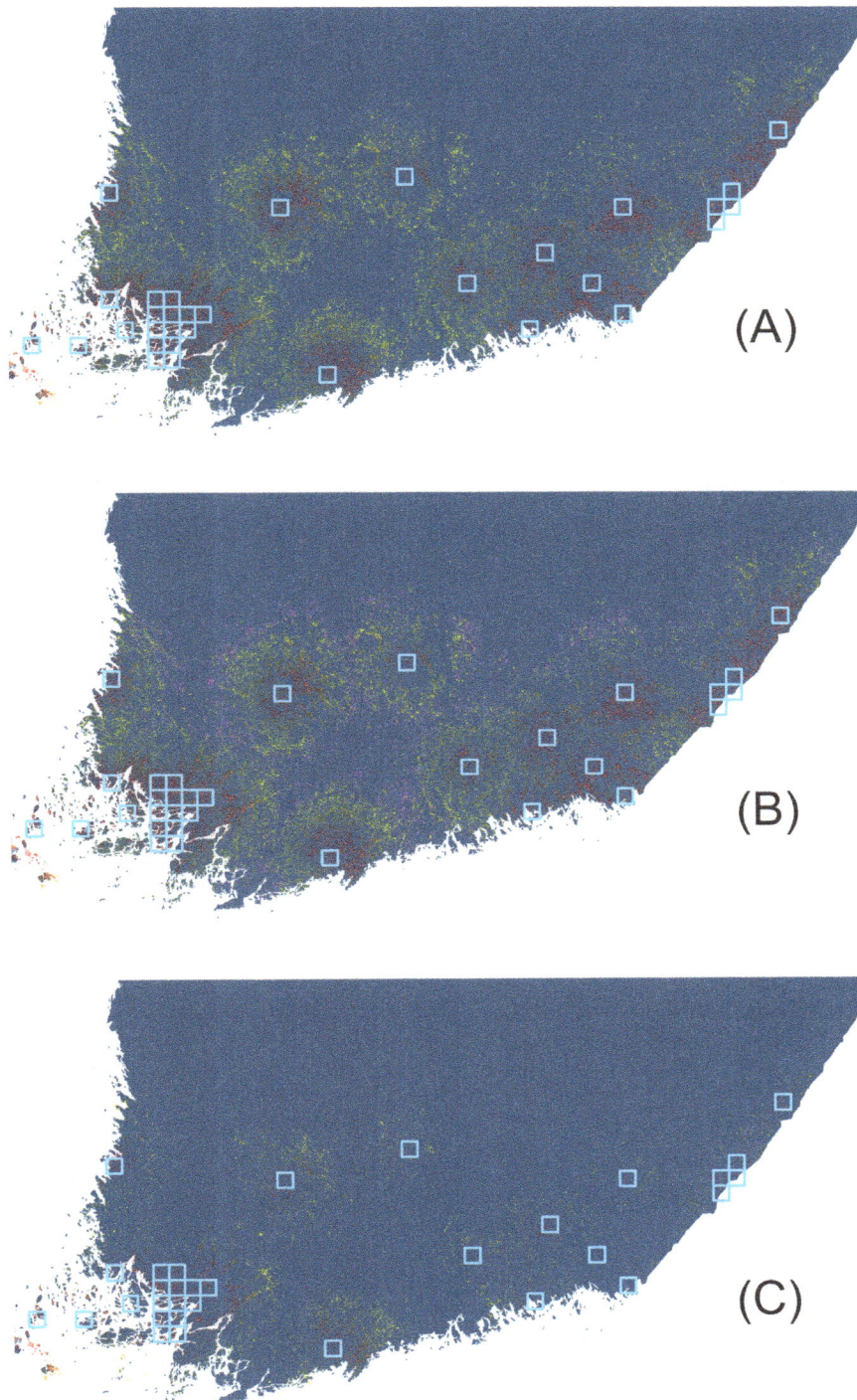

Figure 7. Example output for the simulated range expansion of *Issoria lathonia* in S Finland. Probability of a 200×200 m grid cell to be occupied after a 50-year simulation run is depicted with a colour ramp from red (high) to orange (intermediate) and yellow (low), with areas in dark blue having a probability of zero. Simulations were done using (A) CORINE data and default values for species traits, (B) CORINE data and increased carrying capacity value, and (C) AES – National Survey – SLICES data and default values for species traits. In (B), areas in pink indicate 200×200 m cells projected to have a population only when the higher carrying capacity is assumed.

The differences between the projected maximal range shift between the two land cover data types were also prominent (Figure 6E and 6F). In the CORINE data based results, mean range shifts were often projected to be larger than 20 km, whereas in the AES – National Survey – SLICES data based results only a limited number of individual runs exceeded a shift of 20 km. In both land cover data sets, higher carrying capacity returned the largest projected shifts. These differences are visible in the maps of

Table 1. Summary of the varied life-history traits causing largest change in the three measured species' range expansion measures, i.e. the projected number of butterfly individuals, number of occupied 200×200 m grid cells and mean maximal latitudinal range shift, shown for the two species and two land cover types separately.

| Number of individuals (x 1000) | *Maniola jurtina* | | *Issoria lathonia* | |
	CORINE data	AES – National Survey data	CORINE	AES – National Survey – SLICES data
Lowest	DL1 (1,495±22)	K1 (67±13)	DL1 (461.7±12)	DL1 (296±9.8)
Highest	DL2 (2,350±44)	K2 (222±31)	DL2 (2,324. ±39)	DL2 (1240±39)
Number of occupied 200-m grid cells				
Lowest	K1 (10,283±143)	K1 (416±58)	GR1 (35,132±591)	GR1 (12426±243)
Highest	K2 (16,673±360)	K2 (1,242±146)	K2 (92,356±1,573)	K2 (25124±658)
Maximal latitudinal range shift				
Lowest	K1 (6.3±2.2)	K1 (1.0±1.5)	K1 (10.0±4.3)	DL1 (1.6±4.1)
Highest	K2 (24.1±9.8)	K2 (5.9±7.6)	K2 (38.4±3.0)	K2 (12.0±10.9)

The mean value (+/− standard deviation) from the 100 replicate runs is given for each measure in parenthesis. Species parameterisation abbreviations: *Maniola jurtina* and *Issoria lathonia*: DL1 = 1500 m, DL2 = 5000 m (alternative values for mean distance of long-distance dispersal events); K1 = 200, K2 = 300 (alternative values for carrying capacity); *Issoria lathonia*, GR1 = 1.5 (lower alternative value for population growth rate).

species' occupancy probabilities in the simulations based on CORINE data and default trait parameters vs. increased carrying capacity (Figure 7A and B), particularly as the extended expansion of range margin in certain regions. However, overall the spatial differences are more notable when probability of occupancy values from simulations based on CORINE data are compared with those based on AES – National Survey – SLICES data (Figure 7A and C).

Main traits affecting range expansion

To illuminate which of the four studied life-history traits had the strongest impact of the three measures of the species' range and population dynamics we summarized the most influential traits for the study species in Table 1. This summary shows that altering the carrying capacity typically has the strongest impact, especially on mean maximal range shifting, while the second most important driver is long-distance dispersal ability.

Discussion

Land cover data and species habitat specificity

We have shown that the selection of the land cover data upon which dynamic models are built may have a major effect on the projections of species' range expansion. These findings are important because systematic land cover data from detailed field surveys rarely exist for larger regions. CORINE is one of the few systematically constructed land cover databases covering continent-wide areas and it is commonly used in species distribution modelling [51] [51,52,69]. However, its usefulness with respect to modelling species with strict habitat requirements is insufficiently known [20]. Several niche modelling studies have shown that the projections of species distributions may be substantially affected by the selection of environmental variables, including land cover variables [70,71]. The results from our study illustrate the fact that using insufficient quality landcover data can introduce substantial bias also into the results of dynamic modelling exercises projecting species responses to climate change.

The differences in the projected species population abundance and range dynamics gained using the AES – National Survey data based simulations compared to those based on CORINE data

emerge from certain important sources. The quality of the grassland sites in the National Survey of Rural Biotopes and those managed for biodiversity via AES specific contracts very likely varies less than the CORINE data from the perspective of grassland specialist butterflies which often require managed unimproved grasslands [72–74]. However, spatial cover of more thorough field investigations (including National Survey of Valuable Traditional Rural Biotopes) is often constrained by the limited resources, which may result in the underestimation of habitat availability in insufficiently surveyed areas. This is likely to be one of the reasons behind the very substantial differences in estimated habitat availability in the SW archipelago. In addition, here uptake of AES contracts may also be lower than in the mainland areas. In contrast, the spatial cover of CORINE data is better, but it may be more variable and overestimate the habitat availability, especially for grassland specialists. This is because CORINE data are based on various sources such as other existing land cover databases and satellite imagery.

The recorded occurrences of *Maniola jurtina* in SE Finland are a point of specific interest. Namely, there are some isolated 10×10 km grid cells with records of the species in 2001–2011 that are situated far from the potential source populations (Figure 1). To reach these cells from the earlier known sites would require dispersal over several tens of kilometres within ten years, a situation hardly possible only via the network of AES – National Survey grasslands, as shown by our simulation results. Three factors may play a role here. First, the map of known records for *Maniola jurtina* is inevitably an underestimation of the true distribution because of the spatial variation in survey effort [see 75,76]. Second, it is possible that the grassland network in SE Finland is in many areas insufficient to maintain longer-term populations of grassland specialists and thus regional butterfly populations may be dependent on the constant arrival of immigrants from Russia, where higher quality grasslands are more common, possibly representing a large-scale source-sink system between Russia and Eastern Finland [73]. Thirdly, it is possible that the habitat fidelity of *Maniola jurtina* is in a changing stage. Thus, we may be dealing with changing habitat specificity of a species at its northern range boundary where climate has recently become more favourable. Due to this, the butterfly might now utilise a wider range of grassland habitats than earlier [41].

Impact of uncertainty in life-history traits relative to uncertainty in land cover

As the climate impacts research community increasingly uses dynamic models to project species' future distributions, it is crucial that we begin to gain some insight into the relative importance of different forms of input uncertainty to the uncertainties associated with the outputs from these models. We ran our dynamic model using an illustrative set of values for key life history parameters that were chosen to represent the bounds of uncertainty for those parameters. This allows us to compare how influential uncertainty is in each of those life history parameters as well as comparing how important uncertainty around life history generally is relative to that due to choice of landscape database.

Unsurprisingly, altering the four focal life-history traits had an impact on the projected number of individuals and occupied grid cells of the two study species. Earlier studies have indicated that assuming a higher dispersal ability will allow a faster range expansion and a more successful tracking of changing climate [13,25] and similar results are obtained in our study with more rapid range expansions when the probability of the long-distance dispersal jumps is increased. However, our results are not fully straightforward and the impact of increasing long distance dispersal depends on the landscape context. With *Maniola jurtina*, increasing the mean magnitude of long-distance dispersal events had the strongest impact on projected number of individuals within the CORINE pastures network. Otherwise, the strength of this effect was much reduced. With *Issoria lathonia*, increasing the mean length of long-distance movements typically had a more prominent role than increasing the proportion of long distance dispersal and showed consistent effects on both landscape maps. Increasing the probability of long-distance dispersal events had mainly an intermediate impact on the projected population estimates. This effect was larger than that of the length of long-distance movements for the species' maximal range shifts, being most evident in the results for *Issoria lathonia* and CORINE grassland data.

As expected [77], we found population growth rate impacted the projected population dynamics during range expansion. Theoretical models [5,40] have also highlighted the importance of rapid growth rate for species population persistence under a changing climate. Further, in one study very relevant in the context of our work, Willis et al. [16] showed that the projected rate of expansion of the *Pararge aegeria* butterfly in the UK was especially sensitive to altering population dynamics; a 25% increase in productivity resulted in a 56% increase in range expansion.

Interestingly, we found very substantial sensitivity to uncertainty in carrying capacity showed for both *Maniola jurtina* and *Issoria lathonia*. Classical theory on range expansion has typically stressed the important joint roles of population growth rate and dispersal [67,77–79] and has not highlighted an important role for carrying capacity. However, South [24] showed with a spatially explicit population model that there may be complicated links between dispersal success, dispersal initiation mechanism, patch growth rate and patch carrying capacity, which all ultimately affect population persistence. Moreover, a recent theoretical study by Bocedi et al. [80] using artificial fragmented landscapes and theoretical species, demonstrated that carrying capacity can often be influential. The results we present here lend weight to the suggestion that, at least on fragmented landscapes, we need to pay greater attention to this parameter.

However, it should be noted that what our result highlight most strongly is the substantial uncertainty we find due to the choice of habitat maps relative to the uncertainties due to the demographic parameters. In dynamic modelling studies there is often some discussion of potential uncertainties related to estimation of demographic parameters and especially dispersal, but much less attention is given to uncertainty in the spatial representation of suitable habitat. Our results demonstrate very clearly that uncertainty due to habitat mapping can be at least as great as that due to the demographic parameters. This reinforces the results of a study by Willis et al. [16] that demonstrated that the projected rate of spread on *Pararge aegeria* butterfly was more sensitive to altering habitat availability than variation in demographic factors and seed locations in the simulations. A further study focussing on population viability rather than range expansion [12] has specifically focused the importance of the uncertainties in developing habitat maps for a species, such as potential errors in satellite imagery and georeferencing. They used a spatially explicit model based on two contrasting habitat maps created from remote imagery for a forest dwelling bird species, a 'generous' and 'strict' habitat map. The selection between the two habitat maps caused differences in total population size three times more important than other factors, such as dispersal model type, maximal dispersal distance and bird clutch size. Our study bridges these previous two studies by exploring the extent of uncertainty that arises in projections of range expansion due to the choice of dataset for constructing a habitat map. Interestingly, our results also highlight that the extent of uncertainty due to choice of dataset can be very different between species; we found a much greater effect for *Maniola jurtina* than for *Issoria lathonia*.

Implications for conservation planning

Our modelling results have clear importance for conservation planning because conservation biologists and managers are currently seeking robust tools to project the changes in species' distributions in response to changing climate. Dynamic range expansion models are considered one promising approach to develop improved assessments on the impacts of global changes, potentially providing a sounder basis to allocate the scarce conservation resources than the widely applied bioclimatic envelope models [11]. However, the end-users of dynamic models need to be aware of the limitations in the modelling approaches available. Indeed, our results suggest that dynamic modelling approaches should also be used with caution when applied to real-life nature conservation questions. This is because dynamic range expansion and population models have sources of uncertainty of their own and failing to acknowledge this may invoke a false sense of confidence [12,17,18].

Such uncertainties are centred around three main issues: (1) the scarcity of accurate species and habitat data over larger areas [15], (2) uncertainties in data for species life-history traits critical for dispersal and population dynamics [11,22], and (3) difficulties in determining direct and robust links between the species' habitat requirements and the land cover data available [12]. Under these circumstances, running sensitivity analysis for dynamic species population models before they are used in different applied conservation and management planning questions is essential [18]. The results of this study demonstrate that the sensitivity analysis can indeed provide important insights for the sensitivity of dynamic models to altering species parameters and habitat requirements.

Developing improved conservation planning tools for grassland species is important, because different types of grassland habitats from unimproved semi-natural grasslands to non-cultivated elements such as larger field margins and wooded pastures have faced a drastic decline during the last century [30,34]. This has resulted in a major loss of landscape heterogeneity and habitats favoured by many butterfly species. Our simulation results suggest that the possibilities of both grassland habitat specialist and generalist butterfly species to adapt to projected climate change may be limited. Only in our most generous simulation setting, *Issoria lathonia* in network of CORINE grasslands and abandoned cultivated land and field margins, the mean maximal range shifts were projected to exceed 20 or 30 km kilometres within 50 years. These forecasts are well in line with the recently observed range shifts of grassland butterflies in Finland. Interestingly, they fall much below the observed fastest range shifts in butterflies, generally encountered in forest edge generalist species with high dispersal ability [28].

Moreover, many grassland butterfly species depend on the occurrence of semi-natural grasslands managed by mowing or grazing [73,74], or other similar higher-quality grasslands connected with traditional agricultural practices [81]. Our simulation results for *Maniola jurtina* suggest that the ecological sufficiency of the grasslands included in the National Survey of Rural Biotopes and those managed by specific AES is poor for a strict grassland specialist species depending on high-quality sites. In particular, the likelihood that the grassland specialist species respond to climate change by dispersing into new suitable areas in southern Finland seems low, as the projected maximal range shifts for *Maniola jurtina* were modest at best.

In conclusion, the adaptation and persistence possibilities of grassland specialist species under environmental changes in our study area appear to be very limited; thus, major changes are required to improve the critical situation of these habitats and their species. The two complementary main lines of required future action are: (1) increasing of the area of grasslands managed for biodiversity to mitigate long-term habitat loss impacts [82] and support local population persistence [83], and (2) improving their connectivity to support grassland species range shifting across the landscape [5].

Supporting Information

Figure S1   Shape of emigration probability curve used in the simulations. The calculation of the emigration probability

curve is based on the density-dependent emigration assumption with maximum dispersal probability $D_0 = 0.4$, slope $\alpha = 5.0$ and inflection point $\beta = 1.0$.

Figure S2   The cover of suitable grassland habitat for *Issoria lathonia*, a grassland generalist butterfly, in two exemplary 10×10 km grid cells based on two different sources of spatial grassland data. A and C: summed cover of all kinds of open grassland included in the National Survey, AES and the SLICES database in each of the 200×200 m cells; B and D: summed cover of CORINE classes 'Pastures', 'Natural grassland', 'Land principally occupied by agriculture, with significant areas of natural vegetation', and 'Abandoned arable land', together with the cover of field margins, in each of the 200×200 m cells.

Text S1   Supporting information for the two model species' parameterisation for the RangeShifter dynamic range expansion simulations. More in-depth description of the process how the dispersal and population biological parameters for the two butterfly species, *Maniola jurtina* and *Issoria lathonia*, required to perform the range expansion simulations with RangeShifter algorithm were done based on the following sources: an extensive literature search, the data extracted from long-term butterfly monitoring surveys carried out in Finland such as the Finnish Butterfly Monitoring Scheme, published data on population biological parameters from studies on ecologically similar species, and empirical data -based expert assessments on the general variation of demographic parameters among grassland butterfly species.

Acknowledgments

Kimmo Saarinen kindly provided the NAFI Butterfly Atlas for the purposes of this study.

Author Contributions

Conceived and designed the experiments: RKH GB MK JMJT. Performed the experiments: RKH MK JH NL JP. Analyzed the data: RKH JH JMJT. Contributed to the writing of the manuscript: RKH GB MK JP JMJT.

References

1. Heikkinen RK, Luoto M, Araújo MB, Virkkala R, Thuiller W, et al. (2006) Methods and uncertainties in bioclimatic envelope modelling under climate change. Progress in Physical Geography 30: 751–777.
2. Araujo MB, Peterson AT (2012) Uses and misuses of bioclimatic envelope modeling. Ecology 93: 1527–1539.
3. Vos CC, Berry P, Opdam P, Baveco H, Nijhof B, et al. (2008) Adapting landscapes to climate change: examples of climate-proof ecosystem networks and priority adaptation zones. Journal of Applied Ecology 45: 1722–1731.
4. Sinclair SJ, White MD, Newell GR (2010) How useful are species distribution models for managing biodiversity under future climates? Ecology and Society 15: 1–8.
5. Söndgerath D, Schroder B (2002) Population dynamics and habitat connectivity affecting the spatial spread of populations - a simulation study. Landscape Ecology 17: 57–70.
6. Mustin K, Benton TG, Dytham C, Travis JMJ (2009) The dynamics of climate-induced range shifting; perspectives from simulation modelling. Oikos 118: 131–137.
7. Franklin J (2010) Moving beyond static species distribution models in support of conservation biogeography. Diversity and Distributions 16: 321–330.
8. Hodgson JA, Thomas CD, Dytham C, Travis JMJ, Cornell SJ (2012) The Speed of Range Shifts in Fragmented Landscapes. Plos One 7.
9. Travis JMJ, Delgado M, Bocedi G, Baguette M, Barton K, et al. (2013) Dispersal and species' responses to climate change. Oikos 122: 1532–1540.
10. Keith DA, Akcakaya HR, Thuiller W, Midgley GF, Pearson RG, et al. (2008) Predicting extinction risks under climate change: coupling stochastic population models with dynamic bioclimatic habitat models. Biology Letters 4: 560–563.
11. Huntley B, Barnard P, Altwegg R, Chambers L, Coetzee BWT, et al. (2010) Beyond bioclimatic envelopes: dynamic species' range and abundance modelling in the context of climatic change. Ecography 33: 621–626.
12. Minor ES, McDonald RI, Treml EA, Urban DL (2008) Uncertainty in spatially explicit population models. Biological Conservation 141: 956–970.
13. Anderson BJ, Akcakaya HR, Araujo MB, Fordham DA, Martinez-Meyer E, et al. (2009) Dynamics of range margins for metapopulations under climate change. Proceedings of the Royal Society B-Biological Sciences 276: 1415–1420.
14. Hill JK, Collingham YC, Thomas CD, Blakeley DS, Fox R, et al. (2001) Impacts of landscape structure on butterfly range expansion. Ecology Letters 4: 313–321.
15. del Barrio G, Harrison PA, Berry PM, Butt N, Sanjuan ME, et al. (2006) Integrating multiple modelling approaches to predict the potential impacts of climate change on species' distributions in contrasting regions: comparison and implications for policy. Environmental Science & Policy 9: 129–147.
16. Willis SG, Thomas CD, Hill JK, Collingham YC, Telfer MG, et al. (2009) Dynamic distribution modelling: predicting the present from the past. Ecography 32: 5–12.
17. Bocedi G, Pe'er G, Heikkinen RK, Matsinos Y, Travis JMJ (2012) Projecting species' range expansion dynamics: sources of systematic biases when scaling up patterns and processes. Methods in Ecology and Evolution 3: 1008–1018.
18. Conlisk E, Syphard AD, Franklin J, Flint L, Flint A, et al. (2013) Uncertainty in assessing the impacts of global change with coupled dynamic species distribution and population models. Global Change Biology 19: 858–869.
19. Wilson RJ, Davies ZG, Thomas CD (2009) Modelling the effect of habitat fragmentation on range expansion in a butterfly. Proceedings of the Royal Society B-Biological Sciences 276: 1421–1427.

20. Jimenez-Valverde A, Gomez JF, Lobo JM, Baselga A, Hortal J (2008) Challenging species distribution models: the case of Maculinea nausithous in the Iberian Peninsula. Annales Zoologici Fennici 45: 200–210.

21. Schneider C (2003) The influence of spatial scale on quantifying insect dispersal: an analysis of butterfly data. Ecological Entomology 28: 252–256.

22. Stevens VM, Turlure C, Baguette M (2010) A meta-analysis of dispersal in butterflies. Biological Reviews 85: 625–642.

23. Bonte D, Van Dyck H, Bullock JM, Coulon A, Delgado M, et al. (2012) Costs of dispersal. Biological Reviews 87: 290–312.

24. South A (1999) Dispersal in spatially explicit population models. Conservation Biology 13: 1039–1046.

25. McInerny G, Travis JMJ, Dytham C (2007) Range shifting on a fragmented landscape. Ecological Informatics 2: 1–8.

26. Iverson LR, Schwartz MW, Prasad AM (2004) How fast and far might tree species migrate in the eastern United States due to climate change? Global Ecology and Biogeography 13: 209–219.

27. Menéndez R (2007) How are insects responding to global warming? Tijdschrift voor Entomologie 150: 355–365.

28. Pöyry J, Luoto M, Heikkinen RK, Kuussaari M, Saarinen K (2009) Species traits explain recent range shifts of Finnish butterflies. Global Change Biology 15.

29. Wilson RJ, Maclean IMD (2011) Recent evidence for the climate change threat to Lepidoptera and other insects. Journal of Insect Conservation 15: 259–268.

30. Wenzel M, Schmitt T, Weitzel M, Seitz A (2006) The severe decline of butterflies on western German calcareous grasslands during the last 30 years: A conservation problem. Biological Conservation 128: 542–552.

31. Öckinger E, Smith HG (2007) Semi-natural grasslands as population sources for pollinating insects in agricultural landscapes. Journal of Applied Ecology 44: 50–59.

32. Strijker D (2005) Marginal lands in Europe - causes of decline. Basic and Applied Ecology 6: 99–106.

33. Öckinger E, Eriksson AK, Smith HG (2006) Effects of grassland abandonment, restoration and management on butterflies and vascular plants. Biological Conservation 133: 291–300.

34. Polus E, Vandewoestijne S, Chomtt J, Baguette M (2007) Tracking the effects of one century of habitat loss and fragmentation on calcareous grassland butterfly communities. Biodiversity and Conservation 16: 3423–3436.

35. Warren MS, Hill JK, Thomas JA, Asher J, Fox R, et al. (2001) Rapid responses of British butterflies to opposing forces of climate and habitat change. Nature 414: 65–69.

36. Bocedi G, Palmer SCF, Pe'er G, Heikkinen RK, Matsinos Y, et al. (2014) RangeShifter: a platform for modeling spatial eco-evolutionary dynamics and species' responses to environmental changes. Methods in Ecology and Evolution 5: 388–396.

37. Tscharntke T, Klein AM, Kruess A, Steffan-Dewenter I, Thies C (2005) Landscape perspectives on agricultural intensification and biodiversity - ecosystem service management. Ecology Letters 8: 857–874.

38. Merckx T, Feber R, Parsons M, Bourn N, Townsend M, et al. (2010) Habitat preference and mobility of Polia bombycina: are non-tailored agri-environment schemes any good for a rare and localised species? Journal of Insect Conservation 14: 499–510.

39. Arponen A, Heikkinen RK, Paloniemi R, Pöyry J, Similä J, et al. (2013) Improving conservation planning of semi-natural grasslands: integrating connectivity into agri-environment schemes. Biological Conservation 160: 234–241.

40. Best AS, Johst K, Munkemuller T, Travis JMJ (2007) Which species will succesfully track climate change? The influence of intraspecific competition and density dependent dispersal on range shifting dynamics. Oikos 116: 1531–1539.

41. Oliver T, Hill JK, Thomas CD, Brereton T, Roy DB (2009) Changes in habitat specificity of species at their climatic range boundaries. Ecology letters 12: 1091–1102.

42. Schulman A, Heliölä J, Kuussaari M (2005) Farmland biodiversity on the Åland islands and assessment of the effects of the agri-environmental support scheme. The Finnish Environment 734: 1–210.

43. van Swaay CAM (2003) Butterfly densities on line transects in The Netherlands from 1990–2001. Entomologische Berichten 63: 82–87.

44. Van Swaay CAM, Van Strien AJ, Harpke A, Fontaine B, Stefanescu C, et al. (2010) The European Butterfly Indicator for Grassland species 1990–2009. Wageningen: De Vlinderstichting.

45. Dennis RLH (2004) Butterfly habitats, broad-scale biotope affiliations, and structural exploitation of vegetation at finer scales: the matrix revisited. Ecological Entomology 29: 744–752.

46. Bink FA (1992) Ecologische Atlas van de Dagvlinders van Noordwest-Europa. Haarlem: Schuyt.

47. Maes D, Bonte D (2006) Using distribution patterns of five threatened invertebrates in a highly fragmented dune landscape to develop a multispecies conservation approach. Biological Conservation 133: 490–499.

48. Jonason D, Milberg P, Bergman KO (2010) Monitoring of butterflies within a landscape context in south-eastern Sweden. Journal for Nature Conservation 18: 22–33.

49. Saarinen K, Lahti T, Marttila O (2003) Population trends of Finnish butterflies (Lepidoptera: Hesperioidea, Papilionoidea) in 1991–2000. Biodiversity and Conservation 12: 2147–2159.

50. Luoto M, Heikkinen RK, Pöyry J, Saarinen K (2006) Determinants of biogeographical distribution of butterflies in boreal regions. Journal of Biogeography 33: 1764–1778.

51. Storch D, Konvicka M, Benes J, Martinkova J, Gaston KJ (2003) Distribution patterns in butterflies and birds of the Czech Republic: separating effects of habitat and geographical position. Journal of Biogeography 30: 1195–1205.

52. Titeux N, Maes D, Marmion M, Luoto M, Heikkinen RK (2009) Inclusion of soil data improves the performance of bioclimatic envelope models for insect species distributions in temperate Europe. Journal of Biogeography 36: 1459–1473.

53. Virkkala R, Heikkinen RK, Fronzek S, Kujala H, Leikola N (2013) Does the protected area network preserve bird species of conservation concern in a rapidly changing climate? Biodiversity and Conservation 22: 459–482.

54. Vainio M, Kekäläinen H, Alanen A, Pykälä J (2001) Traditional rural biotopes in Finland. Final report of the nationwide inventory (In Finnish with English summary). Helsinki, Finland: The Finnish Environment 527. Finnish Environment Institute.

55. Maynard-Smith J, Slatkin M (1973) The stability of predator-prey systems. Ecology 54: 384–391.

56. Pöyry J, Leinonen R, Soderman G, Nieminen M, Heikkinen RK, et al. (2011) Climate-induced increase of moth multivoltinism in boreal regions. Global Ecology and Biogeography 20: 289–298.

57. Hovestadt T, Binzenhofer B, Nowicki P, Settele J (2011) Do all inter-patch movements represent dispersal? A mixed kernel study of butterfly mobility in fragmented landscapes. Journal of Animal Ecology 80: 1070–1077.

58. Nathan R, Klein E, Robledo-Arnuncio JJ, Revilla E (2012) Dispersal kernels: review. In: Clobert J, Baguette M, Benton TG, Bullock J, editors. Dispersal Ecology and Evolution. Oxford: Oxford University Press. 187–210.

59. Heliölä J, Kuussaari M, Niininen I (2010) Maatalousympäristön päiväperhosseuranta 1999–2008. (Butterfly monitoring in Finnish agricultural landscapes 1999–2008). The Finnish Environment 2/2010: 1–65.

60. Hanski I (1994) A practical model of metapopulation dynamics. Journal of Animal Ecology 63: 151–162.

61. Dover JW, Clarke SA, Rew L (1992) Habitats and movement patterns of satyrid butterflies (Lepidoptera: Satyridae) on arable farmland. Entomol Gazette 43: 29–43.

62. Öckinger E, Smith HG (2007) Asymmetric dispersal and survival indicate population sources for grassland butterflies in agricultural landscapes. Ecography 30: 288–298.

63. Grill A, Cleary DFR, Stettmer C, Brau M, Settele J (2008) A mowing experiment to evaluate the influence of management on the activity of host ants of Maculinea butterflies. Journal of Insect Conservation 12: 617–627.

64. Ouin A, Martin M, Burel F (2008) Agricultural landscape connectivity for the meadow brown butterfly (Maniola jurtina). Agriculture, Ecosystems & Environment 124: 193–199.

65. Komonen A, Grapputo A, Kaitala V, Kotiaho J, Päivinen J (2004) The role of niche breadth, resource availability and range position on the life history of butterflies. Oikos 105: 41–54.

66. Öckinger E, Smith HG (2008) Do corridors promote dispersal in grassland butterflies and other insects? Landscape Ecology 23: 27–40.

67. Clark JS, Lewis M, Horvath L (2001) Invasion by extremes: population spread with variation in dispersal and reproduction. The American Naturalist 157: 537–554.

68. Travis JMJ, Harris CM, Park KJ, Bullock JM (2011) Improving prediction and management of range expansions by combining analytical and individual-based modeling approaches. Methods in Ecology and Evolution 2: 477–488.

69. Heikkinen RK, Marmion M, Luoto M (2012) Does the interpolation accuracy of species distribution models come at the expense of transferability? Ecography 35: 276–288.

70. Guisan A, Zimmermann NE, Elith J, Graham CH, Phillips S, et al. (2007) What matters for predicting the occurrences of trees: Techniques, data, or species' characteristics? Ecological Monographs 77: 615–630.

71. Syphard AD, Franklin J (2009) Differences in spatial predictions among species distribution modeling methods vary with species traits and environmental predictors. Ecography 32: 907–918.

72. Thomas JA (1995) The conservation of declining butterfly populations in Britain and Europe: Priorities, problems and successes. Biological Journal of the Linnean Society 56: 55–72.

73. Saarinen K, Jantunen J (2002) A comparison of the butterfly fauna of agricultural habitats under different management history in Finnish and Russian Karelia. Annales Zoologici Fennici 39: 173–181.

74. Pöyry J, Lindgren S, Salminen J, Kuussaari M (2005) Responses of butterfly and moth species to restored cattle grazing in semi-natural grasslands. Biological Conservation 122: 465–478.

75. Saarinen K (2004) The national butterfly recording scheme in Finland (NAFI): results in 2004. Baptria 29: 106–114.

76. Pöyry J, Luoto M, Heikkinen RK, Saarinen K (2008) Species traits are associated with the quality of bioclimatic models. Global Ecology and Biogeography 17: 403–414.

77. Skellam J (1951) Random dispersal in theoretical populations. Biometrika 38: 196–218.

78. Hastings A, Cuddington K, Davies KF, Dugaw CJ, Elmendorf S, et al. (2005) The spatial spread of invasions: new developments in theory and evidence. Ecology Letters 8: 91–101.

79. Neubert MG, Caswell H (2000) Demography and dispersal: Calculation and sensitivity analysis of invasion speed for structured populations. Ecology 81: 1613–1628.

80. Bocedi G, Zurell D, Reineking B, Travis JMJ (in press) Mechanistic modelling of animal dispersal offers new insights into range expansion dynamics across fragmented landscapes. Ecography. DOI:10.1111/ecog.01041.

81. Pykälä J (2000) Mitigating human effects on European biodiversity through traditional animal husbandry. Conservation Biology 14: 705–712.

82. Hodgson JA, Thomas CD, Wintle BA, Moilanen A (2009) Climate change, connectivity and conservation decision making: back to basics. Journal of Applied Ecology 46: 964–969.

83. Baguette M, Schtickzelle N (2003) Local population dynamics are important to the conservation of metapopulations in highly fragmented landscapes. Journal of Applied Ecology 40: 404–412.

17

The Effects of Timing of Grazing on Plant and Arthropod Communities in High-Elevation Grasslands

Stacy C. Davis[1]*, Laura A. Burkle[1], Wyatt F. Cross[1], Kyle A. Cutting[2]

1 Department of Ecology, Montana State University, Bozeman, Montana, United States of America, 2 Red Rock Lakes National Wildlife Refuge, US Fish and Wildlife Service, Lima, Montana, United States of America

Abstract

Livestock grazing can be used as a key management tool for maintaining healthy ecosystems. However, the effectiveness of using grazing to modify habitat for species of conservation concern depends on how the grazing regime is implemented. Timing of grazing is one grazing regime component that is less understood than grazing intensity and grazer identity, but is predicted to have important implications for plant and higher trophic level responses. We experimentally assessed how timing of cattle grazing affected plant and arthropod communities in high-elevation grasslands of southwest Montana to better evaluate its use as a tool for multi-trophic level management. We manipulated timing of grazing, with one grazing treatment beginning in mid-June and the other in mid-July, in two experiments conducted in different grassland habitat types (i.e., wet meadow and upland) in 2011 and 2012. In the upland grassland experiment, we found that both early and late grazing treatments reduced forb biomass, whereas graminoid biomass was only reduced with late grazing. Grazing earlier in the growing season versus later did not result in greater recovery of graminoid or forb biomass as expected. In addition, the density of the most ubiquitous grassland arthropod order (Hemiptera) was reduced by both grazing treatments in upland grasslands. A comparison of end-of-season plant responses to grazing in upland versus wet meadow grasslands revealed that grazing reduced graminoid biomass in the wet meadow and forb biomass in the upland, irrespective of timing of grazing. Both grazing treatments also reduced end-of-season total arthropod and Hemiptera densities and Hemiptera biomass in both grassland habitat types. Our results indicate that both early and late season herbivory affect many plant and arthropod characteristics in a similar manner, but grazing earlier may negatively impact species of conservation concern requiring forage earlier in the growing season.

Editor: John F. Valentine, Dauphin Island Sea Lab, United States of America

Funding: The authors are grateful to Montana State University and the U.S. Fish and Wildlife Service for funding this project. The funders had no role in study design, data collection and analysis, decision to publish, or preparation of the manuscript.

Competing Interests: The authors have declared that no competing interests exist.

* Email: stacy.davis1@msu.montana.edu

Introduction

Grazing is a key process in grasslands that has far reaching effects on plant and animal diversity [1,2,3], vegetation structure [4], and ecosystem functioning [5] over multiple spatial scales. Although native ungulate grazers dominated certain landscapes prior to European settlement, many of these grasslands around the world are now grazed by domestic livestock [6]. In 2007, approximately 27% of U.S. land area was classified as grassland pasture and range for livestock production [7]. Current grazing management strategies aim to balance both ecological sustainability and economic considerations [8]. Most conservation practitioners now recognize that grazing may be used as an important land management tool [9,2]. For instance, grazing can be used to modify habitat for species conservation, as demonstrated through improvements of grassland bird habitat [10,11].

Understanding the role of grazing in species conservation requires a thorough assessment of the various components of the grazing regime. Grazing intensity and grazer identity are key components that have received the most attention [12,13,14,15]. In contrast, much less is known about how manipulations of timing of grazing can affect grassland communities. Moreover, although some studies have manipulated grazers at broad temporal scales (grazing across seasons) [16,17,18], very few have examined the effects of timing of grazing at short time scales (within a season) [19]. Timing of grazing management decisions, such as when to initiate grazing, may be especially pertinent for conservation practitioners interested in using grazing as an effective habitat management tool in high-elevation grasslands where grazing only occurs during a contracted growing season.

Altering timing of grazing within a short growing season may have large effects on multiple trophic levels if grazing time periods coincide with key life history stages of organisms in the ecological community (e.g., reproduction, rapid growth) [20]. Livestock can reconstruct grassland bird habitat through grazing-induced modifications to vegetation structure [11]. In this way, timing of grazing may affect higher trophic levels if grazing alters the vegetation structure, either favorably or unfavorably, during key life history phases (e.g., nesting for bird species). Timing of grazing may also affect higher trophic levels through alterations to the arthropod community. For instance, many grassland birds require large amounts of energy and protein, often acquired from arthropod prey, during chick development and adult molting [21,22,23]. Grassland birds may be negatively affected by reductions in the density, biomass, and diversity of arthropods with grazing [3,14,24] if such reductions coincide with birds' key life history stages. Similarly, grassland predators, such as spiders, may be highly sensitive to temporal changes in vegetation structure and prey availability; hence, timing of grazing may have important cascading effects on the structure and stability of invertebrate food webs [25].

In addition to uncertainties with respect to timing of grazing, little is known about how grazing regimes will affect plant and arthropod communities among different grassland habitat types that form the broader vegetation mosaic. Wet meadow and upland grassland habitats often differ in soil moisture, soil type, and major limiting resources [13], which are known to influence community structure and productivity of grassland plants [26,27]. Moreover, the effects of grazing on plant species composition and biomass may be mitigated by favorable soil water conditions [26,28]. Having a clearer understanding of how plant and arthropod communities respond to timing of grazing across wet meadow and upland grassland habitats will be useful for land managers and livestock producers interested in adaptive grazing management.

In an effort to quantify the effects of timing of grazing on plants and arthropods in high-elevation grasslands, we experimentally manipulated timing of cattle grazing over the course of two growing seasons in two grassland habitat types. We asked: (1) how do ungrazed wet meadow and upland grasslands differ in terms of plant, arthropod, and soil moisture characteristics, (2) how does timing of grazing affect plant and arthropod communities in upland grasslands, and (3) what is the range of possible plant and arthropod responses to timing of grazing across wet meadow and upland grasslands? We hypothesized that ungrazed wet meadow grasslands would have greater soil moisture than ungrazed upland grasslands due to the closer proximity of ground water to the soil surface in wet meadows [29]. In addition, we expected greater plant biomass and arthropod densities in wet meadows resulting from the increased soil moisture. While we expected that grazing in upland grasslands would temporarily reduce graminoid and forb biomass and vegetation height, we hypothesized that plots grazed earlier in the growing season would have greater plant biomass and height towards the end of the growing season than plots grazed later in the growing season due to the increased time available for regrowth [30]. Additionally, we hypothesized that arthropod orders (i.e., Hemiptera [true bugs] and Araneae [spiders]) strongly affected by vegetation structure [3,31,32] would respond to grazing in upland grasslands in a similar manner as plants. Finally, we hypothesized that there would be a range of outcomes for plant and arthropod responses to timing of grazing across grassland habitat types. In particular, we expected reduced effects of grazing in wet meadow grasslands versus in upland grasslands due to increased soil moisture and plant regrowth potential [26,17]. Our results provide a more comprehensive understanding of the effects of timing of grazing on plant and arthropod communities, as well as how these effects may differ depending on grassland habitat type. This knowledge is important for understanding how multiple trophic levels are affected by timing of grazing through temporal shifts in vegetation structure and forage availability. Our results are also highly relevant for the conservation and management of high-elevation grasslands that have extremely short growing seasons relative to lower elevations.

Methods

Ethics Statement

This field study was conducted in collaboration with the US Fish and Wildlife Service and The Nature Conservancy; all permissions for site access were granted and no permits were required. This study did not involve any endangered or protected species.

Study Area

We conducted grazing experiments in 2011 and 2012 in the Centennial Valley of southwest Montana (44°40′ N, 111°47′ W, 2030 m elevation) on grasslands leased or owned by Red Rock Lakes National Wildlife Refuge (RRL) and The Nature Conservancy. Climate in the Centennial Valley is characterized by long, cold winters and short, mild summers with highly variable annual precipitation [29]. Mean annual air temperature and precipitation at Lakeview, Montana (located in the southern region of the Centennial Valley; c. 11 km from study sites) are 1.56°C and 500 mm, respectively, with May and June typically being the wettest months [29].

Our 2011 grazing experiment was conducted in wet meadow grasslands, while our 2012 experiment took place in upland grasslands. Wet meadow grasslands occupy over 2,800 hectares at RRL, and vegetation in this habitat is dominated by a dense layer of graminoids and low forb canopy cover [29]. Dominant graminoids include tufted hairgrass (Deschampsia cespitosa), clustered field sedge (Carex praegracilis), and mat muhly (Muhlenbergia richardsonis), while Baltic rush (Juncus balticus) is found in wetter areas [29]. Common wet meadow forbs include Rocky Mountain iris (Iris missouriensis), common dandelion (Taraxacum officinale), and darkthroat shooting star (Dodecatheon pulchellum), among others [29]. Upland grasslands occupy more than 4,900 hectares at RRL, and are largely dominated by graminoids, such as tufted hairgrass (Deschampsia cespitosa), clustered field sedge (Carex praegracilis), basin wildrye (Leymus cinereus), Kentucky bluegrass (Poa pratensis) and Sandberg bluegrass (Poa secunda) [29]. Forb coverage and diversity varies in upland grasslands, depending on soil moisture and type, but silvery lupine (Lupinus argenteus), rosy pussytoes (Antennaria rosea), and common yarrow (Achillea millefolium) are the most common [29]. Both grassland habitat types support diverse breeding bird communities, including long-billed curlew (Numenius americanus), sandhill crane (Grus canadensis), savannah sparrow (Passerculus sandwichensis), western-meadowlark (Sturnella neglecta), and other migratory birds [29].

Current management allows grazing to occur on many of the grasslands owned or leased by RRL, with grazing intensities ranging from 0.31 to 0.85 AUM/acre (where 1 AUM = 1 cow-calf pair per month) between 1994 and 2006 (U.S. Fish and Wildlife Service unpubl. data). Additionally, since the 1950s, 90% of grazing has been initiated after July 10th due to concern over nest trampling of protected avian species (e.g. long-billed curlew and sandhill cranes). Although the long-term grazing history of the specific sites used in this study was unavailable because they were privately owned until 2008, most grazing units (ranging in size from 25 to 3,000 acres) owned or leased by RRL are grazed by cattle on a 3-year rest-rotation [29].

Experimental Design

In both years, we established two cattle grazing treatments that differed in grazing initiation dates: early graze (starting mid-June) and traditional late graze (starting mid-July). We also established ungrazed, control plots (hereafter, ungrazed plots refers to these plots which were not grazed in the study year growing season, but may have been grazed prior to the three year rest-rotation). In 2011, our grazing experiment was conducted at a large spatial scale, using c. 38 hectare (~1120 m by 340 m) experimental plots in heterogeneous, sub-irrigated wet meadow grasslands. We observed a high degree of patchiness in grazing perhaps due to the large amount of heterogeneity in plant community structure and soil moisture in wet meadows. In 2012, we used smaller experimental plots (c. 2 hectare, ~135 m by 150 m) in nearby homogeneous, non-irrigated upland grasslands. Experimental blocks containing control, early-graze, and late-graze plots were established in both years. We had three blocks in 2011, with each block sized ~114 hectares (~1120 m by 1016 m), and four blocks in 2012, each block sized ~6 hectares (~405 m by 150 m). Within each block, plots were randomly selected for one of the three treatments. Due to trespass grazing in early July of 2012, two plots were removed from analyses (one control and one early-grazed), thus reducing the number of replicates for control and early-grazed treatments from four to three. In both years, early- and late-graze treatment plots were grazed by cattle for equal durations (grazed for 2 weeks) and at equal intensities (0.9 AUM/acre).

Vegetation and arthropods were collected concurrently within each plot at six sampling events between mid-June and early September in 2011 and four sampling events between early June and late August in 2012. Sampling event timing and number of events differed between years due to a shorter growing season in 2012 than in 2011. On each sampling event in 2011, ten sampling points were randomly selected for sampling out of 25 sampling points randomly established in each plot. On each sampling event in 2012, ten sampling points that were randomly established east or west of a centrally located north-south transect within each plot were sampled. Since cattle severely trampled the area near water tanks, we created a 30 m buffer zone near the water tanks where no sampling occurred. In both years, we avoided potential "edge effects" between treatments by not sampling in 10 m buffer zones between neighboring plots.

Vegetation Sampling

At each sampling point (N = 10 per plot), above-ground plant biomass was quantified by clipping vegetation to ground level in a 0.03 m^2 quadrat in 2011 and to 5 cm in a 0.25 m^2 quadrat in 2012. On each sampling event, we sampled in different ordinal directions from sampling points to avoid clipping the same vegetation. Quadrats were located at least 1 m away from previously sampled quadrats. Total plant biomass included both live and dead vegetation that represented the previous and current year of growth, with dead vegetation being included because it represents an important structural habitat for arthropods and birds [31,33,34]. Vegetation samples were sorted as graminoids or forbs. In order to further examine grazing effects on vegetation structure, we also sampled vegetation height in 2012. Thus, prior to clipping vegetation, we measured plant height on three out of the four sampling events by estimating the height of 80% of vegetation at four equally-spaced points within the 0.25 m^2 quadrat [35].

Arthropod Sampling

In both years, prior to sampling vegetation at each sampling point, we sampled arthropods using an enclosed 0.25 m^2 plastic barrel that was quickly placed over the collection area, minimizing arthropod escape and restricting the sample to a known area. We then used an inverted leaf blower/vacuum sampler (suction cylinder area of 0.013 m^2; Craftsman XRZ 2000; 30s/sample) to extract arthropods by moving the sampler evenly across the vegetation in the enclosed area [36]. In 2012, we also used sweep

netting in an effort to sample a larger area of the plot. Sweep netting took place during hours of peak arthropod activity (11:00–15:00) on calm, sunny days. We collected a total of 36 sweep net samples on sampling events 2–4 by walking swiftly in the center of each plot in a north-south transect for 50 m (40 sweeps per sample). All arthropods from each sample were frozen until processed. Arthropod density and biomass were determined for a randomly selected set of five out of ten sampling points per plot in 2011 and for all ten sampling points per plot in 2012. In both years, density was determined by separating arthropods greater than 1 mm from vegetation debris, counting individuals, and then identifying individuals to order and size class. Size classes varied depending on the arthropod order in 2011, and in 2012 all arthropod orders were measured to the nearest mm. Biomass was determined for the six arthropod orders that collectively made up over 95% of total composition in terms of abundance (Hemiptera [true bugs], Araneae [spiders], Hymenoptera [mainly ants], Diptera [flies], Coleoptera [beetles], and Orthoptera [grasshoppers]). Total arthropod biomass refers to the sum of these six arthropod orders. In 2011, we dried and weighed arthropods in each size class to obtain dry mass. In 2012, we used more specific size classes (to the nearest mm) and converted the length of each arthropod to dry mass using taxon-specific length/mass regression equations [37]. Total biomass for each arthropod order was calculated as the sum of all size classes in each sample. Individual biomass (a measure of average size) was calculated as total biomass/number of individuals for each sample.

Soil Moisture Sampling

We measured soil moisture in each grassland habitat type to determine how local-scale water availability influenced plant and arthropod responses to grazing treatments. In the wet meadow grassland (2011), we measured soil moisture at each of the 25 sampling points in each plot three times throughout the summer. In the upland grassland (2012), we measured soil moisture at each of the ten sampling points in each plot on all four sampling events. Three soil moisture readings were taken within 30 cm of each other using an Aquaterr 300-T soil probe at a depth of 15 cm. All readings were within 60 cm of the sampling point.

Statistical Analyses

In the wet meadow grassland experiment (2011), we did not detect strong grazing effects because of large variability in soil moisture and plant biomass within experimental plots resulting from sub-irrigation and visible heterogeneity in cattle grazing. Additionally, the discrepancy between the large plot sizes relative to the small quadrats used for plant sampling in 2011 may not have adequately captured the mean effects of the grazing treatments. We therefore restricted our in-depth analysis of timing of grazing effects throughout the growing season to the upland grassland experiment conducted in 2012. In contrast, we used both years of data to examine the range of possible plant and arthropod responses to timing of grazing across grassland habitat types (wet meadow vs. upland grassland). By comparing cumulative (late August = end-of-season) responses, grassland communities had a longer timeframe to develop patterns and reflect grazing treatment differences between grassland habitat types.

Differences Between Ungrazed Wet Meadow and Upland Grasslands. To quantify grassland habitat type differences without the added complexity of grazing, we compared plant, arthropod, and soil moisture metrics from ungrazed plots in wet meadow and upland grasslands. We focused on comparing end-of-season characteristics because this time period represented the cumulative effects over most of the growing season. We first compared mean graminoid biomass, forb biomass, total arthropod density, and total arthropod biomass in each grassland habitat type using paired t-tests.

Histograms of a cumulative soil moisture metric (range 0–200%) were generated to depict end-of-season soil moisture for each grassland habitat type. This cumulative metric, based on the summed average soil moisture values from all August sampling events, was used to better represent moisture conditions in plots at the end of the growing season. To compare soil moisture between grassland habitat type, we limited the sampling points to N = 10 by randomly subsetting ten sampling points in the wet meadow (out of 25 possible) on each sampling event to match N = 10 in the upland grassland. Due to non-normality of distributions, we conducted a Kolmogorov-Smirnov two-sample test to examine inter-habitat differences in soil moisture.

To examine relationships between plant and arthropod variables across habitat types, we calculated Pearson correlation coefficients for end-of-season plant biomass and arthropod density in each grassland habitat type. We also used Pearson correlation coefficients to examine finer-scale relationships between microsite cumulative soil moisture and end-of-season (1) graminoid biomass, (2) forb biomass, (3) total arthropod density, and (4) total arthropod biomass.

Timing of Grazing Effects on Plants and Arthropods Across the Growing Season in Upland Grasslands. To determine the effects of timing of grazing on plant and arthropod characteristics in the upland grassland, we used a repeated-measures analysis of variance (rmANOVA). Treatment and sampling event (sampling event as the repeated measure) were included as independent variables, as well as the interaction term between them. When significance in main effects or interactions were detected (p<0.05), *post-hoc* one-way ANOVAs and Tukey's Honestly Significant Difference (HSD) multiple comparisons were used to determine treatment differences within each sampling event, as well as sampling event differences within each grazing treatment. We did not adjust alpha levels with a Bonferroni correction because doing so may lead to a higher probability of a type II error and a lack of a standard alpha across studies [38]. All plant and arthropod variables were square-root transformed +0.5 to meet assumptions of normality and homogeneity of variances [38].

Because patterns in sweep net sample results were similar to vacuum sample results, we only report the vacuum sample results because they can be expressed on a per area basis. All data were analyzed using R version 2.15.2 [39] and JMP 10 [40].

Timing of Grazing Effects on End-of-Season Plants and Arthropods in Wet Meadow and Upland Grasslands. Our two study years varied in terms of grassland habitat type, but also in accumulated precipitation (averaged 460 mm in 2011 and 400 mm in 2012 [41]), timing of snowmelt (1 month earlier in 2012; data from Natural Resources Conservation Science SNOTEL site at Tepee Creek; c. 24 km from study sites [42]), and length of growing season (defined as the period of time between the last frost of spring and the first frost of fall; 86 days in 2011 and 74 in 2012; data from local MesoWest fire tower data [43]). These differences in climate, along with grassland habitat type differences, allowed us to examine the range of possible plant and arthropod responses to timing of grazing. In order to compare the magnitude and direction of timing of grazing effects on plants and arthropods in two grassland habitat types, we conducted an effect size analysis which allows a standardized (i.e., responses relative to the control) comparison of responses across multiple studies, despite differences in methodology [44]. For each timing of grazing treatment, we calculated the standardized mean

difference between a given plant or arthropod variable in treatment versus control plots using mean log response ratios [45]. We examined end-of-season characteristics, corresponding to late August when plants began to die back, as well as a time when the effects of herbivory on plants are most likely to influence next year's growth [46]. We conducted this analysis on graminoid biomass, forb biomass, and density and biomass of arthropod orders that collectively made up over 95% of total abundance in each study year (Hemiptera, Araneae, Hymenoptera, Diptera, Coleoptera, and Orthoptera). For both years, we used taxon-specific length/mass regression equations [37] using mean lengths based on the larger size class ranges that we established in 2011 in order to standardize methodology in biomass calculations across years. A negative value of the log response ratio indicated lower values for the response variable in the treatment plots compared to the control plots. Effect sizes for each response variable were summarized with a random-effects model using 95% confidence intervals with bias-corrected bootstrapping [44]. The effect size was considered statistically significant if the bootstrapped confidence interval, calculated with 1000 iterations, did not bracket zero. An effect size of 0.2 is considered "small", 0.5 is "medium", 0.8 is "large", and anything greater than 1 is "very large" [45]. All effect size calculations were conducted using MetaWin version 2.1 [44].

Results

Differences Between Ungrazed Wet Meadow and Upland Grasslands

At the end of the season, ungrazed wet meadow and upland grassland habitats differed substantially in many plant and arthropod characteristics. The wet meadow grassland had greater graminoid and forb biomass, as well as total arthropod density (Figure 1a, b, c). In contrast, there were no inter-habitat differences in total arthropod biomass (Figure 1d). Soil moisture differed significantly between grassland habitat types (Kolmogorov-Smirnov test; p<0.05; Figure 2) with greater cumulative soil moisture in wet meadows (mean of 116% and median of 115%) versus uplands (mean of 81% and median of 82%; Figure 2). Total arthropod density was positively correlated with both graminoid (r = 0.06, p = 0.012) and forb biomass (r = 0.55, p = 0.03) in the wet meadow grassland. In contrast, neither graminoid nor forb biomass was significantly correlated with arthropod density in the upland grassland. Cumulative soil moisture was negatively correlated with end-of-season graminoid (r = −0.44, p = 0.02) and forb biomass (r = −0.49, p = 0.007) in ungrazed plots in the upland grassland. All other correlations were non-significant.

Timing of Grazing Effects on Plants and Arthropods Across the Growing Season in Upland Grasslands

Average graminoid and forb biomass in upland grasslands differed significantly across grazing treatments, but these effects depended on the sampling event (Table 1). Both graminoid and forb biomass were similar in all treatments prior to any grazing (Table 2) and remained constant in control plots throughout the growing season (Table 3). Forb biomass was significantly reduced by 73% (by 15.6 g/m^2; Figure 3b; Table 3) with early grazing, whereas both graminoid and forb biomass were significantly reduced with late grazing (by 47% [by 92.67 g/m^2] and by 69% [by 17.51 g/m^2], respectively; Figure 3a, b; Table 3). Average graminoid biomass in both grazed treatments remained at least 30% lower than control plots throughout the rest of the growing season, although neither treatment was significantly different from control plots (Figure 3a; Table 2). The only significant difference

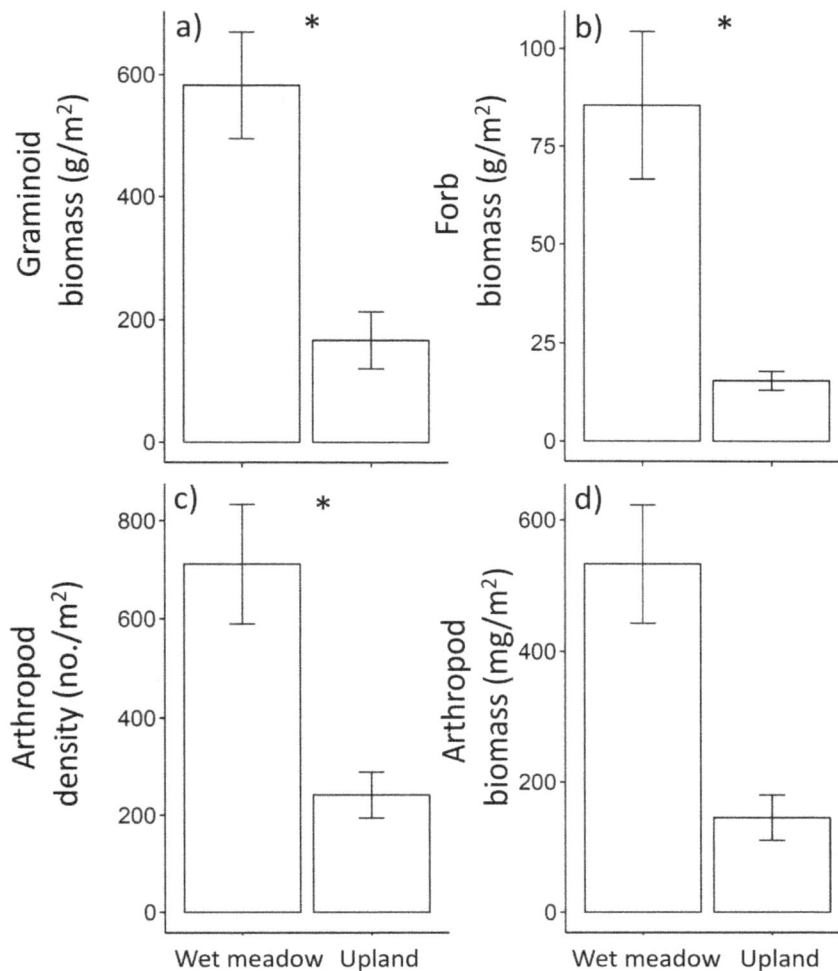

Figure 1. End-of-season comparisons for key grassland variables for ungrazed wet meadow and upland grasslands. End-of-season (late August) values for a) graminoid biomass, b) forb biomass, c) arthropod density, and d) arthropod biomass are untransformed means ± 1 SE. Wet meadow grasslands were sampled in 2011 and upland grasslands in 2012. * Indicates a significant difference between wet meadow and upland grasslands ($p < 0.05$ from paired t-tests, $n = 3$). All variables were square-root transformed +0.5 for t-tests.

among treatments for graminoid biomass was immediately after the early graze in late June, when early-grazed plots had less than half the graminoid biomass of the other plots (Figure 3a; Table 2). There was also a significant difference among treatments for forb biomass immediately after the early graze in late June, as well as between control plots and both early- and late-grazed plots in July and August (Figure 3b; Table 2). Forb biomass in both grazed plots was significantly lower (<50%) than control plots at the end of the growing season (Figure 3b; Table 2).

Average plant height differed significantly across grazing treatments, but these effects depended on the sampling event (Table 1). We assumed there were no pre-treatment differences in plant height because graminoid biomass was similar in all treatments prior to any grazing (Table 1) and post-treatment plant height was strongly correlated with graminoid biomass ($r = 0.72$, $p < 0.0001$, Pearson's correlation). Plant height responses to grazing were similar to graminoid and forb biomass responses, but plant height differed significantly between early- and late-grazed plots at the end of the growing season (Figure 3c; Table 2). Although early- and late-grazed plots had equal graminoid biomass at the end of the growing season (Figure 3a; Table 2),

early-grazed plots had 17% taller vegetation compared to late-grazed plots in late August (Figure 3c; Table 2).

Arthropod density in control plots was largely composed of Hemiptera (57%), Hymenoptera (23%), and Araneae (10%) across all sampling events. The remaining 10% was composed of Coleoptera, Diptera, Orthoptera and six additional low-density orders. Patterns with respect to total arthropod biomass were similar, with the majority of arthropod biomass consisting of Hemiptera (40%), Hymenoptera (19%), Araneae (16%), and Orthoptera (14%).

Average Hemiptera density differed significantly across grazing treatments, but these effects depended on the sampling event (Table 1). Hemiptera density was similar in all treatments prior to any grazing (Table 2) and remained constant in both control and early-grazed plots throughout the growing season (Table 3). Late grazing significantly reduced Hemiptera density by 54% (by ~82 individuals/m²; Figure 3d; Table 3). Early- and late-grazed plots differed significantly immediately after the early graze in late June, when early-grazed plots had less than half the density of Hemiptera than the other treatments (Figure 3d; Table 2). At

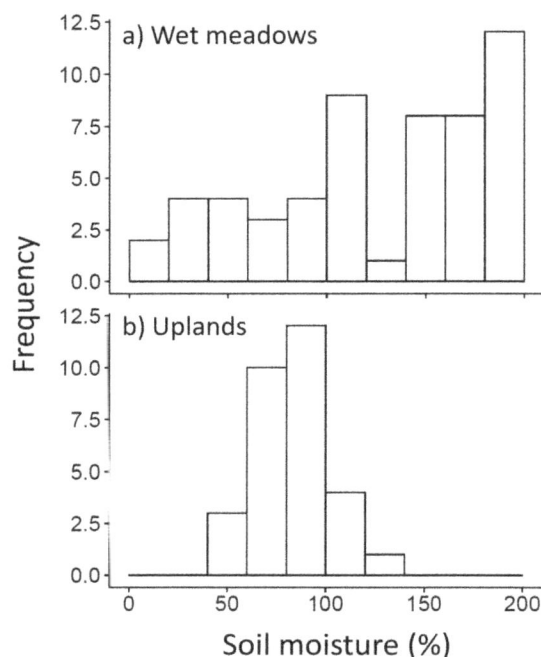

Figure 2. Histograms for end-of-season cumulative soil moisture according to ungrazed grassland habitat type: a) wet meadows and b) uplands. Wet meadow grasslands were sampled in 2011 and upland grasslands in 2012.

the end of the growing season, Hemiptera density in grazed plots did not differ from control plots (Figure 3d; Table 2).

There was no effect of grazing on average body size (i.e., mean individual biomass) of Hemiptera. However, there was a marginally significant effect of grazing on Hemiptera total biomass (p = 0.07; Table 1), but this effect was only evident in late July; late-grazed plots had 37% less Hemiptera biomass than control plots, but only 6% less Hemiptera biomass when compared to early-grazed plots (Table 2). In early June, prior to any grazing, control plots had larger Orthoptera than late-grazed plots (Table 1, Table 3). Grazing did not significantly affect the density or biomass of any other arthropods (Table 1).

Timing of Grazing Effects on End-of-Season Plants and Arthropods in Wet Meadow and Upland Grasslands

Effect sizes of grazing treatments on end-of-season graminoid biomass and forb biomass varied depending on the grassland habitat type. Regardless of timing, grazing had a medium negative effect on end-of-season graminoid biomass in the wet meadow grassland and forb biomass in the upland grassland (Figure 4a, b). Grazing also had a medium to large negative effect on end-of-season arthropod density (driven largely by Hemiptera; Figure 4c and e) and Hemiptera total biomass in both grassland habitats (Figure 4f). Late grazing had a particularly large negative effect (− 1.7) on Hemiptera total biomass in the wet meadow grassland (Figure 4f), as well as end-of-season total arthropod biomass (− 1.1; Figure 4d). Regardless of timing, grazing had a very large negative effect (early = − 1.1, late = − 1.2) on end-of-season Araneae density in the wet meadow grassland (Figure 4g). The effects of timing of grazing on density of other taxa (e.g., Hymenoptera and Coleoptera) were more complex and idiosyncratic, with medium to large positive effects on Hymenoptera density in the wet

meadow grassland and variable effects in the upland grassland (Figure 4h, i).

Discussion

Our grazing experiment in high-elevation, western U.S. grasslands demonstrated that grazing altered plant and arthropod characteristics within a single growing season. In upland grasslands, we found that both timing of grazing treatments generally reduced plant biomass and height, and grazing earlier versus later in the growing season did not result in greater recovery for graminoid or forb biomass. Additionally, the density of the most dominant arthropod order (Hemiptera) was reduced by both grazing treatments. While both grazing initiation dates shared similarities in responses, our results suggest that early grazing has the potential to negatively affect grassland birds that are reliant on forbs and Hemiptera as a main forage source in June. Our results also suggest that the influence of grazers on plant communities may depend on both grassland habitat type and plant functional group, but climatic variation is likely to influence grazing outcomes.

Differences Between Ungrazed Wet Meadow and Upland Grasslands

We found that the ungrazed wet meadow grassland had greater graminoid and forb biomass, as well as total arthropod density, than the ungrazed upland grassland. We expected such differences due to the influence of soil moisture in wet meadows resulting from spring flooding and closer proximity to ground water [29]. Our cumulative end-of-season soil moisture readings suggested that the wet meadow still had greater soil moisture values than the upland grassland at the end of the growing season. However, differences in plant and arthropod communities in varying grassland habitat types may also be due to interannual variation in precipitation. Our study in upland grasslands occurred during a relatively drier year compared to the year we sampled wet meadow grasslands; therefore, precipitation/soil moisture differences between grassland habitat types are likely to represent a wider range of responses than if the two study years had very similar precipitation patterns. Regardless of whether or not soil moisture differences are a result of grassland habitat type or precipitation differences, such soil moisture differences in the broader vegetation mosaic are likely to affect how grazing management between grassland habitat types impacts multiple trophic levels [13]. For instance, total arthropod density was strongly linked to graminoid and forb biomass in wet meadow grasslands, but not in upland grasslands, suggesting that grazing effects on the plant community are likely to indirectly affect the arthropod community in wet meadows.

Effects of Timing of Grazing on Plants and Arthropods Across the Growing Season in Upland Grasslands

Many plant characteristics were temporarily reduced by both grazing treatments, as expected. However, plant height, but not plant biomass, was greater with early grazing than late grazing at the end of the season. Several explanations may help elucidate why plant biomass did not recover to a greater degree with early grazing than late grazing, despite the increased time available for regrowth. First, a plant's ability to regrow after grazing can vary with local conditions, such as the availability of nutrients and moisture [47]. Plant regrowth after grazing may have been limited in the upland grassland because precipitation in 2012 was minimal (47 mm) between June and August, the key growing season in this system. This hypothesis is further supported by the lack of plant growth in control plots across the growing season, suggesting

Table 1. Results of repeated-measures ANOVA for the effect of timing of grazing on plant and arthropod variables in upland grasslands (2012).

Variable	df_N df_D	F	p-value
Graminoid biomass			
Treatment	2,7	1.86	0.2251
Event	3,5	3.67	0.0977
Treatment:Event	6,10	6.88	**0.0041**
Forb biomass			
Treatment	2,7	7.87	**0.0162**
Event	3,5	7.2	**0.0291**
Treatment:Event	6,10	9.74	**0.0011**
Plant height			
Treatment	2,7	31.53	**0.0003**
Event	2,6	0.74	0.5200
Treatment:Event	4,12	21.40	**<0.0001**
Arthropod density			
Treatment	2,6	2.31	0.1805
Event	3,4	1.65	0.3128
Treatment:Event	6,8	1.75	0.2277
Total arthropod biomass			
Treatment	2,6	0.18	0.8417
Event	3,4	0.76	0.5732
Treatment:Event	6,8	1.09	0.4432
Individual arthropod biomass			
Treatment	2,6	4.47	0.0647
Event	3,4	11.91	**0.0184**
Treatment:Event	6,8	0.73	0.6393
Arthropod group total biomass	df_N df_D	F	p-value
Hemiptera total biomass			
Treatment	2,6	0.98	0.4284
Event	3,4	1.26	0.3993
Treatment:Event	6,8	3.12	0.0700
Hymenoptera total biomass			
Treatment	2,6	0.49	0.6344
Event	3,4	1.58	0.3263
Treatment:Event	6,8	0.39	0.8639
Araneae total biomass			
Treatment	2,6	0.45	0.6565

Arthropod group density	df_N df_D	F	p-value
Hemiptera density			
Treatment	2,6	3.18	0.1143
Event	3,4	9.28	**0.0283**
Treatment:Event	6,8	5.92	**0.0125**
Hymenoptera density			
Treatment	2,6	0.91	0.4522
Event	3,4	0.75	0.5763
Treatment:Event	6,8	0.43	0.8376
Araneae density			
Treatment	2,6	0.04	0.9653
Event	3,4	7.45	**0.0409**
Treatment:Event	6,8	1.17	0.4078
Coleoptera density			
Treatment	2,6	0.34	0.7232
Event	3,4	34.87	**0.0025**
Treatment:Event	6,8	1.70	0.2390
Diptera density			
Treatment	2,6	1.06	0.4024
Event	3,4	12.98	**0.0157**
Treatment:Event	6,8	3.31	0.0609
Orthoptera density			
Treatment	2,7	1.49	0.2989
Event	3,4	1.62	0.3192
Treatment:Event	6,8	1.79	0.2188
Arthropod group individual biomass	df_N df_D	F	p-value
Hemiptera ind biomass			
Treatment	2,6	0.29	0.7549
Event	3,4	4.18	0.1004
Treatment:Event	6,8	1.54	0.2800
Hymenoptera ind biomass			
Treatment	2,6	0.85	0.4721
Event	3,4	83.93	**0.0005**
Treatment:Event	6,8	2.13	0.1588
Araneae ind biomass			
Treatment	2,6	0.24	0.7971

Table 1. Cont.

Arthropod group total biomass	df_N df_D	F	p-value	Arthropod group individual biomass	df_N df_D	F	p-value
Event	3,4	11.05	**0.0209**	Event	3,4	1.87	0.2756
Treatment:Event	6,8	2.47	0.1176	Treatment:Event	6,8	0.93	0.5206
Coleoptera total biomass				*Coleoptera ind biomass*			
Treatment	2,6	0.17	0.8465	Treatment	2,6	0.37	0.7051
Event	3,4	8.01	**0.0363**	Event	3,4	0.51	0.6941
Treatment:Event	6,8	1.42	0.3159	Treatment:Event	6,8	0.94	0.5177
Diptera total biomass				*Diptera ind biomass*			
Treatment	2,6	0.45	0.6565	Treatment	2,6	0.24	0.7971
Event	3,4	11.05	**0.0209**	Event	3,4	1.87	0.2756
Treatment:Event	6,8	2.47	0.1176	Treatment:Event	6,8	0.93	0.5206
Orthoptera total biomass				*Orthoptera ind biomass*			
Treatment	2,6	1.13	0.3824	Treatment	2,6	0.92	0.4476
Event	3,4	13.50	**0.0147**	Event	3,4	38.33	**0.0021**
Treatment:Event	6,8	0.53	0.7722	Treatment:Event	6,8	3.83	**0.0422**

Sampling event is the repeated measure. Significant p-values at $\alpha < 0.05$ are in bold; df = degree of freedom.

Table 2. Univariate one-way ANOVAs for each sampling event and Tukey's Honestly Significant Difference multiple comparisons when a significant treatment effect was present within each sampling event for upland grasslands (2012).

Variable	Sampling event	df_N, df_D	F	p-value	Tukey's
Graminoid biomass					
	Early June	2,7	0.10	0.9067	
	Late June	2,7	14.03	**0.0036**	C=L
	Late July	2,7	3.01	0.1141	
	Late August	2,7	2.02	0.2032	
Forb biomass					
	Early June	2,7	0.87	0.4596	
	Late June	2,7	29.23	**0.0004**	C=L
	Late July	2,7	22.68	**0.0009**	E=L
	Late August	2,7	8.67	**0.0128**	E=L
Plant height					
	Late June	2,7	54.18	**<.001**	C=L
	Late July	2,7	16.37	**0.0023**	E=L
	Late August	2,7	40.95	**0.0001**	
Hemiptera density					
	Early June	2,7	0.26	0.7795	
	Late June	2,7	11.02	**0.0098**	C=L
	Late July	2,7	6.53	**0.0251**	C=E, E=L
	Late August	2,7	3.24	0.1008	
Hemiptera total biomass					
	Early June	2,7	0.98	0.4231	
	Late June	2,7	0.97	0.4302	
	Late July	2,7	3.52	0.0874	
	Late August	2,7	2.13	0.1891	
Orthoptera ind biomass					
	Early June	2,7	5.68	**0.0342**	C=E, E=L
	Late June	2,7	1.08	0.3965	
	Late July	2,7	0.30	0.7511	
	Late August	2,7	0.06	0.9418	

Significant p-values at $\alpha < 0.05$ are in bold; df = degree of freedom. The treatment letters sharing a line are not statistically different from one another. Treatments are indicated as follows: C = Control, E = Early, L = Late.

water-limited conditions. Furthermore, the recovery of plant biomass may depend on the phenological stage of plants during grazing. Clipping studies indicate that grasses and forbs are harmed most during the transition from flower stalk formation to seed ripening, which is a time when plants are storing carbohydrates for the dormant period [48]. In our study, it is possible that the majority of grasses and forbs were in this developmental stage during grazing; however, we did not closely track phenological stages of plant species. Finally, there may have been belowground recovery from early grazing that we did not measure; grazers in Yellowstone National Park stimulated root production seven times more than shoot production [49].

We found no significant differences in graminoid biomass at the end of the season between early and late grazing, as well as between grazed and control plots. However, the 40% reduction in mean graminoid biomass in both early- and late-grazed plots (Figure 3a) suggests that graminoid biomass never fully recovered from either grazing treatment, a pattern we may have detected with increased replication. A power analysis using graminoid biomass results from late August indicated that six plots per treatment (instead of four) would be necessary to detect a treatment difference (i.e., at $\alpha = 0.05$; power = 0.29).

Generalizations about effects of grazers on arthropods are elusive [50] due to the large variation in how different arthropod

Table 3. Univariate one-way ANOVAs for each treatment and Tukey's Honestly Significant Difference multiple comparisons when a significant sampling event effect was present within each treatment for upland grasslands (2012).

Variable	Treatment	df$_N$, df$_D$	F	p-value	Tukey's
Graminoid biomass					
	Control	3,8	0.06	0.9829	
	Early	3,8	2.86	0.1044	
	Late	3,8	4.81	**0.0200**	1 = 2, 1 = 3 = 4
Forb biomass					
	Control	3,8	1.31	0.3366	
	Early	3,8	22.22	**0.0003**	2 = 3 = 4
	Late	3,8	9.49	**0.0017**	1 = 4, 3 = 4
Plant height					
	Control	2,6	0.93	0.4458	
	Early	2,6	26.35	**0.0011**	3 = 4
	Late	2,6	30.60	**<0.0001**	3 = 4
Hemiptera density					
	Control	3,8	0.20	0.8967	
	Early	3,8	1.62	0.2598	
	Late	3,8	4.41	**0.0288**	1 = 2 = 4, 1 = 3 = 4
Hemiptera total biomass					
	Control	3,8	0.98	0.4486	
	Early	3,8	5.18	**0.0279**	1 = 2 = 4, 1 = 3 = 4
	Late	3,8	1.26	0.336	
Araneae density	NA	3,8	6.75	**0.001**	1 = 4, 2 = 3 = 4
Coleoptera density	NA	3,8	12.72	**<0.0001**	2 = 3 = 4
Diptera density	NA	3,8	17.30	**<0.0001**	2 = 3 = 4
Hymenoptera individual biomass	NA	3,8	6.99	**0.0008**	1 = 2 = 3, 3 = 4
Araneae total biomass	NA	3,8	9.84	**<0.0001**	2 = 3 = 4
Coleoptera total biomass	NA	3,8	7.35	**0.0006**	2 = 3 = 4
Diptera total biomass	NA	3,8	9.84	**<0.0001**	2 = 3 = 4
Orthoptera total biomass	NA	3,8	5.39	**0.0037**	1 = 2, 2 = 3 = 4
Orthoptera individual biomass					
	Control	3,8	1.09	0.4075	
	Early	3,8	6.21	**0.0175**	1 = 2 = 3, 2 = 3 = 4
	Late	3,8	2.64	0.1019	
Individual arthropod biomass	NA	3,8	3.85	**0.0176**	1 = 2 = 4, 2 = 3 = 4

Significant p-values at α<0.05 are in bold; df = degree of freedom. The sampling event numbers sharing a line are not statistically different from one another. Sampling events are indicated as follows: 1 = Sampling event 1 (early June), 2 = Sampling event 2 (late June), 3 = Sampling event 3 (late July), 4 = Sampling event 4 (late August). If arthropod orders did not vary according to treatment, analysis was done on means across all treatments (variables where treatment = NA).

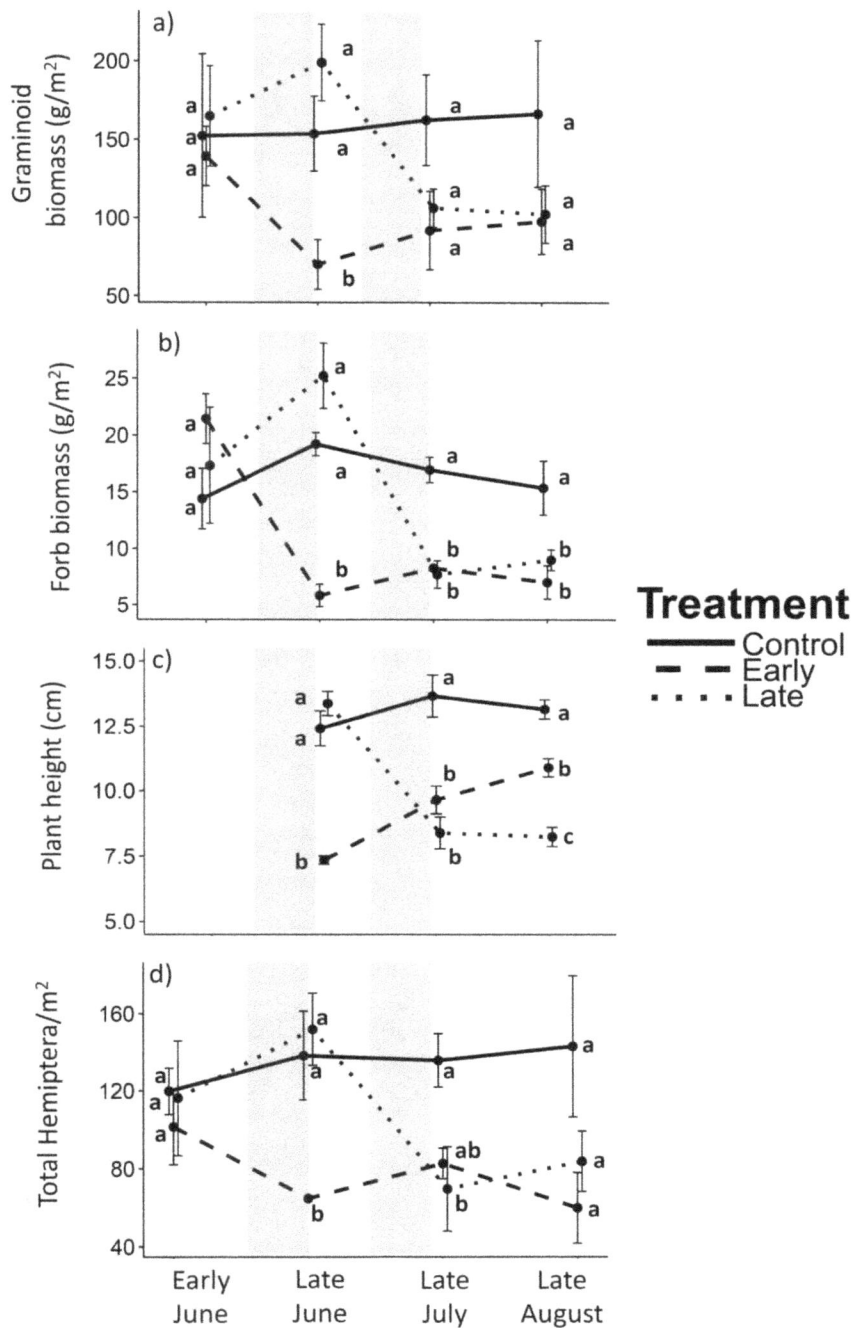

Figure 3. Temporal effects of grazing treatments on plant and arthropod variables for the upland grassland. Values for a) graminoid biomass, b) forb biomass, c) plant height (not sampled in early June), and d) Hemiptera density are untransformed means $\pm$ 1 SE. Duration of early and late grazing treatments are shown by shaded grey boxes. Significant treatment differences within a sampling event are indicated by differing letters. Tukey's HSD results for significant sampling event differences within a treatment are shown in Table 3.

groups respond to grazing-mediated habitat effects [3,12]. Our results showed that Hemiptera density followed a very similar pattern as the plant community to timing of grazing, as expected, whereas spider density was unaffected by grazing. One possible explanation for why spiders did not show a similar response as Hemiptera in our study is that a sufficient level of plant architectural diversity may have been maintained for spider habitat, irrespective of grazing [31]. Additionally, there may be species-specific responses of spiders to timing of grazing, potentially driven by variation in foraging strategy [19]. Even though the effects of timing of grazing on Hemiptera density were temporary (i.e., Hemiptera recovered from both grazed treatments by the end of the growing season), Hemiptera are an important part of this grassland community, comprising over 50% of arthropod density and over 30% of arthropod biomass. Morris and Lakhani [51] likewise found a time-specific response of

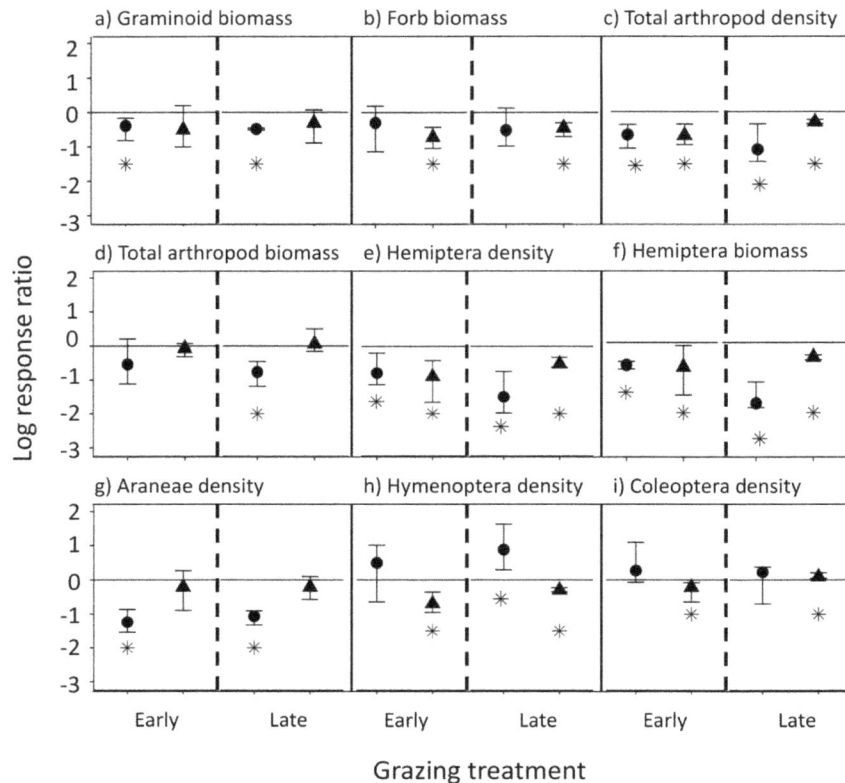

Figure 4. Effect sizes (log response ratio) for end-of-season plant and arthropod variables in wet meadow and upland grasslands. Wet meadows are represented with a circle and uplands with a triangle. The log response ratio compares grazing treatments to control and asterisks (*) denote significant effect sizes at $\alpha = 0.05$. Error bars are bias-corrected 95% confidence intervals. Due to similar results for density and biomass for Araneae, Hymenoptera, and Coleoptera, we only report effect sizes for arthropod density. Diptera and Orthoptera density and biomass in grazed treatments were not significantly different from controls.

Hemiptera to changes in the vegetation community in a lowland grassland; cutting the vegetation early in the growing season (May) reduced the abundance of Hemiptera, but the effects were short-lived.

With the exception of Hemiptera, our results showed that arthropod orders were unaffected by grazing across the growing season in upland grasslands. While some studies that manipulated grazing intensity concur with our results (e.g., [52]), many others have found a negative effect of grazing on arthropods [14,15,53]. These conflicting results may be due to differing habitat types, grazing intensity, grazer identity, and length of experiment. We may have observed stronger responses from the arthropod community with larger sample sizes and if our grazing treatments ran for longer than two weeks, although this short duration of grazing is typical due to the short growing season of our high-elevation study sites.

Timing of Grazing Effects on End-of-Season Plants and Arthropods in Wet Meadow and Upland Grasslands

Regardless of timing of grazing, we found that end-of-season plant biomass was often reduced by grazing, but the effect depended on grassland habitat type and plant functional group. End-of-season forb biomass was unaffected by grazing in the wet meadow grassland, while end-of-season graminoid biomass was unaffected by grazing in the upland grassland. A possible reason for these differences in responses is likely related to the competitive abilities of the two functional groups. Mueggler found that forb

growth benefited more from increased soil moisture than grass growth and suggested that forbs grazed during a wet year would be damaged less than those grazed during a dry year [48]. Our cumulative soil moisture readings indicated that the wet meadow grassland had a different distribution of soil moisture values than in the upland grassland. Additionally, our study in upland grasslands took place in 2012, which was a relatively drier year compared to the 2011 study year in wet meadow grasslands, indicating that our results represent two distinct ends of the abiotic spectrum for this system. These combinations of grassland habitat type and precipitation may have influenced grazing responses, but are representative of realistic conditions and are important factors to consider in adaptive grazing management decisions.

We found that some arthropod orders were unaffected by grazing (i.e., Diptera and Orthoptera), while grazing reduced many other end-of-season arthropod densities in both study years, regardless of timing of grazing or grassland habitat type. End-of-season Hemiptera density was reduced for both grazing treatments in both grassland habitat types, while Araneae density was reduced in the wet meadow grassland. While we did not find as many negative effects of grazing on arthropods *across* the growing season with the rmANOVA, the effect size analysis revealed there were still cumulative negative effects of grazing on several arthropod groups at the *end of the season*, which has important implications for arthropods' ability to recover from a grazing event.

Different plant and arthropod responses in 2011 vs. 2012 may be due to differences in grassland habitat type, but as previously stated, may also be due to interannual variation in climate, such as accumulated precipitation, timing of snowmelt, and length of growing season. It is also possible there were other unmeasured ecological drivers that varied between our two study sites and contributed to the observed responses of plants and arthropods to grazing, such as differences in fine-scale topography [26] or variability in timing of rainfall [16]. Given that the wet meadow grassland was sampled in a wet year and the upland grassland experiment in a dry year, the observed differences between grassland habitat types may be exaggerated compared to other years. Therefore, the effects of timing of grazing on plants and arthropods in upland grasslands remain elusive during a wet year, in which additional soil moisture may benefit an otherwise drier grassland habitat type. Additional studies of wet meadow and upland grasslands in alternative climatic conditions would help to refine grazing management recommendations for these grassland habitat types under variable climate.

Management Implications

Our results demonstrate that the effects of timing of grazing on plant and arthropod communities are not uniform throughout the growing season and such temporal shifts have important implications for species conservation. First, initiating grazing earlier in the growing season in upland grasslands may impact bird species of conservation concern that rely on forbs and Hemiptera as a food source in June. Other studies have similarly suggested that temporal variation in bird forage may have important implications for post-fledgling survival [20]. The temporal shifts we observed in bird forage availability (i.e., Hemiptera) are more likely to affect smaller bird species, such as the savannah sparrow, western meadowlark, and other small-bodied passerines. Larger-sized bird species (long-billed curlew and greater sage-grouse) are less likely to be affected, as they preferably forage on larger-sized arthropod orders, such as Coleoptera [23], that were largely not affected by the grazing treatments in our study. Although we did not measure bird densities as part of this study, previous research has shown that savannah sparrows have a density of 0.52 males/ha (±0.02 SE) in wet meadows at RRL [54]; hence there are higher trophic levels that are likely to be affected by changes in their forage availability as a result of timing of grazing.

Secondly, several plant community variables were affected by timing of grazing and such alterations in vegetation structure have important implications for nesting habitat for bird species of conservation concern. Early grazing in the upland grassland resulted in 17% taller plant height at the end of the season compared to late grazing. Since residual plant height persists into

the next growing season, this information could be used to inform management of grassland birds that require specific vegetation heights for fulfilling life history requirements [11]. All of our grazing treatments, including controls, had less than 15 cm plant height from June to August, indicating that long-billed curlews, with habitat requirements of less than 30 cm plant height for nesting purposes [29], would nest in this area regardless of timing of grazing. However, fine-scale differences in plant height may affect rates of nest predation for other smaller, ground-nesting grassland passerines, such as savannah sparrow and western meadowlark [55].

Proper planning of grazing regimes may aid in reducing possible negative effects of early season grazing on higher trophic levels. Early season grazing could take place outside of core bird nesting and foraging areas in the future, which may help mediate potential early-season impacts on higher trophic levels. One of the most beneficial grazing strategies may include a matrix of grazing plots, with varying grazing initiation dates and years of rest between grazing periods at the landscape level. Doing so would aid certain areas of the grassland in recovery from grazing at any point during the growing season, and provide the necessary structure and forage availability for a broad suite of species with differing life history requirements. A better understanding of how timing of grazing affects plants and arthropods will be useful for conservation grazing management in similar high-elevation grassland systems and in particular, for determining how potential changes in forage availability will affect species conservation.

Acknowledgments

B. Sowell, B. Halstead, and two anonymous reviewers provided helpful comments on earlier versions of this manuscript. We thank B. West of Red Rock Lakes National Wildlife Refuge, and J. Monk and N. Korb of The Nature Conservancy for invaluable logistical support. This project was not possible without the continued partnership and encouragement for rigorous science from our grazing partners, A. Anderson and B. Ulring of J-Bar-L Ranch. We thank W. Running, and S. Muise for assistance in the field, and J. Herrick, T. Schiltz, R. Chaffin, E. Hester, and J. Bernt for help processing lab samples.

Author Contributions

Conceived and designed the experiments: SCD LAB WFC KAC. Performed the experiments: SCD. Analyzed the data: SCD. Contributed reagents/materials/analysis tools: LAB WFC KAC. Wrote the paper: SCD LAB WFC KAC.

References

1. Bakker ES, Ritchie ME, Olff H, Milchunas DG, Knops JMH (2006) Herbivore impact on grassland plant diversity depends on habitat productivity and herbivore size. Ecol Lett 9: 780–788.

2. Collins SL, Knapp AK, Briggs JM, Blair JM, Steinauer EM (1998) Modulation of diversity by grazing and mowing in native tallgrass prairie. Science 280: 745–747.

3. Gibson CWD, Brown VK, Losito L, McGavin GC (1992) The response of invertebrate assemblies to grazing. Ecography 15: 166–176.

4. Davidson AD, Ponce E, Lightfoot DC, Fredrickson EL, Brown JH, et al. (2010) Rapid response of a grassland ecosystem to an experimental manipulation of a keystone rodent and domestic livestock. Ecology 91: 3189–3200.

5. Bai Y, Wu J, Clark CM, Pan Q, Zhang L, et al. (2012) Grazing alters ecosystem functioning and C:N:P stoichiometry of grasslands along a regional precipitation gradient. J Appl Ecol 49: 1204–1215.

6. McNaughton SJ (1986) Grazing lawns: on domesticated and wild grazers. Am Nat 128: 937–939.

7. Nickerson C, Ebel R, Borchers A, Carriazo F (2011) Major Uses of Land in the United States, 2007. EIB-89. U.S. Dept. Agr., Econ. Res. Serv. Dec.

8. Brown JH, McDonald W (1995) Livestock grazing and conservation on southwestern rangelands. Conserv Biol 9: 1644–1647.

9. Pillsbury FC, Miller JR, Debinski DM, Engle DM (2011) Another tool in the toolbox? Using fire and grazing to promote bird diversity in highly fragmented landscapes. Ecosphere 2: 1–14.

10. WallisDeVries MF (1998) Large herbivores as key factors for nature conservation. In: WallisDeVries MF, Bakker JP, van Wieren SE, editors. Grazing and Conservation Management. Dordrecht, the Netherlands: Kluwer Academic Publishers. pp.1–20.

11. Derner JD, Lauenroth WK, Stapp P, Augustine DJ (2009) Livestock as Ecosystem Engineers for Grassland Bird Habitat in the Western Great Plains of North America. Rangeland Ecol Manage 62: 111–118.

12. Sjödin NE, Bengtsson J, Ekbom B (2008) The influence of grazing intensity and landscape composition on the diversity and abundance of flower-visiting insects. J Appl Ecol 45: 763–772.

13. Olff H, Ritchie ME (1998) Effects of herbivores on grassland plant diversity. Trends Ecol Evol 13: 261–265.

14. Dennis P, Young MR, Gordon IJ (1998) Distribution and abundance of small insects and arachnids in relation to structural heterogeneity of grazed, indigenous grasslands. Ecol Entomol 23: 253–264.

15. Cagnolo L, Molina SI, Valladares GR (2002) Diversity and guild structure of insect assemblages under grazing and exclusion regimes in a montane grassland from Central Argentina. Biodivers Conserv 11: 407–420.

16. Pérez-Camacho L, Rebollo S, Hernández-Santana V, García-Salgado G, Pavón-García J, et al. (2012) Plant functional trait responses to interannual rainfall variability, summer drought and seasonal grazing in Mediterranean herbaceous communities. Funct Ecol 26: 740–749.

17. Anderson MT, Frank DA (2003) Defoliation effects on reproductive biomass: importance of scale and timing. J Range Manage 56: 501–516.

18. Morris MG (1973) The Effects of Seasonal Grazing on the Heteroptera and Auchenorhyncha (Hemiptera) of Chalk Grassland. J Appl Ecol 10: 761–780.

19. Lenoir L, Lennartsson T (2010) Effects of timing of grazing on arthropod communities in semi-natural grasslands. J Insect Sci 10: 1–24.

20. Paquette SR, Garant D, Pelletier F, Bélisle M (2013) Seasonal patterns in Tree Swallow prey (Diptera) abundance are affected by agricultural intensification. Ecol Appl 23: 122–133.

21. Sveum CM, Crawford JA, Edge WD (1998) Use and selection of brood-rearing habitat by sage grouse in south central Washington. West N Am Nat 58: 344–351.

22. Peterson JG (1970) The food habits and summer distribution of juvenile sage grouse in central montana. J Wildl Manage 34: 147–155.

23. Drut MS, Pyle WH, Crawford JA (1994) Diets and food selection of sage grouse chicks in Oregon. J Range Manage 47: 90–93.

24. Joern A, Laws AN (2013) Ecological Mechanisms Underlying Arthropod Species Diversity in Grasslands. Annu Rev Entomol 58: 19–36.

25. Schmitz OJ (2008) Herbivory from Individuals to Ecosystems. Annu Rev Ecol Evol Syst 39: 133–152.

26. Milchunas DG, Lauenroth WK, Chapman PL, Kazempour MK (1989) Effects of grazing, topography, and precipitation on the structure of a semiarid grassland. Vegetatio 80: 11–23.

27. Sims PL, Singh JS (1978) The structure and function of ten western north american grasslands: III. Net primary production, turnover and efficiencies of energy capture and water use. J Ecol 66: 573–597.

28. Chase JM, Leibold MA, Downing AL, Shurin JB (2000) The effects of productivity, herbivory, and plant species turnover in grassland food webs. Ecology 81: 2485–2497.

29. United States Fish and Wildlife Service (2009) Comprehensive Conservation Plan, Red Rock Lakes National Wildlife Refuge.

30. Oesterheld M, McNaughton SJ (1991) Effect of stress and time for recovery on the amount of compensatory growth after grazing. Oecologia 85: 305–313.

31. Gibson CWD, Hambler C, Brown VK (1992) Changes in spider (Araneae) assemblages in relation to succession and grazing management. J Appl Ecol 29: 132–142.

32. Kőrösi A, Batáry P, Orosz A, Rédei D, Báldi A (2012) Effects of grazing, vegetation structure and landscape complexity on grassland leafhoppers (Hemiptera: Auchenorrhyncha) and true bugs (Hemiptera: Heteroptera) in Hungary. Insect Conserv Divers 5: 57–66.

33. Wiens JA, Rotenberry JT, Van Horne B (1987) Habitat Occupancy Patterns of North American Shrubsteppe Birds: The Effects of Spatial Scale. Oikos 48: 132–147.

34. Davis SK (2004) Area sensitivity in grassland passerines: effects of patch size, patch shape, and vegetation structure on bird abundance and occurrence in southern Saskatchewan. Auk 121: 1130–1145.

35. Stewart KEJ, Bourn NAD, Thomas JA (2001) An evaluation of three quick methods commonly used to assess sward height in ecology. J Appl Ecol 38: 1148–1154.

36. Kruess A, Tscharntke T (2002) Contrasting responses of plant and insect diversity to variation in grazing intensity. Biol Conserv 106: 293–302.

37. Rogers LE, Buschbom RL, Watson CR (1977) Length-Weight Relationships of Shrub-steppe invertebrates. Ann Entomol Soc Am 70: 51–53.

38. Gotelli NJ, Ellison AM (2004) A Primer of Ecological Statistics. Sunderland, Massachusetts, USA: Sinauer Associates. 492 p.

39. R Core Team (2012) R: A language and environment for statistical computing. R Foundation for Statistical Computing, Vienna, Austria. ISBN 3-900051-07-0. Available http://www.R-project.org.

40. JMP, Version 10. SAS Institute Inc., Cary, NC, 1989–2013.

41. PRISM Climate Group website. Oregon State University. Available: http://prism.oregonstate.edu. Accessed 24 July 2013.

42. United States Department of Agriculture, N.R.C.S. website. (2013) Tepee Creek (813) SNOTEL Site Information. Available: http://www.wcc.nrcs.usda.gov/nwcc/site?sitenum=813&state=mt Accessed: July 24, 2013.

43. Horel J, Splitt M, Dunn L, Pechmann J, White B, et al. (2002) MesoWest: Cooperative mesonets in the western United States. Bull Amer Meteor Soc 83: 211–225.

44. Rosenberg MS, Adams DC, Gurevitch J (2000) MetaWin: Statistical Software for Meta-Analysis. Version 2. Sinauer Associates, Sunderland, Massachusetts.

45. Hedges LV, Gurevitch J, Curtis PS (1999) The meta-analysis of response ratios in experimental ecology. Ecology 80: 1150–1156.

46. Sullivan AT, Howe HF (2009) Prairie forb response to timing of vole herbivory. Ecology 90: 1346–1355.

47. Maschinski J, Whitham TG (1989) The continuum of plant responses to herbivory: the influence of plant association, nutrient availability, and timing. Am Nat 134: 1–19.

48. Mueggler WF (1967) Response of mountain grassland vegetation to clipping in southwestern Montana. Ecology 48: 942–949.

49. Frank DA, Kuns MM, Guido DR (2002) Consumer control of grassland plant production. Ecology 83: 602–606.

50. Bell JR, Wheater CP, Cullen W (2001) The implications of grassland and heathland management for the conservation of spider communities: a review. J Zool 255: 377–387.

51. Morris MG, Lakhani KH (1979) Responses of Grassland Invertebrates to Management by Cutting. I. Species Diversity of Hemiptera. J Appl Ecol 16: 77–98.

52. Mysterud A, Ove Hansen L, Peters C, Austrheim G (2005) The short-term effect of sheep grazing on selected invertebrates (Diptera and Hemiptera) relative to other environmental factors in an alpine ecosystem. J Zool 266: 411–418.

53. Rambo JL, Faeth SH (1999) Effect of vertebrate grazing on plant and insect community structure. Conserv Biol 13: 1047–1054.

54. Newlon K, Warren J (2007) Status Report 2006: Breeding Landbird Monitoring at Red Rock Lakes National Wildlife Refuge.

55. Dion N, Hobson KA (2000) Interactive Effects of Vegetation and Predators on the Success of Natural and Simulated Nests of Grassland Songbirds. The Condor 102: 629–634.

18

Impacts of Diffuse Radiation on Light Use Efficiency across Terrestrial Ecosystems Based on Eddy Covariance Observation in China

Kun Huang[1,2], Shaoqiang Wang[1]*, Lei Zhou[1], Huimin Wang[1], Junhui Zhang[3], Junhua Yan[4], Liang Zhao[5], Yanfen Wang[2], Peili Shi[1]

1 Key Laboratory of Ecosystem Network Observation and Modeling, Institute of Geographic Sciences and Natural Resources Research, Chinese Academy of Sciences, Beijing 100101, China, **2** University of Chinese Academy of Sciences, Beijing 100049, China, **3** Institute of Applied Ecology, Chinese Academy of Sciences, Shenyang 110016, China, **4** South China Botanical Garden, Chinese Academy of Sciences, Guangzhou 510650, China, **5** Northwest Plateau Institute of Biology, Chinese Academy of Sciences, Xining 810001, China

Abstract

Ecosystem light use efficiency (LUE) is a key factor of production models for gross primary production (GPP) predictions. Previous studies revealed that ecosystem LUE could be significantly enhanced by an increase on diffuse radiation. Under large spatial heterogeneity and increasing annual diffuse radiation in China, eddy covariance flux data at 6 sites across different ecosystems from 2003 to 2007 were used to investigate the impacts of diffuse radiation indicated by the cloudiness index (CI) on ecosystem LUE in grassland and forest ecosystems. Our results showed that the ecosystem LUE at the six sites was significantly correlated with the cloudiness variation ($0.24 \leq R^2 \leq 0.85$), especially at the Changbaishan temperate forest ecosystem ($R^2 = 0.85$). Meanwhile, the CI values appeared more frequently between 0.8 and 1.0 in two subtropical forest ecosystems (Qianyanzhou and Dinghushan) and were much larger than those in temperate ecosystems. Besides, cloudiness thresholds which were favorable for enhancing ecosystem carbon sequestration existed at the three forest sites, respectively. Our research confirmed that the ecosystem LUE at the six sites in China was positively responsive to the diffuse radiation, and the cloudiness index could be used as an environmental regulator for LUE modeling in regional GPP prediction.

Editor: Dafeng Hui, Tennessee State University, United States of America

Funding: This research was jointly supported by the CAS for Strategic Priority Research Program (Grant No. XDA05050602), National Basic Research Program of China (973Program) (Grant No. 2010CB833503) and the Key Project in the National Science & Technology Pillar Program of China (Grant No. 2013BAC03B00). The funders had no role in study design, data collection and analysis, decision to publish, or preparation of the manuscript.

Competing Interests: The authors have declared that no competing interests exist.

* Email: sqwang@igsnrr.ac.cn

Introduction

Terrestrial ecosystems play an increasingly important role in global carbon cycle under climate change [1]. Light use efficiency (LUE) was first presented in the context of agricultural ecosystem focusing on the linear relationship between yield and solar irradiance, and gross primary production (GPP) was defined as the overall photosynthetically fixation of carbon per unit space and time [2]. The fact that GPP represents the critical flux component driving the terrestrial ecosystem carbon cycle implies that subtle fluctuations in GPP have substantial implications for future climate warming scenarios [3,4]. With the quantification terrestrial ecosystem GPP for regions, continents, or the globe, we can gain insight into the feedbacks between the terrestrial biosphere and the atmosphere under global change and climate policy-making facilitation [5,6]. Still, GPP predictions at regional scale to global scale are a major challenge due to the spatial heterogeneity [7,8]. Moreover, with the great carbon sequestration potential of the terrestrial ecosystem of China in global carbon budget [9], large

uncertainties exist in terrestrial ecosystem GPP simulation in China.

A number of modeling approaches have been developed for regional/global GPP estimations, including ecological process-based models and light use efficiency models driven by remote sensing data [10]. Among all the models, LUE models encompassing the LUE algorithm proposed by [2] may have the highest potential to identify the spatio-temporal dynamics of regional GPP due to the simplicity of concept and availability of remote sensing data [11]. With this method, GPP was defined as product of photosynthetically active radiation (PAR) absorbed by the vegetation canopy and a conversion factor, LUE [2,12]. Various LUE models have been developed for this purpose, including MODIS GPP algorithm [13], Vegetation Photosynthesis Model (VPM) [14], EC-LUE model [15], Vegetation Index (VI) model [16], C-Fix model [17], Temperature and Greenness Rectangle (TGR) model [18], Temperature and Greenness (TG) model [19] and so on. In order to acquire GPP estimations of high accuracy, the biophysical controls on the ecosystem LUE are significantly important to be fully understood [8,20]. Recent studies indicated that GPP and LUE were affected by both the quantity and composition of the incoming solar radiation [10,21–23]. With a given value of total incoming radiation, LUE of the entire canopy will increase with the increasing fraction of diffuse radiation (FDR) [23–25]. Under cloudy or aerosol-laden skies, incoming radiation was more diffuse and more uniformly distributed in the canopy with a smaller fraction of the canopy that was light saturated [10]. Consequently, canopy photosynthesis was inclined to be more light-use efficient under diffuse sunlight than under direct sunlight condition [10,21,23,24,26,27]. Evidences showed that global secondary organic aerosol in the atmosphere will increase by 36% in 2100 [28]. The aerosol influenced the cloud formation, which was the main contributor to the increment on FDR in the atmosphere [29–31]. Furthermore, an increasing trend of annual diffuse radiation in China has been proved to be $7.03\ MJ.m^{-2}.yr^{-1}$ per decade from 1981 to 2010 [32]. However, few studies on ecosystem GPP predictions took into account effects of the FDR variations of the incoming radiation on LUE based on the LUE models.

Up to now, the eddy covariance (EC) technique provides an alternative way to measure NEE continuously that can be used for GPP calculation by subtracting the modeled ecosystem respiration components [5,33–35]. Multi-sites and continuous eddy covariance (EC) flux and meteorological observation from the China-FLUX network provided a valuable tool for GPP and LUE calculation across ecosystems in China [34]. Therefore, in order to reveal the biophysical controls on measured ecosystem LUE for better regional GPP predictions in terrestrial ecosystems of China which is of high spatial heterogeneity, the impact of diffuse radiation resulting from cloud condition on LUE is of growing concern to be characterized by a uniform proxy. Despite the study that effect of cloudiness change on ecosystem LUE and water use efficiency was detected by the clearness index [36], it was difficult to incorporate the clearness index into LUE model for regional GPP estimates due to the specification of the highest interval of solar elevation angle in each grid. Here we employed an cloudiness index algorithm based on simple inputs [13,22], flux and metrological measurements from six sites of ChinaFLUX encompassing three forest ecosystems and three grassland ecosystems, to address the impact of diffuse radiation on light use efficiency (defined as GPP/PAR) [21,23,26]. The objectives of this study are to: (1) illustrate the seasonal dynamics of the cloudiness index and light use efficiency at different sites; (2) address the influence of fraction of diffuse radiation on ecosystem light use efficiency; (3) identify whether the cloudiness index thresholds favorable for enhancing ecosystem carbon sequestration exist or not.

Materials and Methods

Sites descriptions and measurements

In this study, flux observations were implemented at three forest ecosystems and three grassland ecosystems attached to the Chinese Terrestrial Ecosystem Flux Observational Network (ChinaFLUX). The three forest sites were comprised of the Changbaishan temperate mixed forest (CBS), Qianyanzhou subtropical evergreen needle leaf planted forest (QYZ), and Dinghushan subtropical evergreen broad-leaved forest (DHS). Subject to moosoon-influenced, temperate continental climate, CBS was located in the Jilin province of China, in which growing season ranged from May to September [37]. The QYZ site was located in the subtropical continental monsoon region, in which the mean annual air temperature was 17.9°C [38,39]. Located in the Guangdong province with a subtropical monsoon humid climate, DHS had a wet season from April to September and dry season

from November to March [39]. The three grassland ecosystems were the Inner Mongolia semi-arid *L. chinensis* steppe (NMG) which is C3 grassland, Haibei alpine frigid *P. fruticosa* shrub(HB), and Damxung (DX) alpine meadow-steppe ecosystem with short sparse vegetation(about 10 cm). NMG was located in the Xilin River Basin, Inner Mongolia Autonomous Region of China with a temperate semiarid continental climate. Its growing season lasted from late April to early October [40]. HB was located in the northeast of the Qinhai-Tibet Plateau with a plateau continental climate, which was characterized by lengthy cold winters and very short warm summers. Being situated in a frigid highland, HB receives strong solar radiation, with a mean annual global radiation of up to 6000–$7000\ MJ.m^{-2}$ [36,41]. The DX site was located in the Lhasa City, Tibet, categorized as plateau monsoon climate. Its growing season duration was from May to September. The PAR was usually high, similar to that in alpine meadow area located in eastern Tibetan Plateau and higher than other grassland ecosystems [42]. The locations of six sites were shown in Figure1, and the detailed information of the six sites was provided in Table 1.

Routine meteorological variables were measured simultaneously with the eddy fluxes at each site. Air humidity and air temperature were measured with shielded and aspirated probes (HMP45C, Vaisala, Helsinki, Finland) at different sites. Global radiation and net radiation were recorded with radiometers (CM11 and CNR-1, Kipp & Zonen, Delft, the Netherlands). Photosynthetically active radiation (PAR) above the canopy was measured with a quantum sensor (LI-190Sb, LiCor Inc., USA). All meteorological observations were recorded at 30-min intervals with dataloggers (Model CR10X & CR23X, Campbell Scientific Inc.) [37,39,41–44].

In the study, we only used data measured during the periods of relatively stable leaf area index (LAI) each year from 2003 to 2007 in order to eliminate the potential effect of changing LAI [36]. The LAI of temperate ecosystems (CBS, NMG, HB and DX) remain stable in the mid-growing season. DHS and QYZ were evergreen forest ecosystems, and their LAI did not vary much with season. Therefore, data of mid growing season (June-August) from all the six flux sites were used to analyze the impact of diffuse radiation on ecosystem light use efficiency and photosynthesis.

Ethics statement

Three forest ecosystems (CBS, QYZ and DHS) and three grassland ecosystems (NMG, HB and DX) attached to China-FLUX were maintained by different institutions of Chinese Academy of Sciences (CAS), respectively. The CBS site was maintained by the Institute of Applied Ecology, CAS; the QYZ site and DX site was maintained by the Institute of Geographic Sciences and Natural Resources Research, CAS; the DHS site was maintained by the South China Botanical Garden, CAS; the NMG site was maintained by University of CAS, and the HB site was maintained by Northwest Plateau Institute of Biology, CAS. All necessary permits were obtained for the described field study. The field study did not involve endangered or protected species. Data will be made available upon request.

Eddy flux data

Carbon flux data (GPP and NEP) observed at 6 typical sites from 2003 to 2007 across China were applied to in this study (Figure1). The raw 30-min flux data procedure included: (1) 3D coordinate rotation was applied to force the average vertical wind speed to zero and to align the horizontal wind to mean wind direction, (2) flux data was corrected according the variation of air density caused by transfer of heat and water vapor [45], (3)the storage below EC height was corrected for forest sites [46], and (4)

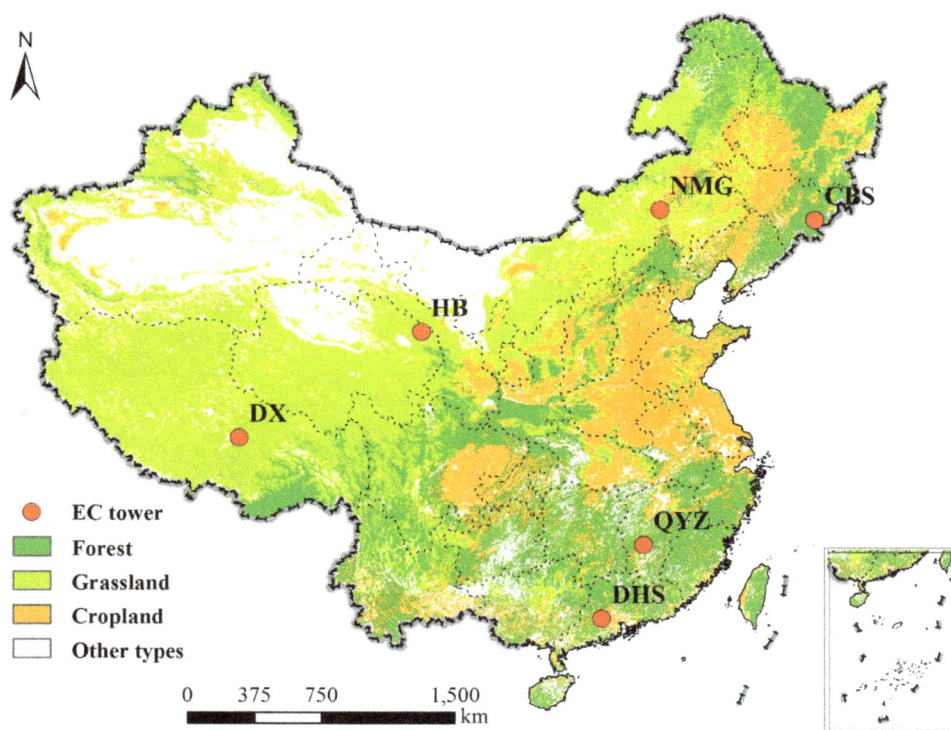

Figure 1. Distribution of the 6 eddy covariance flux sites in China in this study. The background was the MODIS land cover map.

the outlier data were filtered and data gaps were filled by using the look-up table method and mean diurnal variation(MDV) [37,39,47]. In the end, continuous 30 min flux data was performed.

The flux of net ecosystem CO_2 exchange (NEE, mg CO_2 m^{-2} s^{-1}) between the ecosystem and the atmosphere was calculated with equation (1), the net ecosystem productivity (NEP) was assigned to −NEE. Negative NEE values denote carbon uptake, while positive values denote carbon source.

$$NEP = -(\overline{w'\rho'c(zr)} + \int_0^{zr} \frac{\delta\overline{\rho c}}{\delta t} dz), \qquad (1)$$

where the first term on right-hand side is the eddy flux for carbon dioxide or water vapor below the height of observation (z_r), and all advective terms in the mass conservation equation were ignored.

Daily GPP data are partitioned from NEP data measured every 30-min using the eddy covariance technique. GPP was derived from the measured NEP, which was processed using the same method as [36]. Gross primary production (GPP) was calculated employing the following equation:

$$GPP = Re + NEP \qquad (2)$$

Table 1. Site descriptions.

Site (ab.)[a]	Changbaishan (CBS)	Qianyanzhou (QYZ)	Dinghushan (DHS)	Haibei (HB)	Inner Mongolia (NMG)	Damxung (DX)
Location	42°24′N	26°45′N	23°10′N	37°40′N	43°32′N	30°51′N
	128°06′E	115°04′E	112°32′E	101°20′E	116°40′E	91°05′E
Elevation(m)	738	102	300	3293	1189	4333
LAI(m^2m^{-2})	6.1	5.6	4.0	2.8	1.5	1.88
Annual mean precipitation(mm)	600–900	1489	1956	580	350–450	480
Annual mean temperature(°C)	3.6	18.6	21	−1.7	−0.4	1.3
Vegetation type	Mixed forest	Evergreen needle leaf forest	Evergreen broadleaf forest	Alpine frigid shrub	Temperate steppe	Alpine steppe-meadow

[a]Abbreviation for sites.

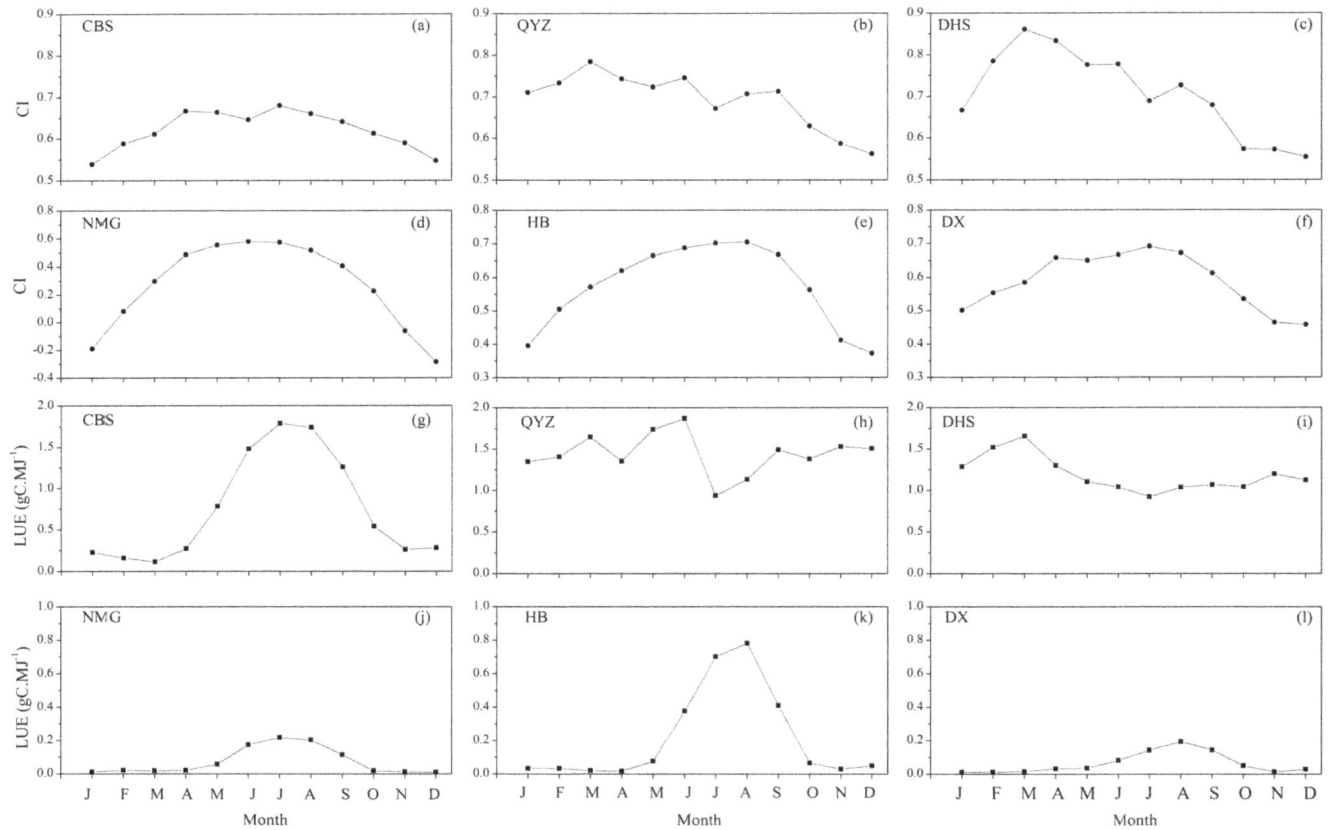

Figure 2. The seasonal variations of monthly mean cloudiness index (CI), monthly mean light use efficiency (LUE) from 2003 to 2007 at the five sites.

NEP was obtained directly from the eddy covariance measurement. Ecosystem respiration (Re) of the seven sites was estimated using the Lloyd-Taylor equation (1994) [41,48]. The nighttime NEP data under turbulent conditions were used to establish Re-temperature response relationship Eq.(3):

$$Re = R_{ref}e^{E_0(1/(T_{ref}-T_0)-1/(T-T_0))} \qquad (3)$$

where R_{ref} represents the ecosystem respiration rate at reference temperature (Tref, 10°C); E_0 is the parameter that determines the temperature sensitivity of ecosystem respiration, and T_0 is a constant and set as $-46.02°C$; T is the air temperature or soil temperature(°C). Eq. (3) was also used to estimate daytime Re.

Calculation of light use efficiency

In this study, LUE (gC.MJ^{-1}) was defined as the ratio of daily GPP (gC.m^{-2}.d^{-1}) to incident PAR (MJ^{-1}.m^{-2}.d^{-1}, using 217 kJ mol^{-1} photons),

$$LUE = \frac{GPP}{PAR} \qquad (4)$$

where PAR was directly measured by the in situ meteorological equipment simultaneous with the flux tower observation.

Cloudiness index

A cloudiness index implemented in CFLUX model was used in our model, since an increase on light use efficiency under overcast conditions at both hourly and daily time steps has been proved in previous studies [22,49]. The cloudiness index was calculated as [22]:

$$CI = 1 - \downarrow PAR/\downarrow PAR_{po} \qquad (5)$$

where CI is the cloudiness index, $\downarrow$ PAR is incident PAR(MJd^{-1}) from daily observation input, $\downarrow$ PAR$_{po}$ is potential incident PAR as a derivation of the algorithm of [50]. With the simple inputs of digital elevation model (DEM) data and readily available parameters, the $\downarrow$ PAR$_{po}$ can be calculated as the global solar radiation at daily time scale in each grid. The spatial resolution of the DEM data was 500 m×500 m, provided by Institute of Geographical Sciences and Natural Resources Research, Chinese Academy of Sciences. More details of the algorithm can be found in the previous literature [50].

The clear sky LUE (LUE$_{cs}$) was specified for each site based on observations of LUE at eddy covariance flux towers. The clear sky LUE was based on the value when $\downarrow$ PAR/ $\downarrow$ PAR$_{po}$ (decreasing cloud cover) approximated 1.0 by a function of LUE under low stress conditions plotted against $\downarrow$ PAR/ $\downarrow$ PAR$_{po}$ [51].

Statistical analysis

The relationships between different variables were fitted with linear and non-linear equations. All analyses were conducted using the origin package v.8.0 (OriginLab Corporation, Northampton, MA, USA). Statistically significant differences were set with P< 0.05 (α = 0.05) unless otherwise stated.

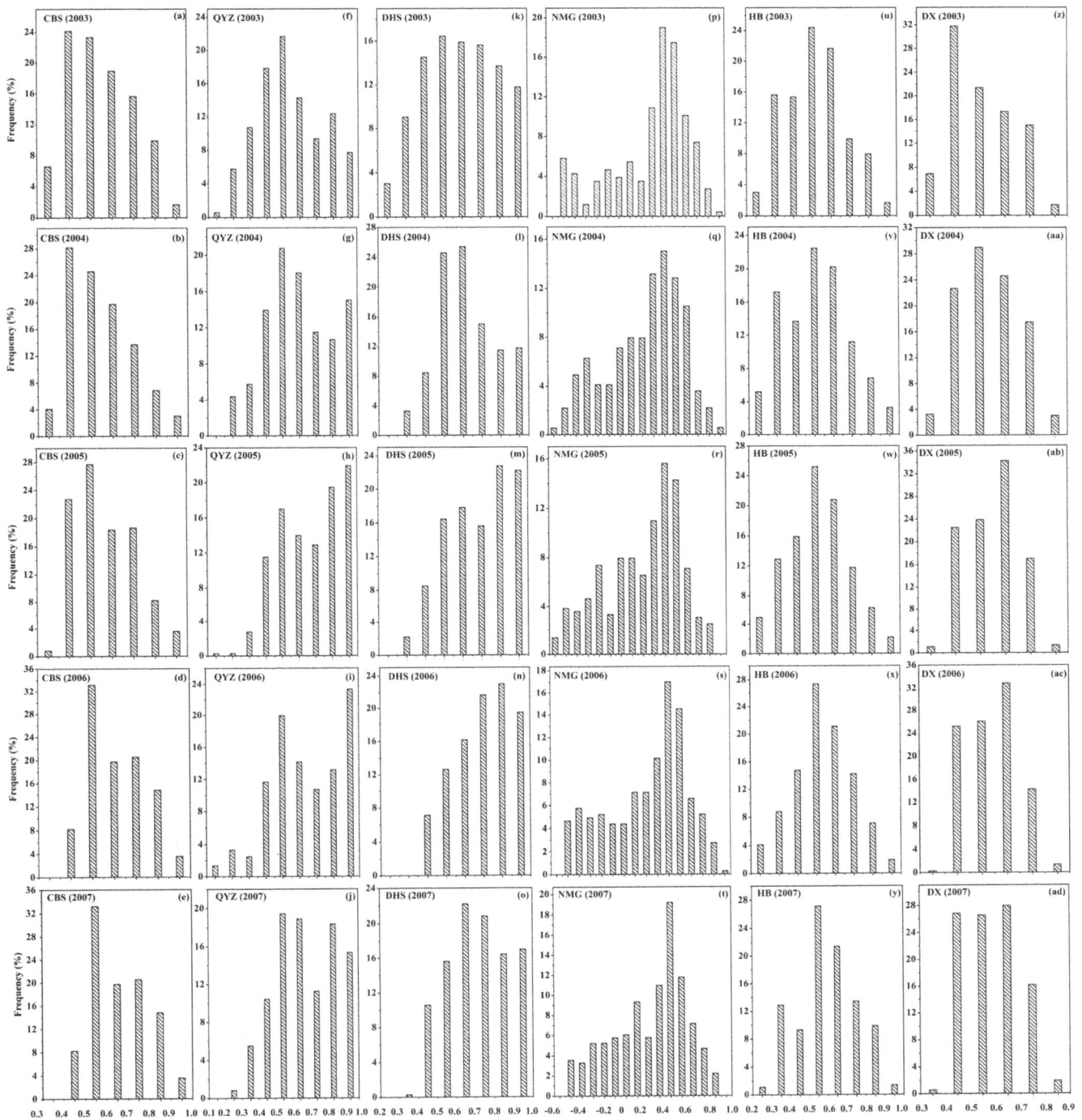

Figure 3. Histograms of the cloudiness index (*CI*) value at the six sites during the mid-growing seasons from 2003 to 2007.

Table 2. Light use efficiency under clear sky in different ecosystems.

Sites	CBS	QYZ	DHS	NMG	HB	DX
$LUE_{cs}(gC.MJ^{-1})$	0.29	0.425	0.569	0.003	0.013	0.009

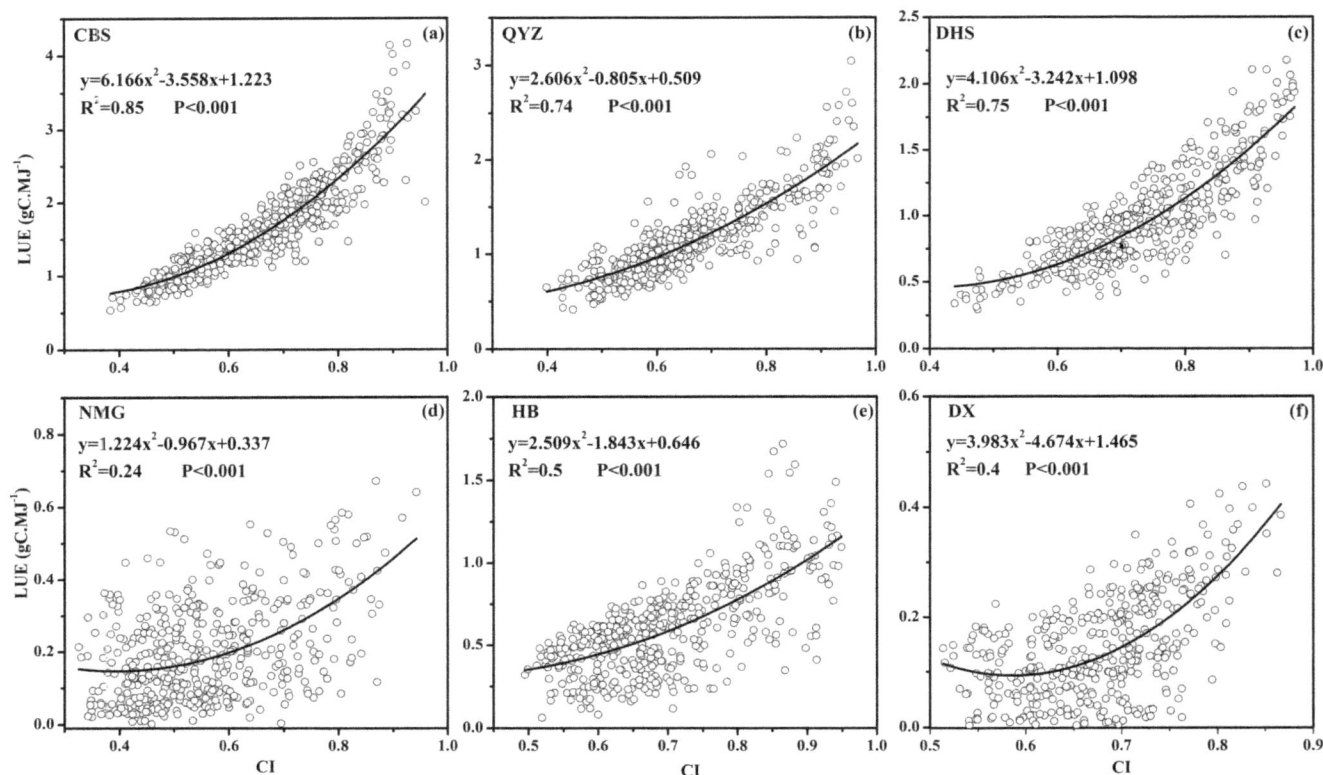

Figure 4. Relationships between LUE and the cloudiness index (*CI*, positive values) during the mid-growing season from 2003 to 2007 at the six sites.

Results and Discussion

Seasonal variation of cloudiness index and light use efficiency across ecosystems

Figure 2 showed the seasonal variations of the cloudiness index and light use efficiency of the six sites from 2003 to 2006. Mostly, cloudiness index (*CI*) was greater at QYZ and DHS than the other temperate ecosystems (CBS, NMG, HB and DX). The *CI* values of subtropical ecosystems (QYZ and DHS) reached the maximum in March, and were higher during the mid-growing season than the two ends of the year (Figure 2b, c). At the temperate ecosystems, the *CI* values peaked during the mid-growing season (Figure 2a, d, e and f), while the *CI* values of the subtropical ecosystems failed to show substantial variations with the seasonal changes. This indicated that sky conditions of two subtropical ecosystem sites were cloudier than those of four temperate ecosystem sites, and cloudy days were more during the mid-growing seasons at the temperate sites. It was also noted that negative *CI* values were found at NMG site, which was in consistency with meteorological observation that the NMG site received stronger solar radiation during the non-growing season. Meanwhile, the forest ecosystems LUE were significantly higher than grassland ecosystem LUE (Figure 2g, h, i, j, k and l). The LUE at subtropical forest sites (QYZ and DHS) failed to show significantly seasonality, while LUE of the temperate ecosystems (CBS, NMG, HB and DX) peaked during mid-growing season. Furthermore, the ecosystem LUE at QYZ site reached its turning point in July during mid-growing season, presented by a sharp fall resulting from the epidemic summer drought [38]. Among grassland sites, the LUE at HB site exhibited apparently higher values than the other two grassland sites and reached its maximal value in August, whereas the ecosystem LUE at NMG and DX site peaked in July and August, respectively (Figure 2j, k and l).

Frequency distribution of cloudiness index value across ecosystems

Apart from the seasonal dynamics of sky conditions (Figure 3a–f), the temporal patterns of cloudiness at the six sites in the mid-growing seasons were showed by the frequency distribution of *CI* values (Figure 3). Despite inter-annual variations resulting from climatic variability, common characteristics of the cloudiness pattern were found to be among the six sites. The *CI* values at CBS site occupied the largest frequency around 0.4 in 2003 and 2004(Figure 3a, b), while the *CI* value frequency took the most part around 0.5 from 2005 to 2007(Figure 3c, d and e). The peaks of *CI* value frequency at QYZ site located around 0.5 (Figure 3f, g and j) and 0.9 (Figure 3h, i). The *CI* value frequency at the DHS peaked between 0.5 and 0.7, except for 2005 and 2006, in which the largest frequency occurred around 0.8 in the mid-growing seasons (Figure 3m, n). As to the NMG site, the largest *CI* value frequency occurred between 0.4 and 0.5. Meanwhile, the *CI* frequency peaked around 0.5 at the HB site (Figure3u–z). The *CI* frequency between 0.5 and 0.7 occupied the largest proportion at DX site, except that peaked around 0.4 in 2003(Figure 3z). Overall, the *CI* frequencies occurred between 0.8 and 1.0 in the subtropical forest sites (QYZ and DHS) were much larger than what in the temperate ecosystems (CBS, NMG, HB and DX), which was verified by the report that spatial patterns of annual diffuse radiation in China showed strong regional heterogeneity, lower in the north but higher in the south [32].

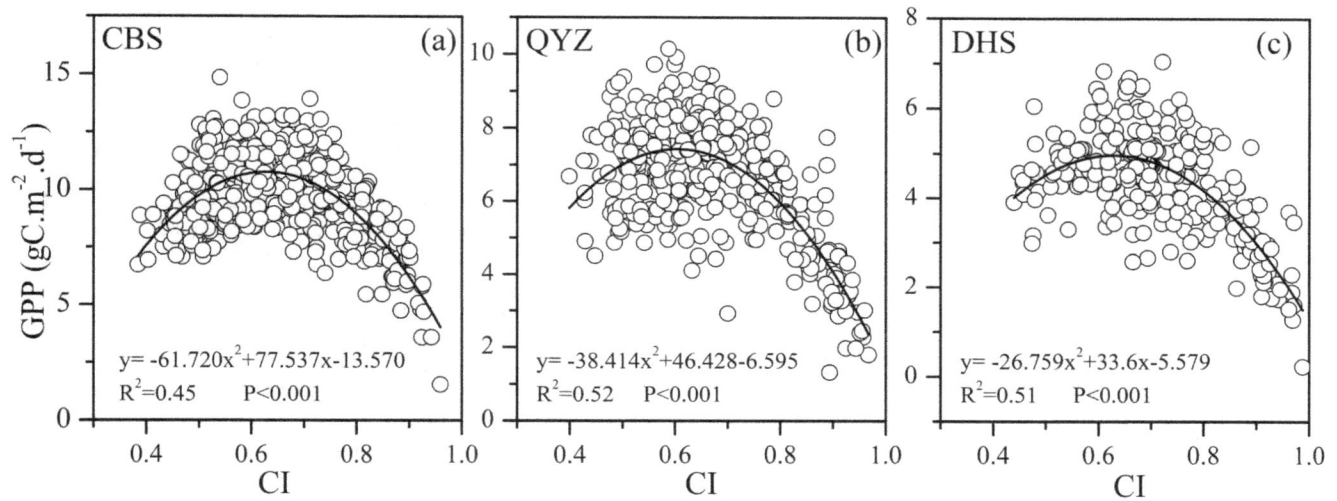

Figure 5. Relationships between ecosystem daily GPP and *CI* during the mid-growing season at three forest ecosystems.

Clear sky light use efficiency in the six ecosystems

The ecosystem LUE was plotted against the ratio of PAR to potential PAR (decreasing cloud cover), and the clear sky LUE (LUE_{cs}) was acquired when PAR/PAR_{po} was around 1.0 [51]. LUE_{cs} values were greater at forest sites than those at grassland sites (Table 2). Among the forest sites, the ecosystem LUE_{cs} was greatest at DHS site, intermediated at QYZ site and lowest at CBS site. The LUE_{cs} of forest ecosystem decreased with the degree of latitude. For grassland sites, the LUE_{cs} peaked at the HB site, followed by the DX site and NMG site, respectively. This was verified by the fact that measured ecosystem LUE was higher than the other two grassland ecosystem sites (Figure 2).

Impacts of diffuse radiation on ecosystem light use efficiency

Figure 4 exhibited the interactive responses of ecosystem LUE to the variation in the diffuse radiation fraction of incoming solar radiation (indicated by the cloudiness index) in different ecosystems. At all sites, significantly quadratic regression relationships were found between the ecosystem LUE and *CI* during the mid-growing season. Once the value of *CI* exceeded a certain one, determined by the minimal value (zero) of the first derived function of each quadratic regression function, the ecosystem LUE increased with *CI* dramatically. LUE of forest ecosystem showed more significantly positive relationship with *CI* ($R^2 \geq 0.74$), compared with three grassland ecosystem sites ($R^2 \leq 0.5$). Also, differences in enhancement on ecosystem LUE induced by the variation of diffuse PAR existed within the ecosystem type across sites. For the forest sites, the ecosystem LUE at CBS site demonstrated stronger increasing trend than the other two subtropical forest sites with the largest correlation coefficient

($R^2 = 0.85$) and quadratic term coefficient (6.166)(Figure 4a, 4b and 4c). The ecosystem LUE at NMG site exhibited least increasing potential ($R^2 = 0.24$, quadratic term coefficient $= 1.224$) with the variation of *CI* among the three grassland ecosystem sites. The expectation that canopy LUE could be enhanced by the diffuse components of solar radiation compared to direct radiation has been reported in previous studies [21,26,52]. Under cloudy skies, incoming radiation was more diffuse and more uniformly distributed in the canopy with a smaller fraction of the canopy that was light saturated [10]. Consequently, canopy photosynthesis was inclined to be more light-use efficient under diffuse sunlight than under direct sunlight conditions [10,21,23,24,26,27]. In addition, differences in canopy structure density across different ecosystem types was presented to contribute to the increasing rate differences of LUE to fraction of diffuse radiation [36], due to its effective penetration to the lower depths of canopy [32,53].

Cloudiness threshold for enhancing forest ecosystem carbon sequestration

Similar significances of quadratic regression relationships between daily GPP and *CI* during mid-growing season were confirmed at three forest sites (Figure 5). The quadratic regression relationships implied that the ecosystem GPP would peak at a certain value of *CI*, and then decreased with increasingly values of *CI*. Specifically, it was noted that the impact of cloudiness on ecosystem carbon exchange process was also dependent on local thermal, moisture and light conditions [36]. At the beginning stage, the forest ecosystems GPP were in positive association with the *CI*. Instead, the forest ecosystems GPP were gradually restrained by the cloudiness when the value of *CI* exceeded a

Table 3. Cloudiness thresholds for enhancing forest ecosystem LUE and GPP.

Cloudiness index	CBS	QYZ	DHS
Lower bounds	0.288	0.154	0.394
Upper bounds	0.628	0.604	0.629

certain one where the symmetric axis of the parabolic curve regression functions located. This phenomenon could partly be ascribed to the decreasing PAR absorbed by the vegetation canopy, based on the radiation conversion efficiency concept of Monteith (1972) [2]. Consequently, the cloudiness thresholds (Table 3) were calculated by a range that began from the value where the symmetric axis of the parabolic curve regression functions (response of LUE to CI) located (Figure 4a, 4b and 4c), and stopped at the point where symmetric axis of the parabolic curve regression functions (response of GPP to CI) located (Figure 5). However, the optimal cloudiness index threshold was not available for the three grassland sites because of the poor quadratic relationship between GPP and CI (P>0.05). The difference in responses of GPP to the variation of diffuse PAR received by the ecosystem between forest sites and grassland sites was likely to result from the difference in canopy structure [36]. The LAI of forest ecosystem at CBS, QYZ and DHS were higher than those of grassland ecosystem (Table 1). Previous studies reported that LUE and GPP of an ecosystem with low LAI, such as grassland and shrubs, did not increase on cloudy days [54,55]. This inconsistency was partly attributed to the differences of climate conditions of the studied ecosystems, including light, water and thermal conditions [36].

Conclusions

Eddy covariance flux observations from six sites encompassing two ecosystem types and the cloudiness index were used to detect the response of LUE and GPP to diffuse radiation during mid-growing season. Results indicated that (1) cloudiness index (CI) was mostly greater at two subtropical forest ecosystem sites (QYZ and DHS) than the other temperate ecosystem sites (CBS, NMG, HB and DX), and LUE in the temperate ecosystem peaked during mid-growing season;(2) LUE under clear sky were greater at forest sites than at grassland sites, and the LUE under clear sky of forest ecosystem decreased with the degree of latitude; (3) significantly quadratic regression relationships were found between the ecosystem LUE and CI during the mid-growing season at all sites;(4) cloudiness thresholds favorable for enhancing ecosystem carbon sequestration existed in forest ecosystem sites.

Due to the large regional heterogeneity existing in terrestrial ecosystem of China, more EC flux sites involved in the future are essential to reveal the impacts of diffuse radiation on terrestrial ecosystem LUE in China. Furthermore, the cloudiness index could be incorporated as an environmental regulator into LUE models for regional GPP simulations in China.

Author Contributions

Conceived and designed the experiments: KH SW. Performed the experiments: KH SW LZ. Analyzed the data: KH SW LZ. Contributed reagents/materials/analysis tools: KH SW LZ HW JZ JY LZ YW PS. Wrote the paper: KH SW. Discussed the manuscript: KH SW LZ HW JZ JY LZ YW PS.

References

1. Nemani RR, Keeling CD, Hashimoto H, Jolly WM, Piper SC, et al. (2003) Climate-driven increases in global terrestrial net primary production from 1982 to 1999. Science 300: 1560–1565.

2. Monteith JL (1972) Solar Radiation and Productivity in Tropical Ecosystems. J Appl Ecol 9: 747–766.

3. Cai WW, Yuan WP, Liang SL, Zhang XT, Dong WJ, et al. (2014) Improved estimations of gross primary production using satellite-derived photosynthetically active radiation. J Geophys Res-Biogeo 119: 110–123.

4. Raupach MR, Canadell JG, Le Quere C (2008) Anthropogenic and biophysical contributions to increasing atmospheric CO_2 growth rate and airborne fraction. Biogeosciences 5: 1601–1613.

5. Wu CY, Munger JW, Niu Z, Kuang D (2010) Comparison of multiple models for estimating gross primary production using MODIS and eddy covariance data in Harvard Forest. Remote Sens Environ 114: 2925–2939.

6. Xiao JF, Zhuang QL, Baldocchi DD, Law BE, Richardson AD, et al. (2008) Estimation of net ecosystem carbon exchange for the conterminous United States by combining MODIS and AmeriFlux data. Agric For Meteorol 148: 1827–1847.

7. Canadell JG, Mooney HA, Baldocchi DD, Berry JA, Ehleringer JR, et al. (2000) Carbon metabolism of the terrestrial biosphere: A multitechnique approach for improved understanding. Ecosystems 3: 115–130.

8. Wang Y, Zhou GS (2012) Light Use Efficiency over Two Temperate Steppes in Inner Mongolia, China. Plos One 7.

9. Piao SL, Fang JY, Ciais P, Peylin P, Huang Y, et al. (2009) The carbon balance of terrestrial ecosystems in China. Nature 458: 1009–U1082.

10. He MZ, Ju WM, Zhou YL, Chen JM, He HL, et al. (2013) Development of a two-leaf light use efficiency model for improving the calculation of terrestrial gross primary productivity. Agric For Meteorol 173: 28–39.

11. Ogutu BO, Dash J (2013) Assessing the capacity of three production efficiency models in simulating gross carbon uptake across multiple biomes in conterminous USA. Agric For Meteorol 174: 158–169.

12. Monteith JL, Moss CJ (1977) Climate and the Efficiency of Crop Production in Britain [and Discussion]. Philosophical Transactions of the Royal Society of London B, Biological Sciences 281: 277–294.

13. Running SW, Nemani RR, Heinsch FA, Zhao MS, Reeves M, et al. (2004) A continuous satellite-derived measure of global terrestrial primary production. BioScience 54: 547–560.

14. Xiao XM, Hollinger D, Aber J, Goltz M, Davidson EA, et al. (2004) Satellite-based modeling of gross primary production in an evergreen needleleaf forest. Remote Sens Environ 89: 519–534.

15. Yuan WP, Liu S, Zhou GS, Zhou GY, Tieszen LL, et al. (2007) Deriving a light use efficiency model from eddy covariance flux data for predicting daily gross primary production across biomes. Agric For Meteorol 143: 189–207.

16. Wu CY, Niu Z, Gao SA (2010) Gross primary production estimation from MODIS data with vegetation index and photosynthetically active radiation in maize. J Geophys Res-Atmos 115.

17. Veroustraete F, Sabbe H, Eerens H (2002) Estimation of carbon mass fluxes over Europe using the C-Fix model and Euroflux data. Remote Sens Environ 83: 376–399.

18. Yang YT, Shang SH, Guan HD, Jiang L (2013) A novel algorithm to assess gross primary production for terrestrial ecosystems from MODIS imagery. J Geophys Res-Biogeo 118: 590–605.

19. Sims DA, Rahman AF, Cordova VD, El-Masri BZ, Baldocchi DD, et al. (2008) A new model of gross primary productivity for North American ecosystems based solely on the enhanced vegetation index and land surface temperature from MODIS. Remote Sens Environ 112: 1633–1646.

20. Garbulsky MF, Penuelas J, Papale D, Ardo J, Goulden ML, et al. (2010) Patterns and controls of the variability of radiation use efficiency and primary productivity across terrestrial ecosystems. Global Ecol Biogeogr 19: 253–267.

21. Gu LH, Baldocchi DD, Wofsy SC, Munger JW, Michalsky JJ, et al. (2003) Response of a deciduous forest to the Mount Pinatubo eruption: Enhanced photosynthesis. Science 299: 2035–2038.

22. Turner DP, Ritts WD, Styles JM, Yang Z, Cohen WB, et al. (2006) A diagnostic carbon flux model to monitor the effects of disturbance and interannual variation in climate on regional NEP. Tellus B 58: 476–490.

23. Mercado LM, Bellouin N, Sitch S, Boucher O, Huntingford C, et al. (2009) Impact of changes in diffuse radiation on the global land carbon sink. Nature 458: 1014–U1087.

24. Oliphant AJ, Dragoni D, Deng B, Grimmond CSB, Schmid HP, et al. (2011) The role of sky conditions on gross primary production in a mixed deciduous forest. Agric For Meteorol 151: 781–791.

25. Roderick ML, Farquhar GD, Berry SL, Noble IR (2001) On the direct effect of clouds and atmospheric particles on the productivity and structure of vegetation. Oecologia 129: 21–30.

26. Gu LH, Baldocchi D, Verma SB, Black TA, Vesala T, et al. (2002) Advantages of diffuse radiation for terrestrial ecosystem productivity. J Geophys Res-Atmos 107.

27. Misson L, Lunden M, McKay M, Goldstein AH (2005) Atmospheric aerosol light scattering and surface wetness influence the diurnal pattern of net ecosystem exchange in a semi-arid ponderosa pine plantation. Agric For Meteorol 129: 69–83.

28. Heald CL, Henze DK, Horowitz LW, Feddema J, Lamarque JF, et al. (2008) Predicted change in global secondary organic aerosol concentrations in response to future climate, emissions, and land use change. J Geophys Res-Atmos 113.

29. Kim SW, Jefferson A, Yoon SC, Dutton EG, Ogren JA, et al. (2005) Comparisons of aerosol optical depth and surface shortwave irradiance and their

effect on the aerosol surface radiative forcing estimation. J Geophys Res-Atmos 110.

30. Feddema JJ, Oleson KW, Bonan GB, Mearns LO, Buja LE, et al. (2005) The importance of land-cover change in simulating future climates. Science 310: 1674–1678.

31. Schiermeier Q (2006) Oceans cool off in hottest years. Nature 442: 854–855.

32. Ren XL, He HL, Zhang L, Zhou L, Yu GR, et al. (2013) Spatiotemporal variability analysis of diffuse radiation in China during 1981–2010. Ann Geophys-Germany 31: 277–289.

33. Baldocchi D, Falge E, Gu LH, Olson R, Hollinger D, et al. (2001) FLUXNET: A new tool to study the temporal and spatial variability of ecosystem-scale carbon dioxide, water vapor, and energy flux densities. B Am Meteorol Soc 82: 2415–2434.

34. Yu GR, Wen XF, Sun XM, Tanner BD, Lee XH, et al. (2006) Overview of ChinaFLUX and evaluation of its eddy covariance measurement. Agric For Meteorol 137: 125–137.

35. Yu GR, Zhu XJ, Fu YL, He HL, Wang QF, et al. (2013) Spatial patterns and climate drivers of carbon fluxes in terrestrial ecosystems of China. Glob Change Biol 19: 798–810.

36. Zhang M, Yu G-R, Zhuang J, Gentry R, Fu Y-L, et al. (2011) Effects of cloudiness change on net ecosystem exchange, light use efficiency, and water use efficiency in typical ecosystems of China. Agric For Meteorol 151: 803–816.

37. Guan DX, Wu JB, Zhao XS, Han SJ, Yu GR, et al. (2006) CO2 fluxes over an old, temperate mixed forest in northeastern China. Agric For Meteorol 137: 138–149.

38. Wen XF, Wang HM, Wang JL, Yu GR, Sun XM (2010) Ecosystem carbon exchanges of a subtropical evergreen coniferous plantation subjected to seasonal drought, 2003–2007. Biogeosciences 7: 357–369.

39. Yu GR, Zhang LM, Sun XM, Fu YL, Wen XF, et al. (2008) Environmental controls over carbon exchange of three forest ecosystems in eastern China. Glob Change Biol 14: 2555–2571.

40. Fu YL, Yu GR, Wang YF, Li ZQ, Hao YB (2006) Effect of water stress on ecosystem photosynthesis and respiration of a Leymus chinensis steppe in Inner Mongolia. Sci China Ser D 49: 196–206.

41. Fu YL, Yu GR, Sun XM, Li YN, Wen XF, et al. (2006) Depression of net ecosystem CO_2 exchange in semi-arid Leymus chinensis steppe and alpine shrub. Agric For Meteorol 137: 234–244.

42. Shi PL, Sun XM, Xu LL, Zhang XZ, He YT, et al. (2006) Net ecosystem CO2 exchange and controlling factors in a steppe - Kobresia meadow on the Tibetan Plateau. Sci China Ser D 49: 207–218.

43. Fu Y, Zheng Z, Yu G, Hu Z, Sun X, et al. (2009) Environmental influences on carbon dioxide fluxes over three grassland ecosystems in China. Biogeosciences 6: 2879–2893.

44. Hu ZM, Yu GR, Fu YL, Sun XM, Li YN, et al. (2008) Effects of vegetation control on ecosystem water use efficiency within and among four grassland ecosystems in China. Glob Change Biol 14: 1609–1619.

45. Webb EK, Pearman GI, Leuning R (1980) Correction of flux measurements for density effects due to heat and water vapour transfer. Quarterly Journal of the Royal Meteorological Society 106: 85–100.

46. Carrara A, Kowalski AS, Neirynck J, Janssens IA, Yuste JC, et al. (2003) Net ecosystem CO_2 exchange of mixed forest in Belgium over 5 years. Agric For Meteorol 119: 209–227.

47. Falge E, Baldocchi D, Olson R, Anthoni P, Aubinet M, et al. (2001) Gap filling strategies for defensible annual sums of net ecosystem exchange. Agric For Meteorol 107: 43–69.

48. Yu GR, Song X, Wang QF, Liu YF, Guan DX, et al. (2008) Water-use efficiency of forest ecosystems in eastern China and its relations to climatic variables. New Phytol 177: 927–937.

49. Turner DP, Urbanski S, Bremer D, Wofsy SC, Meyers T, et al. (2003) A cross-biome comparison of daily light use efficiency for gross primary production. Glob Change Biol 9: 383–395.

50. Fu P, Rich PM (1999) Design and implementation of the Solar Analyst: an ArcView extension for modeling solar radiation at landscape scales. Proceedings of the 19th annual ESRI user conference, San Diego, USA.

51. King DA, Turner DP, Ritts WD (2011) Parameterization of a diagnostic carbon cycle model for continental scale application. Remote Sens Environ 115: 1653–1664.

52. Farquhar GD, Roderick ML (2003) Atmospheric science: Pinatubo, diffuse light, and the carbon cycle. Science 299: 1997–1998.

53. Urban O, Janous D, Acosta M, Czerny R, Markova I, et al. (2007) Ecophysiological controls over the net ecosystem exchange of mountain spruce stand. Comparison of the response in direct vs. diffuse solar radiation. Glob Change Biol 13: 157–168.

54. Letts MG, Lafleur PM, Roulet NT (2005) On the relationship between cloudiness and net ecosystem carbon dioxide exchange in a peatland ecosystem. Ecoscience 12: 53–59.

55. Niyogi D, Chang H-I, Chen F, Gu L, Kumar A, et al. (2007) Potential impacts of aerosol–land–atmosphere interactions on the Indian monsoonal rainfall characteristics. Nat Hazards 42: 345–359.

19

Modeling Pollinator Community Response to Contrasting Bioenergy Scenarios

Ashley B. Bennett[1], Timothy D. Meehan[2], Claudio Gratton[2], Rufus Isaacs[1]*

1 Department of Entomology and Great Lakes Bioenergy Research Center, Michigan State University, East Lansing, Michigan, United States of America, **2** Department of Entomology and Great Lakes Bioenergy Research Center, University of Wisconsin - Madison, Madison, Wisconsin, United States of America

Abstract

In the United States, policy initiatives aimed at increasing sources of renewable energy are advancing bioenergy production, especially in the Midwest region, where agricultural landscapes dominate. While policy directives are focused on renewable fuel production, biodiversity and ecosystem services will be impacted by the land-use changes required to meet production targets. Using data from field observations, we developed empirical models for predicting abundance, diversity, and community composition of flower-visiting bees based on land cover. We used these models to explore how bees might respond under two contrasting bioenergy scenarios: annual bioenergy crop production and perennial grassland bioenergy production. In the two scenarios, 600,000 ha of marginal annual crop land or marginal grassland were converted to perennial grassland or annual row crop bioenergy production, respectively. Model projections indicate that expansion of annual bioenergy crop production at this scale will reduce bee abundance by 0 to 71%, and bee diversity by 0 to 28%, depending on location. In contrast, converting annual crops on marginal soil to perennial grasslands could increase bee abundance from 0 to 600% and increase bee diversity between 0 and 53%. Our analysis of bee community composition suggested a similar pattern, with bee communities becoming less diverse under annual bioenergy crop production, whereas bee composition transitioned towards a more diverse community dominated by wild bees under perennial bioenergy crop production. Models, like those employed here, suggest that bioenergy policies have important consequences for pollinator conservation.

Editor: Shuang-Quan Huang, Central China Normal University, China

Funding: This work was funded by the Department of Energy Great Lakes Bioenergy Research Center (DOE BER Office of Science DE-FC02-07ER64494), DOE OBP Office of Energy and Renewable Energy (DE-AC05-76RL01830), and by a USDA-NIFA grant to CG and RI(2012-67009-20146). The funders had no role in the study design, data collection and analysis, decision to publish, or preparation of the manuscript.

Competing Interests: The authors have declared that no competing interests exist.

* Email: isaacsr@msu.edu

Introduction

Demand for sustainable sources of energy has spurred increasing interest in bioenergy crops as a fuel source. In the United States, government mandates to increase biofuel production to 36 billion gallons per year by 2022 are advancing research into the production and sustainability of both first- and second-generation biofuels [1,2]. First-generation biofuels are produced from annual row crops such as corn, soybean, and canola, while second-generation cellulosic biofuels can be produced from corn stover, switchgrass, or mixed grasslands, a combination of warm-season grasses and forbs [3]. These contrasting options for biofuel cropping systems have the potential to dramatically alter the types and perenniality of vegetative cover in agricultural landscapes, significantly affecting wildlife [4]. Because policies that promote biofuel production have the potential to cause large scale changes in land use [5,6], identifying how bioenergy crops affect biodiversity will be critical to developing sustainable biofuel policies appropriate for regional implementation across the United States.

A growing body of research suggests that bioenergy crops differentially support biodiversity. For example, bird abundance and richness was consistently higher in perennial grassland biofuel plantings compared to annual biofuel plantings, with expanded grasslands predicted to increase bird richness by 12–207% [4,7]. Similarly, predatory arthropods increased in abundance and diversity in perennial grassland biofuel crops compared to corn monocultures [8,9]. Transitioning the landscape into either annual or perennial biofuel crops will affect species diversity and community composition but also the provisioning of valuable ecosystem services, such as arthropod-mediated predation of crop pests [10,11].

Wild bees, which provide $3.1 billion in pollination services annually to agricultural landscapes in the United States [12], are also expected to be affected by the type of bioenergy crops selected. Research at the field level has found bee abundance to be three to four times higher within perennial grasslands than in corn fields [8], while at the landscape level, pollinators respond positively to increasing amounts of natural area and negatively to landscapes dominated by annual agriculture [13,14,15]. The response of pollinators to land-cover change suggests that the selection of bioenergy crops for large-scale production has the potential to positively or negatively impact these organisms. Expanding production of annual bioenergy crops such as corn or soybeans would further simplify the landscape by increasing the proportion of monoculture plantings across landscapes, reducing the availability of food and nesting resources for pollinators. In contrast, expansion of perennial grassland bioenergy crops could benefit pollinators by increasing landscape heterogeneity and augmenting the amount of resource rich-habitat available for foraging and nesting by bees.

With interest in identifying viable bioenergy crops, predictive models have been employed to investigate the effect of different bioenergy crops on species of conservation concern such as grassland birds [4], as well as a range of ecosystem services including biocontrol, carbon sequestration, and phosphorous loading [11,16,17]. Because pollinators provide a valuable ecosystem service and many are experiencing declining populations [18,19], models have also been developed to explore the effects of landscape composition on bee populations [20,21,22]. The conversion of land into more intensive uses is expected to have significant effects on pollinators [23], yet the use of modeling to predict the effects of intensive large scale land change on pollinators is limited [24]. Because pollination is a critical service provided to agricultural crops and to natural plant communities [25,26], bioenergy policies should proactively address how bees can benefit from the development of agricultural systems that advance crop production as well as conservation objectives.

In this research, our aim was to explore the potential effects of two different bioenergy crop production scenarios on pollinator communities using a modeling approach. First, using observations from flower-visiting bees across the state of Michigan, U.S.A, we developed models that related bee abundance, diversity, and community composition to the land cover surrounding study sites. The three community metrics we modeled can indirectly provide insights into how pollination services may be affected by changes in bioenergy production. For example, pollination services tend to increase as flower-visiting bees become more abundant and diverse [27,28,29]. Also, shifts in community composition that result from changes in bioenergy production may affect particular species known to be important pollinators or of conservation concern. Next, we used empirical models to predict the effect of different bioenergy production scenarios on bees. The bioenergy production scenarios tested in this study represented opposite extremes of possible future production scenarios, and assumed a transition to annual bioenergy crops or to perennial grassland bioenergy crops on marginal lands. Production scenarios were limited to marginal lands because a sustainable biofuel policy will likely need to minimize competition between lands devoted to food and biofuel production [30]. Because perennial grassland biofuel production would increase landscape diversity and incorporate more resource-rich habitats into the landscape [21], we predicted that bee abundance, diversity, and community composition would benefit from a biofuel policy that increases perennial grassland production.

Methods

Study Sites

Field sampling was conducted with the permission of land owners in 20 soybean fields located across southern Michigan (Fig. 1) during the summer of 2012. We observed bee visitation at sentinel insect-pollinated plants to estimate bee abundance, diversity, and community composition. Sites were at least 3 km apart and varied in the proportion of annual crops and semi-natural habitats in a 1500 m radius surrounding each site. Land use proportions surrounding study sites were calculated using a geographic pollination system (GIS, ArcGIS, version 10.0, ESRI,

Redlands, CA). The proportion of grassland in the landscape ranged from <1% to 60%, representing a range of possible biofuel production scenarios from sites dominated by annual production to those dominated by perennial grassland production. Bee observations were conducted in soybean fields across a landscape gradient because soybean is a flowering, first-generation biofuel crop that is intensively managed for even plant density and low plant diversity using weed control. This approach reduced variability in bee counts due to variation in flower abundance, plant diversity, and other management practices. However, this approach also requires that inferences about patterns in other crops be viewed with an appropriate degree of caution.

Bee abundance, diversity and community composition

Bee visitation to sentinel sunflowers, *Helianthus annuus*, variety "Sunspot", was measured at each study site to sample the pollinating bee community. Plants were grown in 15.2 cm diameter pots in a greenhouse under 24-h light and a temperature of $26.7 \pm 2°C$. Two sunflowers with open disk flowers were placed at each of two sampling stations located 30 m from field edges and 20 m apart. Bee sampling was conducted simultaneously at each station during a 30-min period with one observer per station. Bees were collected using a hand-held vacuum (Bioquip, Rancho Dominquez, CA) when bee contact was made with disk flowers (i.e., anthers and/or stigmas). Each field (n = 20) was surveyed 3–5 times during the 2012 field season. Unequal sampling across sites was the result of agronomic activities which prevented access to fields on some dates. Sampling occurred on sunny days with temperatures above 24°C, and each field was sampled at least once in the morning and once in the afternoon. Bees were returned to the lab and identified to species using the online key to Bees of Eastern North America at www.discoverlife.org and published species-level keys [31,32].

Bee abundance, diversity, and community composition were quantified for each site. Bee abundance was calculated by averaging the number of bees collected during a 60 min

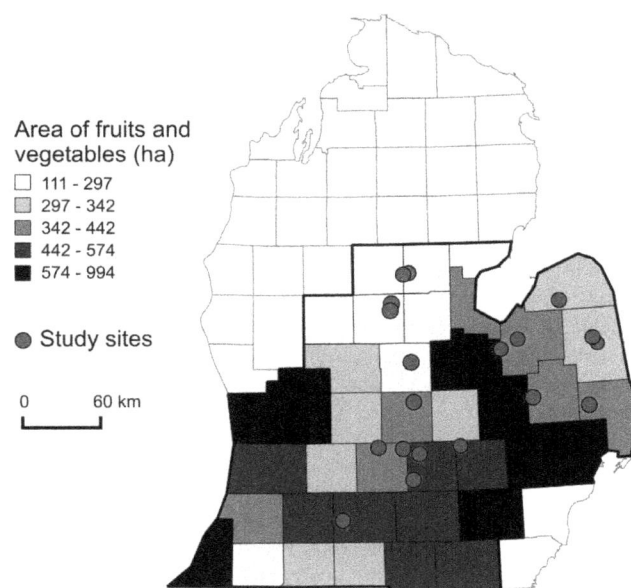

Figure 1. Study sites. Location of study sites sampled for bees across Michigan. Hectares of fruits and vegetables are calculated on a county basis and shown for the lower region of Michigan.

observation period (2 plants/site ×30 min) across all sampling dates for each site, and these values were square-root transformed prior to analysis to improve normality. To avoid obscuring the response of wild bees to landscape change, we excluded *Apis mellifera*, a managed pollinator commonly brought to agricultural landscapes, from calculations of abundance. Because *A. mellifera* was prevalent in agricultural landscapes and influenced composition (see Results), this species was included in calculations of bee diversity and community composition. Bee diversity was quantified for each site using Simpson's diversity index because this index is less sensitive to the degree of sampling effort (Simpson's diversity = 1-D)[33] than species richness and other diversity indices such as Shannon's diversity.

The response of the bee community was then analyzed across sites. First, the abundance of each bee species at each site was averaged across the 2012 season and then square-root transformed and standardized by species and site (i.e., Wisconsin double standardization) [34]. The similarity between sites was then quantified using the Bray-Curtis coefficient. Bee community composition was then ordinated using non-metric multidimensional scaling (NMDS). Next, we explored the relationship between measured landscape variables (e.g., proportions of grassland, forest, wind-pollinated crops, and annual flowering crops; see below) and bee community composition using environmental vector fitting [34]. When viewing the plotted NMDS scores from the community ordination, along with the corresponding environmental vectors (i.e., landscape proportions), we found that most of the variation in bee communities, and most of the association between bee communities and landscape structure, was represented along the first NMDS axis. In order to draw a qualitative link between NMDS scores, bee community composition, and landscape composition, we identified the bee species that contributed most to differences in NMDS scores using a SIMPER (similarity percentages) analysis [34]. Here, the bee community at each site was classified as having a high (>0) or low (<0) NMDS score. The output of the SIMPER analysis elucidated those species that were most highly associated with the two ends of the NMDS spectrum. Given the association between NMDS and landscape gradients, the SIMPER output allowed us to understand how particular bee species responded to landscape composition. Following these preliminary analyses, we used the first NMDS axis values to represent bee composition in subsequent linear regression modeling, described below. NMDS ordinations, environmental vector fitting, and SIMPER analyses were performed using the vegan package [35] written for R statistical computing software [36].

Modeling bee abundance, diversity, and community composition

Multiple linear regression was used to model bee abundance, diversity, and community composition as a function of land cover. Using the 2012 Cropland Data Layer (USDA NASS 2012), the proportion of land cover was calculated in the 1,500 m surrounding sites for nine classes, which accounted for 0.87±0.19 (SD) of land cover: annual wind-pollinated crops, which combined 24 classes of annual crops but was dominated by corn (average proportion of corn = 0.60 in this category, based on all 20 sites); annual flowering crops, which combined 17 classes of annual crops that benefit from pollinators but were dominated by soybean (average proportion of soybean = 0.89); perennial flowering crops (all fruit crops); grasslands (included herbaceous grasslands, old fields, pastures, wildflowers, hayfields, alfalfa fields, and shrublands); forests (combined deciduous, coniferous, mixed forest, and wooded wetlands); wetlands (herbaceous wetlands);

suburbs (areas of low development and open areas dominated by turf); cities (areas of moderate to high development); and other (included water, walnut and Christmas tree farms, and barren). Wind-pollinated crops, annual flowering crops, forests, and grasslands were included as explanatory variables because they accounted for a large proportion of the land cover and (with the exception of forest) are the cover classes that will change under contrasting bioenergy scenarios. Although the proportion of forest does not change under the different bioenergy scenarios, this variable was retained in the model because forests are ecologically important to pollinators, providing nesting habitat and floral resources early in the season [37,38,39].

An information theoretic approach to model selection began with the full model, which included the proportion of wind-pollinated crops, annual flowering crops, grassland, and forest. The full model and all possible subsets of the full model were analyzed using the multimodel inference package, MuMIn, in R [36,40]. The overall best model and all competing models were identified and ranked using bias-corrected Akaike's Information Criterion (AIC_c). Because multiple competing models explained bee abundance, diversity, and community composition, we used model-averaged coefficients from the model set to make predictions about changes in bee communities under the different bioenergy scenarios. Model-averaged coefficients were calculated as weighted averages using model coefficients and Akaike weights, where coefficients were set to zero when a variable was not included in a given model (i.e., a shrinkage coefficient) [41]. Spatial autocorrelation in model residuals was assessed with spline correlograms using the ncf package in R [42]. Statistically significant spatial autocorrelation was not detected in model residuals (Text S1).

Projecting bee abundance, diversity, and community composition

Bee abundance, diversity, and community composition were first estimated across the lower peninsula of Michigan under the current landscape scenario. In GIS, proportional land cover maps were calculated separately for each cover type using a moving window approach. The moving window analysis, which was preformed across Lower Michigan, calculates for each pixel the proportion of cover (e.g., grassland, forest, etc.) in a neighborhood with a radius of 1500 m. To generate predicted values for bees under the current land cover scenario, model-averaged coefficients from the empirical models were multiplied by their respective proportional maps. For example, the model-averaged coefficient for grassland was multiplied by the proportion grassland map and then summed with the products of other terms to produce predicted values for each pixel. The equation used to calculate predicted values for bees in GIS was as follows:

$$Y_{i,j} = b_j + g_j G_i + f_j F_i + w_j W_i + a_j A_i \qquad \text{(eq.1)}$$

where Y is the prediction for the ith pixel for the jth bee community metric (i.e., bee abundance, diversity, or composition), b_j is the intercept for the jth metric, and g_j, f_j, w_j, and a_j are the metric-specific model-averaged parameter values for the proportions of grassland, G, forest, F, wind-pollinated crops, W, and flowering annual crops, A, in the landscape surrounding the ith pixel. The results of this analysis generated maps that estimated bee abundance, diversity, and community composition under current landscape conditions.

The next step was to model abundance, diversity, and community composition under the two contrasting bioenergy scenarios. Because a sustainable bioenergy policy will need to minimize competition with highly productive agricultural land [30], the bioenergy scenarios we developed were focused on marginal land. The U.S. Department of Agriculture's SSURGO database, which lists land capability classes based on soil quality, erosion potential, and water saturation, was used to identify marginal land. Marginal land was defined as cropland with "severe limitations" to "very severe limitations" (land capability classes 3 and 4, respectively) in addition to other lands consider unsuitable for row crop production (classes 5–8).

Of the marginal land in Lower Michigan, we identified 1,200,352 ha in grassland, 550,750 ha in annual wind-pollinated crops (76% corn), and 290,033 ha in flowering annual crops (86% soybean). To keep the number of hectares converted in each scenario consistent, 600,000 ha of marginal lands were converted in each scenario. In the "perennial grassland bioenergy scenario" approximately 360,000 ha of wind-pollinated crops on marginal land and approximately 240,000 ha of flowering annual crops on marginal land were randomly selected and converted into grassland in GIS. In the "annual bioenergy scenario", 600,000 ha of grassland on marginal land was randomly selected and converted into wind-pollinated crops or flowering annual crops based on the proportion of corn (0.58) and soybean (0.41) in the current landscape (CDL 2012). Once the land-cover conversions were completed in GIS, resulting maps represented land cover change under the perennial grassland bioenergy scenario and the annual bioenergy scenario.

The final step was to predict bee abundance, diversity, and community composition under the annual and perennial bioenergy production scenarios. First, we created proportional land cover maps for each cover type using the annual and perennial bioenergy scenario maps. Then model-averaged coefficients from the empirical models were multiplied by their respective proportional land cover maps (Eq. 1). Finally, we calculated the percent change in bee abundance and diversity between the current landscape $(Y_{i,j,c})$ and each bioenergy scenario $(Y_{i,j,s})$ using the equation: percent change = $((Y_{i,j,s} - Y_{i,j,c})/Y_{i,j,c}) \times 100$. Because bee composition used NMDS axis scores, which include both positive and negative values, the difference between the bioenergy landscape and the current landscape was calculated using: difference = $Y_{i,j,s} - Y_{i,j,c}$.

We produced aggregate summaries of the percent change in bee communities for each scenario across the study region in two distinct ways. First, we calculated summary statistics for changes in bee community metrics across all grassland and cropland pixels in the study region. Second, we calculated summary statistics for changes in bee community metrics only for changed pixels and their immediate neighbors across the study region. Hereafter, we refer to results from the former calculations as landscape-level results, and from the latter calculations as local-level results.

Results

Empirical data

During 4,800 min (80 hr) of observation on sentinel plants across all study sites, we observed an average of 0 to 7.25 bee visitors per hour. A total of 38 bee species were identified, and species richness ranged from 1 to 16 species per site, which translated into a range of Simpson's diversity values of 0 to 0.89 per site. The NMDS and vector fitting analyses showed that bee community composition changed along a landscape gradient (Fig. 2, two-dimensional stress = 0.17), where communities in sites with low proportions of grassland and forest had positive NMDS axis scores, while communities associated with high proportions of

grassland and forest had negative NMDS axis scores. The proportion of grassland and forest were significantly negatively correlated with the first NMDS axis (Fig. 2; Grassland: $R^2 = 0.46$, $P = 0.005$; Forest: $R^2 = 0.43$, $P = 0.007$), while the proportion of wind-pollinated and flowering annual crops were positively correlated with the first NMDS axis (Fig. 2; Wind: $R^2 = 0.51$, $P = 0.003$, Flowering annual: $R^2 = 0.45$, $P = 0.006$). In general, the bee communities in landscapes dominated by annual crops were less diverse while those in grassland and forest dominated landscapes were more diverse (Text S2). The SIMPER analysis indicated that *A. mellifera*, *Augochlorella aurata*, and *Halictus ligatus* contributed to the largest difference: 17%, 8%, and 7%, respectively, among sites along the landscape gradient (Text S2). *A. mellifera* and *A. aurata* contributed more to the bee community in agricultural landscapes while *H. ligatus* was increasingly associated with sites having higher proportions of grassland and forest in the surrounding landscape (Text S2).

Modeling pollinators

Bee abundance was best explained by the proportion of forest and grassland in the surrounding landscape (Table 1). Because several competing models were present, model-averaged coefficients calculated with shrinkage were used to predict and map bee abundance in GIS (Table 1, Fig 3A). Overall, forest and grassland had positive model-average coefficients for bee abundance, while wind-pollinated crops and flowering annual crops had negative model-average coefficients.

The best model for bee diversity included only one variable, the proportion of forest in the landscape (Table 1). Again, multiple

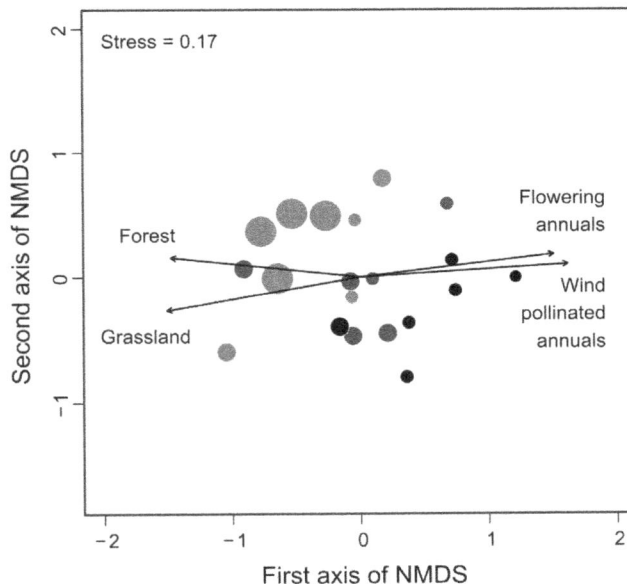

Figure 2. Ordination of bee communities with landscape variables. Ordination of bee community composition using nonmetric multidimensional scaling (NMDS) shows that communities change as the proportion of grassland and forest cover increase in a 1500 m radius surrounding sites (stress = 0.17). Sites with negative NMDS axis one scores are correlated with grassland ($R^2 = 0.46$, $P = 0.0054$) and forest ($R^2 = 0.43$, $P = 0.0071$), while sites with positive NMDS axis one scores are correlated with wind pollinated crops ($R^2 = 0.52$, $P = 0.0025$) and flowering annual crops ($R^2 = 0.45$, $P = 0.0057$). Increasing circle size represents sites with higher bee abundance while decreasing color intensity (black to light gray) indicates sites with higher levels of bee diversity.

(A)   (B)   (C)

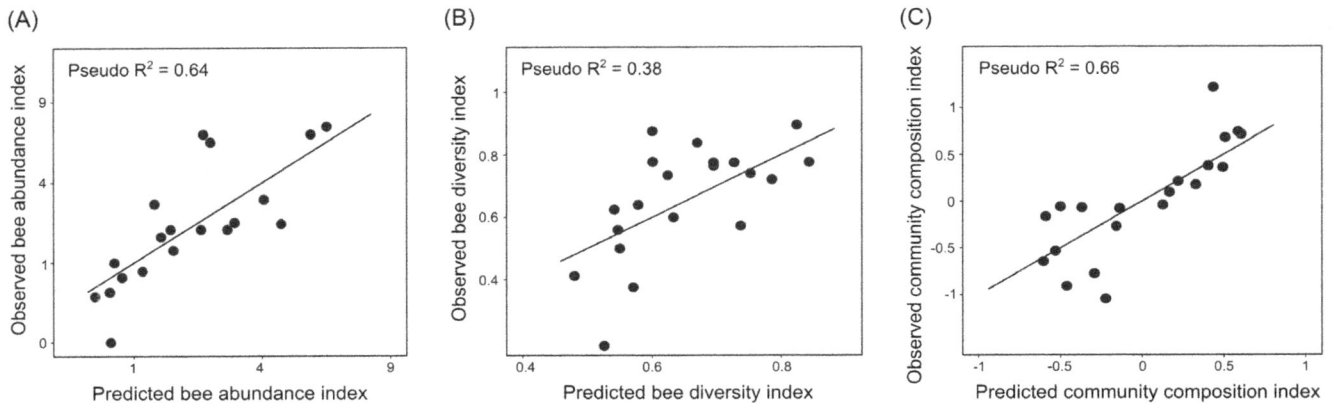

Figure 3. Observed versus predicted bee community metrics. Relationships between observed and predicted values for abundance, diversity, and community composition. Pseudo-R² is derived from regressing observed values versus model-averaged predictions. The graph for abundance displays only 19 points because two data points overlap.

competing models were present in the model set for pollinator diversity and model-averaged coefficients calculated with shrinkage were used to estimate bee diversity in GIS (Table 1, Fig 3B). Model-averaged coefficients for bee diversity were also generally positive for grassland and forest and negative for wind-pollinated crops and flowering annual crops.

Bee community composition, as reflected by the first axis in the NMDS ordination, was best explained by the proportion of grassland and wind-pollinated crops in the surrounding landscape (Table 1). Given the large set of competing models, model-averaged coefficients were used to predict community composition in GIS (Table 1, Fig. 3C). Generally, grassland and forest had negative model-averaged coefficients for this factor, while the model-averaged coefficients for wind-pollinated and flowering annual crops were positive. These results agree with results from vector fitting, which are discussed above in the first paragraph of the results, and have an analogous interpretation.

Projected landscape-level effects of biofuel scenarios on bees

Bee abundance responded positively to the perennial bioenergy scenario and negatively to the annual bioenergy scenario. Under a perennial bioenergy scenario, abundance increased from 0 to 600%, with a landscape-level mean increase of 24% (Fig. 4a). The highest predicted increases in abundance occurred where marginal annual cropland was converted into perennial bioenergy crops (Fig. 5a). Under the annual bioenergy scenario, bee abundance was predicted to decline from 0 to 71%, with a landscape-level mean decrease of 6% (Fig. 4b). The largest loss of abundance is projected to occur in central and western Michigan (Fig. 5a). Areas expected to experience little or no change in abundance, such as pockets of central Michigan and the northwest peninsula of Michigan (Fig. 5A), are regions dominated by prime agricultural soils that are not available for conversion into either perennial or annual bioenergy crops. In contrast, bee abundance is expected to change substantially in regions of Michigan where landscapes support grassland and annual crops on marginal lands. For example, in west Michigan wild bee abundance was projected to experience large declines under the annual scenario, whereas increases were predicted for north-central and southwest Michigan under the perennial scenario.

Similar to the results found for abundance, bee diversity was predicted to increase under the perennial bioenergy scenario and

decrease under the annual bioenergy scenario. Bee diversity was predicted to increase from 0 to 53% (Fig. 4c), with a landscape-level mean increase of 10%, under the perennial bioenergy scenario (Fig. 5b). In contrast, bee diversity was predicted to decrease from 0 to 28% (Fig. 4d), with a landscape-level mean decrease in diversity of only approximately 1% under the annual bioenergy scenario (Fig. 5b). Increases in bee diversity under the perennial scenario were predicted for the counties in the peninsula neighboring Lake Michigan, while declines in bee diversity were not predicted for this area of Michigan under the annual scenario. This region of Michigan is an area of intensive agriculture, meaning that areas of annual crops on marginal soils were available for conversion into perennial bioenergy crops, with potential for increasing pollinator diversity. However, areas of marginal grassland were lacking in this region, preventing land conversion into annual bioenergy crops.

Community composition is also predicted to respond to contrasting bioenergy production scenarios. Change in community composition under the perennial scenario, based on NMDS axis one scores, ranged from 0 to −1.73 under the perennial bioenergy scenario (increasing bee abundance and diversity, Fig. 4E), while changes in scores ranged from 0 to 1.37 (decreasing bee abundance and diversity, Fig. 4F) under the annual scenario. Under the perennial scenario, shifts in composition were predicted to occur in north central, western, and south central Michigan (Fig. 5c). Bee composition scores under the annual scenario were expected to change predominantly in west Michigan (Fig. 5c) where bee communities are expected to experience species declines.

Discussion

Biofuel policies set at the national level are expected to expand biofuel crop production [5,43], causing substantial changes in land cover across the Midwest [44,45]. Policies have the potential to influence crop choice as well as crop placement within the landscape, shaping land cover change in ways that are predicted to affect biodiversity and the provisioning of ecosystem services. Our study shows that bee abundance, diversity, and community composition are sensitive to changes in land cover. We found that bee abundance and diversity were greater where there was a greater proportion of grassland and forest in the landscape and were lower where annual agriculture was more prevalent. These results were used to make predictions about how bees in

Table 1. Model selection.

	Intercept	Grassland	Forest	Wind[a]	FA[b]	ΔAICc	Model weight	Model R^2
Abundance								
Competing model								
1	0.66	1.83	2.98			0.00	0.33	0.63
2	1.35		3.23		−1.50	1.28	0.17	0.60
3	0.83		4.37			1.65	0.14	0.52
Average model[c]								0.64
Parameter value	0.95	1.13	2.84	−0.04	−0.50			
SE	0.46	1.20	1.64	0.49	0.88			
Variable weight		0.60	0.86	0.17	0.37			
Diversity								
Competing model								
1	0.52		0.94			0.00	0.23	0.30
2	0.82			−0.70		0.25	0.20	0.29
3	0.68		0.62		−0.40	0.68	0.16	0.38
Average model								0.38
Parameter value	0.68	0.03	0.45	−0.06	−0.30			
SE	0.16	0.18	0.51	0.02	0.36			
Variable weight		0.19	0.56	0.23	0.54			
Composition								
Competing model								
1	−0.20	−1.46		1.64		0.00	0.25	0.62
2	−0.82			1.63	1.34	0.61	0.18	0.61
3	−0.43	−1.03		1.38	0.85	1.94	0.094	0.65
4	−0.69			2.36		1.97	0.093	0.51
Average model								0.66
Parameter value	−0.33	−0.76	−0.42	1.26	0.62			
SE	0.47	0.90	0.94	0.93	0.83			
Variable weight		0.54	0.29	0.76	0.48			

Summary of model selection statistics for the competing models and model-averaged coefficients predicting bee abundance, diversity, and community composition as a function of land cover variables measured in the 1500 m surrounding study sites.

[a]Wind represents the variable wind pollinated annual crops.

[b]FA represents flowering annual crops potentially visited by bees.

[c]For the averaged model the R^2 is a pseudo-R^2 derived from regressing the observed values versus model-averaged predictions.

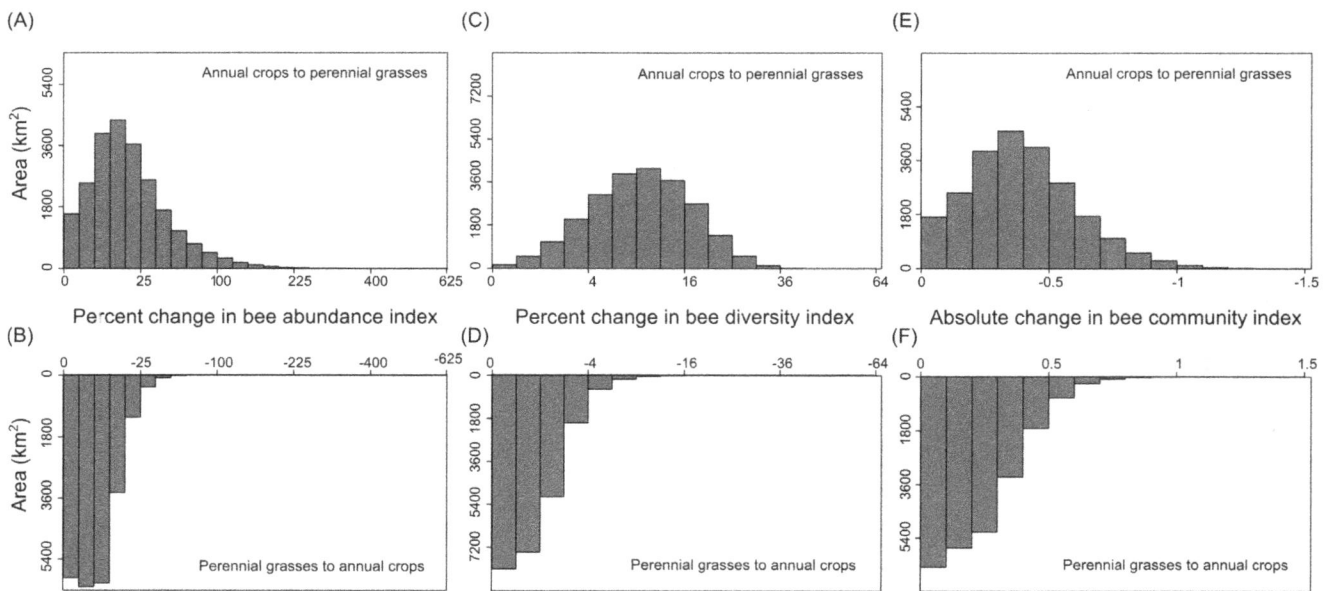

Figure 4. Distribution of percent change for measured bee metrics. The distribution of percent change values calculated under the perennial grassland bioenergy scenario for bee abundance (A), diversity (C), and community composition (E) and under the annual bioenergy scenario for bee abundance (B), diversity (D), and community composition (F). In the annual bioenergy scenario 600,000 ha of marginal grassland was converted into annual bioenergy crops, whereas in the perennial bioenergy scenario 600,000 ha of marginal agricultural land were converted into grassland. Predicted changes in bee communities are only shown for the lower portion of the state where empirical data were collected.

agricultural landscapes would change as annual or perennial bioenergy crops expanded across the region. In a scenario where perennial grassland bioenergy crops are favored, we expect bee abundance and diversity to increase, with shifts to communities that are more dominated by wild bees. In contrast, if policies and markets favor increased adoption of annual bioenergy crops, we predict a reduction in bee abundance and diversity, with community composition moving towards fewer species dominated by generalists such as *A. mellifera*.

In the results reported here, the bee-related metrics were sensitive to landscape-scale land cover change. Mean values reported for increases and decreases in bee abundance and diversity were calculated using all grassland and agricultural pixels across our study region, whether or not they were changed under each scenario. As a result, the magnitudes of the reported percent changes (landscape-level) are smaller than they would be if they were calculated only for the pixels adjacent to those selected for land use change (local-level). At a local level, mean bee abundance increased by 40% under the perennial crop scenario and decreased by 14% under the annual crop scenario. Interestingly, local mean values for bee diversity remained relatively unchanged, with a 9% increase (landscape-scale increase 10%) under the perennial crop scenario and a 2.6% decrease (landscape-scale decrease 1%) under the annual crop scenario. These results suggest that the local effects of bioenergy crop production are more pronounced for bee abundance than for bee diversity.

Forested land played an important role in explaining bee abundance and bee diversity, having the highest variable weight in both sets of models (Table 1). Grassland, however, had the second highest variable weight when explaining bee abundance but the lowest variable weight for bee diversity. The significant relationship between grassland cover and bee abundance may explain why strong local effects are predicted for bee abundance under the perennial and annual land change scenarios but not for bee diversity. The strongest driver of bee diversity was forest cover,

and the proportion of forest in the landscape was not changed under either bioenergy scenario. This might explain why switching energy crops had limited local effects on bee diversity. Although our results suggest that increasing perennial grassland bioenergy production will have local effects on bee abundance, conserving forest habitat will be important for maintaining landscape-level bee diversity.

The response of the bee community to annual and perennial bioenergy scenarios was similar for bee abundance and diversity. In general, the bee community under annual bioenergy production shifted to a community composed of fewer bee species, while the bee community under perennial grassland production transitioned to a more diverse community of wild bees (Text S2). The effect of annual bioenergy crop production on bee community composition is of particular interest in the western counties of Michigan where there is significant production of pollinator-dependent fruit and vegetable crops. Because diverse pollinator communities are expected to provide more reliable pollination services by containing redundant [46] or complimentary [47] pollinator species, a shift to annual biofuel production may lead to a decline in bee diversity, with potential effects on the pollination services provided to fruit growers in this region. The relative location of land used for pollinator-dependent crops and biofuels would be expected to influence the degree to which these changes might affect crop yield.

Changes in bee abundance, diversity, and community composition under the perennial bioenergy scenario highlight where opportunities and challenges for grassland bioenergy production exist across Michigan. Several counties located in north-central and south-east Michigan show little or no change in bee abundance, diversity, or community composition. The lack of change in these counties is due to the presence of fruits, vegetables, corn, and soybeans on prime agricultural soils, yielding few opportunities for perennial grassland bioenergy production on marginal lands. In some cases, small isolated patches of perennial

Perennial grasses to annual crops   Annual crops to perennial grasses

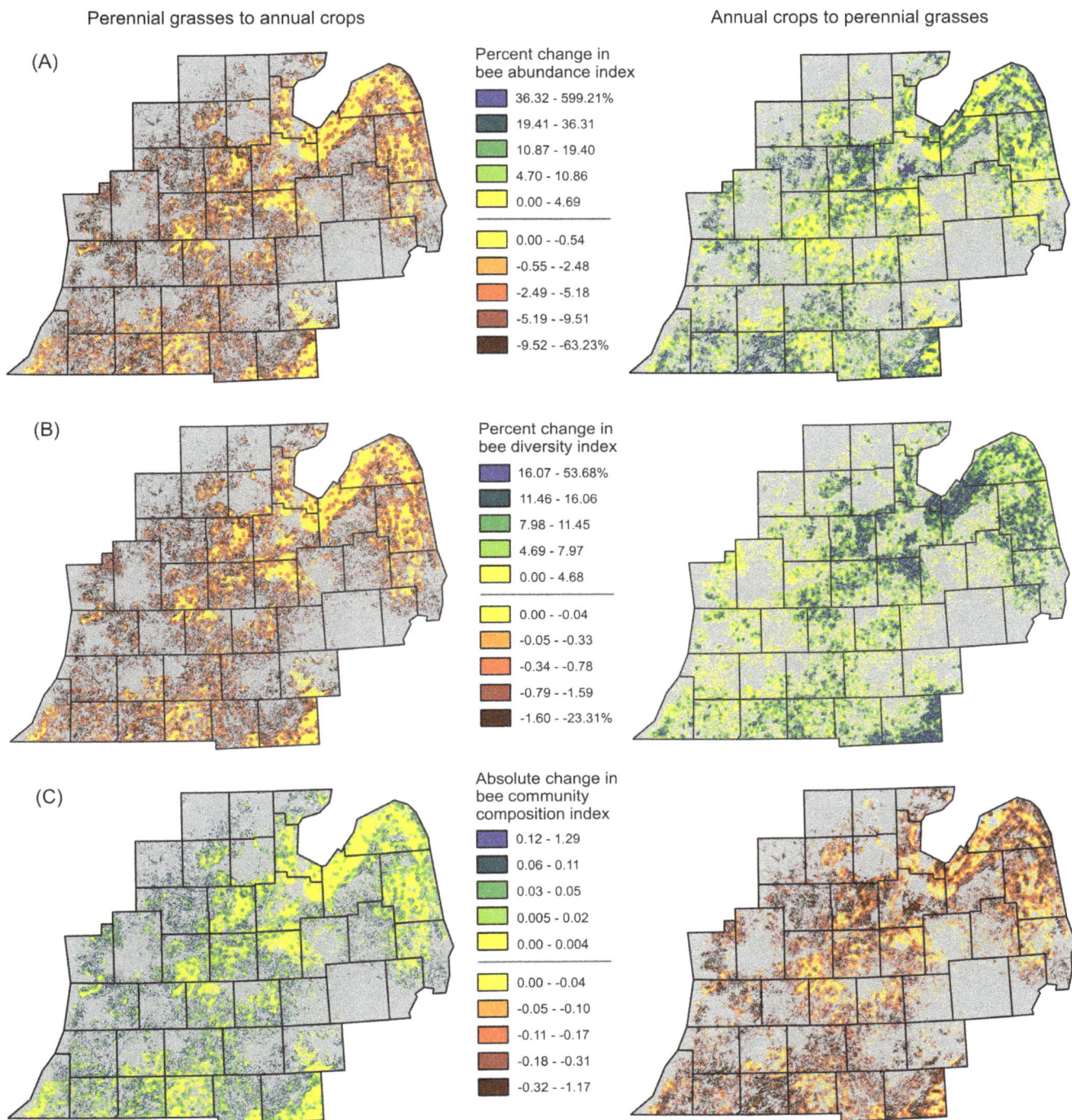

Figure 5. Projected bee metrics. Percent change in bee abundance (A), percent change in bee diversity (B), and difference in community composition (C) predicted for Michigan by an empirical model under annual (left maps) and perennial (right maps) bioenergy production scenarios.

habitats may actually serve as population sinks for bees [48,49], suggesting that one challenge for perennial bioenergy production will be to determine how the size and position of these crops within the landscape will influence pollination services. While placement of bioenergy plantings across the landscape might present challenges, an opportunity exists to target the placement of perennial bioenergy crops near pollinator dependent crops in an effort to increase pollinators and potentially augment pollination services and crop yield.

While the models developed here give insights into the possible effects of future bioenergy production on bees, the resulting maps and our interpretation of these maps depend on several assumptions. First, the models used to predict bee abundance, diversity, and community composition were based on empirical data collected from the lower portion of Michigan and may not extend to other parts of the Midwest. Furthermore, visitation to sunflower measured during the summer may not serve as a good proxy for other pollinator dependent crops, especially early season crops. However, many of the bees we collected, including those in the genus *Bombus* and *Halictus*, are generalists that are present throughout the growing season. Second, conclusions regarding the effect of perennial and annual bioenergy production assume that management practices currently employed for annual crops and perennial grassland habitats do not change substantially with increased bioenergy production. Increasing insecticide use to control emerging pests or annual harvest of bioenergy grasslands could affect bees in ways not reflected in our models. Third, forest was an important variable explaining bee abundance, diversity, and community composition in our study. Under the contrasting bioenergy scenarios developed here, we assumed the proportion of forest remains constant across the landscape, suggesting future loss of forest habitat due to agricultural intensification or urbanization could alter model predictions. Finally, while increasing bee abundance and diversity are generally correlated with higher rates of pollination [28,50,51], we recognize that transitioning the landscape into perennial grassland bioenergy production may not translate into increased pollination services, especially if one effective pollinator species is capable of persisting under both scenarios. However, biologically diverse pollinator communities play an important role in providing stable pollination services [52,53,54], suggesting that land conversion to perennial grassland bioenergy crops can contribute to supporting services in regions where bees have experienced declines.

Conclusions

Using field observations, we generated empirical models and predicted bee abundance, diversity, and community composition across Lower Michigan for two contrasting bioenergy production scenarios. From these analyses, we identified areas where bees are expected to benefit substantially from bioenergy policies that promote perennial grassland production, as well as areas where

further land conversion to annual bioenergy crops is likely to produce significant challenges for the persistence of diverse bee communities. The methods and models developed here have application for the identification of area thresholds required to maximize biodiversity conservation and target areas of the landscape where perennial bioenergy plantings could facilitate pollination services. However, given market values for annual commodity crops, conversion to perennial grassland bioenergy production will likely be limited without policy changes [55]. Policies that acknowledge the value of biodiversity and the services it provides will be necessary for implementing bioenergy production systems that balance trade-offs between crop production and the support of ecosystem services [56].

Acknowledgments

We thank Ashley McNamara, Lindsey Pudlo, Jon Roney, and Laura Maihofer who provided invaluable field assistance, and the landowners who allowed us access to study sites. Thanks to Dr. Jason Gibbs who provided assistance with bee identifications and Dr. David Lusch who provided useful insights into the GIS methodology.

Author Contributions

Conceived and designed the experiments: AB TM CG RI. Performed the experiments: AB TM. Analyzed the data: AB TM. Wrote the paper: AB TM CG RI.

References

1. Tyner WE (2008) The US ethanol and biofuels boom: its origins, current status, and future prospects. Bioscience 58: 646–653.
2. Jarchow ME, Liebman M (2012) Tradeoffs in biomass and nutrient allocation in prairies and corn managed for bioenergy production. Crop Sci 52: 1330–1342.
3. Schubert C (2006) Can biofuels finally take center stage? Nat Biotechnol 24: 777–784.
4. Meehan TD, Hurlbert AH, Gratton C (2010) Bird communities in future bioenergy landscapes of the Upper Midwest. Proc Natl Acad Sci USA 107: 18533–18538.
5. Wright CK, Wimberly MC (2013) Recent land use change in the western Corn Belt threatens grasslands and wetlands. Proc Natl Acad Sci USA 110: 4134–4139.
6. Wiens J, Fargione J, Hill J (2011) Biofuels and biodiversity. Ecol Appl 21: 1085–1095.
7. Robertson BA, Doran PJ, Loomis LR, Robertson JR, Schemske DW (2011) Perennial biomass feedstocks enhance avian diversity. Glob Change Biol Bioenergy 3: 235–246.
8. Gardiner MA, Tuell JK, Isaacs R, Gibbs J, Ascher JS, et al. (2010) Implications of three biofuel crops for beneficial arthropods in agricultural landscapes. Bioenerg Res 3(1): 6–19.
9. Werling BP, Meehan TD, Gratton C, Landis DA (2011) Influence of habitat and landscape perenniality on insect natural enemies in three candidate biofuel crops. Biol Control 59: 304–312.
10. Werling BP, Meehan TD, Robertson BA, Gratton C, Landis DA (2011) Biocontrol potential varies with changes in biofuel-crop plant communities and landscape perenniality. Glob Change Biol Bioenergy 3: 347–359.
11. Meehan TD, Werling BP, Landis DA, Gratton C (2012) Pest-suppression potential of Midwestern landscapes under contrasting bioenergy scenarios. PLOSONE 7(7). E41728. Doi10.1371/journal.pone.0041728.
12. Losey JE, Vaughan M (2006) The economic value of ecological services provided by insects. Bioscience 56: 311–323.
13. Ricketts TH (2004) Tropical forest fragments enhance pollinator activity in nearby coffee crops. Conserv Biol 18: 1262–1271.
14. Ricketts TH, Regetz J, Steffan-Dewenter I, Cunningham SA, Kremen C, et al. (2008) Landscape effects on crop pollination services: are there general patterns? Ecol Lett 11: 499–515.
15. Garibaldi LA, Steffan-Dewenter I, Winfree R, Aizen MA, Bommarco R, et al. (2013) Wild pollinators enhance fruit set of crops regardless of honey bee abundance. Science 339: 1608–1611.
16. Gelfand I, Sahajpal R, Zhang XS, Izaurralde RC, Gross KL, et al. (2013) Sustainable bioenergy production from marginal lands in the US Midwest. Nature 493: 514–ZZZ.
17. Meehan TD, Gratton C, Diehl E, Hunt ND, Mooney DF, et al. (2013) Ecosystem-service tradeoffs associated with switching from annual to perennial energy crops in riparian zones of the US Midwest. PLOSONE 8(11): e89993. doi: 10.137/journal.pone.0080093
18. Potts SG, Biesmeijer JC, Kremen C, Neumann P, Schweiger O, et al. (2010) Global pollinator declines: trends, impacts and drivers. Trends Ecol Evol 25: 345–353.
19. Cameron SA, Lozier JD, Strange JP, Koch JB, Cordes N, et al. (2011) Patterns of widespread decline in North American bumble bees. Proc Natl Acad Sci U S A 108: 662–667.
20. Lonsdorf E, Kremen C, Ricketts T, Winfree R, Williams N, et al. (2009) Modelling pollination services across agricultural landscapes. Ann Bot103: 1589–1600.
21. Kennedy CM, Lonsdorf E, Neel MC, Williams NM, Ricketts TH, et al. (2013) A global quantitative synthesis of local and landscape effects on wild bee pollinators in agroecosystems. Ecol Lett 16: 584–599.
22. Schulp CJE, Lautenbach S, Verburg PH (2014) Quantifying and mapping ecosystem services: demand and supply of pollination in the European Union. Ecol Indic 36: 131–141.

23. Kremen C, Williams NM, Aizen MA, Gemmill-Herren B, LeBuhn G, et al. (2007) Pollination and other ecosystem services produced by mobile organisms: a conceptual framework for the effects of land-use change. Ecol Lett 10: 299–314.

24. Ricketts TH, Lonsdorf E (2013) Mapping the margin: comparing marginal values of tropical forest remnants for pollination services. Ecol Appl 23: 1113–1123.23.

25. Ollerton J, Winfree R, Tarrant S (2011) How many flowering plants are pollinated by animals? Oikos 120: 321–326.

26. Nicholls CI, Altieri MA (2013) Plant biodiversity enhances bees and other insect pollinators in agroecosystems. A review. Agron Sustain Dev 33: 257–274.

27. Hoehn P, Tscharntke T, Tylianakis JM, Steffan-Dewenter I (2008) Functional group diversity of bee pollinators increases crop yield. Proc R Soc B-Biol Sci 275: 2283–2291.

28. Frund J, Dormann CF, Holzschuh A, Tscharntke T (2013) Bee diversity effects on pollination depend on functional complementarity and niche shifts. Ecol 94: 2042–2054.

29. Garibaldi LA, Steffan-Dewenter I, Kremen C, Morales JM, Bommarco R, et al. (2011) Stability of pollination services decreases with isolation from natural areas despite honey bee visits. Ecol Lett 14: 1062–1072.

30. Tilman D, Socolow R, Foley JA, Hill J, Larson E, et al. (2009) Beneficial biofuels-the food, energy, and environment trilemma. Science 325: 270–271.

31. Rehan SM, Sheffield CS (2011) Morphological and molecular delineation of a new species in the *Ceratina dupla* species-group (Hymenoptera: Apidae: Xylocopinae) of eastern North America. Zootaxa 35–50.

32. Gibbs J (2011) Revision of the metallic *Lasioglossum* (*Dialictus*) of eastern North America (Hymenoptera: Halictidae: Halictini). Zootaxa 1–216.

33. Clarke KR, Warwick RM (2001) Change in marine communities: an approach to statistical analysis and interpretation, 2nd ed. Plymouth: PRIMER-E.

34. McCune B, Grace JB (2002) Analysis of ecological communities. MjM software, Gleneden Beach, Oregon, USA: MjM software. 102 p.

35. Oksanen J, Blanchet FG, Kindt R, Legendre P, Minchin PR, et al. (2013) Vegan: community ecology package. R package version 2.0–10. Available: http://CRAN.R-project.org/package=vegan

36. R Development Core Team (2013) R: A language and environment for statistical computing. Vienna, Austria.38

37. Macior LW (1968) *Bombus* (Hymenoptera Apidae) queen foraging in relation to vernal pollination in Wisconsin. Ecol 49: 20–25.

38. Ginsberg HS (1983) Foraging ecology of bees in an old field. Ecol 64: 165–175.

39. Fabian Y, Sandau N, Bruggisser OT, Aebi A, Kehrli P, et al. (2013) The importance of landscape and spatial structure for hymenopteran-based food webs in an agro-ecosystem. J Anim Ecol 82: 1203–1214.

40. Barton K (2013) MuMIn: mulit-model inference. R package version 1.9.5.Avail-.Available: http://CRAN.R-project.org/package=MuMIn

41. Burnham KP, Anderson DR (2002) Model selection and multimodel inference: a practical information-theoretic approach. New York: Springer. 159 p.

42. Bjornstad ON (2013) ncf: spatial nonparametric covariance functions. R package version 1. 1–5. Available: http://CRAN.R-project.org/package=ncf

43. Fargione JE, Cooper TR, Flaspohler DJ, Hill J, Lehman C, et al. (2009) Bioenergy and wildlife: threats and opportunities for grassland conservation. Bioscience 59: 767–777.

44. Wu F, Guan ZF, Yu F, Myers RJ (2013) The spillover effects of biofuel policy on participation in the conservation reserve program. J Econ Dyn Control 37: 1755–1770.

45. Johnston CA (2014) Agricultural expansion: land use shell game in the U.S. Northern Plains. Landscape Ecol 29: 81–95.

46. Winfree R, Kremen C (2009) Are ecosystem services stabilized by differences among species? A test using crop pollination. Proc R Soc B-Biol Sci 276: 229–237.

47. Brittain C, Kremen C, Klein AM (2013) Biodiversity buffers pollination from changes in environmental conditions. Glob Chang Biol 19: 540–547.

48. Fahrig L (2003) Effects of habitat fragmentation on biodiversity. Ann Rev Ecol Evol Syst 34: 487–515.

49. Rosch V, Tscharntke T, Scherber C, Batary P (2013) Landscape composition, connectivity and fragment size drive effects of grassland fragmentation on insect communities. J Applied Ecol 50: 387–394.

50. Albrecht M, Schmid B, Hautier Y, Muller CB (2012) Diverse pollinator communities enhance plant reproductive success. Proc. R Soc B Biol Sci 279: 4845–4852.

51. Holzschuh A, Dudenhoffer JH, Tscharntke T (2012) Landscapes with wild bee habitats enhance pollination, fruit set and yield of sweet cherry. Biol Conserv 153: 101–107.

52. Klein AM, Steffan-Dewenter I, Tscharntke T (2003) Fruit set of highland coffee increases with the diversity of pollinating bees. Proc R Soc B-Biol Sci 270: 955–961.

53. Ebeling A, Klein AM, Schumacher J, Weisser WW, Tscharntke T (2008) How does plant richness affect pollinator richness and temporal stability of flower visits? Oikos 117: 1808–1815.

54. Bartomeus I, Park MG, Gibbs J, Danforth BN, Lakso AN, et al. (2013) Biodiversity ensures plant-pollinator phenological synchrony against climate change. Ecol Lett 16: 1331–1338.

55. James LK, Swinton SM, Thelen KD (2010) Profitability analysis of cellulosic energy crops compared with corn. Agron J 102: 675–687.

56. Landis DA, Werling BP (2010) Arthropods and biofuel production systems in North America. Insect Sci 17: 220–236.

20

Evidence for Frozen-Niche Variation in a Cosmopolitan Parthenogenetic Soil Mite Species (Acari, Oribatida)

Helge von Saltzwedel*, Mark Maraun, Stefan Scheu, Ina Schaefer

Georg-August University, Johann-Friedrich-Blumenbach Institute of Zoology and Anthropology, Dept. Ecology, Göttingen, Germany

Abstract

Parthenogenetic lineages may colonize marginal areas of the range of related sexual species or coexist with sexual species in the same habitat. Frozen-Niche-Variation and General-Purpose-Genotype are two hypotheses suggesting that competition and interclonal selection result in parthenogenetic populations being either genetically diverse or rather homogeneous. The cosmopolitan parthenogenetic oribatid mite *Oppiella nova* has a broad ecological phenotype and is omnipresent in a variety of habitats. Morphological variation in body size is prominent in this species and suggests adaptation to distinct environmental conditions. We investigated genetic variance and body size of five independent forest - grassland ecotones. Forests and grasslands were inhabited by distinct genetic lineages with transitional habitats being colonized by both genetic lineages from forest and grassland. Notably, individuals of grasslands were significantly larger than individuals in forests. These differences indicate the presence of specialized genetic lineages specifically adapted to either forests or grasslands which coexist in transitional habitats. Molecular clock estimates suggest that forest and grassland lineages separated 16-6 million years ago, indicating long-term persistence of these lineages in their respective habitat. Long-term persistence, and morphological and genetic divergence imply that drift and environmental factors result in the evolution of distinct parthenogenetic lineages resembling evolution in sexual species. This suggests that parthenogenetic reproduction is not an evolutionary dead end.

Editor: Xiao-Yue Hong, Nanjing Agricultural University, China

Funding: The author(s) received no specific funding for this work.

Competing Interests: The authors have declared that no competing interests exist.

* Email: hsaltzw1@gwdg.de

Introduction

Parthenogenetic lineages often are successful colonizers of new or disturbed habitats. This success suggests that effective establishment of populations may occur without males and genetic exchange. In parthenogenetic species each individual represents a reproductive unit capable of founding a new population [1–3]. Thelytoky, the exclusive production of daughters from unfertilized eggs, also increases the number of reproductive individuals in a population and thereby population growth. In addition, genotypes that successfully establish in a new habitat are transmitted unchanged to the next generation whereas sexual reproduction potentially breaks up advantageous gene combinations every generation [4]. However, in the long-term, the lack of males and recombination is assumed to result in the accumulation of deleterious mutations [5], [6] and to limit adaptation to changing environments [7], [8]. Therefore, in the long-term parthenogenetic lineages are assumed to be doomed to extinction due to mutational meltdown and competition with sexual sister-taxa.

Among the several hypotheses explaining the ecological and geographical distribution of parthenogenetic and sexual organisms [9] the Frozen-Niche-Variation (FNV) hypothesis [10–13] suggests that widespread parthenogenetic species consist of a number of locally adapted genotypes, each occupying a narrow niche. As parthenogenetic genomes are transmitted in full their genotypes are kept "frozen". In this model asexual individuals arise continuously from sexual populations resulting in genetically diverse populations. Evidence for such specialized genotypes supporting the FNV hypothesis have been found in fishes, frogs, spider mites, shrimps and water fleas [10], [14–19]. On the contrary, spatial and temporal variation of ecological niches may favor the evolution of parthenogenetic genotypes adapted to a wide range of ecological conditions, thereby representing a General-Purpose-Genotype (GPG) [2], [20] with only few parthenogenetic lineages dominating across habitats [21]. In these lineages mutations are the primary source of variation [22], [23] resulting in low genetic diversity within populations contrasting predictions of the FNV hypothesis. Evidence for GPG has been found in fishes, snails, ostracods, oribatid mites and ambrosia beetles [24–29], for a detailed list see [30].

The cosmopolitan thelytokous oribatid mite species *Oppiella nova* (Oudemans, 1902) lives in a variety of habitats including the soils of forests, grasslands, agricultural fields and suspended soils in tree canopies. It can reach high densities (>20,000 ind. m^{-2}) [31–34] and often co-occurs with sexual species of the same genus, such as *O. subpectinata* and *O. falcata* [35]. The existence of sexually reproducing congeneric species suggests that *O. nova* is a parthenogenetic offshoot of the predominantly sexual genus *Oppiella* [36]. However, phylogenetic relationships among *Oppiella* species are unresolved and the sexual sister-taxon of *O. nova* is unknown. The most prominent morphological variation in this

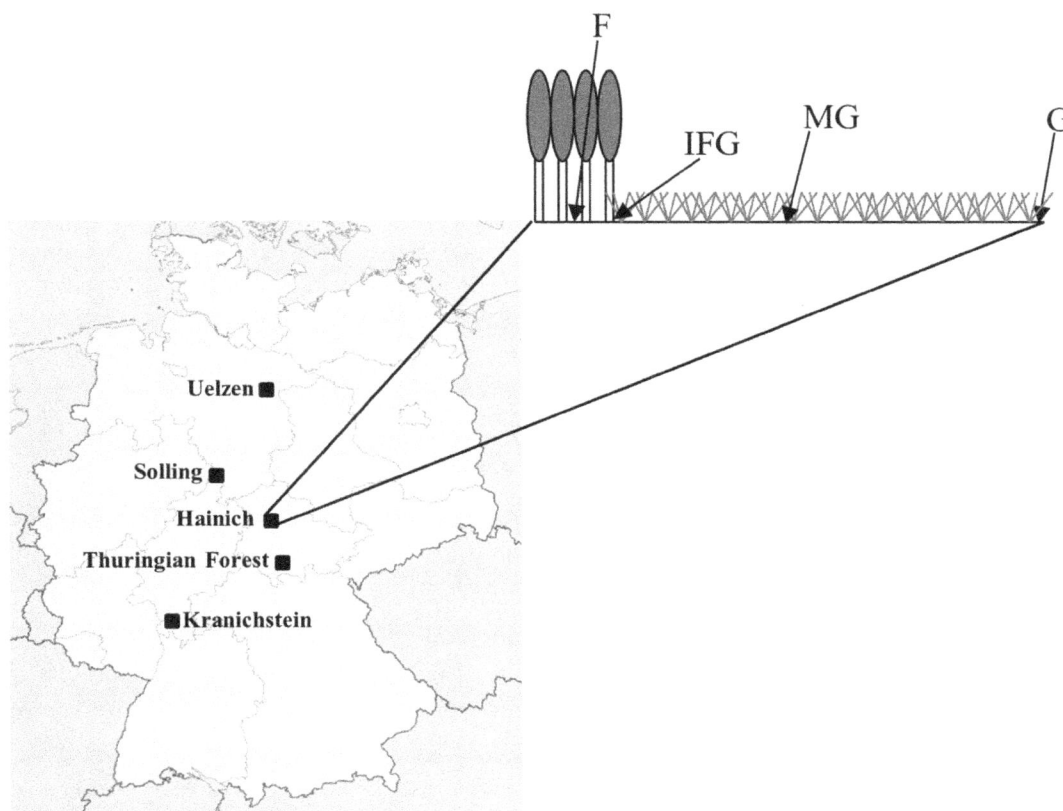

Figure 1. Sampling locations and sampling along a gradient from forest (F) to grassland (G) including transitional habitat types at the intersection between forest and grassland (IFG) and the margin of grassland (MG).

species is body size which ranges from 220 to 320 μm [37]. Due to morphological variation between habitats Woas [38] suggested *O. nova* to comprise different subspecies each adapted to a distinct habitat.

We analyzed the genetic and morphological variance of populations of *O. nova* from grassland and forest soils, forming two distinct soil habitats likely associated with distinct niches, to investigate whether the variation is driven by FNV or GPG processes. Grasslands and forests differ markedly in abiotic and biotic factors, including temperature, humidity, wind, soil structure and fungal community composition. Mites were sampled along a gradient from grassland to forest at five locations spaced at least 50 km from each other. The mitochondrial *COI* gene and the D3 region of the nuclear 28S rDNA were sequenced to identify genetic lineages; the D3 region also served as species marker [39], [40]. According to the FNV hypothesis we expected specimens of the same habitat to cluster together irrespective of sampling locations. In contrast, conform to the GPG hypothesis different habitats (and the associated niches) within the same location were expected to cluster together, i.e. to cluster according to distance. Although oribatid mites are generally poor dispersers, *O. nova* is able to migrate short distances and occasionally disperses long distances by wind [41]. To take dispersal into account, we tested for migration of genotypes between locations and between habitats, i.e. forest and grassland, within locations. Further, we investigated whether body size correlated with habitat type, genetic lineages or sampling location. Similar to haplotype distribution, we expected body size to correlate with habitat type

according to the FNV hypothesis but to correlate with distance of locations according to the GPG hypothesis.

Materials and Methods

Ethics statement

Permission for sampling at Kranichstein was given by the forestry office Darmstadt, permission at Hainich was issued by the state environmental office of Thüringen (§ 72 BbgNatSchG). All other sampling sites were outside Nature Reserve Areas and no permission for soil samples was required. The field study did not involve any endangered or protected species.

Sampling and study sites

A total of 147 individuals of the oribatid mite species *O. nova* were collected from five locations in Germany: Hainich (HA), Kranichstein (KW), Solling (SO), Thuringian Forest (TW) and Uelzen (UE) (**Table 1, Figure 1**). We restricted the analysis to the parthenogenetic species because the sexual sister-taxon is unknown. Individuals were sampled from soil and litter of adjacent grassland and forest along a gradient, including the habitat types forest (F) and grassland (G) and two transitional habitats, grassland margin (MG) and intersection of forest and grassland (IFG). MG was located in grassland but close to the forest edge which formed a sharp boundary, IGF samples were taken where tree litter and grassland vegetation mixed (**Figure 1**). The maximum distance between F and G sampling sites was 100 m, MG and IFG sampling sites were 15–20 m apart; sampling locations were 56–350 km apart. From each habitat three samples of 15×15 cm were

Table 1. Sampling locations, habitat type, number of collected individuals, number of sequences for *COI* and D3 and respective GenBank accession numbers of *Oppiella nova* analyzed in this study.

Location	Habitat type	n individuals	n sequences (*COI*)	GenBank acc. no.	n sequences (D3)	GenBank acc. no.
Kranichstein Forest, near Darmstadt (KW)	F	17	12	KF293419-26, 35–38,	14	KF293529, 33–41, 51–54
	G	23	19	KF293427-28, 39–55	21	KF293530, 42–44, 55–71
	MG	26	20	KF293415-16, 29–32, 56–69	24	KF293545-48, 72–90
	IFG	4	4	KF293417-18, 33–34	3	KF293532, 49–50
Solling Forest, near Neuhaus (SO)	F	12	7	KF293470-76	8	KF293591-98
	G	1	–	–	1	KF293599
	MG	7	4	KF293477-80	5	KF293600-04
Hainich Forest, near Weberstedt (HA)	F	4	4	KF293402-05	4	KF293514-17
	IFG	12	9	KF293406-14	11	KF293518-28
Thuringian Forest, near Ilmenau (TW)	F	3	3	KF293481-83	3	KF293605-07
	G	2	2	KF293484-85	1	KF293608
	IFG	1	–	–	1	KF293609
Uelzen Forest, near Uelzen (UE)	G	4	3	KF293486-88	3	KF293610-12
	MG	7	3	KF293489-91	7	KF293613-19
	IFG	24	20	KF293492-511	20	KF293620-39

Individuals were collected along a gradient from forest (F) to grassland (G), covering the intersection of forest and grassland (IFG) and margin of grassland (MG).

taken, including litter and the uppermost 5 cm of the soil. Invertebrates were extracted by heat [42] and collected in 75% EtOH. *O. nova* was separated using a dissecting microscope, and morphological identification was confirmed by light microscopy [37].

DNA extraction and PCR

Genomic DNA was extracted from single individuals using the DNeasy Blood and Tissue Kit (Qiagen; Hilden, Germany) following the manufacturer's protocol for animal tissue. Purified DNA was eluted in 30 µl buffer AE and stored at −20°C until further preparation. All PCR reactions for sequencing were performed in 25 µl volumes containing 12.5 µl HotStarTaq Mastermix (Qiagen; Hilden, Germany) with 1 µl of each primer (10 pM), 1 µl of MgCl$_2$ (25 mM) and variable volumes of template DNA (5 µl for D3 and 8 µl for *COI*) and H$_2$O (4.5 µl for D3 and 1.5 µl for *COI*). A 709 bp fragment of the *COI* gene was amplified using the primers LCO1490 (forward) 5′-GGT CAA CAA ATC ATA AAG ATA TTG G-3′ and HCO2198 (reverse) 5′-TAA ACT TCA GGG TGA CCA AAA AAT CA-3′ [43]. Amplification consisted of one initial activation step at 95°C for 15 min, followed by 35 amplification cycles of denaturation at 94°C for 30 s, annealing at 40°C for 60 s, elongation at 72°C for 60 s and a final elongation step at 72°C for 10 min. Amplification of the 356 bp fragment of the D3 region of the 28S rDNA was performed using the primers D3A (forward) 5′-GAC CCG TCT TGA AAC ACG GA-3′ and D3B (reverse) 5′-TCG GAA GGA ACC AGC TAC TA-3′ [44]. The PCR protocol for D3 consisted of an initial activation step at 95°C for 15 min, followed by 35 amplification cycles of denaturation at 94°C for 30 s, annealing at 54°C for 45 s, elongation at 72°C for 60 s and a final elongation

step at 72°C for 10 min. PCR products were purified with the QIAquick PCR Purification Kit (Qiagen; Hilden, Germany) following the manufacturer's protocol. Sequencing in both directions (forward and reverse strands) of *COI* fragments was done by Macrogen Inc. (Seoul, South Korea). The D3 fragments were sequenced at G2L (Institute for Microbiology and Genetics, Laboratory for Genomic Analyses, University of Göttingen). All nucleotide sequences are available at GenBank (www.ncbi.nlm. nih.gov/genbank; KF293402 - KF293513 for *COI* and KF293514 - KF293641 for D3).

Data analysis

Sequences were edited, ambiguous positions were corrected by hand, aided by the respective chromatograms, and nucleotide sequences were translated into amino acid sequences using the invertebrate mitochondrial code implemented in Sequencher 4.9 (Gene Codes Corporation, USA). Consensus sequences were assembled in BioEdit 7.0.1 [45] and aligned with ClustalX v1.81 [46] using multiple alignment parameters: 10.0 (gap opening) and 0.1 (gap extension) for the nucleotide and default settings for the amino acid dataset. In total, three different alignments were generated which included two individuals of *Berniniella hauseri* as outgroup. The D3 alignment included 126 individuals of *O. nova* (**Alignment S1**) and the *COI* alignment 110 individuals (**Alignment S2**). The nucleotide alignments were 356 bp (D3) and 709 bp (*COI*) long; the protein alignment of *COI* had 235 positions.

Phylogenetic trees were calculated with RAxML v8.0.2 [47], MrBayes v3.1.2 [48] and BEAST v1.7.4 [49]. Phylogenetic optimality criterion was maximum likelihood for RAxML, Bayesian inference for MrBayes and BEAST. The best fit model of sequence evolution was estimated with jModeltest 2.1.4 [50], [51], according to the AIC the best model was GTR+I+Γ [52], [53] for both nucleotide alignments. The MCMC chain was run for ten million generations and sampled every 1,000[th] generation in MrBayes. In BEAST, the MCMC chain ran for 100 million generations and sampled every 10,000[th] generation, the majority consensus trees were generated with a burnin value of 2,500 (25%). In RAxML 8,000 bootstrap replicates were calculated for statistical node support. A median-joining haplotype network for the nucleotide dataset of COI was generated with Network 4.6 (Fluxus Technology, Suffolk, Great Britain).

A strict molecular clock was performed with BEAST v1.7.4, BEAUti v1.7.4 and TreeAnnotator v1.7.4 [49] with a fixed substitution rate of 0.0115 which corresponds to the common invertebrate rate of COI of 0.023 substitutions per site per million years [54], [55] for the COI nucleotide alignment. The site model was GTR+I+Γ and as tree prior we used "Yule Process" [56] to allow higher rate variation among branches in this parthenogenetic species than coalescent tree priors do. The Yule.birth rate prior had uniform distribution; all priors were estimated by the software. The MCMC chain ran for 20,000,000 generations, every 2,000th generation was sampled and a burnin of 2,500 was applied and convergence of the MCMC was confirmed using Tracer v1.4 [57].

To test for potential migration between forest and grassland and between sampling locations, three models of migration were tested with grassland and forest specific COI haplotypes using Bayesian inference in MIGRATE-N 3.2.16 [58]. The three models included (1) panmixis among all locations (50 individuals from forest and grassland) assuming a single population, (2) migration between forest (26 individuals) and grassland (24 individuals), and (3) migration between the five sampling locations. This analyses included 4 individuals from HA, 31 from KW, 7 from SO, 5 from TW and 3 from UE. The models were tested in several independent runs. The following parameters deviated from default settings: 10,000 record steps in chain: heating set to on, static heating; 4 chains sampling at every 10th interval using the temperature scheme suggested with the character #; Theta prior distribution, uniform, 0 (minimum) 1 (maximum) 0.1 (delta); migration prior distribution, uniform, 0 (minimum) 10000 (maximum) 1000 (delta); running multiple replicates set to YES, 4 independent chains; number of long chains to run set to 2. To identify the best-fit model, marginal likelihoods of the three runs were compared by calculating the log Bayes Factor (LBF) and model probability (MP) by substracting the largest log likelihood from every other log likelihood, exponentiating the difference and summing up the results. The exponential elements were divided by this product; results indicate which model is most likely relative to the others [59].

Two independent analyses of molecular variance (AMOVA) were calculated in ARLEQUIN 3.5 [60] to investigate between and within population structure based on p-distances, selecting (1) habitat (forest and grassland) and (2) sampling locations as group. Populations represented by less than three individuals were excluded. According to the FNV hypothesis we expected higher variance within sampling locations (i.e., high variance between habitats and associated niches) than between sampling locations, whereas according to the GPG hypothesis we expected variance within habitats to be similar or lower than between sampling locations. Isolation by distance was tested by Mantel test implemented in ARLEQUIN using 10,000 permutations and straight-line (Euclidian) distances. Haplotype and nucleotide diversity were also calculated in ARLEQUIN. To distinguish between divergent selection and neutral drift the distribution of

synonymous and non-synonymous substitutions between locations and between habitat types was compared using the McDonald Kreitman test [61] in DnaSP v5.10.1 [62].

For morphological variation body length and width of 147 individuals were measured from dorsal pictures, taken with AxioCam HRm and processed with the image analyzing software AxioVision 4.8.2 (Zeiss, Göttingen, Germany) by quantifying pixels. Differences between mean values of body length of O. nova were analyzed in R 3.1 (R Development Core Team 2014) using the linear mixed effects model (nlme package) [63]. Locations were set as random variable and body-sizes were compared between habitat types (F, G, IFG, MG) and additionally between individuals with forest and grassland specific COI genotypes. Post hoc multiple comparisons of means were made using Tukey's honestly significant difference test (multicomp package) [64] with p<0.001 as threshold for significance.

Results

Densities of O. nova varied between zero and 30 individuals per sample. To obtain equal numbers of individuals per habitat type, the three samples of each sampling location were pooled for further analysis. Numbers of individuals at the four habitat types were 36, 30, 40 and 41 for F, G, MG and IFG, respectively. Molecular variation of the D3 fragment was low; only nine positions of the 354 bp fragment varied in 126 analyzed individuals (positions 59–61, 114 and 120–124). Accordingly, the phylogenetic tree had no structure and habitats and locations were mixed (**Figure 2a, Figure S1–2**). Amino acid sequences of the COI fragment were almost identical in the 110 individuals sequenced. Only 16 specimens had one or two variable sites with non-synonymous substitutions (**Table S1**) and the overall genetic distances between protein sequences were low (<0.5%). In each of the phylogenetic trees O. nova was monophyletic and separated with high support from the outgroup taxon B. hauseri. Trees (MrBayes, BEAST and RAxML) based on the COI nucleotide alignment were similar (**Figure S3–4**) and COI haplotypes generally clustered according to habitat type, irrespective of sampling locations (**Figure 2b**). Applying a mitochondrial substitution rate of 2.3% per million years, F and G lineages diverged between 6 and 16 mya (**Figure 2b**).

The COI haplotype network also showed a strong habitat related structure (**Figure 3**). Individuals from several sampling locations had identical or closely related haplotypes. However, individuals from forest (F) and grassland (G) had distinct haplotypes, irrespective of sampling locations, i.e., individuals from F and G always clustered separately. Haplotype from IFG either clustered with individuals from G or F, individuals from MG either clustered with individuals from G or IFG (except for two individuals from Solling that formed an isolated clade).

In total, 37 haplotypes were sampled and one haplotype was very common with a total of 21 individuals, 11 from IFG, 9 from MG and one from G. Haplotypes from IFG commonly occurred in more than one habitat type, F and IFG shared five, G and MG shared four common haplotypes, whereas IFG and MG as well as G and IFG shared only one haplotype. In contrast, no haplotypes were shared between F and MG as well as between F and G. Haplotype (Hd) and nucleotide (Π) diversity showed similar patterns (**Table S2**). Within sampling locations Hd of the four habitat types was similar, being between 0.8–0.96. In HA, both Hd and Π were lowest, in SO haplotype diversity was highest (Hd = 0.95) but nucleotide diversity was only intermediate (Π = 0.1). Haplotypes in other locations and in F and G were more different from each other.

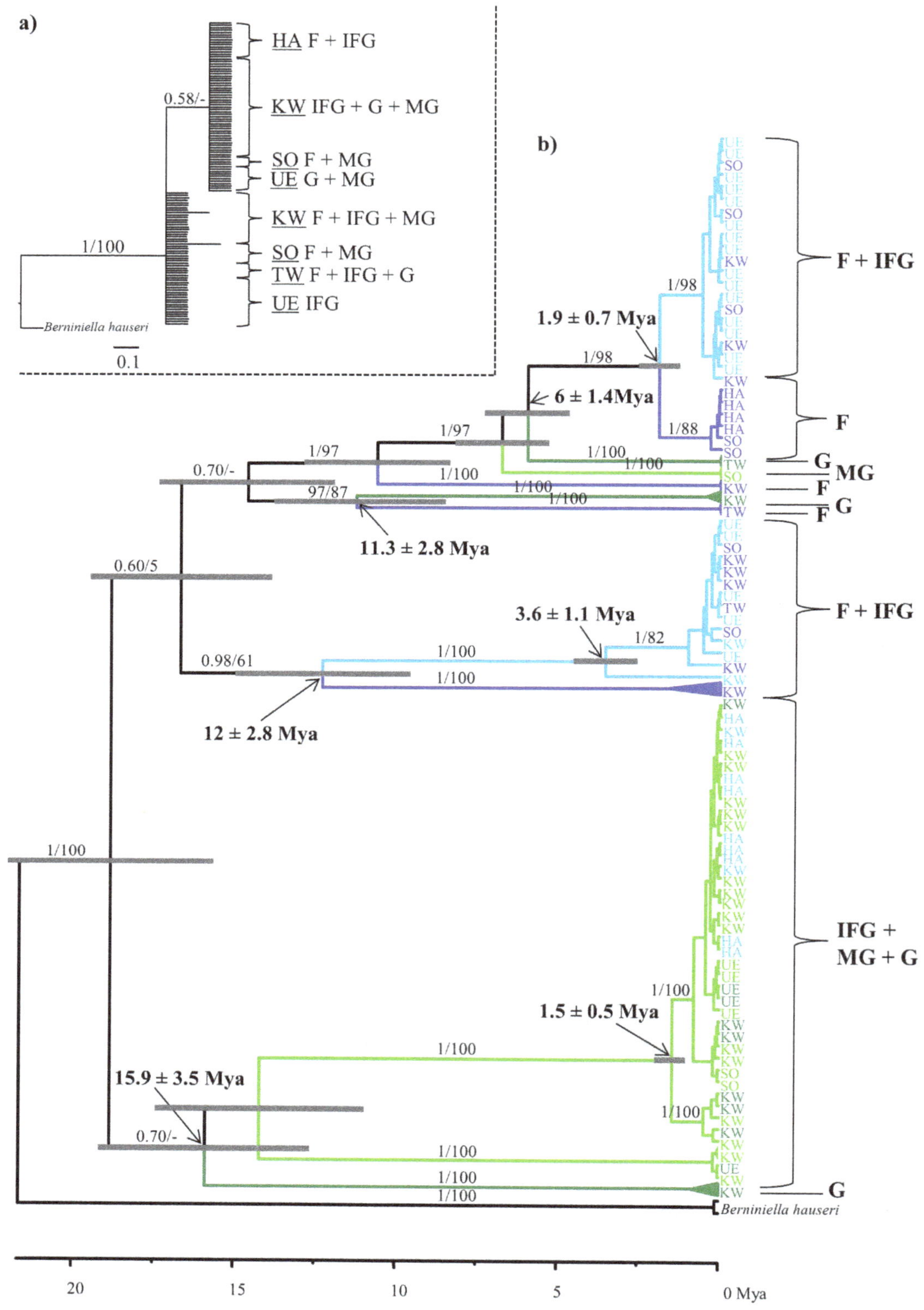

Figure 2. Bayesian phylogenetic trees showing the relatedness among individuals of *Oppiella nova* from different habitats based on (a) nuclear (D3 region of the 28S rDNA, 126 individuals) and (b) mitochondrial markers (*COI*, 110 individuals). Different colors indicate *COI* haplotypes of forest (F, blue), grassland (G, green), intersection of grassland and forest (IFG, light blue), and margin of grassland (MG, light green). Numbers on nodes represent posterior probabilities and bootstrap values, bold numbers are median estimated divergence times ±95% HPD calculated in BEAST using a strict molecular clock and grey bars on nodes indicate 95% HPD intervals. UE (Uelzen), SO (Solling), HA (Hainich), TW (Thuringian Forest) and KW (Kranichstein Forest) refer to the locations of the five forest – grassland gradients studied.

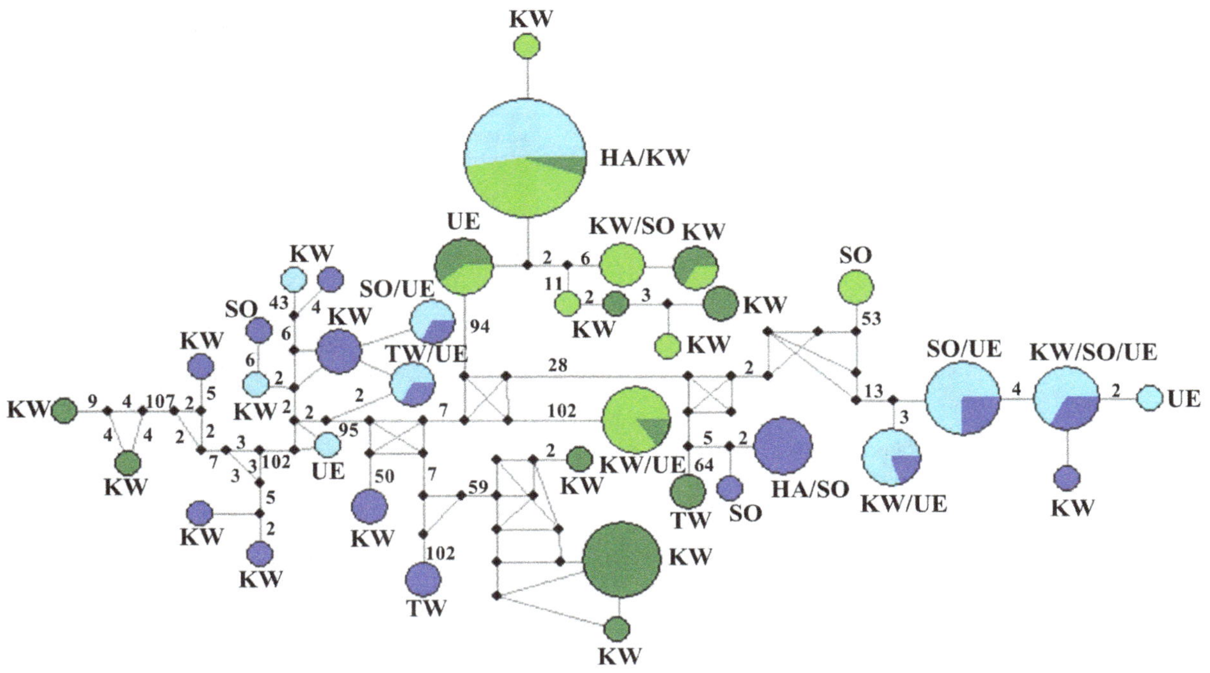

Figure 3. Median Joining Haplotype Network of 110 *COI* sequences of *Oppiella nova* collected from forest (blue), intersection of grassland and forest (light blue), margin of grassland (light green) and grassland (green) from five sampling locations (UE, Uelzen; SO, Solling; HA, Hainich; TW, Thuringian Forest; KW, Kranichstein Forest). The size of circles is proportional to the number of sequences per haplotype. Numbers on lines represent mutation steps separating the haplotypes; no number indicates a single mutation step. Haplotypes from forest and grassland are always separated by many mutation steps, but either haplotypes from forest or from grassland are closely related, even from distant sampling locations.

As indicated by AMOVA, genetic variance was generally high, being highest within samples (58%) and lower within locations (43%) and lowest within habitat types (35%) (**Table 2**). The negative variance component among locations resulted from low or nearly absent genetic structure. If the expectation of the estimator is zero, AMOVA can generate slightly negative variance components.

Among the three models tested with MIGRATE-N, migration between locations (model 3; log marginal likelihood = −5066, LBF = 0, MP = 1) was most likely. Substantially less likely were migrations between F and G (model 2; log marginal likelihood = −5317, LBF = −502, MP = 9.8E-110) and panmixis (model 1; log marginal likelihood = −5259, LBF = −386, MP = −1.5E-84). Isolation by distance was rejected being not significant (r(Y) = −0.24, P(rY) = 0.98). The McDonald Kreitman test was not significant for all comparisons as non-synonymous substitutions were not fixed within habitat types or locations.

Body length of *O. nova* (**Table S3**) in the different habitats ranged from an average of 251 to 275 μm with individuals from G being 24 μm longer than those from F ($F_{3,139} = 23.83$, p<0.001 for habitat type; **Figure 4**). Accordingly, body length of individuals with forest and grassland specific genotypes differed significantly ($F_{1,22} = 22.06$, p<0.001). Body size of individuals from IFG and F was similar; MG and IFG were in between that of individuals from F and G.

Discussion

The results indicate that *O. nova* differs both genetically and morphologically between forest and grassland. In agreement with the FNV hypothesis, haplotypes of forest and grassland were distinct and formed well-supported grassland and forest clades. Although individuals from both habitats were always distinct, some haplotypes also occurred in the transitional habitat types IFG and MG. This suggests niche-related environmental filtering between forest and grassland haplotypes with forest and grassland haplotypes coexisting in transitional habitats. Notably, forest and grassland haplotypes significantly differed in morphology with body size gradually increasing with distance from forest reaching a maximum in grassland specimens.

Considerable molecular variance was found in each of the locations and habitat types, suggesting independent colonization by different lineages rather than by a single locally adapted lineage. High molecular variance within sampling locations suggests that different lineages exist in neighboring habitat types at each sampling location. The results indicate that forest and grassland habitats are associated with certain niches selecting for specific genotypes with both niches being present in transitional habitats, which is consistent with the FNV model. Notably, haplotypes present in more than one habitat type also occurred at different locations. These widespread haplotypes predominantly colonized transitional habitats but haplotype diversity in these habitats was generally lower than in forests and grasslands. Environmental conditions in transitional habitats probably favor more generalist genotypes.

According to ecological niche theory, interspecific competition favors the evolution of species occupying separate niches. Species performance therefore is limited by environmental conditions and genetic adaptation, restricting geographic distribution. Similarly, intraspecific differentiation also can be linked to divergences in environmental conditions or resources. Niche differentiation typically is manifested in morphological differentiation, but may

Table 2. AMOVA of the *COI* gene of *Oppiella nova* on the variance among and within locations and among and within habitat types.

source of variation	d.f.	sum of squares	variance components	percentage of variation	fixiation indices	
Among locations	4	874	−0.40 Va	−1	Fct	−0.01
Among habitats	3	1,124	3.20 Va	6	Fct	0.06
Among samples within locations	8	1,609	22 Vb***	43	Fsc	0.43***
Among samples within habitats	9	1,308	19 Vb***	35	Fsc	0.39***
Within samples	97	2,898	30 Vc***	58	Fst	0.42***

Within samples variance was identical for both analyses; asterisks indicate significant differences at p<0.001; d.f. are degrees of freedom.

also be cryptic and only recognizable at physiological, genetic or transcriptomic levels [65–68]. Difference in body size is a common feature that separates individuals along a single resource dimension [69], whereas genetic differentiation is usually correlated with reproductive or geographic isolation [70–72].

In *O. nova* isolation by distance was not significant and differences in body size correlated with separation into forest and grassland, indicating that niche specific size-dimorphism is due to habitat specific adaptations rather than geographical differentiation. Variation in body size likely reflects niche differentiation, which often is induced by resource shifts and differential exposure to predators [73–75]. Stable isotope data from oppiid species indicate that *O. nova* lives as predator or scavenger [76], [77] and size dimorphism therefore may reflect adaptation to prey of different body size. However, differences in habitat structure and different predator communities in grasslands and forests may also

be responsible for the observed variations in body size. Adult oribatid mites typically are well protected from predation by morphological and chemical defenses [78–80]. However, Schneider and Maraun [81] demonstrated that gamasid mites, the most vigorous predators of soil microarthropods, prey heavily on *O. nova*. Gamasid mites preferentially prey on oribatid mite species of a body size of 200–300 µm [81], indicating that larger and smaller species live in size refuges. Large oribatid mites are heavily sclerotized while smaller ones and juveniles typically are weakly sclerotized but colonize pore space inaccessible for predators such as gamasid mites. For *O. nova*, which is small, weakly sclerotized and mobile, top-down control by gamasid mite predators is likely to be important with larger individuals suffering less from predation by gamasid mites than smaller ones. Differences in body size in forest and grassland therefore may reflect body size related differences in predation by gamasid mites. Unfortunately,

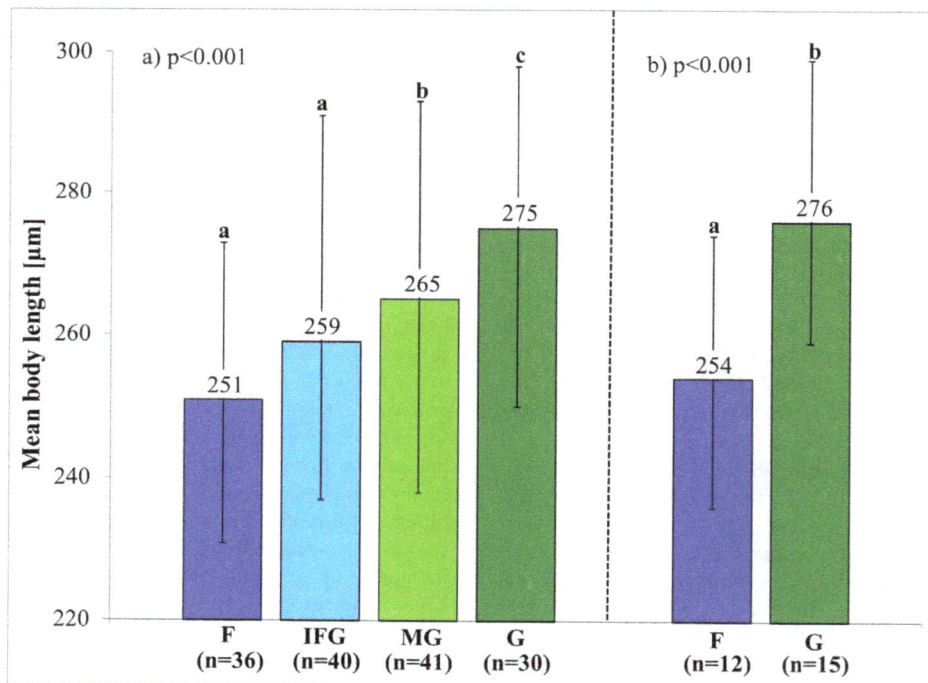

Figure 4. Body length of *Oppiella nova* from forests, grasslands and transitional habitat types between forests and grassland. (a) Differences between all collected individuals (147); forest (F), grassland (G), intersection of forest and grassland (IFG), and margin of grassland (MG). (b) Differences in body size between individuals with forest and grassland specific genotypes. Numbers in brackets refer to the number of individuals included in the analyses; error bars indicate standard deviation. P-values correspond to Tuckey's HSD test.

little is known on the control of oribatid mites by gamasid mite predators in the field and whether this differs between forest and grassland.

Overall, our data indicate ecological differentiation of a parthenogenetic lineage into discrete genetic and morphological entities. The gradual change in haplotype composition and body size between forest and grassland indicates adaptation to specific environmental conditions, i.e. a shift in ecological niches. Further, the results suggest that in addition to haplotypes from both forest and grassland, transitional habitats are colonized by widespread genotypes with lower haplotype diversity than forest and grassland. In contrast to forests and grassland, oribatid mites of transitional habitats may be less affected by predation but rather by abiotic forces due to more variable climatic conditions. Despite the distinctness of forest and grassland lineages and non-synonymous substitutions in the *COI* gene, no indications for divergent selection were found. This may be due to the large population size of *O. nova* as genetic drift and fixation probability of mutations decrease with increasing population size. *Oppiella nova* is among the most abundant oribatid mite species in grasslands and forests and can reach densities of thousands of individuals per square meter [82], [83], [34]. This suggests that extinction rates and bottlenecks are of minor importance explaining why genetic variance of the *COI* fragment within populations is high. Despite separation of shallow clades by long branches in the mitochondrial dataset, which may indicate a cryptic species complex, low D3 variance suggests that *O. nova* may best be treated as single (parthenogenetic) species. High intraspecific *COI* variance is common in arthropods [84], [85], especially in those living in soil [86–88], including parthenogenetic oribatid mites [28] and bdelloid rotifers [89].

In contrast to *O. nova*, haplotype diversity in the parthenogenetic oribatid mite *P. peltifer* suggested a general purpose genotype [28]. *Oppiella nova* is a fast reproducing [90] weakly sclerotized r-strategist [91] presumably feeding on living resources and therefore subject to co-evolutionary adaptations [76]. In contrast, *P. peltifer* reproduces slowly and is strongly sclerotized, characters typical for K-strategists. It predominantly feeds on dead organic matter suggesting that co-evolutionary processes between consumer and (dead) food resource are non-existing [91], [77], thereby facilitating more generalist genoptypes.

Our age estimations suggest that lineages of *O. nova* from grassland separated from those of forests during the Middle and Late Miocene (16-6 mya). The substitution rate of parthenogenetic species may differ from the general rate of *COI* established for arthropods. Still, age estimates and high genetic distances between forest and grassland lineages suggest long-term separation and persistence of lineages, contradicting the commonly held view that parthenogenetic lineages are short-lived evolutionary dead ends. Speciation of parthenogenetic lineages has been assumed to be responsible for the formation of large phylogenetic clusters in bdelloid rotifers [92–94] and certain groups of oribatid mites [40], [95]. The age of grassland lineages correlated well with the expansion of grasslands in the Miocene [96], [97] indicating long-standing adaptation to this habitat. Present day occurrence of grassland and forest lineages in managed European grasslands and forests, respectively, suggests recurrent establishment of lineages due to environmental filtering, i.e. grassland and forest lineages remained bound to the respective habitats.

Our results suggest that, as in sexual species, environmental filters and biotic interactions contribute to the evolution of parthenogenetic species. High genetic variability presumably is maintained by adaptation of certain genotypes to environmental settings as suggested by the FNV hypothesis. Habitat partitioning and coexistence of parthenogenetic lineages at local scales suggest that speciation may occur sympatrically.

Supporting Information

Figure S1   Maximum Likelihood tree showing the relatedness among individuals of *Oppiella nova* from different habitats based on nuclear marker (D3 region of the 28S rDNA, 126 individuals). UE (Uelzen), SO (Solling), HA (Hainich), TW (Thuringian Forest) and KW (Kranichstein Forest) refer to the locations of the five forest – grassland gradients studied (see Figure 1).

Figure S2   Bayesian phylogenetic tree showing the relatedness among individuals of *Oppiella nova* from different habitats based on nuclear marker (D3 region of the 28S rDNA, 126 individuals). UE (Uelzen), SO (Solling), HA (Hainich), TW (Thuringian Forest) and KW (Kranichstein Forest) refer to the locations of the five forest – grassland gradients studied.

Figure S3   Maximum Likelihood tree showing the relatedness among individuals of *Oppiella nova* from different habitats based on mitochondrial marker (*COI*, 110 individuals). UE (Uelzen), SO (Solling), HA (Hainich), TW (Thuringian Forest) and KW (Kranichstein Forest) refer to the locations of the five forest – grassland gradients studied.

Figure S4   Bayesian phylogenetic tree showing the relatedness among individuals of *Oppiella nova* from different habitats based on mitochondrial marker (*COI*, 110 individuals UE (Uelzen), SO (Solling), HA (Hainich), TW (Thuringian Forest) and KW (Kranichstein Forest) refer to the locations of the five forest – grassland gradients studied.

Table S1   Non-synonymous amino acid substitutions among *COI* sequences of Oppiella nova from F, G, IFG and MG. Non-synonymous substitutions are highlighted in red and positions in the *COI* fragment are indicated; individuals affected are from Kranichstein Forest (KW), Thuringian Forest (TW) and Uelzen (UE).

Table S2   Number of individuals (n Ind), number of haplotypes (n haplo), haplotype (Hd) and nucleotide diversity (π) of habitats and locations of Oppiella nova.

Table S3   Body length [μm] of all individuals of *Oppiella nova* sampled for this study and are included in Fig. 4a.

Alignment S1   Alignment of the 28S rDNA (D3 region; 356 bp) including 126 individuals of *O. nova*.

Alignment S2   Alignment of the *COI* gene (709 bp) including 110 individuals of *O. nova*.

Acknowledgments

We thank Georgia Erdmann for help with species determination and Bernhard Klarner and Christoph Digel for help with statistical analyses.

Data Accessibility

DNA sequences: Genbank accession numbers KF293402 - KF293641.

References

1. Glesener RR, Tilman D (1978) Sexuality and the components of environmental uncertainty: clues from geographic parthenogenesis in terrestrial animals. Am Nat 112: 659–673.
2. Lynch M (1984) Destabilizing hybridization, general-purpose genotypes and geographical parthenogenesis. Q Rev Biol 59: 257–290.
3. Suomalainen E, Saura A, Lokki J (1987) Cytology and evolution in parthenogenesis. Bocca Raton: CRC Press.
4. Birdsell JA, Wills C (2003) The evolutionary origin and maintenance of sexual recombination: A review of contemporary models. Evol Biol 33: 27–137.
5. Muller HJ (1964) The relation of recombination to mutational advance. Mutat Res 1: 2–9.
6. Kondrashov AS (1988) Deleterious mutations and the evolution of sexual reproduction. Nature 336: 435–440.
7. Fisher RA (1930) The Genetical Theory of Natural Selection. In: Bennet JH, editor. The Genetical Theory of Natural Selection. Oxford: Oxford University Press.
8. Bell G (1982) The Masterpiece of Nature: The Evolution and Genetics of Sexuality. London: Guildford and King's Lynn.
9. Butlin R (2002) Evolution of sex: The costs and benefits of sex: new insights from old asexual lineages. Nature Rev Genet 3: 311–317.
10. Vrijenhoek RC. 1979 Factors affecting clonal diversity and coexistance. Amer Zool 19: 787–797.
11. Vrijenhoek RC (1984a) The evolution of clonal diversity in *Poeciliopsis*. In: Turner BJ, editor. Evolutionary Genetics of Fishes. New York: Plenum Press. 399–429
12. Vrijenhoek RC (1984b) Ecological differentiation among clones: the Frozen-Niche Variation model. In: Wöhrmann K, Loeschke V, editors. Population Biology and Evolution. Heidelberg: Springer-Verlag. 217–231.
13. Wetherington JD, Schenck RA, Vrijenhoek RC (1989) The origins and ecological success of unisexual *Poeciliopsis*: the Frozen Niche-Variation model. In: Snelson FF Jr, editor. Ecology and evolution of livebearing fishes (Poeciliidae). New Jersey: Prentice Hall, Englewood Cliffs. 259–275.
14. Lima NRW (1998) Genetic analysis of predatory efficiency in natural and laboratory made hybrids of *Poeciliopsis* (Pisces: Poeciliidae). Behaviour 135: 83–98.
15. Gray MM, Weeks SC (2001) Niche breadth in clonal and sexual fish (*Poeciliopsis*): a test for the frozen niche variation model. Can J Fish Aquat Sci 58: 1313–1318.
16. Hotz H, Guex GD, Beerli P, Semlitsch RD, Pruvost NBM (2007) Hemiclone diversity in the hybridogenetic frog *Rana esculenta* outside the area of clone formation: the view from protein electrophoresis. J Zoolog Syst Evol Res 46: 56–62.
17. Groot TVM, Janssen A, Pallini A, Breeuwer JAJ (2005) Adaptation in the asexual false spider mite *Brevipalpus phoenicis*: evidence for frozen niche variation. Exp Appl Acarol 36: 165–176.
18. Browne RA, Hoopes CW (1990) Genotype diversity and selection in asexual brine shrimp. Evolution 33: 848–859.
19. Pantel JH, Juenger TE, Leibaold MA (2011) Environmental gradients structure *Daphnia pulex* x *pulicaria* clonal distribution. J Evolution Biol 24: 723–732.
20. Baker HG (1965) Characteristics and Modes of Origin of Weeds. In: Baker HG, Stebbins GL, editor. The Genetics of Colonizing Species. New York: Academic Press. 147–172.
21. Lynch M, Bürger R, Butcher D, Gabriel W (1993) The mutational meltdown in asexual populations. J Hered 84: 339–344.
22. Lynch M (1985) Spontaneous mutations for life history characters in an obligate parthenogen. Evolution 39: 804–818.
23. Lynch M, Gabriel W (1987) Environmental tolerance. Am Nat 129: 283–303.
24. Schlosser IJ, Doeringsfeld MR, Elder JF, Arzayus LF (1998) Niche relationships of clonal and sexual fish in a heterogeneous landscape. Ecology 79: 953–968.
25. Myers MJ, Meyer CP, Resh VH (2000) Neritid and thiarid gastropods from French Polynesian streams: how reproduction (sexual, parthenogenetic) and dispersal (active, passive) affect population structure. Freshw Biol 44: 535–545.
26. Van Doninck K, Schön I, De Bruyn L, Martens K (2002) A general purpose genotype in an ancient asexual. Oecologia 132: 205–212.
27. Van Doninck K, Schön I, Martens K, Backeljau T (2004) Clonal diversity in the ancient asexual ostracod *Darwinula stevensoni* assessed by RAPD-PCR. Heredity 93: 154–160.
28. Heethoff M, Domes K, Laumann M, Maraun M, Norton RA et al. (2007) High genetic divergences indicate ancient separation of parthenogenetic lineages of the oribatid mite *Platynothrus peltifer* (Acari, Oribatida). J Evolution Biol 2: 392–402.
29. Andersen FA, Jordal BH, Kambestad M, Kirkendall LR (2011) Improbable but true: the invasive inbreeding ambrosia beetle *Xylosandrus morigerus* has generalist genotypes. Ecol Evol 2: 247–257.
30. Schoen I, Martens K, van Dijk P (2009) *Lost Sex: The Evolutionary Biology of Parthenogenesis*. Heidelberg: Springer-Verlag.
31. Hutson BR (1980) Colonization of industrial reclamation sites by acari, collembola and other invertebrates. J Appl Ecol 17: 255–275.
32. Wanner M, Dunger W (2002) Primary immigration and succession of soil organisms on reclaimed opencast coal mining areas in Eastern Germany. Eur J Soil Biol 38: 137–143.
33. Lindberg N, Bengtsson J (2005) Population responses of oribatid mites and collembolans after drought. Appl Soil Ecol 28: 163–174.
34. Siira-Pietikäinen A, Penttinen R, Huhta V (2008) Oribatid mites (Acari: Oribatida) in boreal forest floor and decaying wood. Pedobiologia 52: 111–118.
35. Erdmann G, Scheu S, Maraun M (2012) Regional factors rather than forest type drive the community structure of soil living oribatid mites (Acari, Oribatida). Exp Appl Acarol 57: 159–167.
36. Cianciolo JM, Norton RA (2006) The ecological distribution of reproductive mode in oribatid mites, as related to biological complexity. Exp Appl Acarol 40: 1–25.
37. Weigmann G (2006) Hornmilben (Oribatida) In: *Die Tierwelt Deutschlands*. Keltern: Goecke & Evers.
38. Woas S (1986) Beitrag zur Revision der Oppioidea sensu Balogh, 1972 (Acari, Oribatei). Andrias 5: 21–224.
39. Cruickshank RH (2002) Molecular markers for the phylogenetics of mites and ticks. Syst Appl Acarol 7: 3–14.
40. Maraun M, Heethoff M, Schneider K, Scheu S, Weigmann G et al. (2004) Molecular phylogeny of oribatid mites (Oribatida, Acari): evidence for multiple radiations of parthenogenetic lineages. Exp Appl Acarol 33: 183–201.
41. Lehmitz R, Russell D, Hohberg K, Christian A, Xylander WER (2012) Active dispersal of oribative mites into young soils. Appl Soil Ecol 55: 10–19.
42. Kempson D, Lloyd M, Ghellardi R (1963). A new extractor for woodland litter. Pedobiologia 1: 1–21.
43. Folmer O, Black M, Hoeh W, Lutz R, Vrijenhoek R (1994) DNA primers for amplification of mitochondrial cytochrome c oxidase subunit I from diverse metazoan invertebrates. Mol Mar Biol Biotechnol 3: 294–299.
44. Litvaitis MK, Nunn G, Thomas WK, Kocher TD (1994) A molecular approach for the identification of meiofaunal turbellarians (Platyhelminthes, Turbellaria). Mar Biol 120: 437–442.
45. Hall TA (1999) BioEdit: a user-friendly biological sequence alignment editor and analysis program for Windows 95/98/NT. Nucleic Acids Symp 41: 95–98.
46. Thompson JD, Gibson TJ, Plewniak F, Jeanmougin F, Higgins DG (1997) The CLUSTAL_X windows interface: flexible strategies for multiple sequence alignment aided by quality analysis tools. Nucleic Acids Res 25: 4876–4882.
47. Stamatakis A (2014) RAxML Version 8: A tool for Phylogenetic analysis and Post-Analysis of large Phylogenies. Bioinformatics.
48. Ronquist F, Huelsenbeck JP (2003) MrBayes 3: Bayesian phylogenetic inference under mixed models. Bioinformatics 19: 1572–1574.
49. Drummond AJ, Suchard A, Xie D, Rambaut A (2012) Bayesian Phylogenetics with BEAUti and the BEAST 1.7. Mol Biol Evol 29: 1969–1973.
50. Darriba D, Taboada GL, Doallo R, Posada D (2012). jModelTest 2: more models, new heuristics and parallel computing. Nat. Methods 9: 772.
51. Guindon S, Gascuel O (2003). A simple, fast and accurate method to estimate large phylongeies by maximum-likelihood. Syst Biol 52: 696–704.
52. Lanave C, Preparata G, Saccone C, Serio G (1984). A new method for calculating evolutionary substitution rates. J Mol Evol 20: 86–93.
53. Ziheng Y (1994). Maximum likelihood phylogenetic estimation from DNA sequences with variable rates over sites: Approximate methods. J Mol Evol 39: 306–314.
54. Brower AVZ (1994) Rapid morphological radiation and convergence among races of the butterfly *Heliconius erato* inferred from patterns of mitochondrial DNA evolution. Proc Natl Acad Sci 91: 6491–6495.
55. Avise JC (1994) Molecular Markers, Natural History and Evolution. New York: Chapman & Hall.
56. Gernhard T, Hartmann K, Steel M (2008) Stochastic properties of generalised Yule models, with biodiversity applications. J Math Biol 57: 713–735.
57. Drummond AJ, Rambaut A (2007) BEAST: Bayesian evolutionary analysis by sampling trees. BMC Evol Biol 7: 214.
58. Beerli P (2009) How to use MIGRATE or why are Markov chain Monte Carlo programs difficult to use? In: *Population Genetics for Animal Conservation* (ed. Bertorelle G, Bruford MW, Hauffe HC, Rizzoli A, Vernesi C), 42–79. Cambridge University Press, Cambridge, UK.
59. Beerli P, Palczewski M (2010) Unified framework to evaluate panmixia and migration direction among multiple sampling locations. Genetics 185: 313–326.
60. Excoffier LG, Laval G, Schneider S (2005) Arlequin ver. 3.0: An integrated software package for population genetics data analysis. Evol Bioinform Online 1: 47–50.

Author Contributions

Conceived and designed the experiments: IS. Performed the experiments: HvS. Analyzed the data: HvS. Contributed reagents/materials/analysis tools: SS. Contributed to the writing of the manuscript: HvS MM SS IS.

61. McDonald JH, Kreitman M (1991) Adaptive protein evolution at the Adh locus in *Drosophila*. Nature 351: 652–654.

62. Librado P, Rozas J (2009) DnaSP V5: A software for comprehensive analysis of DNA polymorphism data. Bioinformatics 25: 1451–1452.

63. Pinheiro J, Bates D, DebRoy S, Sarkar D and the R Development Core Team (2013) nlme: Linear and Nonlinear Mixed Effects Models. R Package version 3.1–111.

64. Hothorn T, Bretz F, Westfall T (2008) Simultaneous Inference in General Paramatric Models. Biomet J 40: 346–363.

65. Christensen B (1980) Constant differential distribution of genetic variants in polyploid parthenogenetic forms of *Lumbricillus lineatus* (Enchytraeidae, Oligochaeta). Hereditas 92: 193–198.

66. Posthuma L (1990) Genetic differentiation between populations of *Orchesella cincta* (Collembola) from heavy metal contaminated sites. J. Apll. Ecol. 27: 609–622.

67. Heethoff M, Etzold K, Scheu S (2004) Mitochondrial COII sequences indicate that the parthenogenetic earthworm *Octolasion tyrtaeum* (Savigny 1826) constitutes of two lineages differing in body size and genotype. Pedobiologia 48: 9–13.

68. Janssens TKS, Roelofs D, van Straalen NM (2009) Molecular mechanisms of heavy metal tolerance and evolution in invertebrates. Insect Sci 16: 3–18.

69. Hutchinson GE (1959) Homage to Santa Rosalia or why are there so many kinds of animals? Amer Nat 93: 145–159.

70. Hansen MM, Mensberg KLD (1998) Genetic differentiation and relationship between genetic and geographical distance in Danish sea trout (*Salmo trutta* L.) populations. Heredity 81: 493–504.

71. Whitaker RJ, Grogan DW, Taylor JW (2003) Geographic barriers isolate endemic populations of hyperthermophilic Archaea. Science 301: 976–978.

72. Ramachandran S, Deshpande O, Roseman CC (2005) Support from the relationship of genetic and geographic distance in human populations for a serial founder effect originating in Africa. Proc Natl Acad Sci 102: 15942–15947.

73. Brönmark C, Pettersson LB, Nilsson PA (1999) Predator-induced defense in crucian carp. In: Tollrian R, Harvell CD, editors. The Ecology and Evolution of Inducible Defenses. New Jersey: Princeton University Press. 203–217.

74. East TL, Havens KE, Rodusky AJ, Brady MA (1999) *Daphnia lumholtzi* and *Daphnia ambigua*: population comparisons of an exotic and native cladoceran in Lake Okeechobee, Florida. J Plankton Res 21: 1537–1551.

75. Cooper Jr WE, Stankowich T (2010) Prey or predator? Body size of an approaching animal affects decisions to attack or escape. Behav Ecol 21: 1278–1284.

76. Schneider K, Migge S, Norton RA, Scheu S, Langel R et al. (2004) Trophic niche differentiation in soil microathropods (Oribatida, Acari): evidence from stable isotope ratios (^{15}N/^{14}N). Soil Biol Biochem 36: 1769–1774.

77. Maraun M, Erdmann G, Fischer BM, Pollierer MM, Norton RA et al. (2011) Stable isotopes revisited: Their use and limits for oribatid mite trophic ecology. Soil Biol Biochem 43: 877–882.

78. Sanders FH, Norton RA (2004) Anatomy and function of the ptychoid defensive mechanism in the mite *Euphthiracarus cooki* (Acari: Oribatida). J Morphol 259: 119–154.

79. Peschel K, Norton RA, Scheu S, Maraun M (2006) Do oribatid mites live in enemy-free space? Evidence from feeding experiments with the predatory mite *Pergamasus septentrionalis*. Soil Biol Biochem 38: 2985–2989.

80. Heethoff M, Raspotnig G (2012) Expanding the 'enemy-free space' for oribatid mites: evidence for chemical defense of juvenile *Archegozetes longisetosus* against the rove beetle *Stenus juno*. Exp Appl Acarol 56: 93–97.

81. Schneider K, Maraun M (2009) Top-down control of soil microarthropods – Evidence from a laboratory experiment. Soil Biol Biochem 41: 170–175.

82. Maraun M, Scheu S (2000) The structure of oribatid mite communities (Acari, Oribatida): patterns, mechanisms and implications for future research. Ecography 23: 374–783.

83. Penttinen R, Siira-Pietikäinen A, Huhta V (2008) Oribatid mites in eleven different habitats in Finland. In: Bertrand M, Kreiter S, McCoy KD, Migeon A, Navajas M, Tixier MS, Vial L, editors. Integrative Acarology. Montpellier: Proceedings of the 6th European Congress of the EURAAC. 237–244.

84. Schäffer S, Koblmüller S, Pfingstl T, Sturmbauer C, Krisper G (2010) Contrasting mitochondrial DNA diversity estimates in Austrian *Scutovertex minutus* and *S. sculptus* (Acari, Oribatida, Brachypylina, Scutoverticidae). Pedobiologia 53: 203–211.

85. Edmands S (2001) Phylogeography of the intertidal copepod *Tigriopus californicus* reveals substantially reduced population differentiation at northern latitudes. Mol Ecol 10: 1743–50.

86. Rosenberger M, Maraun M, Scheu S, Schaefer I (2013) Pre- and post-glacial diversifications shape genetic complexity of soil-living microarthropod species. Pedobiologia 56: 79–87.

87. Torricelli G, Carapelli A, Convey P, Nardi F (2010) High divergence across the whole mitochondrial genome in the "pan-Antarctic" springtail *Friesea grisea*: Evidence for cryptic species? Gene 449: 30–40.

88. Boyer SL, Baker JM, Giribet G (2007) Deep genetic divergences in *Aoraki denticulata* (Arachnida, Opiliones, Cyphophthalmi): a widespread "mite harvestman" defies DNA taxonomy. Mol Ecol 16: 4999–5016.

89. Fontaneto D, Boschetti C, Ricci C (2008) Cryptic diversification in ancient asexuals: evidence from the bdelloid rotifer *Philodina flaviceps*. J Evolution Biol 21: 580–587.

90. Kaneko N (1988) Life history of *Oppiella nova* (Oudemans) (Oribatei) in cool temperate forest soils in Japan. Acarologia 29: 215–221.

91. Norton RA (1994) Evolutionary Aspects of Oribatid Mite Life Histories and Consequences for the Origin of the Astigmata. In: Houck MA, *editor*. Mites. Ecological and Evolutionary Analyses of Life-History Patterns. New York: Chapman & Hall. 99–135.

92. Mark Welch D, Meselson M (2000) Evidence for the evolution of bdelloid rotifers without sexual reproduction or genetic exchange. Science 288: 1211–1215.

93. Normark BB, Judson OP, Moran NA (2003) Genomic signatures of ancient asexual lineages. Biol J Linnean Soc 79: 69–84.

94. Birky Jr CW, Wolf C, Maughan H, Herbertson L. Henry E (2005) Speciation and selection without sex. Hydrobiologia 546: 29–45.

95. Laumann M, Norton RA, Weigmann G, Scheu S, Maraun M et al. (2007) Speciation in the parthenogenetic oribatid mite genus *Tectocepheus* (Acari, Oribatida) as indicated by molecular phylogeny. Pedobiologia 51: 111–122.

96. Retallack GJ (2001) Cenozoic expansion of grasslands and climatic cooling. J Geol 109: 407–426.

97. Osborne CP, Beerling DJ (2006) Nature's green revolution: the remarkable evolutionary rise of C$_4$ plants. Phil Trans R Soc A 361: 173–194.

Permissions

The contributors of this book come from diverse backgrounds, making this book a truly international effort. This book will bring forth new frontiers with its revolutionizing research information and detailed analysis of the nascent developments around the world.

We would like to thank all the contributing authors for lending their expertise to make the book truly unique. They have played a crucial role in the development of this book. Without their invaluable contributions this book wouldn't have been possible. They have made vital efforts to compile up to date information on the varied aspects of this subject to make this book a valuable addition to the collection of many professionals and students.

This book was conceptualized with the vision of imparting up-to-date information and advanced data in this field. To ensure the same, a matchless editorial board was set up. Every individual on the board went through rigorous rounds of assessment to prove their worth. After which they invested a large part of their time researching and compiling the most relevant data for our readers.

The editorial board has been involved in producing this book since its inception. They have spent rigorous hours researching and exploring the diverse topics which have resulted in the successful publishing of this book. They have passed on their knowledge of decades through this book. To expedite this challenging task, the publisher supported the team at every step. A small team of assistant editors was also appointed to further simplify the editing procedure and attain best results for the readers.

Apart from the editorial board, the designing team has also invested a significant amount of their time in understanding the subject and creating the most relevant covers. They scrutinized every image to scout for the most suitable representation of the subject and create an appropriate cover for the book.

The publishing team has been an ardent support to the editorial, designing and production team. Their endless efforts to recruit the best for this project, has resulted in the accomplishment of this book. They are a veteran in the field of academics and their pool of knowledge is as vast as their experience in printing. Their expertise and guidance has proved useful at every step. Their uncompromising quality standards have made this book an exceptional effort. Their encouragement from time to time has been an inspiration for everyone.

The publisher and the editorial board hope that this book will prove to be a valuable piece of knowledge for researchers, students, practitioners and scholars across the globe.

List of Contributors

William T. Bean
Humboldt State University, Arcata, California, United States of America

Robert Stafford
California Department of Fish and Game, Los Osos, California, United States of America

H. Scott Butterfield
The Nature Conservancy, San Francisco, California, United States of America

Justin S. Brashares
Department of Environmental Science, Policy and Management, University of California, Berkeley, California, United States of America

Eduardo Carneiro, Olaf Hermann Hendrik Mielke and Mirna Martins Casagrande
Laboratório de Estudos de Lepidoptera Neotropical, Zoology Department, UFPR. Curitiba, Paraná Brasil

Konrad Fiedler
Division of Tropical Ecology & Animal Biodiversity, University of Vienna, Vienna, Austria

Jian Sun and Shuli Niu
Synthesis Research Centre of Chinese Ecosystem Research Network, Key Laboratory of Ecosystem Network Observation and Modelling, Institute of Geographic Sciences and Natural Resources Research, Chinese Academy of Sciences, Beijing, China

Xiaodan Wang, Genwei Cheng, Jianbo Wu and Jiangtao Hong
The key laboratory of mountain surface processes and eco-regulation, Institute of Mountain Hazard and Environment, Chinese Academy of Sciences, Chengdu, China

Peter J. Blank and Monica G. Turner
Department of Zoology, University of Wisconsin, Madison, Wisconsin, United States of America

David W. Sample
Wisconsin Department of Natural Resources, Madison, Wisconsin, United States of America

Carol L. Williams
Wisconsin Energy Institute, University of Wisconsin, Madison, Wisconsin, United States of America

Claire D. Stevenson-Holt
Centre for Wildlife Conservation, University of Cumbria, Ambleside, Cumbria, United Kingdom

Kevin Watts
Centre for Ecosystems, Society and Biosecurity, Forest Research, Farnham, Surrey, United Kingdom

Chloe C. Bellamy
Centre for Ecosystems, Society and Biosecurity, Forest Research, Roslin, Midlothian, United Kingdom

Owen T. Nevin
School of Medical and Applied Sciences, Central Queensland University, Gladstone, Queensland, Australia

Andrew D. Ramsey
School of Biological and Forensic Sciences, University of Derby, Derby, Derbyshire, United Kingdom

Rosemary L. Sherriff
Department of Geography, Humboldt State University, Arcata, California, United States of America

Rutherford V. Platt
Department of Environmental Studies, Gettysburg College, Gettysburg, Pennsylvania, United States of America

Thomas T. Veblen, Tania L. Schoennagel and Meredith H. Gartner
Department of Geography, University of Colorado, Boulder, Colorado, United States of America

Bing Wang, Fenxiang Wen, Jiangtao Wu and Xiaojun Wang
College of Environmental Science and Resources, Shanxi University, Taiyuan, China

Yani Hu
Library, Hebei University of Science and Technology, Shijiazhuang, China

Dan Liu, Wenwen Cai, Jiangzhou Xia, Wenjie Dong, Yang Chen,
Haicheng Zhang and Wenping Yuan
State Key Laboratory of Earth Surface Processes and Resource Ecology, Beijing Normal University, Beijing, China

Guangsheng Zhou
Chinese Academy of Sciences, Institute of Botany, State Key Laboratory of Vegetation and Environmental Change, Beijing, China
Chinese Academy of Meteorological Sciences, Beijing, China

Xuan Fang and Guoan Tang
Key Laboratory of Virtual Geographic Environment, Ministry of Education, School of Geography Science, Nanjing Normal University, Nanjing, China

Bicheng Li
Research Center of Soil and Water Conservation and Ecological Environment, Chinese Academy of Sciences, Yangling, Shaanxi, China

Ruiming Han
School of Geography Science, Nanjing Normal University, Nanjing, China

Fei Peng, Quangang You, Manhou Xu, Jian Guo and Tao Wang, Xian Xue
Key Laboratory of Desert and Desertification, Chinese Academy of Sciences, Cold and Arid Regions Environmental and Engineering Research Institute, Chinese Academy of Sciences, Lanzhou, China

Huhe, Shinchilelt Borjigin, Nobukiko Nomura, Toshiaki Nakajima Toru Nakamura and Hiroo Uchiyama
Graduate School of Life and Environmental Sciences, University of Tsukuba, Tsukuba, Ibaraki, Japan

Yunxiang Cheng
State Key Laboratory of Grassland Agro-Ecosystems, College of Pastoral Agriculture Science and Technology, Lanzhou University, Lanzhou, China

Jens Schirmel, Martin Alt, Isabell Rudolph and Martin H. Entling
Institute of Environmental Science, University of Koblenz-Landau, Landau, Germany

Yang Fu, Haicheng Zhang and Wenjie Dong
State Key Laboratory of Earth Surface Processes and Resource Ecology, Beijing Normal University, Beijing, China

Wenping Yuan
State Key Laboratory of Earth Surface Processes and Resource Ecology, Beijing Normal University, Beijing, China
State Key Laboratory of Cryospheric Sciences, Cold and Arid Regions Environmental and Engineering Research Institute, Chinese Academy of Sciences, Lanzhou, Gansu, China

Pénélope Lamarque and Sandra Lavorel
Laboratoire d'Ecologie Alpine, Unité Mixte de recherche 5553, Centre National de la Recherche Scientifique, UniversitéJoseph Fourier, Grenoble, France

Patrick Meyfroidt
Fonds de larecherche scientifique (F.R.S.-FNRS), Louvain-La-Neuve, Belgium
Earth and Life Institute, Georges Lemaître Centre for Earth and Climate Research, University of Louvain, Louvain-la-Neuve, Belgium

Baptiste Nettier
Irstea centre de Grenoble, uniteé de recherche Deéveloppement des territoires montagnards, Grenoble, France

Jie Zhao, Xunyang He and Kelin Wang
Key Laboratory of Agro-ecological Processes in Subtropical Region, Institute of Subtropical Agriculture, Chinese Academy of Sciences, Changsha, Hunan, China
Huanjiang Observation and Research Station for Karst Ecosystems, Chinese Academy of Sciences, Huanjiang, Guangxi, China

Shengping Li and Lu Liu
Key Laboratory of Agro-ecological Processes in Subtropical Region, Institute of Subtropical Agriculture, Chinese Academy of Sciences, Changsha, Hunan, China
Huanjiang Observation and Research Station for Karst Ecosystems, Chinese Academy of Sciences, Huanjiang, Guangxi, China
Graduate University of Chinese Academy of Sciences, Beijing, China

Risto K. Heikkinen, Mikko Kuussaari, Janne Heliölä, Niko Leikola and Juha Pöyry
Finnish Environment Institute, Natural Environment Centre, Helsinki, Finland

Greta Bocedi and Justin M. J. Travis
Institute of Biological Sciences, University of Aberdeen, Aberdeen, United Kingdom

Stacy C. Davis, Laura A. Burkle and Wyatt F. Cross
Department of Ecology, Montana State University, Bozeman, Montana, United States of America

Kyle A. Cutting
Red Rock Lakes National Wildlife Refuge, US Fish and Wildlife Service, Lima, Montana, United States of America

Shaoqiang Wang, Lei Zhou, Huimin Wang and Peili Shi
Key Laboratory of Ecosystem Network Observation and Modeling, Institute of Geographic Sciences and Natural Resources Research, Chinese Academy of Sciences, Beijing 100101, China

Kun Huang
Key Laboratory of Ecosystem Network Observation and Modeling, Institute of Geographic Sciences and Natural Resources Research, Chinese Academy of Sciences, Beijing 100101, China
University of Chinese Academy of Sciences, Beijing 100049, China

Yanfen Wang
University of Chinese Academy of Sciences, Beijing 100049, China

Junhui Zhang
Institute of Applied Ecology, Chinese Academy of Sciences, Shenyang 110016, China

Junhua Yan
South China Botanical Garden, Chinese Academy of Sciences, Guangzhou 510650, China

Liang Zhao
Northwest Plateau Institute of Biology, Chinese Academy of Sciences, Xining 810001, China

Ashley B. Bennett and Rufus Isaacs
Department of Entomology and Great Lakes Bioenergy Research Center, Michigan State University, East Lansing, Michigan, United States of America

Timothy D. Meehan and Claudio Gratton
Department ofEntomology and Great Lakes Bioenergy Research Center, University of Wisconsin - Madison, Madison, Wisconsin, United States of America

Helge von Saltzwedel, Mark Maraun, Stefan Scheu and Ina Schaefer
Georg-August University, Johann-Friedrich-Blumenbach Institute of Zoology and Anthropology, Dept. Ecology, Göttingen, Germany

Index

A

Arthropod Communities, 135, 189-190, 194, 201

B

Bee Abundance, 212-220

Beginning Of The Growing Season (bgs), 138

Bioenergy, 30-34, 36-39, 174, 212-216, 218-221

Biomass Measure, 23

Biomass Yields, 30-31, 33-34, 36-38

Bird Communities, 30-31, 38, 190, 220

C

Carbon Sequestration, 147, 149, 154, 203-204, 209-210, 213

Catchment Scale, 68-69, 73, 76, 78

Clipping, 105-108, 110-111, 113-114, 116-118, 120, 191, 197, 202

Cloudiness Index, 203-204, 206-210

Community Composition, 11, 124-125, 127, 167, 169, 172, 212-220, 223

Community Structure, 9-11, 13, 17-18, 20, 28, 32, 117, 119, 124, 127, 169, 174, 191, 202, 230

Cosmopolitan Parthenogenetic Soil Mite, 222

D

Data Selection, 175-176

Denaturing Gradient Gel

Electrophoresis (dgge), 120

Diffuse Radiation, 203-204, 208-211

Diversity, 8-11, 19-20, 30-31, 36-38, 92, 95, 97, 100-104, 117, 119-127, 129-131, 133-137, 151, 153-156, 158-159, 162-165, 171, 173-174, 186, 189-190, 199, 201-202, 212-222, 225, 227, 229-231

E

Ecosystem Carbon Fluxes, 105, 117-118

Ecosystem Respiration (er), 105-106, 109-115

Ecosystem Service Values (esvs), 92-95, 98, 100-102

Ecosystem Services (es), 150

Elevational Gradients, 10-13, 16, 18-20

Endangered Rodent, 1-2, 9

F

Frozen-niche Variation, 222

G

General-purpose-genotype, 222

Global Validation, 80

Grazing Exclusion (m3), 21, 24

Grazing Regimes, 21, 134, 190, 201

Gross Ecosystem Production (gep), 105-106, 109-111

Gross Primary Production, 80, 82, 87-88, 90, 105, 149, 203, 205, 210-211

Growing Degree Day Model (gdd), 141

H

Habitat Suitability Modelling, 40-41, 50

High-elevation Grasslands, 189-190

Hilly-gully Loess Plateau, 68

Historical Range Of Variability (hrv), 52

Hydrologically Contrasting, 105, 116-117

I

Invasive Grey Squirrel, 40

Invertebrates, 129-130, 136-137, 187, 202, 224, 230-231

L

Land Use Pattern, 92-93, 95, 97-100, 103

Least-cost Connectivity Models, 40

Light Use Efficiency (lue), 203, 206

Log Bayes Factor (lbf), 225

M

Mean Nearest Neighbor Distance (mntd), 14

Mean Pairwise Distance (mpd), 14

Microclimate, 29, 105, 111

Model Probability (mp), 225

Montane Forests, 51-55, 60-63, 65-66

Multi-scale Distribution Model, 1

N

Net Ecosystem Exchange (nee), 105-106, 109-115

Nitrogen-fixing Bacteria, 119, 124-127

Non-equilibrium Populations, 1

Normalized Difference Vegetation

Index (ndvi), 9, 82

P

Permanent Grazing (m1), 21

Phenology Models, 138-139, 141-148

Phospholipid Fatty Acids (plfa), 169

Phylogenetic Community Structure, 10, 13, 17-18, 20

Plant Traits, 21-22, 28, 165

Pollinator Community, 212

Potential Bioenergy Grasslands, 30
Process-based Model, 80

Q
Quantitative Pcr (qpcr), 119, 122

R
Resource Limitation, 1

S
Seasonal Grazing (m2), 21, 24
Secondary Vegetation, 166-167, 171-172
Skipper Butterflies, 10-11, 18
Soil Biota Composition, 166
Soil Nutrients, 21, 23, 27-28, 173
Soil Water Content, 28, 68, 74-75, 77-79, 89, 106, 167, 169-170, 172
Species' Range Expansion, 175, 184, 186

T
Temporal Variations, 92, 111
Total Bacteria, 119, 122-124, 127
Traditional Flood Irrigation, 129-130, 133-134
Trait Parameterisation, 175

U
Ungrazed Wet Meadow, 190-194
Upland Grasslands, 189-195, 197-198, 200-201

V
Vertical Profiles, 68-73, 75

W
Warming, 28, 105-111, 113-114, 116-118, 148-149, 176, 187, 203
Wildfire Severity, 51, 65

Y
Yule Process, 225

www.ingramcontent.com/pod-product-compliance
Lightning Source LLC
Chambersburg PA
CBHW080517200326
41458CB00012B/4246